Elements of Pelagos Biology

Genuario Belmonte

Elements of Pelagos Biology

With Focus on the Mediterranean Sea

Springer

Genuario Belmonte
Biological & Environmental Science & Technologies
University of Salento
Lecce, Italy

ISBN 978-3-031-74279-8 ISBN 978-3-031-74280-4 (eBook)
https://doi.org/10.1007/978-3-031-74280-4

This Springer imprint is published by the registered company Springer Nature Switzerland AG
The registered company address is: Gewerbestrasse 11, 6330 Cham, Switzerland

Preface

Oceans cover more than 70% of the planet's surface, with an average water layer thickness of approximately 3800 m. Mankind is largely excluded from this enormous world, which is consequently largely unfamiliar to them.

In addition, the lifestyles in a world of 3D water, darkness, and without any connection with a solid bottom imposes a psychological limit on the human mind. The consequence is the existence of a large part of the living planet (its majority) still unknown, foreign, and scarcely proposed even in university and/or specialized courses.

This textbook stems from the experience of more than 10 years of teaching in the frame of the master's degree "Coastal and Marine Biology and Ecology" proposed at the University of Salento (Italy), one of the first international degrees in marine biology proposed by an Italian university. The geographic focus of the lessons is the Mediterranean Sea, but general principles and biological aspects are continuously considered.

Species and communities are presented in detail because the basic level of taxonomy, life cycles, and rules in ecological functioning of the pelagos is not typically learned in basic courses of biology and natural sciences, which often focus on terrestrial or freshwater models and, in the marine environment, mostly on benthic organisms. The pelagic lifestyle, in addition, is not the main part of marine biology textbooks. This proposal aims to revert the point of view and focuses on the pelagic lifestyle of living organisms, practically disregarding the benthos.

This book aims to encourage students to project their future studies in a foreign and unknown world, such as the marine pelagos. It should be and probably it will be in the next future, one of the most attractive targets for knowledge makers, thus stimulating technology and engineering, other than attracting naturalists' interest.

This book aims to support basic knowledge ("elements") of pelagic lifestyles, and it does not propose in detail synthesis and/or models of resource use, food chain functioning, physiology, or management for the protection or restoration of nature, because ecology, oceanography, physiology, and biochemistry are covered in separate, independent courses within the same master degree's program, taught by other teachers. The aim of this book is, consequently, to provide a basis and/or to stimulate curiosity on the particular and unknown world of pelagos.

As a didactic strategy, species, geographic sites, details, and subjects are not isolated in single chapters but are repeatedly proposed at each occasion in the text to give them maximum exposition and possibly enhance their permanence in student minds.

Illustrations are mostly line drawings just to give evidence of the main features of each organism and to suggest to the readers an exercise (the drawing) incredibly disregarded but extremely useful to learn anatomy of marine organisms, understand differences between species, and remember details. My personal line drawings are sometimes associated with spectacular drawings of ancient naturalists, whose artistic representations of microscopic and pelagic organisms still act as explanatory tools to understand their anatomy.

In the past, drawings were indispensable for describing anatomy. However, in more recent times, drawings have become crucial for illustrating schemes and processes. This dual utility of drawing attitude is attributed to A. Gennari, whose representations described the anatomy of single organisms, scenarios where species thrive, or schemes of ecosystem functioning, for the illustrated calendars of the Natural History Museums of the University of Salento in the last 20 years.

I was also assisted by and I would like to thank and acknowledge:

Prof. Otto Larink, who kindly made available a collection of 33 photographic plates (comprising approximately 300 photographs taken under the compound microscope) of plankton organisms, thus enriching the textbook with an atlas annexed at the end of the main text.

Dr. Anna Maria Miglietta, who graphically realized all the remakes and newly proposed graphics and schemes of the textbook.

Dr. Giacomo Milisenda, who authored the chapter on pelagic fisheries.

References and literature are cited in the text only if actually necessary. I have avoided adding all references to the main text to facilitate the lecture for university students. However, each chapter has its proper literature list, which includes citations from the text as well as recommendations for further reading.

With contributions of (in alphabetic order):

Alberto Gennari, Natural history illustrator (drawings of species, habitats, and processes)

Otto Larink, Technische Universität Braunschweig, 38106 Braunschweig, Germany (plankton photographic atlas)

Anna Maria Miglietta, Museum of Marine Biology, University of Salento, 73100 Lecce, Italy (graphics and schemes)

Giacomo Milisenda, Stazione Zoologica Anton Dohrn, 90142 Palermo, Italy (pelagic fishery, Sect. 9.2)

Lecce, Italy Genuario Belmonte

Contents

Introduction

1

1.1 Terminology, Geographic Frame, and Main Water Mass Circulation

Pelagos is the aquatic world where life is independent from a solid bottom and/or a coastal world. Although the term can also be used by limnologists for fresh waters (lakes), its etymology is Greek and undoubtedly refers to the sea (Peleus, father of Achilles, was a personage linked to marine world and he was the husband of Thetis, also a goddess of the sea) (Fig. 1.1).

The term was coined to indicate open sea waters and was in opposition to the coastal realm. At sea, interactions with the bottom are inversely proportional to the distance from the coast; thus, pelagic existence will be enhanced by the distance from the coast. Pelagos, however, is also referred to as a lifestyle of living organisms; hence, even confined shallow waters can host pelagic species if these, generally small, spend their life in the water column without (or with negligible) interaction with the bottom. All situations are considered in terms of the relative size of each organism and its water world. Thus, even a drop of water is a pelagic environment for a bacterium.

Due to an increasing conditioning of bottoms on the water column, in progressive vicinity to coasts, the water column over the continent shelf (until approximately 200 m below sea level, where the continental slope starts) is characterized by the adjective **neritic**. This last situation is distinguished from the actual open sea, far from the coastline, and from the continental shelf, which is better described by the adjective **pelagic**.

The first layer of marine waters, from the surface down to 200 m below sea level, has the name **epipelagos** both in the neritic domain and in the pelagic domain (Fig. 1.2).

This upper layer of the pelagos domain is characterized by the most intense interactions with the atmosphere and anthropic pressure, which progressively diminish with depth. From an ecological point of view, the presence of sunlight

G. Belmonte, *Elements of Pelagos Biology*,
https://doi.org/10.1007/978-3-031-74280-4_1

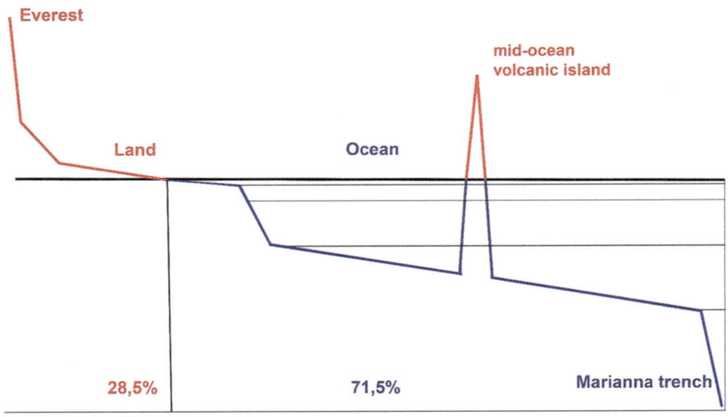

Fig. 1.1 Scheme (hydro-clino-gram) representation of the planet surface. Average altitude of lands = 800 m above sea level. Average depth of oceans = 3800 m below sea level. (Courtesy G. Belmonte, graphic composition of A. M. Miglietta)

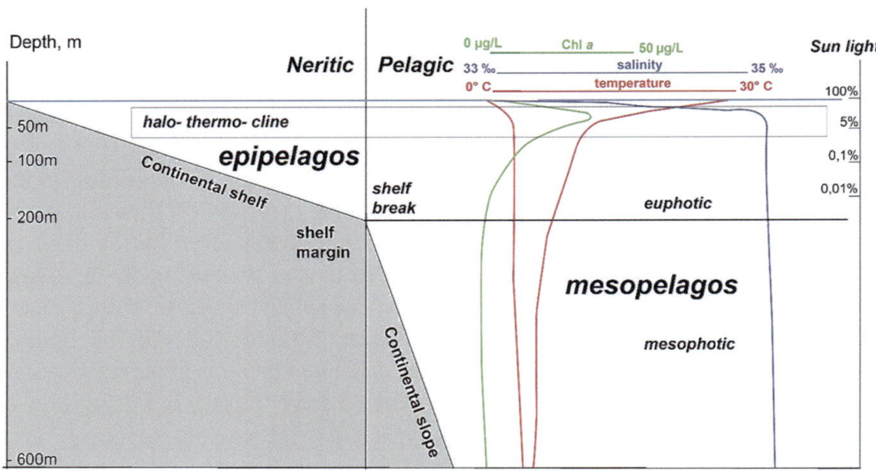

Fig. 1.2 Partitioning of the water column (only the first 600 m below the sea surface) with principal terms adopted in the present volume. (Courtesy G. Belmonte, graphic composition of A.M. Miglietta)

(variable in intensity and duration) is fundamental for productivity based on **photosynthesis** carried out by **autotrophic** organisms (mostly grouped in the **phytoplankton**). This productive pelagic layer is also defined as **euphotic** and is responsible for biomass formation and the sustainment of the whole trophic net in the pelagos. Among the abiotic conditions characterizing the epipelagos, we have to mention some typical features generally absent in the remaining water column, such as possible salinity variations (due to freshwater input from rain or rivers interesting the more surface layer) and temperature variations that follow the seasonality of atmospheric climate.

Regardless of the surface, salinity and temperature generally reach a constant value below a certain depth. The layers of this condition change (even intense according to the geographic site and/or the season) are called the **halocline** and **thermocline**, respectively, and they generally occur in the first 100 m below sea level (Fig. 1.2).

In addition, large geographic portions of epipelagos can be affected by the presence of ice at the level of polar regions.

A biological characteristic of epipelagos is the presence of lunged animals (those obliged to stay close to the surface because they are obliged to breathe air) that are generally large sized species and very representative of marine pelagic life (whales, dolphins, seals, turtles, and penguins).

Plankton, in general, are abundant and well represented in the epipelagos and, consequently, are also well represented the largest size classes of pelagic organisms (termed **nekton**), depending on such a food resource.

If plankton and nekton are also represented in deeper pelagic layers, what is exclusive to epipelagos is the **neuston**, a world of very surface organisms. A final, important characteristic of the epipelagos is the presence of the maximum ecological diversity of pelagic communities. This maximum generally occurs at the level the shelf break (in the water column over the margin of the continental shelf).

Below the epipelagos, the presence of an intermediate water layer between −200 and −1000 m is recognized, which is the **mesopelagos**, a quasi-dark world with a twilight undetectable from the human eye. This layer also represents the maximum possibility of interaction with the fishery and human pressure on the natural oceanic environment. Due to the very negligible presence of light, this layer is also called **disphotic**. The small quantity of sunlight reaching the mesopelagos, in fact, is not enough to fuel the process of photosynthesis. The abiotic conditions (e.g., temperature, salinity, pH, absence of light) are quite stable and are not interested in variations such as those typically occurring in the epipelagos. Due to the absence of light and photosynthesis, living organisms are only **heterotrophic,** and many of them perform daily vertical migrations to feed in the epipelagos at night. The absence of light can determine hypoxic or even anoxic conditions in those basins where an exchange between epi and mesopelagic waters does not occur. This is the case in the Black Sea, where brackish epipelagic waters (at salt concentrations of 18–22 g/L), even if cold, do not reach density values responsible for their sinking and exchange with deep, saltier waters (at 35 g/L of salinity). This absence of remixing, together with the absence of photosynthesis in the mesopelagic layer, produces an absence of oxygen below a depth of 200 m and a general **anoxia** that characterizes the whole mesopelagic Black Sea.

Below the mesopelagos, a third layer is recognized, the **bathypelagos**, between −1000 and −3000 m (Fig. 1.3). Here, the abiotic conditions are definitively invariable, and the water density is at its maximum possible (value of 1000) at a temperature of 3 °C, which is largely homogeneous in any geographic partitions of the planet. The maximum density of (pure) water is at 4 °C, but the salt content of oceanic water (average, 35 g/L) modifies the physical characteristics of this liquid material, moving the transition points between liquid and gas at more than 100 °C and that between liquid and solid at less than 0 °C.

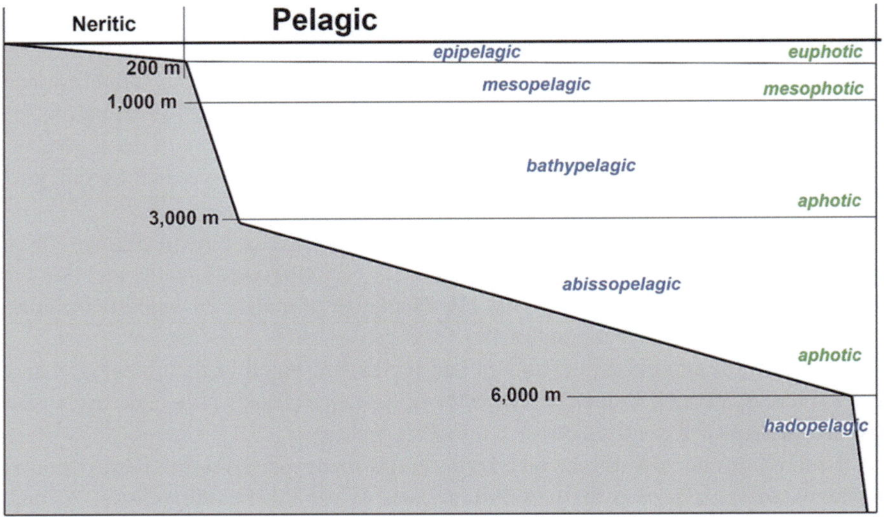

Fig. 1.3 Schematic representation of the pelagos vertical partitioning. (Courtesy G. Belmonte, graphic composition of A. M. Miglietta)

The salt water warmer or cooler than 3 °C is, for this physical principle, less dense than that having 3 °C and floats above this denser water. The cooling of surface water masses at 3 °C, at any point on the planet, provokes the reaching of its maximum density and its consequent sinking to deeper layers. At different points on the planet, this descending, dense water mass moves deep water masses that upwell to the surface somewhere else.

Sun light cannot reach this deep layer, but **bioluminescence** is diffuse among living organisms and justifies the presence of **photoreceptors** (even true eyes) in bathypelagic species.

Etymology of scientific terms
Auto = independent *Cline* = passage, transition *Epi* = upper *Eu* = good, true *Halo* = saline, salt *Nekton* = swimming community *Nero* = water *Neustos* = sailing at surface *Pelagos* = sea *Photos* = light *Phytos* = plant *Planktos* = going around *Syn* = together *Thermo* = heat *Trofe* = food

The physical environment of the bathypelagos is also hostile to organisms with calcareous structures (such as tests, shells and thecae of some plankton) because the water pressure provokes the dissolution of $CaCO_3$ in the water at more than 300 atm (those corresponding to 3000 m depth), as sugar in a cup of tea (on land, at 1 atm).

Below the bathypelagos (between 3000 and 6000 m below sea level) is the world of the **abyssopelagos,** the largest habitat of the planet, over the oceanic plains, which extends from the base of the continental slope at these depths (Fig. 1.3).

This world is dark (absence of sunlight) and constant in temperature (approximately 3 °C) and salinity (approximately 35%). It extends for approximately 50% of the planet's surface, being at the level of the average depth (reported as 3800 m) of the oceans. Such a large extension, however, is not accessible to humans, and our knowledge of the abyssopelagic is practically feasible and due to pioneer scientific explorations.

Below 6000 m, the pelagic world is called **hadopelagic** and classifiable as "unknown".

If some data have been obtained (due to scientific explorations) from the bottoms at these depths, the knowledge of pelagos is considered inexistent. The enormous pressure due to more than 6000 m of water on living mechanisms and biological functions was considered (until a few decades ago) incompatible with the chemical life of proteins (these molecules were considered denatured at such a pressure value) and, consequently, with life as a whole. Some life forms, however, exist under these conditions and have been reported from a maximum of more than 10,000 m below sea level. This is the world of oceanic trenches (whose maximum depth has been measured in the **Mariana Trench at −10,911 m**), and most of them are isolated basins that should represent independently evolving units that are not interconnected.

Regarding the **circulation** of principal water masses, the solar energy that differently hits the earth surface produces gradients of heat that are able to provoke changes in the water mass position. The planet's rotation somehow dictates where wind and waters are directed, and land distribution drives these movements according to the continent's interposition along the main current routes.

At the equator, where the solar energy arrives with its maximum power, evaporation produces a rise in humidity, which, on reaching the high (and cold) atmospheric layers, generates rain that falls down. The continuous ascension of humidity at the equator obliges the pre-existing cold air layer, deprived of humidity by rain, to move apart (south and northwards) and to fall down (because cold and hence denser) to the ground at the level of the tropics. The continuous ascension of humid air at the equator attracts air from the tropics in a circulation cell that is very regular at most of the equator length.

Abyssus = unlimited, unmeasurable (depth) *An* = without, deprived *Bathos* = deep *Bios* = life; *Lumen* (L) = light *Dis* (L) = contrary, oppositional *Hado* = *Ade* = the underground world for souls after death *Heteros* = different *Mesos* = intermediate *Neustos* = sailing at surface *Pelagos* = sea *Photos* = light *Planktos* = going around *Recipio* (L) = receive

Due to the planet rotation (eastwards directed), each fluid moving at its surface is deviated by the Coriolis force on the right (in the northern hemisphere) or on the left (in the southern hemisphere) of its movement direction. This deviation produces a constant westward direction for all winds converging from the tropics to the equator (the so-called **westerlies**).

1.2 Pelagic Biogeography

The geographic subdivision of the pelagos is somewhat impossible if based on land subdivision criteria. The pelagos, in fact, is a fluid water volume interested in vertical and horizontal remixing processes. On land, the biodiversity distribution is linked to the soil or even anchored to it (if the vegetation part is considered), but managed by climate and a moving and mixing atmosphere. In the ocean, the moving and mixing medium contemporaneously contains biodiversity, which, consequently, is not simply affected by but participates in and characterizes moving water masses.

Each homogeneous water mass evolves with time and modifies (more or less intensely) the extension and margins of each recognized biological zone. With this consideration, only large and approximate areas of the pelagos have been classified and given a geographic name.

Box 1.1 The Main Oceanographic Currents

Due to the different heat storage capabilities of materials distributed at the surface of the planet (essentially rocks of continents and water of oceans), different atmospheric heat is produced over them (Fig. 1.4).

Differences in atmospheric temperatures cause air movements (winds) from high atmospheric pressure locations (colder) towards low atmospheric pressure locations (warmer). Where such a wind force is intense and persistent enough, wind-driven currents are also generated in the surface water masses of the ocean. The movement of water masses at the surface finds obstacles or is deviated by the interposition of continent masses along such current routes. In addition to these horizontal circulations, different temperatures at different latitudes determine the vertical movements of water masses due to changes in density values (dependent on temperature). Thus, at the poles, the ocean water reaches its maximum density and sinks to the bottom, attracting new and less dense water at the surface from surrounding temperate areas. This flow towards the pole (cold areas) offers the possibility for deep water to re-emerge in different points of the planet (areas of **upwelling**) with the redistribution of nutrients accumulated on the ocean's deep layers and deriving from the digestion of organic material confluent from any part of the planet.

Such a complex scenario of horizontal and vertical flow of ocean water masses produces relatively constant currents that receive a name and could be geographically identified.

As a general rule, the most constant currents are produced by winds that blow from approximately 30° latitude north and south towards the equator, with a NE–SW direction in the northern hemisphere and a SE–NW direction in the southern hemisphere. The water masses pushed towards the equator and westward turn northwards and southwards when they encounter a continent, sustaining a counter current (at higher latitudes) that closes the circuit. When the westwards flow drifts apart the surface ocean from a continent, an occasion is created there for the upwelling of deep, nutrient-rich water. In general, such a scheme of surface water flow produces a latitudinal enlargement of warm water (and climate) on the west border of the oceans and its narrowing on the eastern border. The subtraction of surface water from East Ocean borders also creates the occasion for the upwelling of deep waters and the reinforcement of cold water situations.

Generally, in the neritic environment (over the continental platform), pelagos and benthos are more closely interdependent and share the same geographic partitioning. In the offshore ocean pelagos, however, the interdependence with benthos is minimized or null, and the zonization of the two domains does not overlap.

One possible generalization is that the upper ocean layer is more affected by atmospheric climate and can be subdivided in more parts than the deeper layers, which are more stable and homogeneous at increasing depths.

Robert G. Bailey (1998) proposed a partitioning of the surface ocean waters into three main **domains**: polar (two areas), temperate (two areas), and one warm intertropical area, arranged according to the latitude between poles and the equator. The 3 domains (ecoregions) of Bailey were further partitioned into 14 **divisions** on the basis of latitude positioning, salinity, wind regime, and precipitation-evaporation balance (Fig. 1.5).

Physical geographical oceans, however, do not coincide with biological oceans. These are the consequences of the ancient (cultural) separation of the students of two science fields. For this reason, the terminology used by physical geographers to describe and subdivide the world has to be substituted with terms of

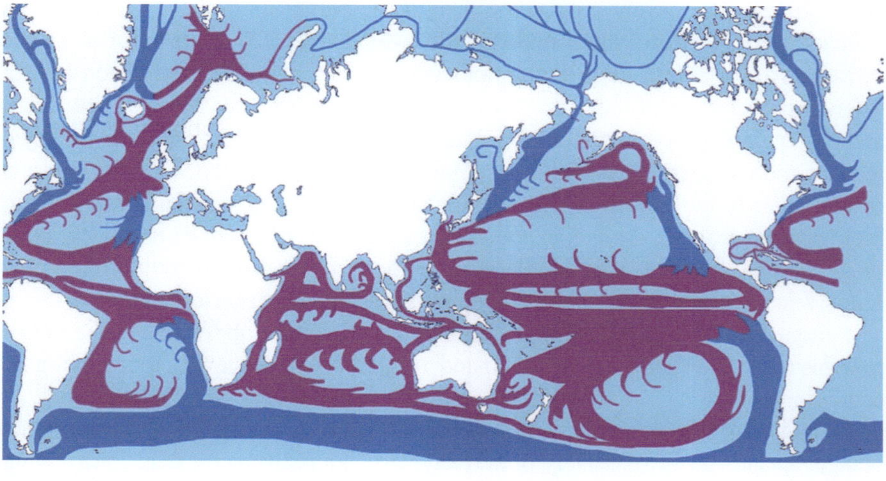

Fig. 1.4 Schematic representation of surface ocean circulation. Warm currents (warmer than the average temperature of the geographic latitude) are indicated in violet. Cold currents (colder than the average temperature of the geographic latitude) are indicated in blue. (Courtesy G. Belmonte)

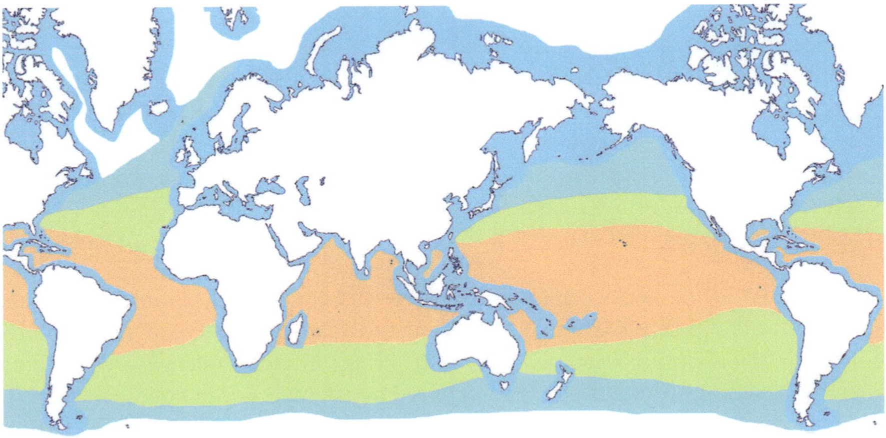

Fig. 1.5 Climatic subdivision of epipelagos according to three climate domains. (Courtesy G. Belmonte)

biogeographers, and we have to continuously specify that "region" and "province" are not geographic terms but biogeographic terms (they will always be used with the prefix "bio" in the present text).

Even at corresponding latitudes, and hence with corresponding oceanographic situations, the biological world in different oceans could be characterized by different assemblages, and different oceanographic units can host homogeneous biological assemblages, thus not coinciding with oceanographic space units.

As an answer to Bailey's (1998) proposal (3 domains and 14 divisions), Spalding et al. (2007) defined terms and biologically subdivided the coastal, epi-continental seas into **12 coastal domains** (realms), **62 provinces**, and **232 ecoregions**. Spalding et al. (2012) successively proposed a geographic partitioning of the open sea (the pelagos proper), identifying 4 **ocean realms** divided into **37 bioprovinces** for the surface layer of 200 m thickness (epipelagos). Such a complex subdivision clearly offers the dimension of the distinctness of neritic areas (the coastal ones between 0 and 200 m depth) based on species richness and typology, depicting the coastal systems as more diversified (also because more studied) and more fractioned than the open ocean. In both partitioning of epipelagos, however, it was undebatable that, biologically speaking, the Indian and Pacific oceans are different names of the same biogeographic unit and that the Eastern Pacific is distinguished by the remaining Indo-West Pacific, notwithstanding being geographically internal to the Pacific Ocean.

Sutten et al. (2017) also attempted to give a geographic subdivision of the **mesopelagos** (the water layer between 200 and 1000 m below sea level), recognizing 33 ecoregions.

In analysing species richness, however, it is useful to take into consideration that pelagic species move (passively or actively) over long distances, thus contributing to extending and intertwining the different bioareas.

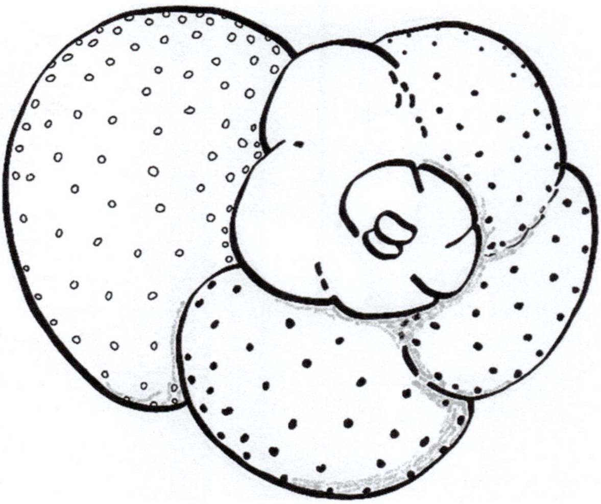

Fig. 1.6 *Globigerina glutinata*. The eukaryote species most distributed in the world's oceans (present in 578 cells of the 2065 which partition the whole ocean according to Costello et al. (2017)). Only the calcareous, multilocular shell is represented. (Courtesy G. Belmonte)

Generally, the microbiota of plankton is composed of species among the most widespread in the whole planet's oceans, and the eukaryotic species with the largest geographic distribution in the oceans is planktonic *Globigerina glutinata* (Foraminifera), reported from 28% of the 2065 50° cell partitioning of the oceans of the world (Fig. 1.6).

Large species are able to migrate long distances and have long life durations. Such megafauna, together with typical plankton, contribute to maintaining the large extension of biopelagic areas.

More than ocean gyres, currents, and continent interpositions, the study of Costello et al. (2017), based on the geographical distribution of 65,000 species, suggested the existence of bioassemblages relatively independent from physical oceanography and sometimes more confident with palaeogeography data. For example, the Mediterranean and Red Seas (separated by the Suez isthmus) were diverse up to the point they are included in different realms. On the other hand, the Caribbean Sea in the West Atlantic and East Pacific are separated by the Panama isthmus but belong to the same realm. This is probably due to the different periods of isthmus emersion for tectonic reasons and the time necessary for a different evolution of marine communities on the two opposite sides (in fact, the Panama isthmus is geologically younger) (Fig. 1.7).

Box 1.2 Terminology for Marine Biogeography

According to Spalding et al.'s terminology relative to epipelagos (Spalding et al. 2012), a **Realm** is "*the largest scale unit of phylogenetic organization in the oceans ('regions' of other authors) across which biotas are internally coherent at higher taxonomic levels as a result of a shared and unique* *evolutionary history. Realms have high levels of endemism, including* ***unique taxa at the genus and family levels in some groups. Driving factors beyond the development of such unique biotas include temperature, historical and broad-scale isolation, and the proximity of the benthos***."

A **Province** (here a **Bio-Province**) is "*a large area defined by large-scale, spatially and temporally stable (or seasonally recurrent) oceanographic drivers (ocean gyres, equatorial and other upwellings, semienclosed pelagic basins). Provinces present distinct biotas that have at least some cohesion over evolutionary time frames. Provinces will hold some level of endemism, principally at the level of species. Although historical isolation will play a role, many of the distinct biotas have arisen as a result of distinctive abiotic features that circumscribe their boundaries.*"

A **Biome** is "*a grouping of provinces sharing oceanographic processes (boundary current systems, mid-oceanic gyres, etc.). They may be separated by large physical distances and characterized by very different evolutionary histories. They would therefore be expected to host ecosystems with comparable structural and functional properties but would not host the same species.*"

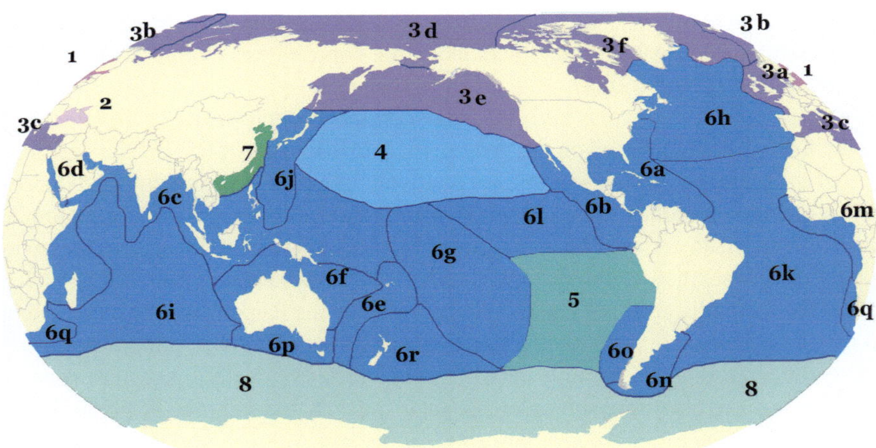

Fig. 1.7 Epipelagic biogeography proposed by Costello et al. (2017) on the basis of the distribution of 65,000 species in 2065 oceanic cells delimited by 5° of latitude and longitude. Geographic units (Bioprovinces) of the same Biogeographic Realm have the same colour. (From Costello et al. 2017, modified by G. Belmonte)

The proposal of Costello et al. (2017) recognizes **30 geographic units** (seas, or bioprovinces) nested in **8 biogeographic realms**.

Realms can be small (e.g., the Black and/or Baltic Seas) or largely extended (e.g., Realm 6, Warm Temperate seas, with 18 bioprovinces) but in any case they have similar biodiversity assemblages characterized by an elevated level of endemism.

Realm 1. Inner Baltic Sea.

No subareas; 458 species distribution examined; endemism, 63%.

The realm is small, and biological associations are heavily conditioned by freshwater (in the Gulf of Bothnia, waters have a salinity less than 5‰), by a short history (the Baltic Sea started to exist after the last glacial melt completion, approximately 15,000 years ago), and by an enclosed geographic configuration that impedes complete communication with the northeast Atlantic.

Realm 2. Black Sea.

No subareas; 192 species distribution examined; endemism, 84%.

The realm is small, and biological associations are conditioned by brackish water (salinity less than 22‰ at any point on the surface), by a young historical communication with the Mediterranean Sea (which has started to be connected with the Black Sea through the Bosphorus after the last glacial melt completion, approximately 15,000 years ago), and by an enclosed geographic configuration that impedes complete communication with the Mediterranean Sea. Biological assemblages are possible only in the first 200 m from the surface because below such a depth, light cannot support the photosynthetic derivation of oxygen, and this is scarce or absent at all.

Realm 3. Cold and temperate areas.

n.6 biogeographic units (seas).

3.a North East Atlantic; 7117 species distribution examined; endemism, 27%.

All the seas around the Britannic archipelago and the North Sea are part of such a biounit. The biological connection existing with the North Ocean is probably derived from the action of colonization of northern areas during ice cap retirement between 30,000 and 12,000 years ago from the Northeast Atlantic. During the last glacial period, the North Sea and the England Channel were also absent due to partial submersion below the northern ice cap of that period and to the emersion of sea bottoms as subaerial prairies with cattle and mammoths.

3.b Norwegian Sea (part); 1345 species distribution examined; endemism, 43%.

It is a cold temperate biounit that extends between 60° and 80° latitude or more due to the absence of ice (at least for a part of the year). This situation is due to the ultimate effect of the Gulf Current coming from the western tropical Atlantic. At the same latitudes, on the opposite side of the Atlantic, close to the American continent, ice and icebergs are known to produce troubles and problems in navigation.

3.c Mediterranean Sea; 3096 species distribution examined; endemism, 45%.

It is a cold temperate biounit at the boundary with a subtropical area. The ongoing warming climate is currently producing the so-called **tropicalization** of the Mediterranean Sea. Dried or strongly reduced from **7 to 5 million years** ago, it has successively been invaded by North East Atlantic fauna. Since 1869, a communication (**the Suez Canal**) has also been opened with the Red Sea, which today is responsible for most of the tropicalization of the Mediterranean biota.

3.d Arctic; 1907 species distribution examined; endemism, 19%.

The ice cap impedes the penetration of sunlight in the liquid ocean, thus provoking the absence of autotrophic eukaryotes and bacteria. The ice-free area enlarges seasonally, thus allowing the migration of pelagic autotrophs from temperate oceans. Such colonization is probably responsible for the smallest endemism found among all the biounits of the Costello et al. subdivision.

3.e North Pacific; 5535 species distribution examined; endemism, 27%.

Part of the Pacific Ocean from the Bering Sea to the La Perouse Strait south of Sakhalin Island on the west side and to the extreme south US coast on the east side.

3.f North America subarctic; 1492 species distribution examined; endemism, 31%.

The biounit contains Iceland and Greenland, other than all the northern Canadian archipelago. Due to surface ocean circulation, ice here is more present than at the corresponding latitudes on the opposite side of the North Atlantic (Europe).

Realm 4. Subtropical.

Mid-Tropical North Pacific; 2859 Species Distribution Examined; Endemism, 47%.

This is the only biounit of this realm. It does not touch continents and is oceanographically characterized by the large oceanic gyre existing at these latitudes.

Realm 5. South East Pacific Ocean.

South East Pacific Ocean; 1618 Species Distribution Examined; Endemism, 59%.

A single important biounit in the southeast of the Pacific Ocean interested, on the extreme east, by the Humboldt current of cold waters coming from the south, and the upwelling area connected.

Realm 6. Warm temperate areas.

n.18 biogeographic units (seas).

6.a Caribbean Sea and Gulf of Mexico; 13,281 species distribution examined; endemism, 30%. Area where the **Gulf Stream** is generated. Notwithstanding its positioning at low latitudes in the Atlantic Ocean, its biological community shows similarities with that of the Eastern Tropical Pacific, on the other side of the **Panama isthmus**. The age of such a geological structure is approximately 3 million years, and this geologically short time is responsible for the similarity still existing between the two subunits and their belonging to a unique Realm.

6.b Gulf of California; 3279 species distribution examined; endemism, 30%. The name does not describe the whole extension. The biounit, in fact, spread from the Gulf of California, on the north, to the Ecuadorian coast, on the south. The biounit contains the Galapagos Archipelago and has a biological community closely similar to biounit 6a (Caribbean Sea and Gulf of Mexico) in the Atlantic Ocean.

6.c Indo-West Pacific extended; 16,505 species distribution examined; endemism, 31%. The **largest biounit** of the planet's oceans. Characterized by coastal (thus not properly pelagic) communities with a high level of species richness. Coastal epicontinental seas and true pelagic areas are equally represented.

6.d Red Sea; 997 species distribution examined; endemism, 74%. A small sea (less than 5 geographic units of the Costello et al. subdivision based on 5° squares), characterized by intense evaporation and a small communication with the Indian Ocean only from approximately 15,000 years, when the melting of the ice produced a sea level lift of approximately 120 m useful to establish such a communication at level of the strait **Bab el Mandeb** (here the marine bottom has a threshold at approximately −85 m). This isolation, however, has been a repeated situation in past glacial oscillations and produced an intermittent injection of biological propagules from the Indian Ocean, with intermittent possibilities of independent evolution during the isolated periods. Since **1869**, it has been connected with the Mediterranean Sea throughout the artificial **Suez Canal**.

6.e Tasman Sea; 1468 species distribution examined; endemism, 57%. Part of the southwest Pacific between Australia and New Zealand.

6.f Coral Sea; 10,349 species distribution examined; endemism, 33%. Characterized by the presence of a madreporic reef along the north-east coast of Australia.

6.g., South Tropical Pacific; 2818 species distribution examined; endemism, 44%. It extends from New Guinea, on the west, to French Polynesia, on the east. The Hadal environment is particularly present (with principal oceanic trenches of the planet).

6.h North Atlantic; 7591 species distributions examined; endemism, 26%. A large pelagic area exists between North East America and West Europe, and the American side is interested in the Gulf Stream of tropical waters, which arrives to mitigate West Europe. This is the biological unit responsible for the biological replenishment of the Mediterranean basin, which started 5 million years ago.

6.i Indian Ocean; 3486 species distribution examined; endemism, 43%. The open ocean part of the Indian Ocean, characterized by the absence of islands and/or archipelagos.

6.j West Pacific; 4678 species distributions examined; endemism, 40%. The Japanese archipelago and other minor islands close to oceanic trenches.

6.k South Atlantic; 5512 species distributions examined; endemism, 33%. One of the most extended ocean units, it stretches from the north tropic line (Tropic of Cancer) to the extreme south capes of South America and South Africa.

6.l Mid East Pacific; 1217 species distribution examined; endemism, 36%. Far from continents.

6.m Gulf of Guinea; 992 species distribution examined; endemism, 57%. Coastal biounit, from Cabo Verde to Angola.

6.n Rio de la Plata; 1651 species distribution examined; endemism, 45%. Extended from the mar del Plata to the Chilean sea, it contains the Malvinas archipelago, and Cape Horn (the southernmost geographic point of South America).

6.o Chile; 584 species distribution examined; endemism, 68%. Although geographically coastal, it does not receive water from rivers, and contains very deep oceanic bottoms. This allows the resuspension of cold and nutrient waters due to the occurring water circulation.

6.p South Australia; 2158 species distribution examined; endemism, 40%. A small biounit in the southern hemisphere.

6.q South Africa; 6700 species distribution examined; endemism, 45%. All the waters around South Africa either in the Atlantic Ocean or in the Indian Ocean.

6.r New Zealand; 3126 species distribution examined; endemism, 33%. A small biounit around the New Zealand islands.

Realm 7. North-West Pacific.

7. North-West Pacific. A coastal, internal sea between Corea (north) and Vietnam (south) in a mostly epicontinental situation.

Realm 8. Southern (Antarctic) Ocean.

8. Southern Ocean. Homogeneous, it is characterized by currents that do not encounter continents and thus perform complete circumnavigation of the planet at the same latitude.

Most of the zooplankton in pelagic waters below 200 m (mesopelagos, bathypelagos, abyssopelagos, and adopelagos) migrate up every day between different pelagic layers for alimentary reasons.

Regardless of the temperature at the surface, in the deepest part of the mesopelagos, it reaches a constant value of approximately 3 °C (the temperature corresponding to the maximum density value for marine water). The sunlight that sustains the chlorophyll-based photosynthesis in the epipelagos produces only a twilight effect in the upper mesopelagos before the passage to complete darkness. This is also the frontier for fishery devices, and fishermen collect biological resources at not more than 800–1000 m below sea level but in any cases trawling close to the bottom.

Sutton et al. (2017) are the only scientists who attempted a **geographical partitioning of the mesopelagos**. Their attempt was based on the physical-chemical characteristics of mesopelagic water masses (Fig. 1.8) and showed the surprising

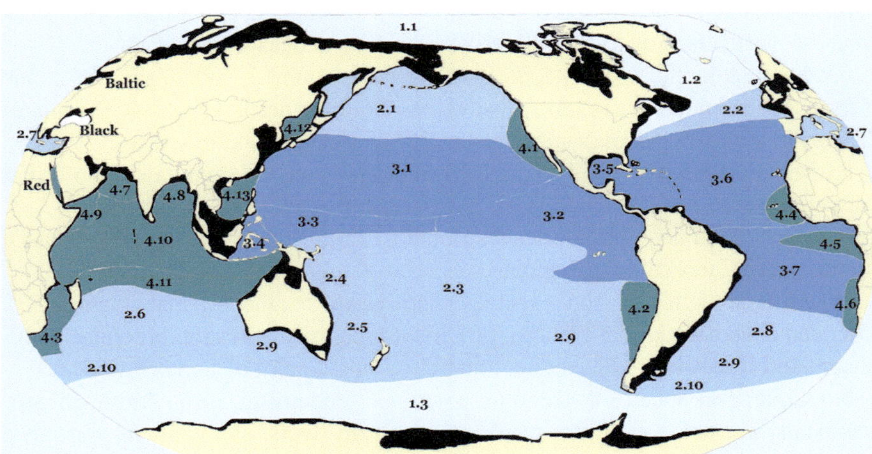

Fig. 1.8 Biogeographic partitioning of the mesopelagos on the basis of the water mass characteristics. (From Sutton et al. 2017, modified by G. Belmonte, licensed under CC BY 4.0)

presence of 33 bioregions (corresponding to the biounits of Costello et al. 2017) nested in 4 biomes (corresponding to the realms of Costello et al. 2017) notwithstanding the absence, from the list, of shallow areas from the comparison and of some deep basins for absence.

In the attempt to uniformize terminologies and geographic names between the two approaches, the list of Sutton et al. (2017) ecoregions is presented here, with numbers and names remembering the epipelagos partitioning of Costello et al. (2017).

Mesopelagic Realm 1. Inner Baltic Sea.

Absent in mesopelagos biogeographic partitioning, probably for the extreme reduction of the appropriate layer (200–1000 m below sea level).

Mesopelagic Realm 2. Black Sea.

Absent in mesopelagos biogeographic partitioning, in any case characterized by a substantial absence of eukaryotic life below the epipelagos (200 m below the sea surface) due to persistent anoxia.

Mesopelagic Realm 3. Polar Biome.

Geographically corresponding to Realm 3 Cold and Temperate areas and Realm 8 Southern (Antarctic) Ocean. It contains a total of three ecoregions against the seven biounits of the epipelagos.

Arctic. Cold, low salinity water, circulation restricted by continental masses. Twenty percent of zooplankton species are endemic.

Sub-Arctic NW Atlantic. Deep areas directly connected with the Arctic Ocean.

Antarctic/Southern Ocean. Cold waters delimited by the polar front. Seasonally highly productive plankton, with high endemism among zooplankton. Mesopelagic fish community with low species richness and absence of endemism.

Mesopelagic Realm 4. Westerly Winds Biome.

Geographically corresponding to parts of different Realms of the epipelagos. Contains a total of ten ecoregions from any geographic ocean and ecoclimatic province. Apart from the first two ecoregions, maximum algal production (bloom) occurs in the winter, and nutrient limitation occurs the rest of the year.

Sub-Arctic Pacific. There is an important spring phytoplankton bloom and a secondary bloom in summer–autumn. Sunlight and/or nutrient limitation in the other periods.

North Atlantic Drift. As in the ecoregion Sub-Arctic Pacific, pulses of algal blooms alternate to low productivity periods (strong seasonal oscillations).

South Central Pacific.

Coral Sea.

Tasman Sea.

Southern Indian Ocean.

Mediterranean Sea.

South Atlantic Ocean.

Circumglobal tropical front.

Sub-Antarctic seas.

Mesopelagic Realm 5. Trade Winds Biome.

Geographically corresponding to parts of different realms of the epipelagos. Seasonality is weak, and primary production is low. A deep chlorophyll maximum is a persistent feature. Contains a total of seven ecoregions.

Northern Central Pacific.

Eastern Tropical Pacific.

Equatorial Pacific.

South East Asian Pocket.

Gulf of Mexico.

Central North Atlantic.

Tropical/West Equatorial Atlantic.

Mesopelagic Realm 6. The Neritic Biome Far from the Coastline.

Geographically corresponding to parts of different realms of the epipelagos. Oceanic circulation is modified by continental topography and associated local winds. Algal blooms are produced by different oceanic effects (terrestrial runoff, upwelling, and tides). The biome contains a total of 13 ecoregions. Ecoregions 1–6 are coastal sites of pulsed production and upwelling. Ecoregions 7–11 exhibit large amplitude conditioning by monsoon reversals. Ecoregions 12–13 are "marginal", and each is characterized by unique features due to the surrounding continental masses.

California current.

Perú/Humboldt current.

Agulhas current.

Mauritania/Cabo Verde.

Guinea/East Equatorial Atlantic.

Benguela current.

Arabian Sea.

Gulf of Bengal.

Somali Sea.

Northern Indian Ocean.

Mid Indian Ocean.

Sea of Japan.

China Sea.

Mesopelagic Realm 7. Red Sea.

Mesopelagos are completely isolated and not connected with other ecoregions. The isolation, the deep temperature at 22 °C, and the salinity of 42‰ realize unique evolutionary conditions for the Red Sea mesopelagic fauna, which are still largely unknown.

Mesopelagos have a total extension that is smaller than that of epipelagos. This would justify a simplification of its biogeography. It is characterized, in addition, by a narrower range of environmental variables (up to the assumption that they are constant), justifying the reduction of possible habitats to just one. Neither the space extension nor the environmental variability could justify the diversified biogeography found by Sutton et al. (2017). This suggests that the isolation that large communities experience for a long time under constant conditions could produce the evolution of diversity over geological time. Such a hypothesis waits to be clarified, but some recent data (Vannini et al. 2021) on mesopelagic zooplankton suggest that their diversification is higher than that measurable in surface plankton at the geographic scale of the Mediterranean Sea.

The bathy-pelagos (1000–3000 m below sea level) are even more unknown than the meso-pelagos, and only some scientific expeditions have reported data from the area. In any case, it must be considered that some large lunged carnivores (Cetacea Odontoceti) regularly perform apnoea dives to a depth of 1000–3000 m to obtain feeding resources from immense prey populations.

As in the case of such apnoea performances among Mammalia (porpoises and sperm whales), which allow us to suspect an abundant availability of food at bathy-pelagic depths (1000–3000 m), the presence of eyes in all the animals also suggests that darkness should not be complete where sunlight is absent. Most likely, bioluminescence (diffused among all phyla) is an intense and widespread activity in deep-world organisms.

Large species distributions favoured by pelagic environments should not only connect distant populations but should also reduce intraspecific genetic variability and slow speciation rates, thus determining a broader compartmentation than in shallow waters or on land. Paradoxically, rates of speciation can also be as high for pelagic taxa as for shallow marine and terrestrial species, suggesting that high dispersal and large population size have not impeded species formation at sea. Biogeographic ranges of living and fossil species provide evidence that the distributions of pelagic species are set more by environmental limits on reproduction and population maintenance than by limits to dispersion. Clearly, the distributions of some well-known fish species are controlled not by failure to disperse but by factors that limit successful reproduction or population growth outside their core ranges.

In conclusion, we must consider that our ignorance of pelagic biodiversity is probably the main driver of the approximate and vague depiction of pelagic zonization that we humans have, probably distant from an actual situation.

1.3 The Palaeogeography of Oceans

The affinity of flora and fauna and their geographic distributions on continents allowed the naturalists to propose a geographic subdivision of living organisms on the land (biogeography) that did not agree with the canonical physical geography. To justify some intercontinental bio-connections, continental bridges were proposed, and the flooding of ancient, interposed continents (such as *Atlantis* in the middle of the Atlantic Ocean) was proposed, although never proven. The continental drift theory of Alfred Wegener (1915-1919) resolved the problem, unveiling an incredible reality: continents continuously move their positions and, over long periods of times, collide or separate from each other. This proven theory has the consequence of indirectly declaring that oceans also change their size and contours with time.

The new ocean floor is slowly but continuously produced at the level of **mid-ocean ridges,** where crustal plates drift apart from each other and fluid magma emerges from the break points. The newly produced ocean floor must be consumed elsewhere to maintain the equilibrium of the total earth surface extension. The areas where the ocean floor is consumed are called **trenches,** and the ocean floor returns to the magma of the subsurface lithosphere. The problem of production–consumption concerns just and only the ocean floor. This part of the lithosphere has a **basaltic** composition and is denser and heavier than the continental lithosphere, which is **granitic**. Thus, continents are modified in geography and position but are always at the surface, while ocean floors are continuously renewed. After the formation at the level of an ocean ridge, the ocean floor is on the top of a submarine mountain crest (mainly volcans) at approximately 2000–3000 m of depth. With the production of new ocean floors, the pre-existing floors drifted apart along a submarine slope and until the ocean's deep plains, at some hundred km from the ridge and at 4000–6000 m of depth. The increased water column weight and the sedimentation process (proportional to the age of the bottom and to the height of the water column) contribute to the compaction of the oceanic mud floor until a depth where the melting of the deepest basaltic layers, in contact with beneath mantle magma, allows the submersion of the ocean floor below the continental crust (at the level of trenches). At this level, the continent tends to move because the ocean crust tends to push it, and sinks in the fluid magma below the continent margin. The continent at this point starts to return towards the centre of the ocean (an ocean closure period) due to the opposite pushing force of a new forming ocean floor (on one side) and the contemporary cessation of the opposite action of the ancient ocean floor (on the opposite side). A complete cycle of ocean opening (formation) and closure seems to last approximately 500 million years. Thus, nowhere, on the planet's surface could we find an ocean floor older than 200–250 My (the ocean floor of 200–250 My is that which starts to sink in the magma, thus allowing the start of the closure process of the ocean).

Finally, the changing position of continents must be carefully considered for the consequences it has on oceanic main surface currents. Today, in fact, an equatorial surface current that is westerly directed finds obstacles at the level of continents and is turned southwards and northwards by them with implications for the global climate (e.g., the formation of the Gulf Stream in the tropical Atlantic, which mitigates the climate of northern Europe). Such a situation is where we live, but an uninterrupted oceanic equatorial westward current was present until 10 M y ago because continental masses were not positioned to impede the westward equatorial oceanic circulation.

In summary, we must consider that oceans (and continents) are not geographically stable. Two large continental masses existed (**Laurasia** in the Northern Hemisphere and **Gondwana** in the Southern Hemisphere), and a large, global ocean (**Panthalassa**) was present with a narrow extension just where continents were separated. From that moment on, the Atlantic Ocean started to open, thus dividing the northern continent into two main parts, and successively, this also occurred at the level of the southern continent, which gave origin to South America, Africa, India,

Australia, and Antarctica, contemporaneously circumscribing new oceanic areas. The complete opening of the Indian Ocean (today between Africa, India, and Australia) was parallel with the closure of the Thetis, whose bottoms are today uplifted in mountain chains of the Alpine orogenesis from the Pyrenees to the Himalaya.

Box 1.3 Palaeogeographic Maps of Oceans and Continents

See Fig. 1.9.

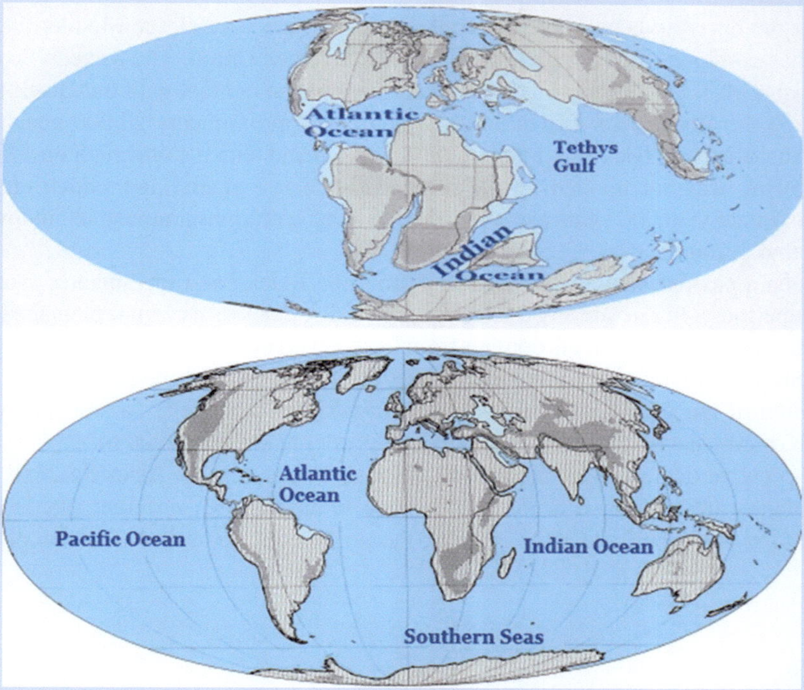

Fig. 1.9 World maps from 60 (top) and 5 (bottom) million years ago. (From Smith et al. 1994 modified by G. Belmonte)

1.4 Oceanography and Physical Geography of the Mediterranean Sea

The Mediterranean Sea is the largest "internal" sea on the planet and extends approximately **2,500,000 km²**. It is part of the cold temperate marine realm and one of the seas most historically affected by the presence of human settlements and activities along its coasts. It is sometimes inserted with the Black and Azov Seas in the Ponto-Mediterranean province (in terrestrial biogeography, overall), but neighbouring Black and Azov Seas (for a total of **469,000 km²**) are distinct and exceptional environments representing an isolated marine realm (Costello et al. 2017) .

The Ponto-Mediterranean area is more extended according to longitude (a total of 48°, from 5° W to 43° E) than according to latitude (a total of 16°, from 30° to 46°N) in the temperate zone of the planet, thus showing an approximately homogenous climate (Fig. 1.10).

The Mediterranean Sea is connected with the Atlantic Ocean from 5 million years via the **Gibraltar Strait**, which varied in widening up to only 13 km today and with a submerged threshold of −365 m.

At the opposite geographic extremity, in Egypt, since 1869, the Mediterranean Sea has communicated with the Red Sea via the **Suez Canal**. The Aegean Sea (a Turkish-Greek Gulf of the Mediterranean) has been in contact with the Black Sea through the **Bosphorus–Dardanelles** system for approximately 20,000 years. In fact, the Bosphorus passage, today wide only 1 km and with a submerged threshold at −50 m, was not flooded by the sea during the last glacial age, which ended approximately 30,000 years ago when the sea level was approximately 120 m lower than that of today.

Such a passage was a freshwater exit from the Black Sea (a freshwater lake in those periods). In fact, during the ice age, the Black Sea was a basin replenished by the freshwater of large rivers coming from Europe and the Sarmatic plains. The lack of communication between the deep waters of the two basins is responsible for the isolation of the deep **Black Sea,** which is considered practically "**azoic**" below a depth of 200 m and until the maximum depths are greater than 2000 m.

Typical vertical migrations of zooplankton do not exist in the Black Sea, and the epipelagos are organized according to three biolayers with different planktonic predators: the uppermost layer is under the predaceous control of the ctenophora

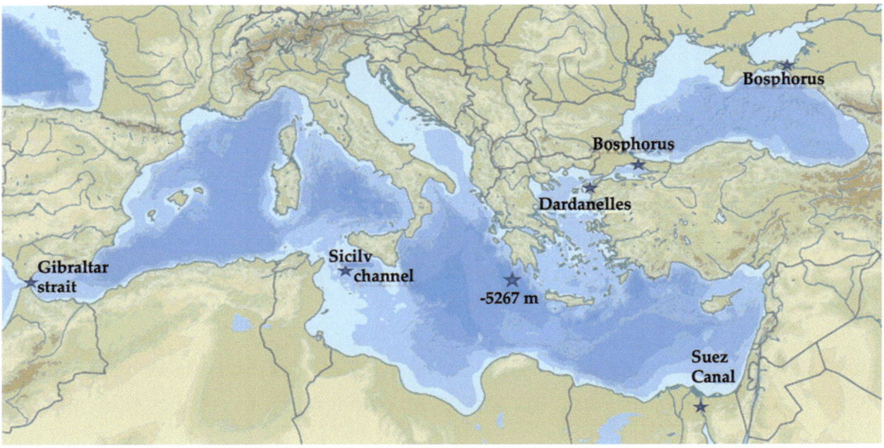

Fig. 1.10 Physical map of the Mediterranean Basin. Stars indicate critical points. (Courtesy G. Belmonte)

Pleurobrachia pileus; the intermediate layer is characterized by the Copepoda *Calanus finmarchicus*, and the third layer is under the control of the chaetognatha *Sagitta setosa* (Fig. 1.11).

The **Azov Sea** is the smallest portion of the Ponto-Mediterranean area. It is a shallow and brackish area that communicates with the North Black Sea through the Cimmerian **Bosphorus** which is only 3 km wide. The Azov Sea does not have a mesopelagic layer because its maximum depth is 14 m.

This means that the basin closely resembles a large lagoon due to the very brackish conditions (salinity, 10 g/L).

Differently from the Black and Azov Seas, the Mediterranean is geographically subdivided into a western, northernmost part, and an eastern, southernmost one, separated by the Italian peninsula and Sicily, and communicating at the level of the **Sicily Channel**, where a threshold is present at a depth of **427 m**. The neritic (epicontinental) areas of the Mediterranean are mainly in the central-northern Adriatic and in the Tunisian Sea.

The Mediterranean Sea is never deeper than 6000 m; hence, it does not have the Hadal portion of ocean life (maximum depth, close to Cape Matapan, Greece, **−5267 m**). The average depth of the western basin is 1650 m, and that of the eastern basin is 1570 m.

Due to the threshold existing at Gibraltar (−365 m), the deep and cold waters of the Atlantic Ocean cannot penetrate into the Mediterranean basin. The deep waters of the Mediterranean, consequently, have a minimum **temperature** of **12–13 °C**, which is the minimum reachable in its northern part during winter (thus being the denser water masses produced inside the basin).

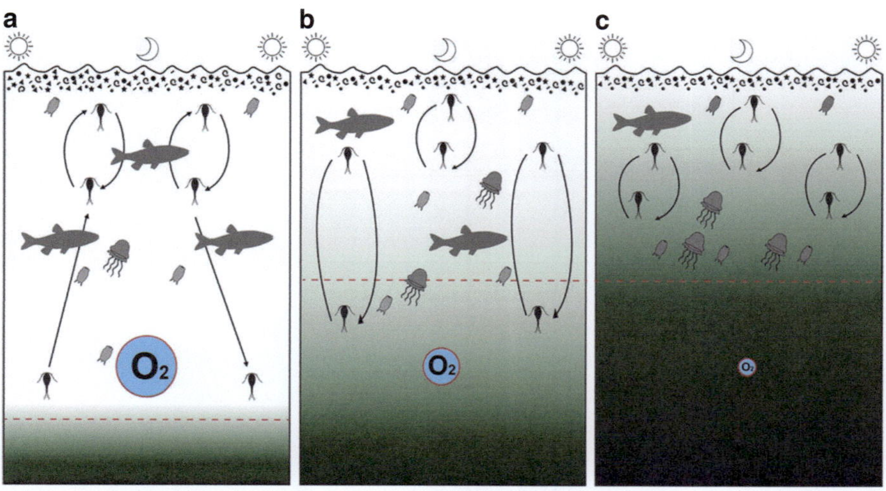

Fig. 1.11 Zooplankton vertical migration in the Black Sea. (**a–c**) Decreasing concentrations of dissolved oxygen (graphic composition of A. M. Miglietta)

From Gibraltar to the easternmost portion of the Mediterranean Sea, the **salinity** of the surface layer increases from **35‰ to 39‰** due to the increase in evaporation consequent to the temperate-tropical and semiarid climate of those regions.

The Mediterranean Sea represents the remaining western part of the **Tethys Ocean,** which has been closed with the progressive opening of the Indian Ocean in a tectonic scenario that has modified the geography for more than 180 million years. The biological appurtenance of such an eastern ocean (Tethys) characterized the Paleo-Mediterranean until 7 million years ago, when the closure of the communication on the east (at the level of the Hatay-Eufrates valley) and on the west (at the level of Gibraltar) provoked the almost complete drying of the basin (Fig. 1.12).

The intense evaporation of that period was responsible for the reduction of approximately 2000 m of the water level and an excess salt concentration of the remaining water in the deeper parts of the basin.

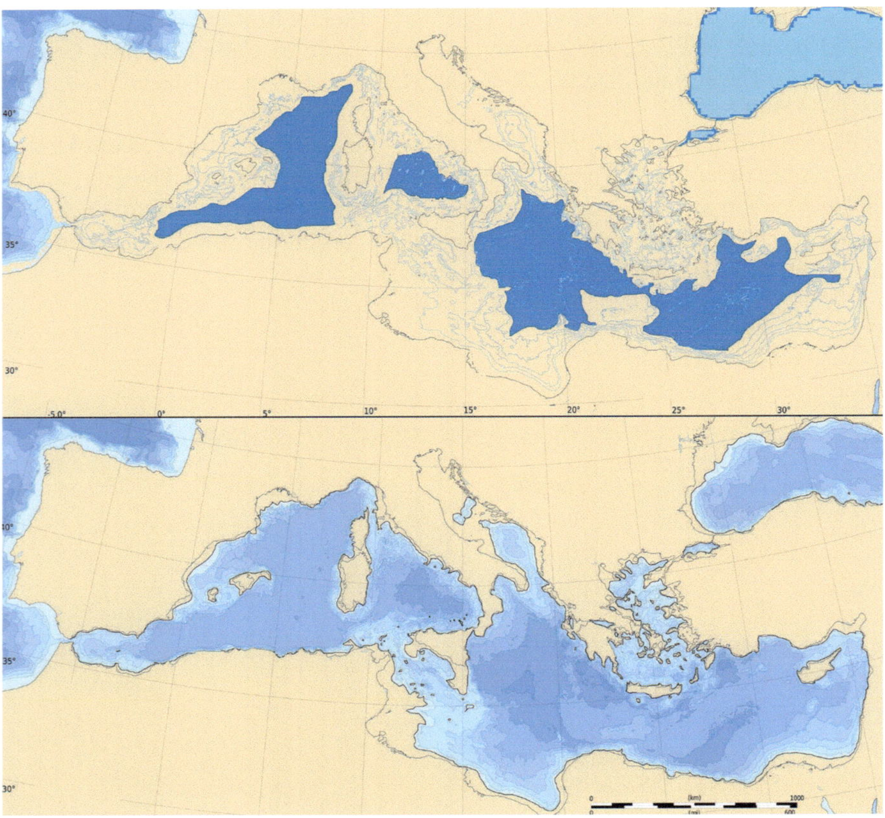

Fig. 1.12 Mediterranean maps of the Messinian period (top, 7–5 million years ago) and of the glacial period (bottom, approximately 35,000 years ago); Black and Marmara seas were isolated and probably freshwater lakes in both periods. (Courtesy G. Belmonte)

Azoic = A, without; zoon, animal life *medium* (L) = in the middle *pontos* = the sea *Mediterranean* = *medi* (in the middle); *terra* (land)

This ecological crisis (the so-called **Messinian crisis**) provoked the disappearance of the living forms that were the heritage of the Tethys ocean. Approximately 5 million years ago, a connection with the Atlantic Ocean opened at the western extremity (a little bit south of Gibraltar), and that ocean replenished the dried basin with its water and biological contents. For such a geologic story, today, the Mediterranean does not have a memory of the Tethys Ocean, but it is an entity biologically derived from the Atlantic Ocean.

Geographers have long adopted methods and proposed solutions for the representation of a sphere surface (or a curved part of it) on the two dimensions of a map chart.

This effort produced proposals and libraries of maps, which gave birth to the scientific discipline of cartography (inside the science of geography).

Most of the Earth's surface (the ocean surface), however, is geographically designed only with approximation (precise limits between seas are still not proposed by anyone).

The whole map of oceanic bottoms has become a reality only in the last 70 years after being a military (navy) secret for a long time. What is on the road, today, is the continuous effort to recognize boundaries in the sea, and the morphological-physical approach, basic for land geography, is paralleled by biology-based compartmentation in the sea.

The Mediterranean Sea is without doubt a uniform geographic unit, being well delimited by the emerged land masses of three continents (Europe, Africa, and Asia), but inside it, many other basins received a name since from ancient times due to the intense navigation of many peoples in its waters (Pelasgians, Minoans, Phoenicians, Greeks, Romans, Arabs, and Vikings, just to mention the most famous).

These different basins inside the Mediterranean, however, are limited only to their coastal sides; but they do not have a precise limit in the open sea. The Adriatic Sea, for example, is a very well delimited sea in the central-north Mediterranean, but the only communication it has with the northern Ionian Sea is promptly a problem. In fact, sailors, environment managers, Navy officers, and common people living on the coasts have different opinions on where it starts (or finishes) more or less at the level of the Otranto Channel.

When the Marine Biology Italian Society started with the project to individuate zones differently characterized by the biological communities (overall those of the marine coastal benthos), it was immediately evident that the historically defined "Italian seas" did not correspond to the "biogeographical" subdivisions recognizable on the basis of the assemblages of organisms living along the Italian coasts.

That subdivision divided the Italian coastline into **nine different biological areas** in a geographic map where "only" five seas have a geographic name.

Thetis = Greek goddess of the sea *Messinian* = geological period at the end of the Miocene, from the Italian city of Messina

Box 1.4 An Example of Territorial Competence, the Italian Seas

In the case of Italy, which is largely surrounded by sea in the middle of the Mediterranean, the geographic atlases report, among other geographical characteristics, a detailed **coastline** extension of **7914** km and a **surface area of 301,000 km²**. The political limits, however, are not those at the line where land immerges below the sea. The Italian government, as in every other country, individuates a **baseline** around the coasts, which represents the very border of its territorial extension and contains "internal" marine areas even very far from the coastline.

Such a baseline, in addition, is the line from which the "territorial waters" have to be calculated (at 12 nautical miles, NM, towards the open sea).

This last bend of 12 NM, which extends from the baseline into the open sea, is undoubtedly under the political competence of the country that it surrounds. This political appurtenance is officially recognized by the international community.

Finally, each country even recognizes its own economic competence on the sea until 200 km from the coastline.

This additional marine area competence produces, as a result, a total competent surface (from the political-economic point of view) for each country, abundantly larger than that suggested by geographic maps. In the case of Italy, a peninsula with 2 large islands in the center of the Mediterranean Sea, the surface area of the politically competent sea (**587,000 km²**) doubles the extension of the land territory (**301,000 km²**) (Fig. 1.13).

Even in such a deep effort to justify a sea subdivision, the open-sea subdivision appears to be an impossible enterprise. Some scientists, however, have tried to obtain suggestions based on the distribution (although at a large geographic scale) of pelagic species, fish overall, molluscs, and crustaceans.

The necessary data for such a new point of view was (and still is) the fishery world and its enormous amount of information regarding the fish market species.

One of the first attempts was carried out by Garibaldi and Caddy (1998), who proposed a biological subdivision of the Mediterranean based on the geographical distribution of 536 different species of economic importance over 3 main layers: 0–200 m (epipelagos), 200–1000 m (mesopelagos), and more than 1000 m (bathypelagos).

A more recent attempt was that of Berline et al. (2014), who used an oceanographic distance matrix as a basis for hierarchical clustering and recognized **22 areas** within a more or less stable net of boundaries in the whole basin of the Mediterranean. Areas were discussed in light of existing ecoregionalization and available knowledge on plankton distributions (Fig. 1.14).

Very recently, the realization of the **Suez Canal** (official opening, **1869**) by the French engineer **Ferdinand de Lesseps** reopened a connection with the eastern world (the Red Sea, in this case, which is a "mediterranean" extension of the Indian Ocean).

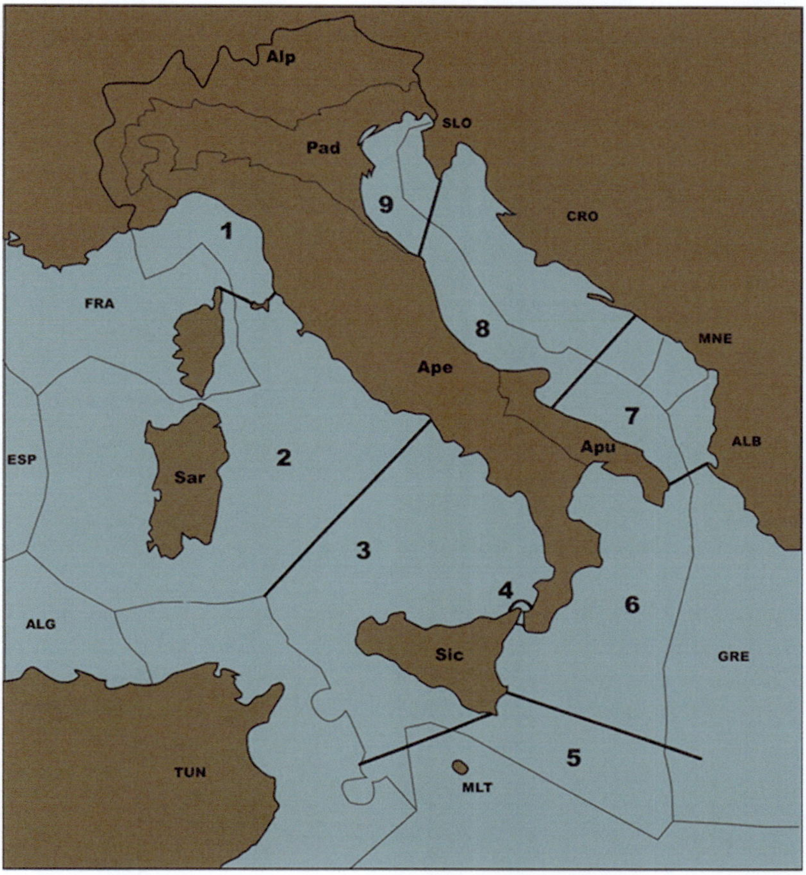

Fig. 1.13 Map of Italian seas. Alp, Ape, Pad, Apu, Sar, and Sic indicate six faunal districts on land. Numbers 1–9 indicate the marine biogeographical districts. (From Belmonte 2018, modified by G. Belmonte)

The connection between the Red and Mediterranean Seas already existed in the historical past, when the Persian Emperor Darius, and successively the Egyptian Pharaoh Ptolemy, connected the Red Sea with the easternmost branch of the Nile River delta via an intermediate bitter lake.

That canal (sized for the small ships of seventh to eighth centuries BC) was not a free connection between the two marine worlds because the interposition of a high salinity situation (the "bitter" lakes) and a long trait of freshwater (the Nile Riverway) realized an ecological barrier to the free exchange of marine life between the two worlds. The excavation of the modern **Suez Canal**, on the contrary, avoided the use of a Nile branch and provoked a progressive reduction in salinity in the bitter lakes (arrived at approximately 43% in the middle of the twentieth century). The Aswan dam on the Nile, in addition, eliminated the seasonal freshwater flushes at the Port Said level (the Mediterranean termination of the Suez Canal) since the same period (half of the twentieth century). The great commercial success of the canal

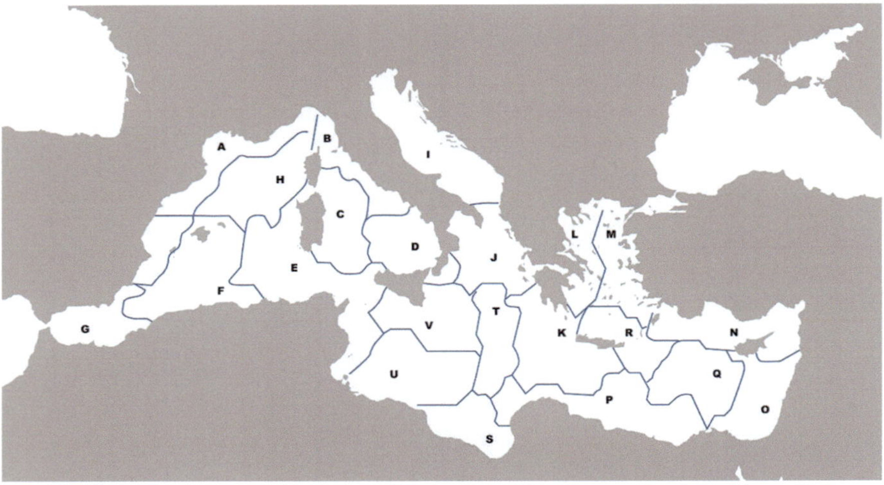

Fig. 1.14 Map of Mediterranean partitioning. Twenty-two biological districts are identified, eight of which are around Italy. (Based on Berline et al. 2014, redrawn and modified by G. Belmonte)

encouraged interventions of widening and depth increase to favour an always increasing traffic demand. This situation led to the modification of **164 km** of the canal length, passing from an average depth and width of **8** and **56 m,** respectively (1869), to the current average values of **24 m** and more than **200 m** (Galil et al. 2017).

Box 1.5 The Suez Canal

On 17 November 1869, an artificial canal connecting the Red Sea with the Mediterranean Sea was opened. Ferdinand de Lesseps was the chief of that enterprise derived from the project of Eng. Luigi Nardelli. That canal was 164 km long and 8 m deep. Today, the length has been extended to 194 km (with the duplication of some traits), and the average depth is 24 m, with traffic of more than 50 ships per day. What was realized to serve as a connecting way between East Asiatic and European markets has also functioned as a connection between the marine ecosystems of the Red Sea and Mediterranean Sea (Fig. 1.15).

The elimination of the salinity barriers (bitter lakes and Nile delta) and, overall, the widening of the communication between the Red and Mediterranean Seas created conditions for biological invasion, which received the name of **Lessepsian migration** (Por 1978).

For a long time, the bitter lakes (salinity >60%) in the middle of the road and the Nile freshwater outflow at the north extremity of the canal impeded the free circulation of marine organisms. After 1950, however, both of these conditions decreased to a level that allowed biological exchange to start. At present, more than 400 species of Red Sea biota have been registered as arrived and affirmed in the Mediterranean (**Lessepsian migration**). The easternmost part of the Mediterranean is obviously the most interested in such a biotic invasion. No more than ten species passed from the Mediterranean Sea to the Red Sea (**anti-Lessepsian migration**) in an unbalanced exchange, which attracted the attention of many scientists.

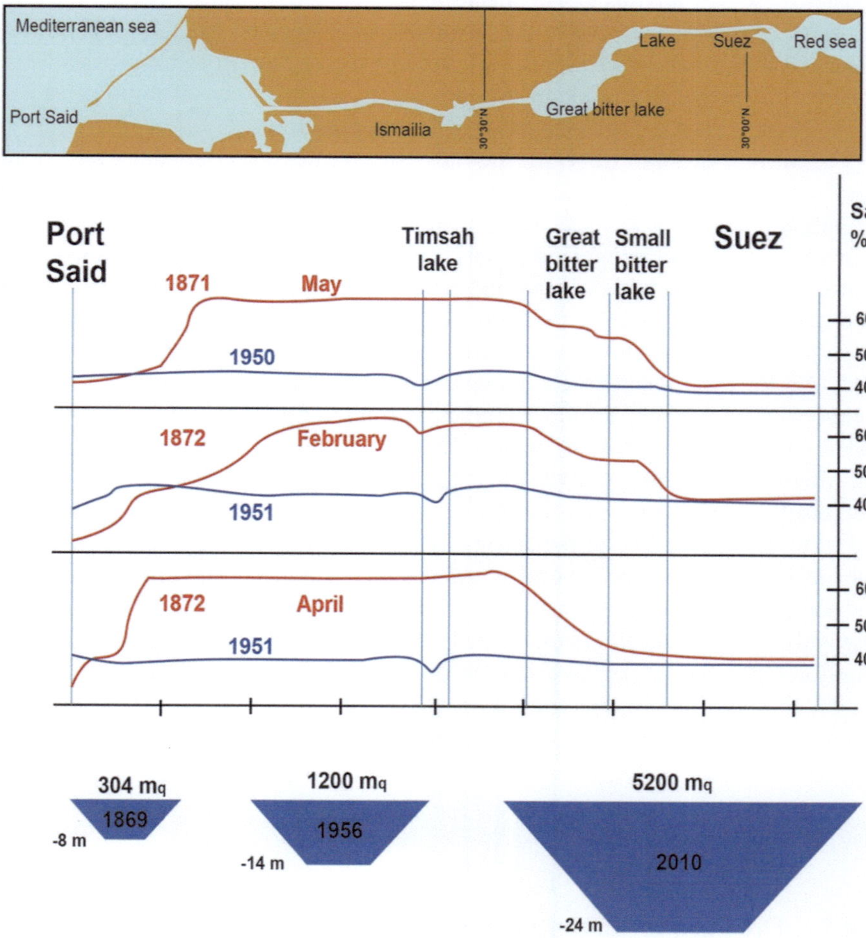

Fig. 1.15 The Suez Canal. Up, the map of the total length (164 km), with North (Mediterranean Sea) on the left. Centre, the evolution of salinity along the canal measured in 3 months (May, February, April) of years 1871–1872 (in red) and 1950–1951 (in blue). Down, the section of the canal in 1869, 1956, and 2010 (graphic composition of A. M. Miglietta)

Such geographic communication is most responsible for the arrival of "alien" species (or **nonindigenous species, NIS**) recently recorded in the Mediterranean. In the eastern basin, the new arrivals from the Red Sea amounted to **225 species** in 2009 (considering only fish, Mollusca, and Crustacea, which were noted due to their importance in the fishery market) (Rilov and Galil 2009). The recording of NIS coming from the Red Sea via the Suez Canal progressively diminishes westwards, and in the western Mediterranean, there are no more than **27 species**. A recent list updates these new arrivals to **443 species** (all kinds of organisms, not only the taxa of fishery importance), with the strongest involvement of the easternmost Mediterranean countries (Egypt, Israel, and Lebanon). The list of Lessepsian NISs contains either coastal benthic or pelagic species, overall if we take into consideration the smallest-sized ones (as those typical of plankton).

The **anti-Lessepsian migrants** are those species that pass the Suez Canal from Port Said to Suez, i.e., from the Mediterranean to the Red Sea. The "Lessepsian migration" (involving 443 species) is a unidirectional phenomenon because fewer than ten species are recognized as anti-Lessepsian. This unbalanced situation is probably due to the level of specialization typical of tropical organisms coming from the Red Sea in comparison with the comparatively more generalist character of Eastern Mediterranean species. In the new Mediterranean habitat, the specialists from the Red Sea result in competitive winners because they find occasions of "empty niches" where to exploit resources. In contrast, the Mediterranean generalist species find more competition and problems for the conquering of the Red Sea habitat, where specialists are well positioned at their ecological places.

Ante-Lessepsian species in the Mediterranean, finally, are those species somehow linked to and descending of the ancient Tethys *taxa*. In fact, the destruction of the tethydean fauna during the Messinian period was not complete. Few species survived, and some rare and archaic species of the current fauna belong to this group. One representative is *Aphanius fasciatus*, a small halotolerant Cyprinidae, and an entire fish family could be that of Triglidae (sea robins), interesting models of adaptation to very benthic lifestyles, with ventral spines used for walking on sand and very large pectoral fins used for visual communication more than for swimming (Fig. 1.16).

ante = before a certain period *anti* = contrary, opposite

1.4.1 The Case of *Mnemiopsis leidyi* Invasion

The management of NISs is also a problem for every environment and for the human economy. The history of NISs started to receive attention and scientific diffusion, probably with the invasion of the ctenophore ***Mnemiopsis leidyi*** in the Black Sea (Fig. 1.17).

This comb jelly is not a Lessepsian migrant but derives from the North American Atlantic coast, where it typically inhabits bays and confined coastal areas.

The comb jelly (*Mnemiopsis leidyi*) was first recorded in the Black Sea in 1982, and in 1988, a population explosion started in the Azov Sea (Zaitsev and Ozturk 2001). By 1994, the total biomass of *Mnemiopsis* was estimated at 100 million tons,

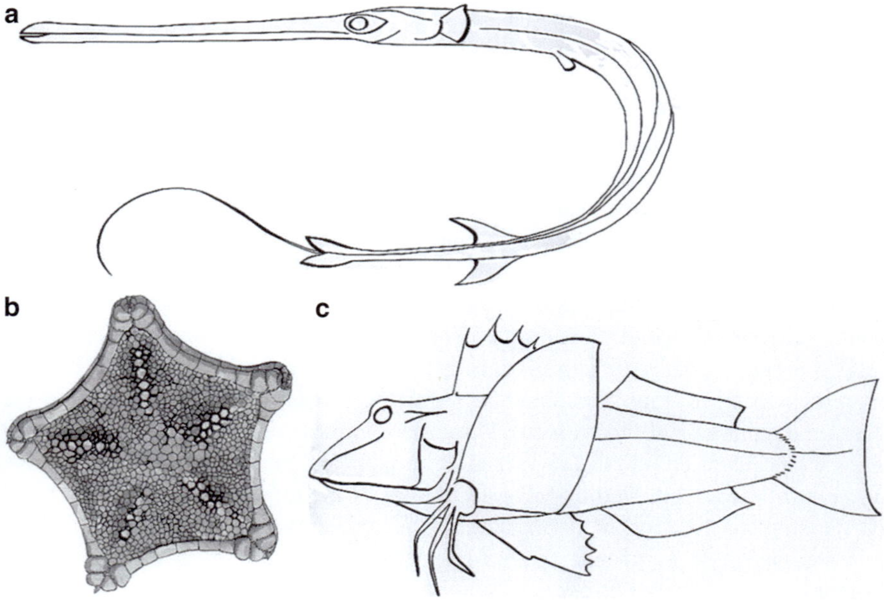

Fig. 1.16. (**a**) Lessepsian species *Fistularia commersonii* (Osteichthyes, Fistulariidae); (**b**) anti-Lessepsian species *Peltaster placenta* (Echinodermata, Asteroidea); (**c**) Ante-Lessepsian species, *Chelidonichthis* (Osteichthyes, Triglidae). (Courtesy G. Belmonte)

Fig. 1.17 The lobated ctenophore *Mnemiopsis leidyi* is a nonindigenous species that causes heavy alteration to the ecological equilibrium of the Black Sea pelagos. The larger *Beroe ovata*, a predaceous ctenophore, is responsible for the restoration of the ecological equilibrium by diminishing the *Mnemiopsis* population (Drawing of A. Gennari)

and 30 million tons were recorded in the Azov Sea in 1989. The *M. leidyi* population in the Black Sea had an increase in offshore biomass from 225 g wet wt m^{-2} in early 1988 to 2000 g m^{-2} in 1989. A second biomass peak was recorded in 1995 that surpassed the level of the first peak (2700 g m^{-2} in 1995) in Russian offshore waters.

Since 1995, the abundance of *Mnemiopsis* has begun to decrease and stabilize at 300–800 g m^{-2} in the Black Sea and at 500–600 g m^{-2} in the Sea of Azov.

One possible explanation for the quick expansion of *Mnemiopsis* in the 1980s may be its wide tolerance to salinity and temperature. It can live and reproduce between 1.3 and 32 °C and between 3.4‰ and 75‰. After its invasion, the structure of the planktonic communities in coastal waters and the open sea changed significantly. The abundance of mesozooplankton declined by 2–2.5 times or more.

A pronounced decrease (approximately 2–10 times) in planktonic larvae in summer also occurred, thus also showing a grazing impact of *Mnemiopsis* upon the benthic populations of those larvae. Three main impacts of *Mnemiopsis* on the fisheries were identified: (1) direct predation on fish eggs and larvae; in shelf waters, *Mnemiopsis* was estimated to graze up to 70% of the total ichthyoplankton stock; (2) feeding on the food of fish larvae and adults, thus causing starvation in fish populations; and (3) further acceleration of ongoing ecological change due to **eutrophication**.

A drastic decrease in fish production occurred, e.g., a four- to five-fold decrease in kilka (*Clupeonella* spp.) and more than a ten-fold decline in anchovy. The annual losses attributed to the *Mnemiopsis* plague were calculated to be ca. 200 million US dollars in the Black Sea and 30–40 million US dollars in the Sea of Azov.

The **euriecios** lifestyle of the species allows it to support large salinity and temperature intervals, rather than very low Oxygen concentrations in the water. This adaptability to even extreme conditions allowed the species to be a winner competitor against most of the local Black Sea species.

Although *Mnemiopsis* is a **holoplankter**, its presence also had consequences on benthic life (generating a 30% reduction in the pre-existing benthic biomass), not only due to the subtraction of benthos deriving larvae (**meroplankton**) but also with the subtraction of Oxygen for the decomposition of decaying jelly biomass accumulated on the bottom when they died.

The direct subtraction of larvae and juveniles for future populations was even aggravated by the subtraction of small-sized food for the support of high-level consumers, and top predators (such as tuna fish) completely disappeared from the Black Sea or were severely reduced in number in that period.

The number of commercially important species and their availability in the market declined from 26 to 5, and total fishing in the NW area of the Black Sea (one of the most productive in Europe) passed from 700,000 tons per year to only 80,000.

The arrival of another Ctenophora species in late 1997 (***Beroe ovata***) was interpreted as responsible for the collapse of the *Mnemiopsis* population. *Beroe*, in fact, acted as a competitor or even a predator of *Mnemiopsis*, whose population, from that moment onwards, was put under biological control.

In 1999, *M. leidyi* and the jellyfish of Black Sea origin *Aurelia aurita* were recorded in the middle Caspian Sea. In 2001, *M. leidyi* greatly increased in

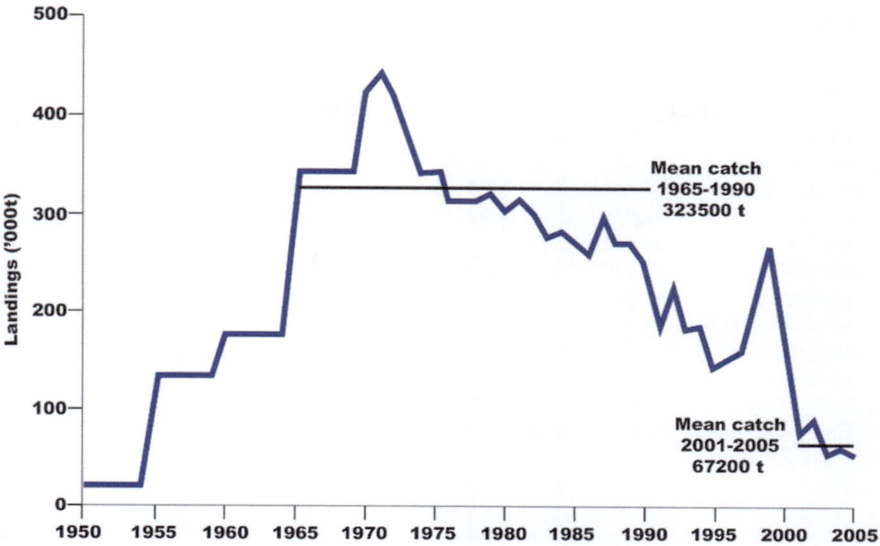

Fig. 1.18 Caspian Sea *Kilka* landings. (From Daskalov and Mamedov 2007, graphically modified by A. M. Miglietta)

population size, particularly in the southern Caspian, where its abundance doubled the maximum values recorded in the Black Sea in 1989 and reached values of 1000 ind. m^{-3}. The consequences of the arrival of *M. leidyi* in the Caspian Sea were severe and even disastrous. In 2000, zooplankton biomass decreased by 5–20 times in all groups when compared to previous years, and this trend continued. The near collapse of the formerly intensive kilka fishery was also registered (Fig. 1.18).

Kilka is an important food reserve for beluga and other sturgeon species (family Acipenseridae), as well as the Caspian seal. A decrease in *kilka* abundance causes nutritional deficiency, which may lead to seal diseases and fecundity decreases. An increase in phytoplankton biomass was recorded due to a decline in zooplankton grazing pressure, giving an example of the trophic cascade effect.

More recently, *Mnemiopsis* also appeared in the Baltic and Eastern Mediterranean Seas but without serious consequences for coastal ecosystems.

This widening geographic distribution started its story just off the south coast of Crimea, in the Black Sea, where it arrived (probably) in the ballast waters of cargo ships from North America. The long travel (even 10 days) is not a problem for *Mnemiopsis,* which can also support starvation for weeks without consequences but only a size diminution.

eury = good performer (not conditioned by) *eu* = good, *throphe* = food *holo* = complete, total *mero* = partial *indigenous* = *ein* (in), *deimos* (house), *gens* (L) (people)

References

Bailey RG (1998) Ecoregions. The Ecosystem geography of the oceans and continents. Springer-Verlag, New York, 176 pp

Belmonte G (2018) Calanoida (Crustacea: Copepoda) of the Italian fauna: a review. Eur Zool J 85:273–289

Berline L, Rammou AM, Doglioli A, Molcard A, Petrenko A (2014) A connectivity-based eco-regionalization method of the Mediterranean Sea. PLoS One 9:e111978. https://doi.org/10.1371/journal.pone.0111978

Costello MJ, Tsai P, Wong PS, Cheung AKL, Basher Z, Chaudhary C (2017) Marine biogeographic realms and species endemicity. Nat Commun 8:1057. https://doi.org/10.1038/s41467-017-01121.2

Daskalov GM, Mamedov EV (2007) Integrated fisheries assessment and possible causes for the collapse of anchovy kilka in the Caspian Sea. ICES J Mar Sci 64(3):503–511

Galil B, Marchini A, Occhipinti-Ambrogi A, Ojaveer H (2017) The enlargement of the Suez Canal—Erythraean introductions and management challenges. Manag Biol Invasions 8:141–152

Garibaldi L, Caddy JF (1998) Biogeographic characterization of Mediterranean and Black Seas faunal provinces using GIS procedures. Ocean Coast Manag 39:211–227

Por FD (1978) Lessepsian migration: the influx of Red Sea biota into the Mediterranean by way of the Suez Canal. Ecological studies, vol 23. Springer-Verlag, Berlin, 228 pp

Rilov G, Galil B (2009) Marine bioinvasions in the Mediterranean Sea—history, distribution and ecology. In: Biological invasions in marine ecosystems, Ecological studies. Springer, Berlin, pp 549–575

Smith AG, Smith DG, Funnel BM (1994) Atlas of mesozoic and cenozoic coastlines. Cambridge University Press, New York

Spalding MD, Fox HE, Allen GR, Davidson N, Ferdaña ZA, Finlayson M, Halpern BS, Jorge MA, Lombana A, Lourie SA, Martin KD, McManus E, Molnar J, Recchia CA, Robertson J (2007) Marine ecoregions of the world: a bioregionalization of coastal and shelf areas. Bioscience 57:573–583

Spalding MD, Agostini VN, Rice J, Grant SM (2012) Pelagic provinces of the world: a biogeographic classification of the world's surface pelagic waters. Ocean Coast Manag 60:19–30

Sutton T, Clark MR, Dunn DC, Halpin PN, Rogers AD, Guinotte J, Bograd SJ, Angel MV, Perez JAA, Wishner K, Haedrich RL, Lindsay DJ, Drazen JC, Vereshchaka A, Piatkowski U, Morato T, Błachowiak-Samołyk K, Robison BH, Gjerder KM, Pierrot-Bults A, Bernal P, Reygondeau G, Heino M (2017) A global biogeographic classification of the mesopelagic zone. Deep-Sea Res I Oceanogr Res Pap 126:85–102

Vannini J, Yebra L, Granata A, Guglielmo L, Moscatello S, Belmonte G (2021) Horizontal distribution of deep sea microplankton: a new point of view for marine biogeography. Water 13:1263

Wegener AL (1915-1919) Die Entstehung der Kontinente und Ozeane. Mit handschriftlichen Bemerkungen von Alfred Wegener, Notizen und Briefen sowie neu erstelltem Index. Hrsgg. vom Alfred-Wegener-Institut für Polar- und Meeresforschung. Vieweg, Braunschweig 1915, 1929, Borntraeger, Berlin 2005 (Repr. d. 4.Aufl.)

Zaitsev Y, Ozturk B (eds) (2001) Exotic species in the Aegean, Marmara, Black, Azov and Caspian seas. Turkish Marine Research Foundation, Istanbul, 264 pp

Plankton: Why Do So Many Species Co-exist in the Same Water Volume?

2

2.1 Terminology and Conceptual Grouping of Pelagic Organisms

All organisms living in the water column have to resolve the physical problem of being denser (and heavier) than the water where they stay suspended and have to adopt strategies to oppose or retard sinking. Solutions are many and encompass anything, from any devices to diminish the body density (gas bubbles, oil droplets) to the development of swimming ability to actively regain superior layers of the water column (generally, swimming is not used to sink).

Smaller, unicellular organisms are somehow facilitated to avoid sinking in the water thanks to their size and the salt concentration, which elevates the water density, thus favouring the organism suspension.

The difference in anatomical devices between unicellular and multicellular organisms could be the first reason to distinguish them according to size groups. This is, to tell the truth, the main scheme of **plankton** groups proposed in the textbooks.

Box 2.1 Subdivision of Plankton Organisms According to Average Size (Fig. 2.1)

Femto plankton, size < 0.2 μm; **virus.**

Pico plankton, 0.2 < size < 2.0 μm; **Prokaryota** (Bacteria and Archaea), **Chrysophyta** and smallest **Eukaryota.**

Nano plankton, 2.0 < size < 20.0 μm; mainly **Protozoa** and **Chromista**; generally **autotrophic** and photosynthetic organisms (the so-called unicellular algae, or phytoplankton). Components of the smallest micro- and mesoplankton can be included (e.g., early developmental stages of tiny metazoans).

Micro plankton, 20.0 < size < 200.0 μm; the largest Protozoa and Chromista, the smallest **Metazoa**. Generally, autotrophy is restricted to the unicellular components (phytoplankton), and heterotrophy is present in the unicellular

G. Belmonte, *Elements of Pelagos Biology*,
https://doi.org/10.1007/978-3-031-74280-4_2

world but more efficaciously characterizes the multicellular components (**zooplankton**).
Meso plankton, 200.0 (0.02) < size < 20,000.0 μm (20.0 mm); exclusively heterotrophic organisms (animals, the zooplankton), representing most of the Metazoa **Phyla**.
Macro plankton, 20,000.0 (20 mm) < size < 200,000.0 μm (200.0 mm); exclusively heterotrophic organisms (animals), not able to efficaciously swim to change their positioning in a water mass. The largest portion was catchable with common plankton nets.
Megalo plankton, size > 200.0 mm (20 cm); isolated and visible floating individuals or colonies, generally transparent (with a high content of water) and jelly.

Apart from size groups, the functional and/or ecological role of plankton organisms can also be responsible for their grouping.

Bacterio-plankton. **Detritivores** are prokaryote organisms generally feeding on detritus generated by the death of living organisms or the decay of their parts. Prokaryotes are also autotrophic blue algae, but these functionally belong to the phytoplankton (Fig. 2.2).

Auto = self, *trophe* = food (able to produce energetic food autonomously) *Chrome* = colour, *istos* = living matter (organisms with (photosynthetic) pigments) *Eu* = good, *Karion* = *nucleus* (equipped with a true nucleus) *Femto* (DAN) = from the Danish Femten (15); 1 Fm = 1×10^{-15} m *heterotrophic* = differently feeding *Macro* = large enough to be visible *Megalo* = more than visible 1,000,000 times larger the 1 μm *Meso* = intermediate (between micro and megalo) *Meta-zoa* = final-animals (the last ones appeared in the evolution) *Micro* = small, below the visual limit of the human eye; 1 million times smaller than 1 m *Nano* = dwarf; 1 billion times smaller than 1 m *Photo-synthetic* = synthetized by using the sun light as energy *Phylum* = race, stock *Phyto-plankton* = Vegetal-Plankton *Pico* (E) = Spanish word to indicate a "bit"; 1 trillion times smaller than 1 m *Pro* = before, *karyon* = nucleus (stage preceding true nucleate cells) *Proto-zoa* = primitive-animals *Zoo-plankton* = animal plankton

Phytoplankton. Prokaryotic blue algae (Bacteria) are a component, but the category is mostly composed of unicellular eukaryotes that are able to synthesize organic molecules by using the sunlight energy captured thanks to special pigments (many molecules, the most cited of them being chlorophyll), which give them colouration (for this reason, they belong to the kingdom **Chromista**). The whole process is called photosynthesis. For this role, they are ecologically classified as **producers**.

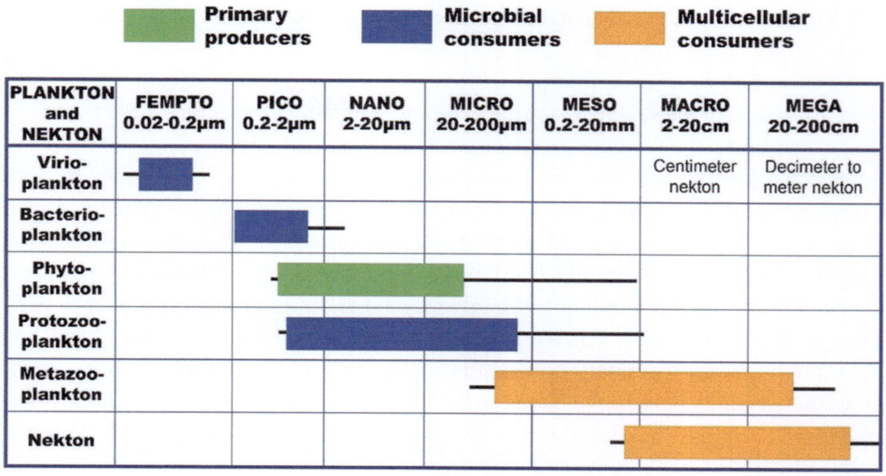

Fig. 2.1 Plankton size groups (graphic composition of A. M. Miglietta)

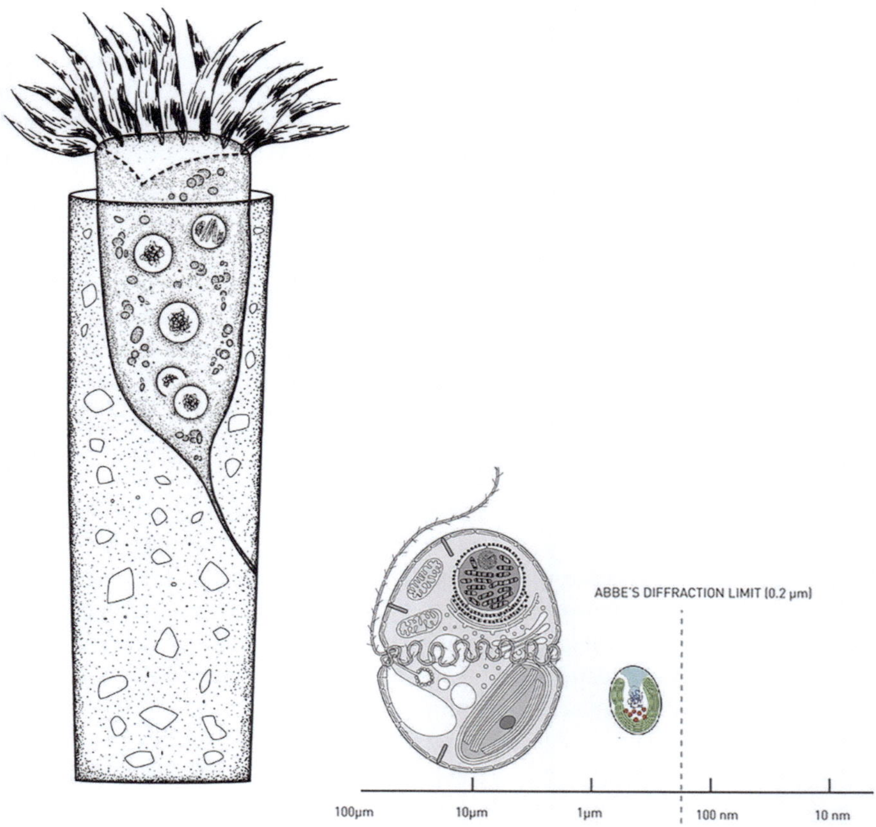

Fig. 2.2 Protozoa, Chromista, and Bacteria in size comparison. (**a**) A ciliate Tintinnina; (**b**) a dinoflagellate; (**c**) a blue alga (graphic composition of A. M. Miglietta)

Commonly also described as (unicellular) **algae**, they belong to the vegetal world (autotrophic organisms).

Phytoplankton organisms stay in different size groups, pico-, nano-, and micro-plankton, and are the sustainers of the ecological pyramid, producing food for themselves and being the food for all heterotrophic organisms (Atlas Plates 1–6).

Zooplankton. They comprise unicellular and multicellular eukaryotes that are distributed among Protozoa and Metazoa and are generally defined as heterotrophic because they obtain food in different ways than photosynthesis. The zooplankton stay in different size categories, from micro to meso and larger plankton. Organisms of zooplankton feed directly or indirectly on phytoplankton products (entire cells or parts of them) and can be divided into primary consumers (if directly dependent on phytoplankton) or secondary consumers (if dependent on phytoplankton consumers). Such a food chain in the ocean can be very long, with many levels of consumers where each level hosts carnivores preying on animals of the preceding level and being contemporaneously predated by animals of the next level (Atlas Plates 7–33).

The above-described categories suggest that we can simply identify the bacterioplankton as the ecological compartment of **detritivores**, the phytoplankton as the compartment of **producers**, and the zooplankton as the compartment of **consumers**.

The reality, however, is not so simple because, in each *taxon*, overall those of unicellular organisms, we can find either producer species or consumer ones. In addition, single species and even single individuals among Chromista are called **mixotrophers** because they have the ability to shift feeding behaviour from detritivorous to planktivorous habitus and from heterotrophy to autotrophy depending on environmental conditions. The frame is further accomplished by the existence of symbiosis, generally between organisms belonging to different kingdoms, which allows animals to be sustained in obtaining energy from sunlight because they host endosymbiotic algae and allow them to perform photosynthesis inside their bodies.

Additional categories for subdividing the plankton are available on the basis of body composition and consistency (mineralized, jelly), life cycle part passed in the water column (mero- and holo-plankton, ichthyo-plankton), and absence of specific adaptation to the plankton life (**pseudoplankton**, **tychoplankton**).

2.2 Species Richness and Species Abundance

Pelagos and deep oceans have not received the same attention that Science has dedicated to freshwater and terrestrial habitats. For this reason, many (if not all) rules that address species assemblages and their abundance historically came from terrestrial and freshwater situations, and they simply have to be tested before they can be considered applicable in ocean environments.

Hutchinson (1959) justified the embarrassing abundance of species on the planet with the existence of complex relationships (**trophic nets**) among living organisms (Fig. 2.3). According to this proposal, small-sized species, characterized by a short life span and the absence of multiple developmental stages, can concentrate themselves upon narrower feeding specializations, typical of each separate species, thus

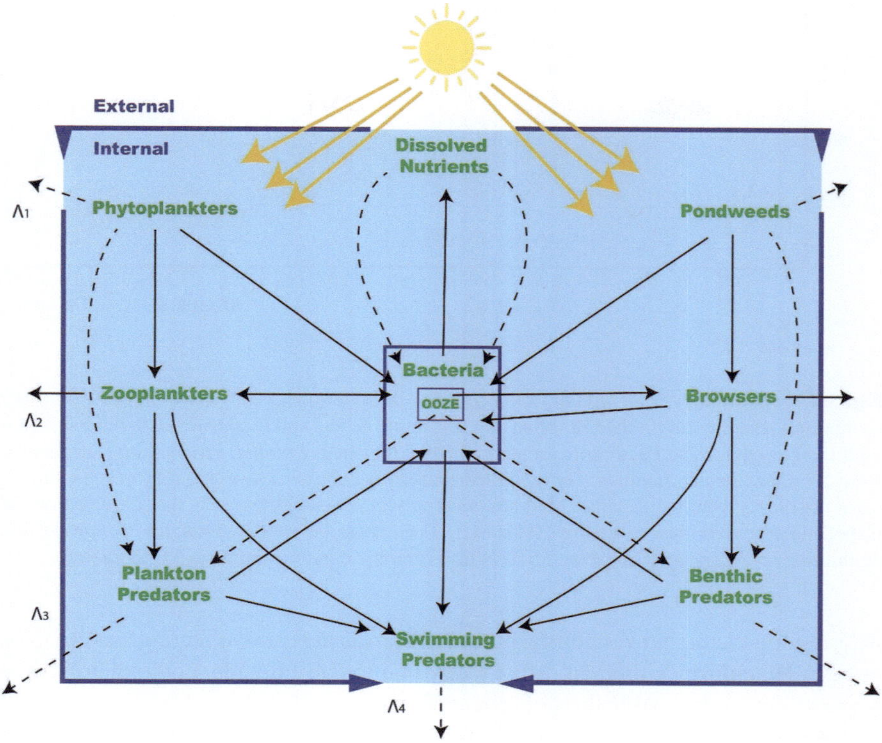

Fig. 2.3 Trophic relationships (net) modified from (Hutchinson 1959) (graphical composition of A. M. Miglietta)

favouring species coexistence in the same site, without competition. The complexity of community composition was intended to be directly derived from the **specialization** of living organisms, with such a specialization being the possible result of evolution under **stable environmental conditions**.

This assumption indirectly played a role in time: environmental stability (intended as time passed under invariant conditions) allows the existence of complex community compositions.

This implicit admission, however, was never evident; in fact, Hutchinson did not develop it in the future. Only 2 years later, Hutchinson (1961) was convinced that the stability (homogeneity) of the water environment could not explain the high biodiversity in the plankton, and he suggested seasonality (time **instability**, i.e., variability) as the actual reason.

In accordance with the principle of competitive exclusion (constant conditions lead to the affirmation of only one actor in each ecological role) (Gause 1934), only variation allows the coexistence of ecologically similar species.

In the water of the pelagic habitat, it is not the variability of conditions in the space that allows microalgae of phytoplankton to coexist, and competitive exclusion is probably avoided by the condition variability existing in time. Time was a

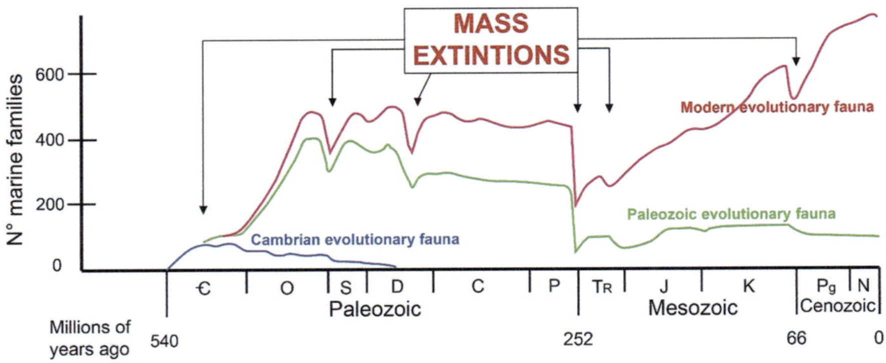

Fig. 2.4 Sepkowsky (1984) graphic of increasing taxa (families) of marine animals from the Cambrian to the present. In the Cambrian, the first fauna is blue and is considered to have survived until the Devonian. The Palaeozoic fauna dominated the ancient planet, and it is still represented today by a few archaic families. The so-called modern fauna has roots in the very past and dominates the planet from the Permian. Red arrows represent catastrophic events that occurred on the planet and produced mass extinctions (at least 5). From these mass extinctions, the life started, and dominance levels changed between different taxa (graphic composition of A. M. Miglietta)

determining factor, but variability, not stability, was the cause affecting species richness, in Hutchinson's thought. Most likely due to these double, contradictory time implications in the justification of species richness, derived from Hutchinson's suggestion, successive ecologists did not rely on the time role and concentrated more on the space-scale variability of conditions in the environment. To tell the truth, the role of evolutionary time in the biological diversification (enrichment) of the planet is probably given as obvious, and, avoiding discussion, scientists simply wanted to demonstrate how many and what causes were involved, in addition to the time factor.

Sepkowsky (1984) demonstrated that the number of *taxa* in the fossil record has a positive, although discontinuous, increasing trend in the whole Earth story (Fig. 2.4).

On the other hand, Gould (1994) pointed out that after the abrupt appearance of animal life approximately 550 million years ago (the so-called **Cambrian explosion**), only extinctions occurred on the Earth regarding *phyla*, and biodiversity growth was related to exaggerated speciation inside one or a few *phyla*.

Chloro = green, *phyllos* = leaf *Detritus* = organic litter, *vore* = eat (organism that feeds on *detritus*) *mixo* = mixed, *trophe* = food (organism or feeding regime able to shift from auto to heterotrophy) *ichthyos* = fish, *plankton* (fish plankton stages) *Photo* = light, *Syntetos* = put together *pseudo* = false, *plankton* = (sessile organisms on rafting objects) *Syn* = together; *bios* = life *tychos* = accidental, *plankton* (benthic organisms temporarily suspended in the water column by disturbance events)

Over evolutionary time, the biorichness in terms of *phyla* was considered impoverished, notwithstanding the evident growth of the species number.

Coming back to the Gause principle (Gause 1934), just to certify that time for ecologists is not the same entity that it represents for evolutionists, it implicitly admits an impoverishment of the species number with time (the ecological exclusion destroys locally the diversification if conditions become constant). After Hutchinson (1959, 1961), in the 1960s of the twentieth century, the scene was monopolized by the model of MacArthur and Wilson (1967), where the species number possibly existing in a restricted area (an island) depends on the availability of space (area) and on the distance from the source of the arriving specimens (generally the closest continent), a more enriched natural situation. Both of these problems are attained with the space scenario.

The role of time in that model is limited to what would be necessary to reach the equilibrium point (to replenish the available space on an island).

The model applies only to newly formed islands, e.g., for oceanic islands. What could change on an oceanic island at its biological equilibrium, when new species arrivals equal the resident species extinctions, is the type of species present, not their number. The time necessary for the biological replenishment of newly formed islands is relatively short. As the species number reaches an equilibrium, the number of species apparently stops growing. This number, however, seems to grow with a slow rhythm over a longer time, and only great age differences among islands can unveil this time effect.

The **Krakatau** "son" (an island remnant from a volcanic explosion of 1883) has been considered to have been recolonized by species coming from neighbouring islands (Indonesia), with a re-proposition of the previous situation after no more than 50 years. Three years after the eruption, an expedition of botanists found mosses, algae, flowering plants, and 11 species of fern on the newly formed substratum. Further colonization of the barren island then began to accelerate. Butterflies, beetles, flies, and a single monitor lizard were found after 6 years in 1889. Fifty years after the eruption of 1883, the island was home to 171 species of plants, and the original habitat of the equatorial forest was reestablished. The "colonizers" came with the wind (40%), rafted on the sea (30%), and most of the remainder were brought by moving animals.

Simberloff and Wilson (1970) demonstrated the restoration of the initial species number (insects) in experimentally deserted islets of the Florida Keys after only 2 years (Fig. 2.5). The number of species continued to rise in the more distant islands, thus also confirming the dependence of the colonization on the proximity of the potential source of propagules. Species immigrations and extinctions have continued at a high rate, but the total number of species has remained approximately constant.

These facts (one natural and one experimental) sustain the theory of island biogeography on a complete ecological basis.

At the birth of an oceanic island, generally a volcano emerging from the sea, all the arrivals represent biological novelties. Time by time, with the successful settlement of species, new arrivals have a greater probability of belonging to an already established species, and the probability of the arrival of "new" species progressively diminishes with time (in a continuing arrival scenario). In contrast, the extinction rate of resident species on the new island is completely non-existent at the beginning of the story (when species are not present) and can only grow with the progressive increase in species number.

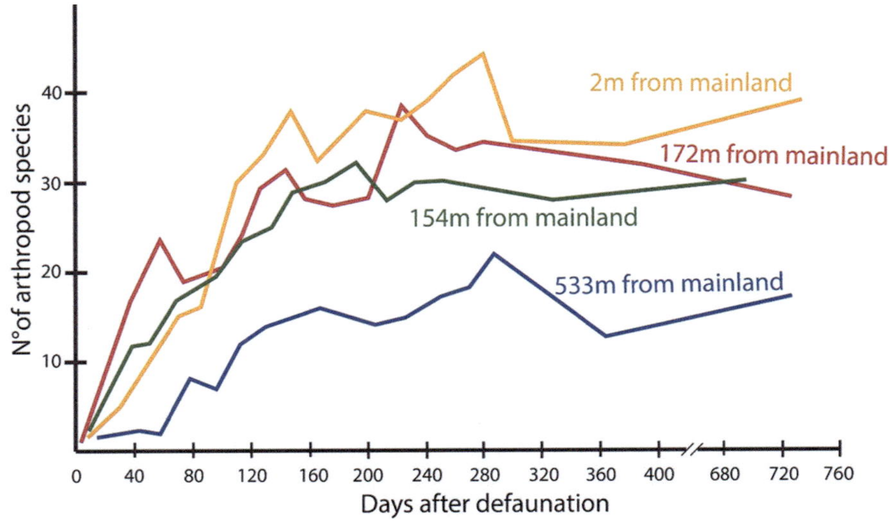

Fig. 2.5 Faunal recovery in four islets close to the Florida mainland. The number of species before defaunation was inversely proportional to the distance from the mainland. The same number of species was approximately reached after 2 years from defaunation. (From Simberloff and Wilson 1970; redrawn by A. M. Miglietta)

In the MacArthur and Wilson (1967) theory, the curve of new species arriving on an island (progressively diminishing with time) encounters the curve of extinct local species (progressively increasing with time) at a point that represents the maximum number of species coexisting on the island (Fig. 2.6). The species assemblage composition can also vary with time, but the number "at equilibrium" should remain constant, and it is typical of each island.

The species abundance on each island is affected by the size of the island and by distance from the closest continent. These two conditions are undoubtedly responsible for the total number of coexisting species and their successful arrival from elsewhere.

The entire theory is based on the processes of arrival and extinction of species already existing elsewhere (in source areas or on the same island). The matter is completely in the field of ecology and geographic dispersal strategies.

McArthur and Wilson's theory does not predict the presence of endemic species on islands. If they are endemic, by definition, they cannot be arrived at from elsewhere. Or, in any case, if a species arriving on an island has the time to evolve into another species, this happens within a restricted, isolated population, thus producing the result of a species addition without any involvement of arrivals.

A number of studies tending to demonstrate the validity of the McArthur and Wilson rule on a general scale were realized in the successive 20 years, considering the mountain tips, lakes, and island submerged coasts as isolated biotas subjected to the same rule. The consideration of temporary ponds (freshwater islands on the land) as islands also added **a time point of view** to the problem. According to Ebert

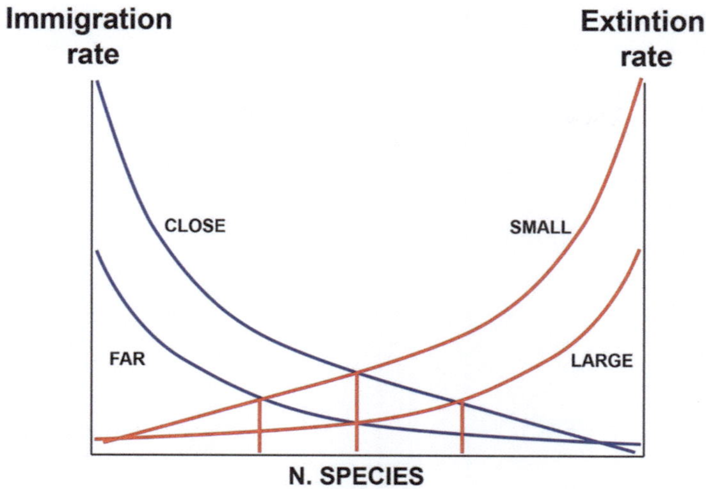

Fig. 2.6 McArthur and Wilson (1967) theory graphically represented. Red curves represent the size of islands (small/large); blue curves represent their distance from the continent (close/far). Red vertical lines indicate the number of species (on the x-axis) at the cross of lines. The minimum number of species is for small islands, far from the continent. The maximum number of species is for large islands close to the continent. The y-axis indicates the immigration rate (number of possible new species arrivals) on the left and the extinction rate (number of possible extinction risks for species present on the island) on the right (graphic composition of A.M. Miglietta)

and Balko (1987), the length of the periodical existence of temporary ponds (the so-called **hydroperiod**) affects the number of species possibly present in each pond independently of the basin size (Fig. 2.7). The authors proposed a time-based reinterpretation of the MacArthur and Wilson space-based model, with the area dimension corresponding to the period length of the pond's existence and distances from the continent corresponding to periods of water absence.

The time they considered, however, was an ecological factor because it was related to something like the seasonal absence/presence cycles and was not useful for the realization of evolutionary modifications (normally requiring a long time).

Rosenzweig (1995) described small populations in the periphery of the geographic distribution of large populations (those that cannot realize a **pangamy**) as differentiating, generating new species over time. The process, intrinsic to population dynamics, is completely disconnected from environmental diversification and occurs in homogeneous habitats in the absence of recruitment from elsewhere.

The hypothesis of Rosenzweig admits a spontaneous numeric increase in species due to mechanisms intrinsic to population genetics. Species richness for Rosenzweig, in opposition to Hutchinson's thought based on **allopatric speciation**, is also funded by **sympatric and parapatric** speciation processes.

The situation described by Rosenzweig justifies more species than those admitted by the equilibrium situation due to the model of MacArthur and Wilson, even in the absence of immigration and environmental variability.

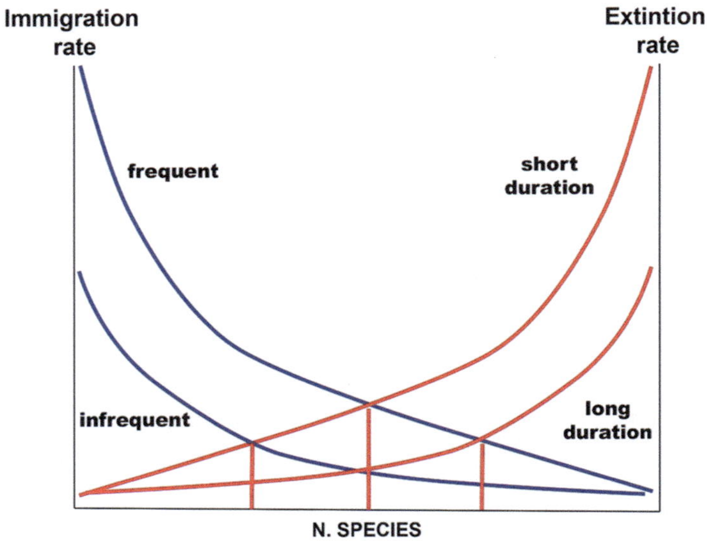

Fig. 2.7 Scheme of McArthur and Wilson (1967), interpreted by Ebert and Balko (1987) on the basis of the frequency (blue lines) and hydroperiod (red lines) of temporary ponds (graphic composition of A. M. Miglietta)

Notwithstanding this evolutionary (time-based) approach to the problem of species richness, Rosenzweig officially adhered to the hypothesis of Terborgh (1973), who gave **land extension** responsibility for the geographic distribution and abundance of species on the whole Earth. In fact, even if only time can lead to the realization of genetic differences and their affirmation, for Rosenzweig, this can be possible only in large populations, hence for those populations that are distributed in large geographical areas.

In addition, more extended areas contain a larger number of rare species because each lonely (rarely findable) individual in each site is potentially connected with other scattered individuals distributed in a larger space. In contrast, a lonely individual on an island (a limited space) cannot maintain such a connection with the rest of its own conspecifics, and the species will extinguish on that island with the death of lonely individuals (Fig. 2.8).

This situation suggests an interpretation of the species-number dynamics in each isolated system. In fact, when a continental island is formed (e.g., as a consequence of sea level uplift that separates a land portion from the rest of the continent), the number of species on the island can only diminish (for the extinction of those represented by small numbers of specimens) from that moment onwards.

The new situation of equilibrium reached on the newly formed island will host a number of species lower than the number of species hosted on a similar extended area on the continent. This is just the opposite case of the species dynamics on oceanic islands, where their number grows from the birth of the island (a barren volcano emerged out of the sea) until the final equilibrium.

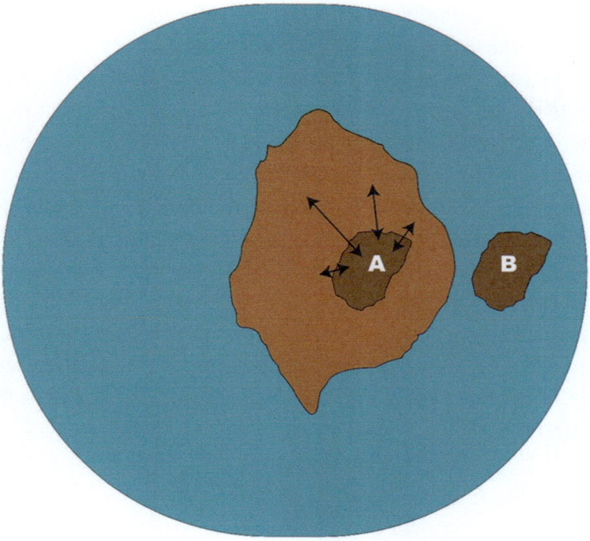

Fig. 2.8 Hypothesis of species number growth in a delimited area. (**a**) The area is inside a continent. Individuals inside the area are possibly connected with other conspecifics living in the rest of the continent. (**b**) An area of equivalent extension of that in (**a**), but isolated in an ocean. Individuals on the island do not have the possibility to maintain a genetic connection with those of the continent; the species on the island is reduced to a few individuals and can become extinct in short time. (Courtesy G. Belmonte)

This role of time appeared in a context of great success for ecology, which was accepted without problems by Connell's (1978) theory of intermediate disturbance and the evident descent of the continuous instability proposal of Hutchinson as the main factor responsible for species richness and/or ecological diversity in each geographic site.

Hence, Rosenzweig opposes his evolutionary theory (large areas have higher species numbers because large populations can exist, and this also occurs under constant conditions) against the ecological theory of Connell (there are more species where the environment is variable and subjected to intermediate disturbance).

One thesis affirms that species richness is independent of environmental conditions, and speciation also occurs in mature and stable situations; the other thesis affirms that species richness is favoured by variations that maintain the system in a state of immaturity (Belmonte 2012).

Observing pelagic diversity, Norris (2000) indicated that morphological taxonomy underestimated the number of pelagic species among fish, zooplankton, and phytoplankton. Species may be broadly distributed but fail to maintain large populations in the water masses into which they are known to disperse.

Such cases strongly suggest that biogeographic distribution ranges are determined more by where pelagic species can maintain viable populations than by their dispersal ability.

Therefore, speciation models that rely on limiting dispersal, such as vicariant and **allopatric** hypotheses, are probably not as satisfactory for explaining pelagic biodiversity as **sympatric** and **parapatric** models are. Speciation in the open ocean shares important similarities with that present in widely dispersed insects, in which changes in season or location of mating are key to establishing reproductive isolation and conditions for new species formation.

Major palaeoceanographic events, such as the closure of the Central American Seaway (approximately three million years ago), contribute to speciation by altering the vertical structure, seasonality, and ecology of the water masses in which **viable reproductive populations can thrive**.

The increasing recognition of numerous cryptic species within clades that have traditionally regarded as low-diversity groups suggests that **open-ocean *taxa* may have narrower tolerance limits and more circumscribed niches** than have been generally admitted.

Hydro = water *Pan* = complete, without obstacles; *Gamos* = fertilization *Para* = close *Syn* = together, *Patria* (L) = homeland

2.2.1 Isolated Environments and the Role of Time and Stability on Species Richness

Lynch (1988) summarized the ***refugia* hypothesis** from which the present planet's biodiversity could derive. During a period of *habitat* fragmentation, species were separated into many noncommunicating populations that had the time to accumulate genetic novelties through the mechanism of genetic drift (in conditions of **allopatry**). The same interpretation (fragmentation of the habitat leading to isolation of many populations) has also been proposed for the flock speciation of hundreds of species of Cichlidae fish in African rift lakes (Sturmbauer 1998).

What is evidently unmanageable with an ecological approach is just the situation of the existing ancient lakes (water islands). In these situations, ecologically based predictions on the number of species they should host are often disobeyed.

The ancient lakes currently existing on Earth are all characterized by an elevated number of species coupled with elevated endemism.

Lakes such as Bajkal, Tanganika, Malawi, Victoria, Biwa, and Ohrid seem to show a species richness directly proportional to their own age, more than to their geographic position, their size, and/or episodes of habitat fragmentation, which could have succeeded (and probably succeeded) during their existence. The dependence of the species number on time (on an evolutionary scale) is evident because sites with high species numbers (the mentioned lakes) also show an elevated **endemism** grade (directly correlated with their presumed age) (Fig. 2.9).

Being endemics, species present in such environments did not arrive there from elsewhere, and they appeared in situ according to a mechanism (**speciation**) that

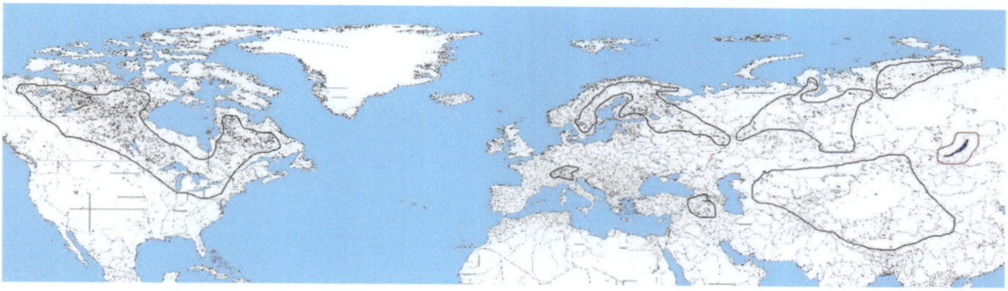

Fig. 2.9 Lake Bajkal, evidenced with a red ring, in the geographic frame of boreal lakes (mainly of postglacial origin, in large areas circumscribed with black lines). The biodiversity of the lake is astonishingly elevated in comparison with other lakes, probably due to its long-lasting existence (3–20 Million years). This hypothesis also justifies its high level of endemism (over 80% for some taxa). (Courtesy G. Belmonte)

requires time, and, according to the Rosenzweig hypothesis, this simply needs large initial populations.

The evolutionary phenomenon of speciation produces endemism, and it requires, in lakes, probably more than 10,000 years because postglacial lakes (formed from 10,000 years ago) in the boreal hemisphere have common fauna and not endemics among their species.

Allo = foreign, *Patria* (L) = homeland

The necessary time for evolution is, on the other hand, completely compatible with the volcanic islands of the MacArthur & Wilson model.

For example, the Galapagos or Hawaii archipelagos, well known for studies on the presence of endemics, have ages of some million years (with some islands being younger than others).

Howarth (1990) admitted time as the main factor responsible for the high species richness of **Hawaiian Drosophilidae** (at least 1000 species recognized), giving importance not only to biota fragmentation (the Hawaiian archipelago) but also to the presence of similar conditions on all archipelago islands over millions of years.

Briggs (1995) reported that the marine species richness from the two volcanic islands **Saint Helen and Ascension**, notwithstanding that they should be colonized according to MacArthur and Wilson's model, shows numbers compatible with the different ages of the two islands and not with their distance from the continent or their size (Fig. 2.10).

Although Ascension is closer than St. Helena, to the nearest continent (1500 km from Africa coasts against 1850 km), it has fewer fish species in its coastal habitats. If we look at the position of the two islands relative to the *Mid*-**Atlantic ridge**, we discover that Ascension is closer to this fracture line where volcanic activity is intense; hence, it is the youngest among the two islands: it is the last that has been formed, being the closest to the middle ocean ridge.

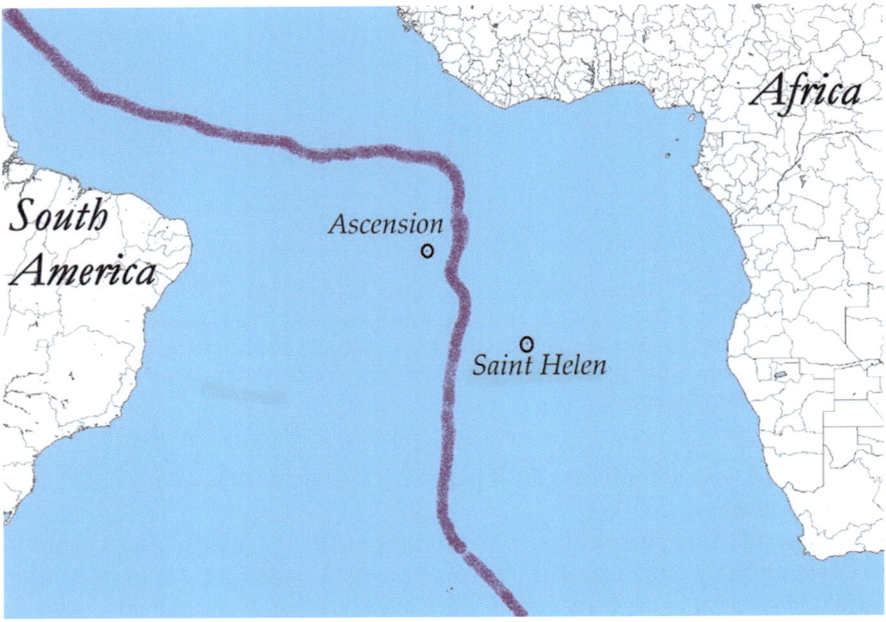

Fig. 2.10 Geographic position of Ascension and Saint Helen islands in the middle Atlantic Ocean. Ascension is closer to the continent (Africa) than Saint Helen (and the same situation is also verifiable with South America), but Saint Helen has more species (reef fishes). The species richness of Saint Helen can be explained by its distance from the middle Atlantic Ridge (this island is older than Ascension). The violet stripe represents the position of the middle Atlantic Ridge. (Courtesy G. Belmonte)

Marine dominance and other environments also add data to the time-species model. One of these situations is the **continental slope between 500 and 2000 m** deep. Some pioneering data from benthonologists have astonished the scientific community, showing the diversification of such an environment in the NE American slope (Grassle and Maciolek 1992) (Fig. 2.11). In those environments, not only are species numerous (862 Metazoa species on a total sampled area of only 21 m^2), but they sometimes reach maxima of possible abundance, with no species more abundant than another. Additionally, here, the reason was searched among ecological cues, and the intermediate disturbance of Connel (1978) was considered the only possible reason. Nobody has ever identified the intermediate disturbance that determines such a situation, but the continental slope has been considered inevitably interested in turbid falls of strength, extension, and frequency suitable to sustain such species diversification.

Continental slopes are, however, among the most ancient marine habitats of the planet, and depths below 1000 m are those where temperature is invariably fixed at approximately 3 °C, sunlight is completely absent, differences in pressures are relatively small at different layers, and turbulence is probably weak.

Due to continental drift, which separates the North Atlantic Ocean floor on the two sides of the central ridge, the slope of North-East America has drifted from North-West Africa for approximately **190 million years**. While surface coastal

Fig. 2.11 A map of the North Atlantic Ocean with the dark blue spot representing the sampling area from which Grassle and Maciolek (1992) described an unattended high species richness. (Courtesy G. Belmonte)

areas suffer from the variability of seasons and climates, the cited slope (1000–2000 m depth) represents the zone of maximum age where present conditions have been realized first (and they have remained unaltered).

Time seems to be the most important factor responsible for species richness, according to Rosenzweig's hypothesis of spontaneous formation of species from a large initial population, and the strongest alliance for this result comes from the stability of conditions that should remain there.

This is completely opposite to the ecological monopoly of reasons to explain species richness (space or time variability, intermediate disturbance, fractioning of habitats, and so on).

Unfortunately, the short time scale (1–50 years) that our studies can use to establish a rule has no efficacy from an evolutionary point of view.

The area of an environment, or its variability can of course be the cause of a temporary enrichment of species coming from elsewhere. However, all biogeography can be better understood if interpreted in terms of environmental age.

2.2.2 Latitudinal and Altitudinal Gradients

Classic environmental gradients (latitudinal and longitudinal) are strictly correlated with the age of environments because cold environments (at high latitudes and/or high altitudes) have obligatorily fewer years than warm environments because ice expansion during the last glacial era somehow impeded biodiversity flourishing and terminated approximately 12,000 years ago.

The polar cap retraction at high latitudes and their rising at high altitudes of mountain glaciers are the certification of the young age of currently periglacial isolated environments.

Dodson (1991, 1992) confirmed the ecological rule area/species number on crustacean species abundance in lakes of the northern hemisphere (Fig. 2.12). However, this author considered mostly postglacial lakes, which (as the majority of boreal lakes) formed in the postglacial era, hence having a comparable age (20,000–10,000 years). In the study of 32 European lakes, the regression line on the area/species log plot shows an evident scattering distribution of *y* values around the *x* value (lake area) of approximately 1×10^5 km^2. This *x* value (area) shows the youngest lakes (Finstertaler, a dam; Port Bielh, at 2700 m above sea level; and

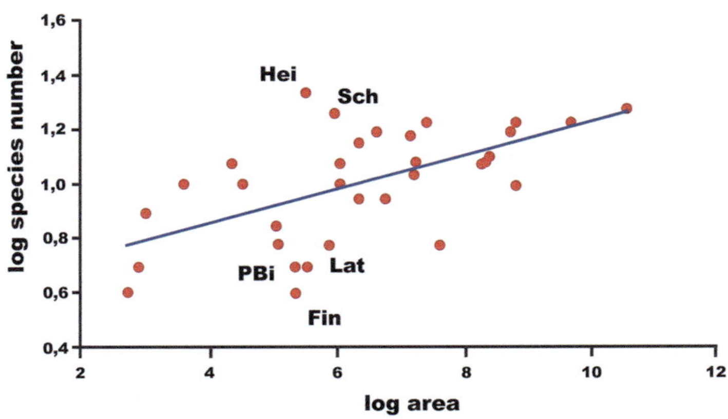

Fig. 2.12 The Dodson's (1991) correlation area/species (modified by G. Belmonte) as reported from 32 European lakes. Lakes with an area (*x*) of approximately 1×10^5 km^2 are visibly scattered around the regression line. Among the four sites with the minimum numbers (*y*) three are "young" lakes: **Fin**, the Vorderer Finstertaler, that is a recent dam (less than 100 years old), **PBi** is Port Bielh, an alpine lake at more than 2200 m above sea level, and **Lat**, the Latnjajaure (near the Arctic polar circle), both sites less than 5000 years old, and seasonally glaciated

Latnjajaure, near the Arctic polar circle) as those with the lowest *y* values (species number). The less biodiverse lakes were all evidently "young" because of a newly built dam or postglacial basins.

A different hypothesis could be attempted with lakes arranged according to age vs. species plot. The growth of species numbers with time is obvious at the start of the story of each island (and lake).

Islands (isolated environments) of millions of years of age can offer the possibility of easily evaluating the correlation of species numbers with the system age. Such a challenge has been accepted by limnologists more than land island biogeographers. The **ancient lakes** allow us to explain the species richness, avoiding searching for a solution to the dispersal/immigration scenario and paying more attention to in situ speciation.

2.2.3 The Case of Lake Aral

The Aral Sea (today reduced by approximately 90% of its past extension) had a surface area approximately 20 times that of Lake Ohrid in the Balkans. This notwithstanding, and contrary to the rule that the area correlated with the species number, Aral had an evident low species richness (maximum 136 species according to Arrow Aladin (1995)) compared with the "small" Lake Ohrid (approximately 1200 species according to Albrecht and Wilkie (2008)). Many suggestions make it possible to hypothesize that Lake Aral did not exist in ancient historical times, and its story has only lasted some centuries (Fig. 2.13).

The expert Arab geographers of the Middle Ages did not indicate a great lake in the position of the Lake Aral, and the lake has been reported on maps only from Russians after the conquest of those steppe lands in 1620. The history tells about the Sultan of Khiwa, who turned the Oxus river into the desert to eliminate a riverway for Czar soldiers coming from the Caspian coastline, thus hoping to impede Russians from individuating the city and his kingdom. Oxus is very possibly what is now called **Amu Daria**, the southern tributary of the Aral, and it possibly arrived in the Caspian Sea in the past, from where Czar soldiers could easily reach Khiwa. This is probably only a hypothesis to be confirmed, but the present retraction of the Aral extension left dried lake bottoms where, in 2004, an ancient city was recognized with archaeological remnants of the thirteenth to fourteenth centuries.

Aral, in the Kirghiz language, means "island". It is now evident that the Aral Sea was a young water environment and, although exaggeratedly large, the hosted species were only those that arrived from elsewhere, according to the process described by MacArthur and Wilson, and no one of them had the time to undergo the genetic process of speciation.

The case of the Aral also suggests that, unlike what was proposed by Hutchinson (1961), at least for long time intervals (millions of years), the species richness is probably directly correlated with the **stability** of the system (Connell 1978). Speciation, according to Rosenzweig's model, does not require the variability of conditions. The **age vs. species curve**, hence, has to be corrected into a sum of

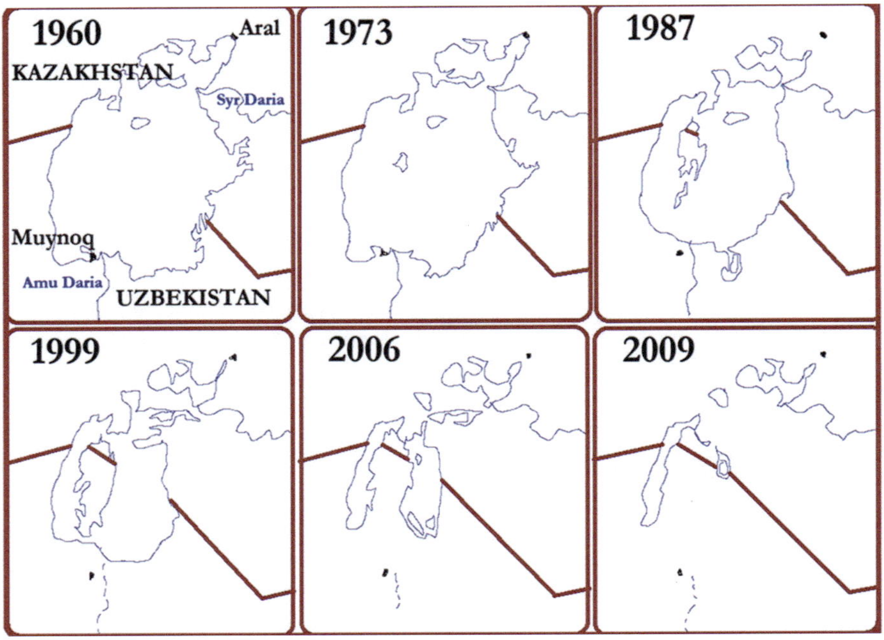

Fig. 2.13 Maps of the Lake Aral area from 1960 to 2009 (graphic composition of A. M. Miglietta)

curves acting each for a different time duration, and nothing can be said about the existence of a superior limit to the total number of species in a particular environment.

According to the model of Hubbell (2001) on a neutral theory of species geographic distribution, a species richness value is typical for each environment, which depends not only upon MacArthur and Wilson's model but also on speciation and evolution. This value, however, in Hubbel's model obeys a sort of **zero-sum game** where the number of species cannot exceed the **carrying capacity** of the system.

If the carrying capacity is a biomass value, the number of species can grow in combination with a decrease in the number of individuals per species and/or a decrease in the size of each species.

However, this model still needs an appropriate set of data to be completely accepted. It proposes an upper limit, in any case, and somehow a fixed total number of species, which is not what is suspected. At this point, Whittaker (1972) suggests an interpretation: "more species existing, more opportunities to additional species to exist".

Time undoubtedly has an effect on the number of species that each environment can host. This effect, however, is not simply that indicated by MacArthur and Wilson on oceanic islands due to the immigration and extinction rates of species. The role of time in species richness starts to be recognizable only in isolated environments where contamination from neighbours is limited or null and, after many years, allows evolution to occur.

This action is so slow that it cannot be recognized in terms of ecological time, and even a limit (a maximum number of species) is probably difficult to find if it exists. Time, however, elegantly explains the disobeying faunal situation of St. Helena and Ascension in the middle of the Atlantic Ocean, the high species richness of ancient lakes, the latitudinal gradient of the species number, the species diminution with increasing altitude, and the astonishing species richness of benthic fauna on the continental slope of the NW Atlantic Ocean.

The temptation to redescribe each pattern and geographic situation is great. In Italy, for example, a series of faunal geographic areas have been recognized on the ecological basis of the climate.

If we look at the lakes, however, we can note that the Padano-Venetian district has large lakes at low altitudes, formed at the beginning of glacier retraction (approximately 35,000 years ago), whereas the Alpine district hosts small lakes over the tree altitudinal limit and hence is ice-free since only 5000–7000 years. Along the Italian peninsula (Apennine district) are concentrated several volcanic calderas replenished with water from hundreds of thousands of years (those of the volcan quiescence), and at the extreme south and on islands (insular district), most of the lakes have been realized only recently with dam buildings (ages lower than 100 years).

In the brief example of the geographic distribution of species in lakes, the biogeographic subdivision of the territory could be based not on ecology but on environmental age, and it proposes the age of environments as an alternative key in the interpretation of species distribution, other than species richness (Connell 1978).

Suggestions derived from this review on long-aged environments allow us to predict that many species wait for a discovery in deep oceans (either in benthos or plankton) due to the constancy of the environmental conditions (over an evolutionary time span) coupled with the enormous available space and population sizes.

2.3 Abundance and Representativeness of Species

In each environment, trophic level, and/or high level taxon, it is possible to verify a general rule that distributes the individuals according to their species appurtenance (Fisher et al. 1943). Only very few species are represented by many individuals (or only a few high-level taxa are represented by many species) because the majority of species are represented by a small number of individuals.

This affirmation appears as Nature's law, and before naturalists, mathematicians geometrically described such an affirmation with a **"Jota" distribution** in a plot with N (number of specimens belonging to the same species) on the y axes and S (number of species) on the x-axis, where correlation between individuals and species is described by a hyperbolic function.

The utility of such a graphical description induced the proposal of a number transformation to minimize otherwise extremely long axes. Numbers (both specimens and species) were transformed into a log base or grouped according to the rank of importance (Preston 1948). These transformations produced similar and comparable curves in the **"rank-abundance" graphs**, which were typically **right-skewed** (Fig. 2.14).

The curve with abundances (individuals per species) grouped in a few ranks (on the *x*-axis) was mostly vertical and quite parallel to the *y*-axis. The curve with abundances distributed in many ranks was quite horizontal and parallel to the *x*-axis.

In ecology, the first curve (number of specimens vs. few ranks of species importance in each habitat) is the description of a **geometric sequence**, typical of extreme habitats where few species coexist and generally only one is dominant and the others are rare (e.g., a hyperhaline environment). The opposite situation (that with many ranks of importance) is typical of a habitat with very high biodiversity.

The more common distribution curves were called **log-normal** because, in a rank-distribution graph, the curve resembles the **normal distribution** of values around a central average value, although it is **deprived of the "left part"** of the distribution. The absence of the "left part" was considered a typical consequence of the data affected by the **sampling effort**. In other words, the nonperfect design of a normal distribution was interpreted as a lack of enough data from the environment. Additionally, the Jota distribution was interpreted as the right part of a normal distribution with a veiled left part.

Some long-term collection series (e.g., insects collected for years at the same site) demonstrated that the "veiled left part" of the normal distribution actually exists, thus suggesting that species less represented (the rare ones, on the left side of the *x*-axis) in a community are not the majority. This new rule, in addition, suggests

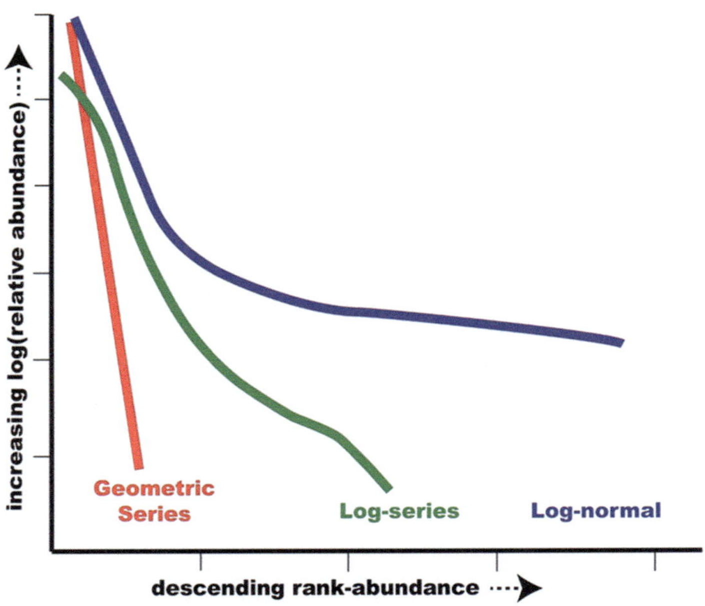

Fig. 2.14 Curves of species abundance–distribution. The steepness of the curve is inversely related to the species richness (steepest curves describe poorest communities). (Courtesy G. Belmonte)

that more intense sampling efforts ensure better descriptions of natural communities and biodiversity (Fig. 2.15).

More recently, it emerged that the number of species does not follow a log-normal distribution (this means that either rare or dominant species are scarcely present in comparison with a large number of species with an intermediate number of individuals). Robust databases and historical studies suggested (Hubbell 2001) that distributions in nature show **a long tail of rare species** (not well represented in terms of individuals) **on the left of the average**, which is not forecasted by the log-normal distribution. Hubbel had the merit of being the first to propose additional reasons for the only ecological ones in the numerical description of biodiversity. Distributions with a long tail of very rare species were proposed as a byproduct of a "zero-sum" **multinomial distribution** (of relative species abundances) that recognizes a sort of "carrying capacity" to the community that could not host more than a fixed number of individuals (or could not sustain more than a fixed value of biomass). Community

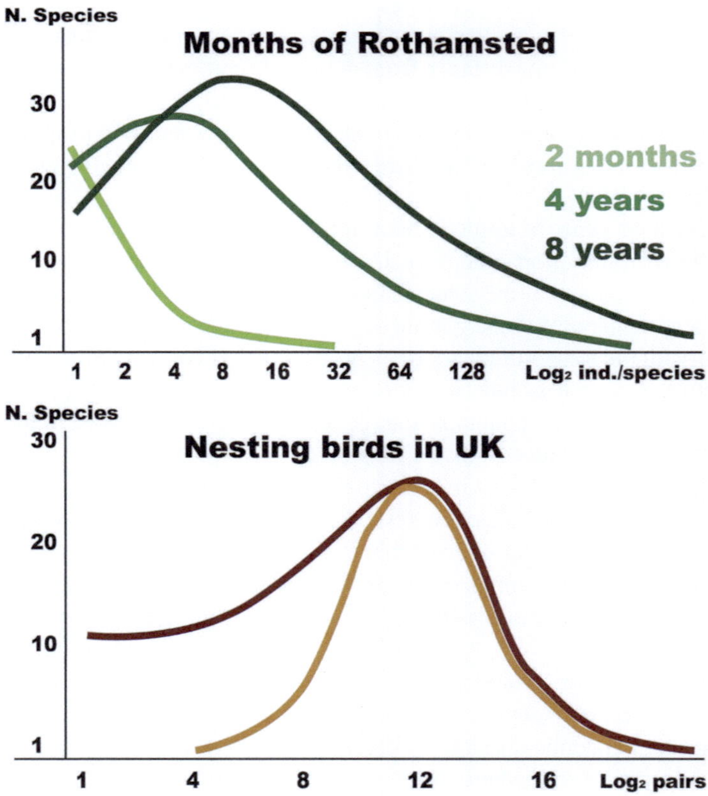

Fig. 2.15 The unveiled left side of the log distribution curves. Above, specimens collected with different time-efforts (2 months, 4 years, and 8 years) gave different rank-abundance graphics. Below, the historical data on nesting birds in Great Britain are distributed in a rank-abundance plot, showing a prevalence of species with intermediate importance (Williamson and Gaston 2005). (Courtesy G. Belmonte)

diversification, at this point, depends on how many species can be represented by that fixed number of individuals (or biomass value).

The focus on a limitation due to the environmental capability of host species is an "ecological" point of view that does not consider the role of time in the richness of species assemblages.

Whittaker (1972) noted that "*evolutionary time is difficult to measure but is important as the dimension through which an increase in diversity occurs. For terrestrial plants and insects, the* ***increase of species diversity****, ...,* ***is a self-augmenting evolutionary process without any evident limit***".

The absence of the factor "evolutionary time" in the evaluation of biodiversity represents the most important obstacle to the universal application of some of the most important theories in the distribution of living organisms on Earth.

The successive arrival of colonizers on a newly formed oceanic island is obviously dependent on the distance of the island from the continent, the space that it can offer to the colonizer assemblage, and from then on, the time necessary to occupy all the possible ecological roles.

The **distance** acts as a filter on the propagules launched by species from the continents. The greater the distance from the source continent, the lower the species number on the sink island.

The **island area** is also an obvious element that plays a role in shaping the final number of coexisting species. The more island surfaces there are, the more coexisting species there are.

The so-called **capture angle** is also important. This is the width of the island profile offered to the propagule diffusion point (the continent). For islands having the same extension and at the same distance from the continent, the island with the longer diameter parallel to the continent's coast will receive more propagules than the island with the longer diameter perpendicular to the continent's coastline. From this point of view, flat islands will also receive fewer propagules (and ultimately host fewer species) than mountain islands. In this latter case, however, the variability in altitude also produces a diversification of climatological levels, thus enhancing the availability of different habitats to the arriving propagules.

Considering all these rules, the system just needs time to allow propagules to arrive in the new situation and affirm themselves. This could represent an additional reason for inter-island differences in species numbers, and some studies on newly formed islands have demonstrated that time plays a limited role in colonization history.

We can easily note, however, that very isolated environments (mostly oceanic islands) host **endemic species**, sometimes at a high percentage of the total species assemblage. An endemic species, by definition, is not present elsewhere; hence, it has not arrived from elsewhere, and this raises the problem of completeness and universality in Mac Arthur and Wilson's theory.

Some cases (isolated environments) strongly suggest that new species can be formed in situ by evolutionary processes because they do not come from elsewhere. These endemic species normally do not simply substitute some preexisting relatives but, at least for a certain time during the **sympatric speciation** process, coexist with

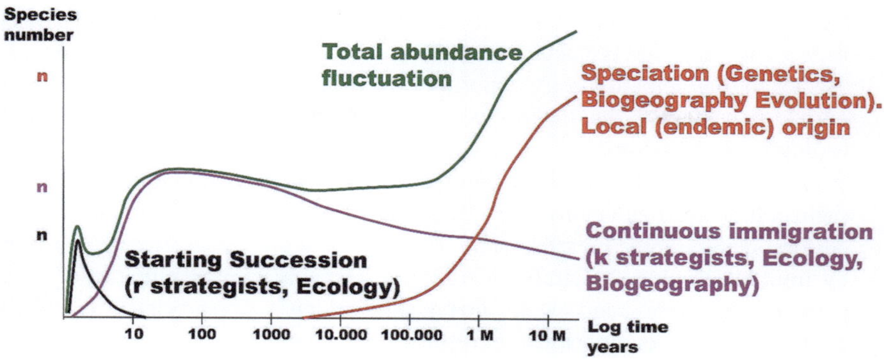

Fig. 2.16 The species number on each island grows accordingly with different phases. (1) Early colonization, arrival of pioneer species (r-strategists) that sustain successive immigration (ecological succession). (2) Immigration from neighbours (MacArthur and Wilson (1967) model) up to a constant number (the equilibrium point) of species (k-strategists) where new arrivals equal extinctions. (3) Speciation and species flock appearances. The species number grows for intrinsic, evolutionary reasons and independently from the ecological ones. (From Belmonte 2012; redrawn by A. M. Miglietta)

them on the same island. This process results in the growth of the species number at an isolated geographic site. From this simple consideration, it is deduced that species numbers in isolated environments can grow over evolutionary time, even in the absence of colonization from elsewhere.

This is the case for the already mentioned Lake Ohrid (on the Albania–North Macedonia border), which shows endemism of approximately 50% of its species (Albrecht and Wilkie 2008) (i.e., half of the existing species do not exist elsewhere) and has an estimated age of 3 million years.

The same explanation fits well with the exceptional case of Lake Bajkal. This large lake (not the largest of the world, however) is in the middle of East Siberia and has been studied by the same scientific community that studied Lake Aral (which was twice as large as Bajkal). The results of research on Bajkal diversity estimate the existence of approximately 2600 species, with some taxa at 80–90% endemism (Timoshkin 2001). The exceptional species richness of Bajkal obliged Dodson (1991) to eliminate it from the area vs. species plot (Fig. 2.12) that he obtained from all the lakes of the northern hemisphere because they were sensibly different. The other lakes are considered well described by the corresponding species–area plot because they are of postglacial formation (35,000 years) and are thus homogeneous from the geologic age point of view.

The only datum in accordance with the species richness of Bajkal (please remember that Aral had 136 species at its maximum flooded period) is the age of the lake, which ranges between 3 and 20 million years.

In conclusion, the number of species existing in an isolated geographical area varies according to three-time steps depending on the age of the environment (Fig. 2.16):

1. New arrivals of **pioneers** at the very birth time of the area (island); this process lasts a few years and prepares the situation for successive species assemblages; species of this step are mostly opportunistic, grouped under the ecological category of **r-strategists**, and the process of species arrival is described by experts of **ecological successions**.
2. Newly arrived species are those that require a certain maturation of the system (hence, they need an environment "prepared" by opportunistic colonizers) and develop demographic pressure on it. The system is consequently characterized by limited resource availability, defining a **carrying capacity** that drives the population dynamics, and species are type **K strategists.** The process is described by the ecology of **competition** for resource utilization.
3. New species are formed inside the system (**speciation**). The species are endemics that do not substitute preexisting species but add themselves to the total species assemblage. The formation of new species requires evolutionary time that depends on genome variability, which is different for any taxon, but in any case, it requires a high number of generations and an advanced age for the isolated environment. The zero-sum game of Hubbell (2001) seems to not be respected by ancient environments where the species number grows continuously (on an evolutionary time scale).

2.4 Pelagic Lifestyles, Feeding Habits, and Trophic Relationships

Here, a pelagic lifestyle is considered to be as any behaviour of an organism living surrounded by liquid water. A blue-fin whale has to be considered in a large water volume, but very small organisms, such as bacteria or viruses, can conduct a pelagic lifestyle even in just a drop of water. For this reason, "pelagos" is a relative concept and should not correspond to simply the open sea.

Box 2.2 Viscosity and Sedimentation

$$\text{sedimentation rate Vg} = \frac{d^{2*}(\Delta\rho) * g}{18^{*}\eta}$$

***d*, Diameter of the particle:** in the formula, d is reported as d^2. The greater d is, the greater the sedimentation rate of the particle.

$\Delta\rho$ Difference of density between particle and liquid: more $\Delta\rho$ corresponds to a higher sedimentation rate.

η dynamic viscosity: less η corresponds to a higher sedimentation rate.

Many plankton species belong to the ecological category of "**primary producers**", or **autotrophs**, which are able to use the energy coming from sunlight to synthesize molecules for life energetic requirement (**photosynthesis**). Typically, in an ecological pyramid, the primary producers reach the maximum value of living biomass and stay at the base of the building to sustain, with the production of energetic molecules, the whole set of nonautotrophic organisms. The pelagic marine pyramid shows a base of unicellular, very small autotrophic organisms (**phytoplankton** of the size range of picoplankton), which are food for a large variety of **filter feeders**. Due to their size, small phytoplankton cells are of either eukaryotic or prokaryotic types. The latter comprise **Cyanobacteria** (or blue algae), which are **Prokaryota** and are extremely diffuse. ***Prochlorococcus*** is a genus of Cyanobacteria that, in terms of the number of individuals, is the most abundant photosynthetic organism in the whole biosphere. Autotrophic organisms that convert sunlight energy into chemical links are also **Eukaryota** of the Kingdom **Chromista** (see Atlas Plates 1–6).

They are probably the best known among phytoplankton taxonomists and are grouped in the phyla of **Ochrophyta** (**Bacillariophyceae** or diatoms), **Miozoa** (Dinophyceae or dinoflagellates), and **Haptophyta** (**Coccolithophycea** or coccolithophorids).

All these organisms are unicellular and reproduce themselves by cell division (**mitosis**) without any recombination of genetic nuclear material.

Their size ranges from **2** to **200 μm** among Bacillariophyceae and from **7** to **500 μm** among Dinophyceae and are all characterized by cell capsules of different materials. Bacillariophyceae have a **siliceous theca** (*frustule*), Dinophyceae have a **cellulosic theca**, and Coccolitophyceae have the cell protected by rounded **calcareous plates**.

Bacillarius (L) = shape of a small rod *Caryos* = nucleus *Chrome* = colour *Chloros* = green; *Phyllos* = leaf *Cyano* = blue *Dynamos* = strength, power *Phycus/Fucus* = plant, alga *Pro* = previous, precedent; *Chloro* = green; *Coccus* = rounded

The sexual reproduction (which performs a recombination of genetic material) is carried out by many species under unfavourable environmental conditions and very commonly can be associated with the production of a resting stage (a **cyst**) that waits for the return of favourable conditions. All phytoplankton organisms have a molecule of pigment (chlorophyll) that absorbs energy from light wavelengths corresponding to blue and red ($\lambda = 400$ and 700 nm) from sunlight (Fig. 2.17).

Apart from the common "chlorophyll *a*", present in most photosynthetic organisms, each taxon is characterized by additional pigments of different colours (i.e., absorbing sunlight at different wavelengths), which enlarge the light spectrum useful for **photosynthesis**.

Bacillariophyceae (diatoms) are unicellular algae typically deprived of cilia or flagella (but a flagellum exists in male gametes) and thus unable to swim. Phytoplankton species are likely to remain suspended in the water column because

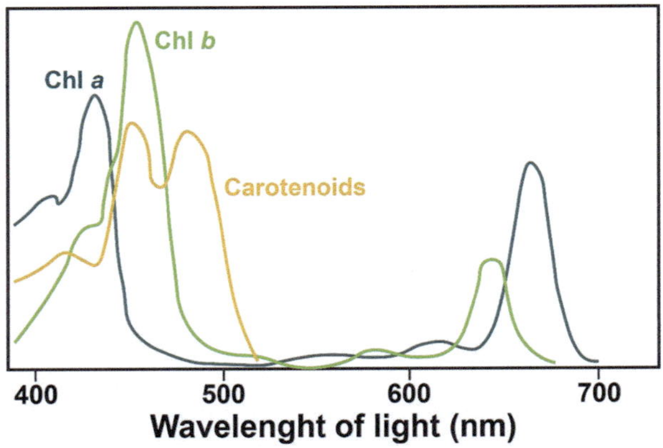

Fig. 2.17 Chlorophyll *a* and *b* and carotenoid absorbance spectra (graphic composition of A.M. Miglietta)

their small size retards sinking, notwithstanding the presence of a siliceous skeleton (frustule). Diatom cells are typically **diploid**.

Among Dinophyceae (**dinoflagellates**), chromosomes are always condensed and never associated with protamine molecules. The mitotic plate is not formed during cell division, and the organism is typically **aploid**. Dinoflagellates also represent the intermediate lifestyle between autotrophs and heterotrophs, being in many cases **mixotrophs** (thus choosing the feeding regime according to the environmental conditions), and some species are completely heterotrophic, performing the feeding style of **phagotrophy** in engulfing other, smaller, unicellular organisms.

The larger of them (up to 0.5 mm) are visible to the human eye, but many microscopic species can manifest their presence also by colouring the water during episodes of intense cellular multiplication and consequently high cell concentration in the water (the so-called **red tides** in coastal confined areas) or emitting light during night (the phenomenon of **bioluminescence**) (Fig. 2.18).

2.4.1 The Microbial Food Web

Prokaryote microbes are the most important consumers of dissolved organic matter (DOM) in the ocean. Their respiration is the major biogeochemical loss for oceanic dissolved organic Carbon (DOC). It should be noted that most Bacteria benefit from DOM, and only a few of them stay on particles of particulate organic matter (POM). The microbial communities of the sea contain a complete set of roles and specializations (primary producers, herbivores, carnivores, mixotrophs, scavengers, and parasites) among microbes with sizes of approximately 5 μm. This complex (and large) community is responsible for nutrient regeneration in ocean environment(s). Bacteria incorporating organic matter serve as food for bacterivorous organisms (mostly mixotrophs unicellular Eukaryota), which are regenerators of inorganic nutrients due to their extruded material. Bacterivores are food for mesozooplankton

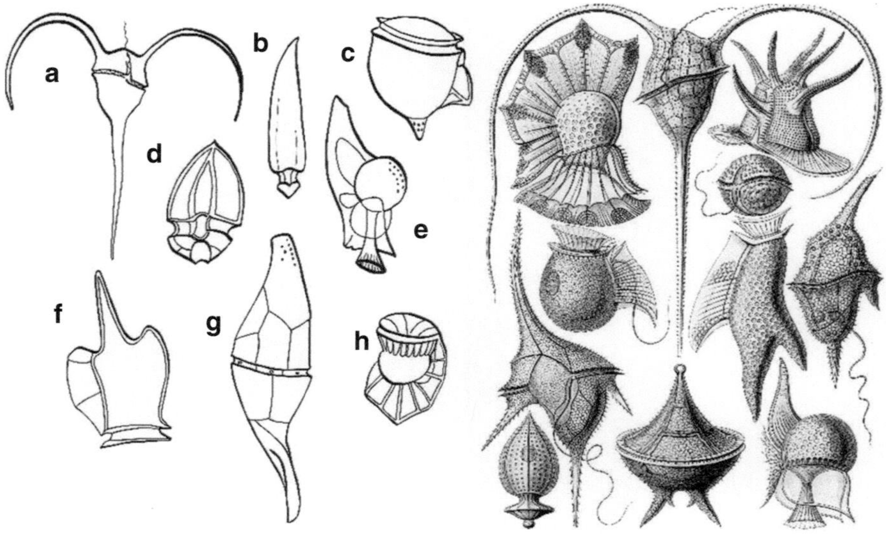

Fig. 2.18 Left, sketches representing the diversity in Dinophyceae morphology. (**a**) *Ceratium*; (**b**) *Oxytoxum* sp1; (**c**) *Phalachroma*; (**d**) *Oxytoxum* sp 2; (**e**) *Histioneis*; (**f**) *Dinophysis*; (**g**) *Centrodinium*; (**h**) *Ornitocercus*. (Courtesy G. Belmonte). Right, more details (and some different species) in drawings illustrating the Haeckel's (1904) atlas

(larval and tiny Metazoa), which, contrary to what was assumed in the past, are only marginally herbivores (on phytoplankton). Such an alternative pathway to the classic phytoplankton-zooplankton-nekton trophic chain has been identified as more important and is called a **microbial loop** (Azam et al. 1983) or, more recently, a **microbial food web** (Fig. 2.19).

Bacteria use approximately 35–50% of their photosynthetic production for their growth and metabolism, and then a large fraction of molecules obtained with photosynthesis are expelled in the water as small carbohydrates (molecules not subject to nutrient limitations because they do not contain N or P). Such "exudates" realize an extracellular jelly cover in many unicellular phytoplankton. This jelly cover (Transparent Extracellular Polymers, TEP) is also useful, at least in eukaryotic phytoplankton, to favour floatability (if containing opportune ions), but it also hosts and rears Bacteria for the production of useful molecules (vitamins).

Additionally, 16–37% of the Carbon present in phytoplankton ingested by herbivores passes quite immediately in the water as DOC. The unicellular herbivorous Eukaryota do not completely digest the ingested phytoplankton, and residues present in their vacuoles return to the water. In addition, many herbivorous zooplankton lose parts of the phytoplankton cells that they capture during chewing or manipulation before ingestion. Additional DOC is liberated from the faecal pellets of these Metazoa plankton.

This large availability of DOM has been a surprise for ecologists 50 years ago, who measured sea productivity only in terms of the chlorophyll content of water masses. The oligotrophy of the ocean, as a consequence, is a poor production

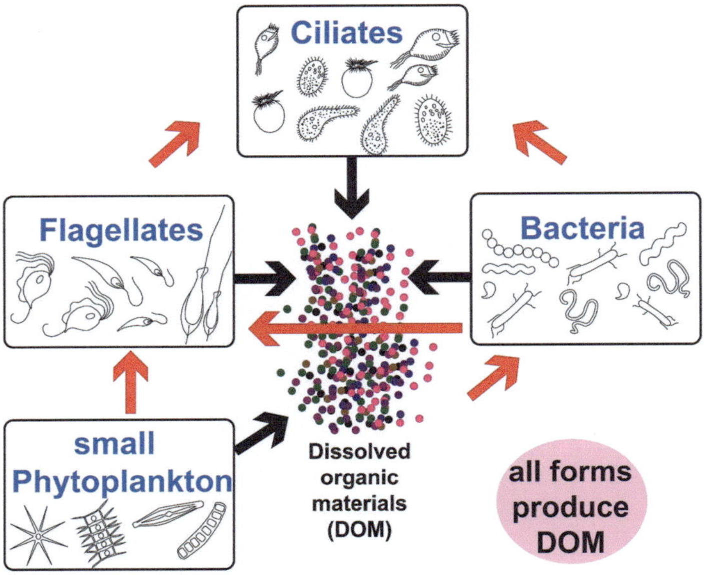

Fig. 2.19 The microbial food web. Red arrows, ingested food; black arrows, produced material (graphic composition of A. M. Miglietta)

situation only if photosynthesis and autotrophy are considered. The microbial food web must be considered a fundamental way to recycle materials, either to photosynthetic or heterotrophic organisms.

If we consider that most of the ocean is deprived of the possibility of performing photosynthesis (because it is in the dark), we can easily understand the importance of such alternative ways for communities of living organisms to obtain energy for their lives.

Coccolithophyceae are unicellular algae, exclusively marine. They show alternate diploid and aploid generations and are the most important producers of Ca carbonate in the pelagos.

Among unicellular Eukaryota, the largest planktonic organisms are generally heterotrophic and belong to the kingdom **Chromista** and the phyla Ciliophora and Retaria.

Ciliophora are heterotrophic, large unicellular organisms (20–200 µm) with the evident characteristic of the cell surface being ornamented with short and numerous cilia.

Cilia can be distributed on the whole surface of the cell, can be concentrated in a single part, or can be associated with forming cirri and tufts in precise parts of the cell. In any case, ciliary movement produces food currents other than movement for the cell.

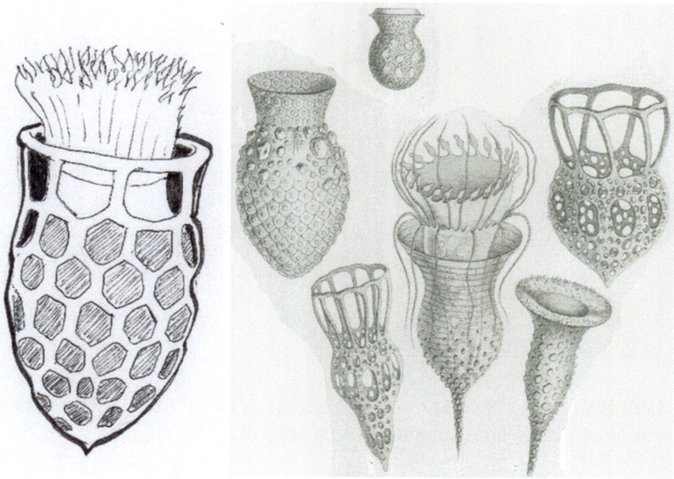

Fig. 2.20 Tintinnina. Left, a sketch of *Dyctiocysta* (original). Right, diversity of Tintinnina techae from the Haeckel's (1904) atlas

In the plankton, there are Ciliophora equipped with chitin theca with a wide opening from where the cell protrudes its ciliary apparatus. Such a theca is often shaped as a small bell and gives the name to the typical plankton group of **Tintinnina** (Fig. 2.20; see also Atlas, Plate 8).

Ciliates have two types of nuclei, one **micro** (diploid, for genetic and reproductive purposes) and one **macro** (generally polyploid, for functional purposes). Sexual reproduction implies the exchange of genetic material (**conjugation**) of micronuclei between two different individuals, and it is generally followed by the formation of a resting stage (the **cyst**) useful to avoid life in non-suitable conditions.

Asexual reproduction is a binary scission with a separation plane transverse to the long axis of the cell.

Aplo = a (not) plus (plural) *Coccus* (L) = rounded; *Lithos* = stone; *Forein* = bring *Cyst* = case with biological material inside *Diplo* (L) = double *Globigerus* (L) = rounded, globe shaped *Phago* = engulf, introduce; *Trophe* = food *Plasma* = liquid life material *Radius* (L) = ray *Rete* (L) = net *Tintinnulus* (L) = small bell

Retaria is a phylum of Chromista of unicellular organisms (Foraminifera, Radiolaria, Acantharia) living in all aquatic environments (not only plankton) and reaching the largest size among planktonic cells (40–300 μm) (Fig. 2.21; see also Atlas, Plate 7).

Foraminifera are 4000 species with some holo-plankton components (family Globigerinidae) that are very useful in identifying the geologic facies of ancient oceans. These organisms (as all the Foraminifera) produce a calcareous shell divided in successively larger chambers to host the progressively growing cell. **Globigerinidae** of the Miocene period (23–5 My), for example, sinking to the

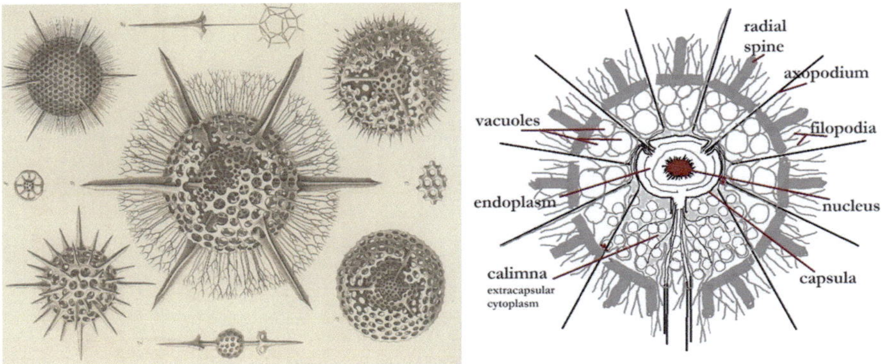

Fig. 2.21 Left, Radiolaria (Retaria) from the Haeckel's (1904)'s atlas. Right, scheme of the cellular organization of a radiolarian (graphic composition of A. M. Miglietta)

Fig. 2.22 The formation of sea bottom muds containing *Globigerina* calcareous spherules. After some million years, the marine bottom mud becomes rock. Such a limestone, excavated from quarries, gives the material used in the realization of Baroque buildings typical of the seventeenth century in Lecce (Italy) (Drawing of A. Gennari)

bottom when dead in the Tethys Ocean, produced mud whose particles (the globular spheres constituting the shell) were successively compacted in **limestone**, a tender stone that gave material for the realization of Lecce's baroque of the fifteenth to sixteenth centuries (Fig. 2.22).

Calcareous, multichambered shells are typically perforated. Each forum allows the internal cytoplasm to extend outside in rays and reticular connections, where most food capture occurs.

Rays of cytoplasm are sustained by the microtubular axis and individuate, at the internal cytoplasm, a double verse of material circulation (centrifugal and centripetal). The external cytoplasm net can also host "**photosymbionts**".

Foraminifera have the characteristic of alternating "**aplo-biont**" with to **diplobiont** generations.

In contrast to Foraminifera, **Radiolaria** are mostly planktonic and have a more accentuated radial symmetry. Their skeleton is composed of siliceous spines radiating in any direction, which give the organism an extended diameter (40–500 μm).

Extensions of cytoplasm (pseudopodia) stay along the spines, and the cytoplasm somehow "walks" on these axial elements, expanding or contracting along them. The cytoplasmic volume is so large that many nuclei can be present (or a single polyploid nucleus).

Generally, the central part of the cytoplasm, with nuclear material, is protected inside a perforated theca that separates the internal part (**endoplasm**) of the cell from the external part (**ectoplasm**). In the ectoplasm, many vacuoles are also present to allow flotation. Being heterotrophic, Radiolaria can be found at any depth in the ocean. The siliceous skeleton does not dissolve at high pressures; hence, unlike calcareous shells (which suffer such a physical phenomenon below 3000 m), Radiolaria can thrive well below 3000 m of depth.

Radiolaria are today distinguished in **Polycystina**, **Phaeodaria**, and **Acantharea** on the basis of their skeleton architecture. Acantharea, in detail, has a very unique skeleton of Strontium sulfate.

Among microzooplankton (mostly unicellular eukaryotes), tiny Metazoa (multicellulars) can also be considered, other than the small developmental stages of meso-plankton and benthos (meroplankton). The smallest of these tiny Metazoa are **Rotifera**. They are not the most important *taxon* of microzooplankton in terms of representativeness, but in confined coastal waters, and seasonally, they can reach important concentrations of specimens.

Rotifera are **pseudocelomates** characterized by a crown of cilia used either for swimming or for generating alimentary currents.

Box 2.3 Wonderful Symmetries and Microscopic Complexity

Radiolaria are among the unicellular organisms with the largest and most complex cells. The size and richness of cytoplasmic organelles correspond to a highly complex morphology and function. Such a morphological complexity of the cell is also associated with a high variability of **cytoskeleton** elements (spines, thecae, baskets, capsules), which gave the *taxon* incredible success among naturalistic illustrators and artists. The most celebrated among them is **Ernst Haeckel** (1834–1919). He was an eminent biologist of the nineteenth century. He imposed on the Science terms as Heterochrony, Phylogeny, Phylum, Protista, and Ecology. He created historical statements as "ontogeny recapitulates phylogeny" and proposed the existence of putative creatures (Blastea, Gastrea, etc.) as ancestors of modern Metazoa, or the existence of human ancestors in the southeast Asia (Indonesia), where important remains were effectively discovered 10 years after his suggestion. He did not agree with the Darwinian theory on the evolution of living organisms and was a supporter of racial ideology, which was used to support Nazism. These errors in the Haeckel scientific doctrine did not impede his fame from being enormous even more than 100 years after his death. In fact, he was a great naturalist (he described and gave the name to hundreds of new species), and the artistic aspect of his production was never debated. He illustrated marine unicellular organisms and invertebrates (Radiolaria, overall). Such a work also arrived to contaminate the *Art Nouveau* style. His illustrated books still represent a "must" for scientific illustration and the diffusion of nature forms among the large public (Fig. 2.23).

At the centre of the **ciliary crown**, Rotifera show the presence of a mouth preceding a **chitinous** and chewing pharynx (the **mastax**) able to crash the diatom frustules. The aboral extremity is generally equipped with a telescopic foot to clasp the substratum (Fig. 2.24).

Thus, they are plankto-benthic and primary consumers.

They do not have larvae and very commonly show the reproductive phenomenon of **parthenogenesis**, with entire populations of only females (diploid). When they shift to a sexual mate, they complete egg meiosis without waiting for fertilization and originate a segmentation process that gives aploid individuals (males) producing spermatozoans without meiosis. The male gametes are now available, and the fertilized aploid eggs originate a final product, diploid and resting (**cyst**), which passes the adverse period under lethargic conditions.

The **meso-zooplankton** is an ensemble of plankton organisms collectable with a 200 μm mesh net. They correspond to all the larger sizes of larvae (from the micro-zooplankton) and to the smaller non-Rotifera Metazoa.

The size of single specimens is inversely related to their abundance. Consequently, the plankton net for meso-zooplankton must be towed for a longer distance to filter larger water volumes and obtain larger numbers of specimens. Generally, the

Fig. 2.23 Radiolaria from the Haeckel's illustrations used for the decoration of a window in the Natural History Museum of Vienna (photograph, G. Belmonte)

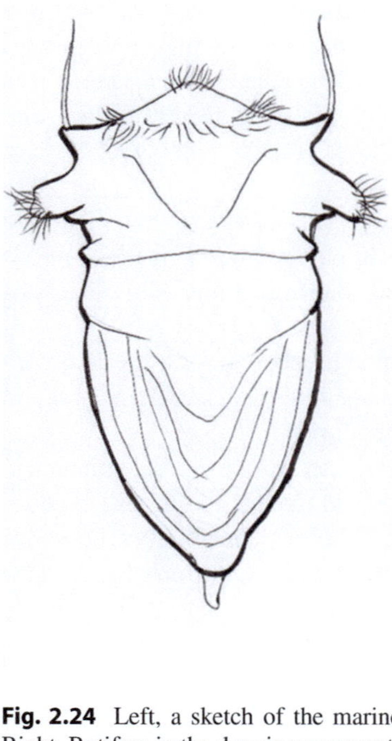

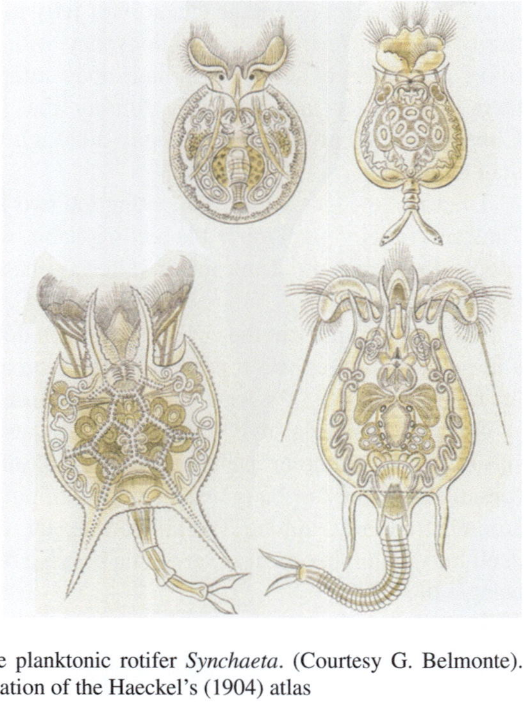

Fig. 2.24 Left, a sketch of the marine planktonic rotifer *Synchaeta*. (Courtesy G. Belmonte). Right, Rotifera in the drawing representation of the Haeckel's (1904) atlas

volume unit used to refer **to microplankton** collections is 1-10 L, while for mesoplankton, 1 m^3 is preferred.

On these bases, the net used for collecting **macroplankton** generally has a mesh size of 500 μm, and the net used to catch n**ekton** has a minimum mesh size of 3.0 cm (it is a fishing net).

Macroplankton comprise all the organisms visible to the human eye, and these can also be truly macroscopic, as in the case of some Scyphozoa medusae (**Cnidaria**).

Small individuals (larvae and juveniles) of nekton (the so-called **ichthyoplankton**) are generally considered part of the meso- and macro-plankton, thus representing a connection between two size categories and two different lifestyles (planktonic and nektonic) in the pelagos.

Among Cnidaria, Scyphozoa and Cubozoa medusae are generally evident, but Hydrozoa medusae are microscopic and generally ascribable to the category of mesoplankton (Fig. 2.25; see also the atlas, Plates 9–12).

Acantha = spine *Chitin* (Semitic) = covering *Cnide* = stinging *Coelum* (L) = void space (sky) *Ecto/endo* = esterior/interior *Phaeo* = gray *Pseudo* = false *Rota* (L) = wheel

Nekton is typically formed by fish (Osteichthyes and Chondrichthyes), and the largest species are not captured with standard fishing nets but with alternative devices. This large-sized category of marine organisms also contains other Vertebrata (Mammalia, Reptilia, Aves) and is the largest among invertebrates (Mollusca Cephalopoda).

All medusas represent the category of **jelly plankton**, a group of organisms that resolve the problem of negative buoyancy with the presence of high percentages (more than 97%) of sea water in their body composition. This characteristic makes them transparent and jellylike. Among the jelly plankton, **Ctenophora** and **Chordata Tunicata** are also present, although with a smaller number of species than **Cnidaria**.

The most evident difference among the two large taxa (Cnidaria and Chordata Tunicata) of jelly plankton is the feeding habit, which in medusae is predaceous on zooplankton and in Tunicata is filtering based on unicellular organisms of phytoplankton.

Cnidaria also present the complication of a double lifestyle in each species, with a life cycle that proposes a partial existence on a substrate (as a sessile polyp) and the remaining part in the water column (as a free medusa).

Not a few Cnidaria are holoplanktonic; in these species, the tiny ciliate planula is substituted by complex superorganisms (**colonies**) equipped with a gas bladder (**pneumatophore**) to buoy and/or with pumping elements (**nectocalyx**) to swim around. Similarly, among Trachimedusae, the ciliated larva ***actinula*** transforms itself into a pumping medusa (avoiding passage from the polyp stage) to continue its pelagic life.

The destiny of ciliated larvae in meroplanktonic species, is to sink to the bottom where they continue growth and avoid resolving problems linked to their stay in the

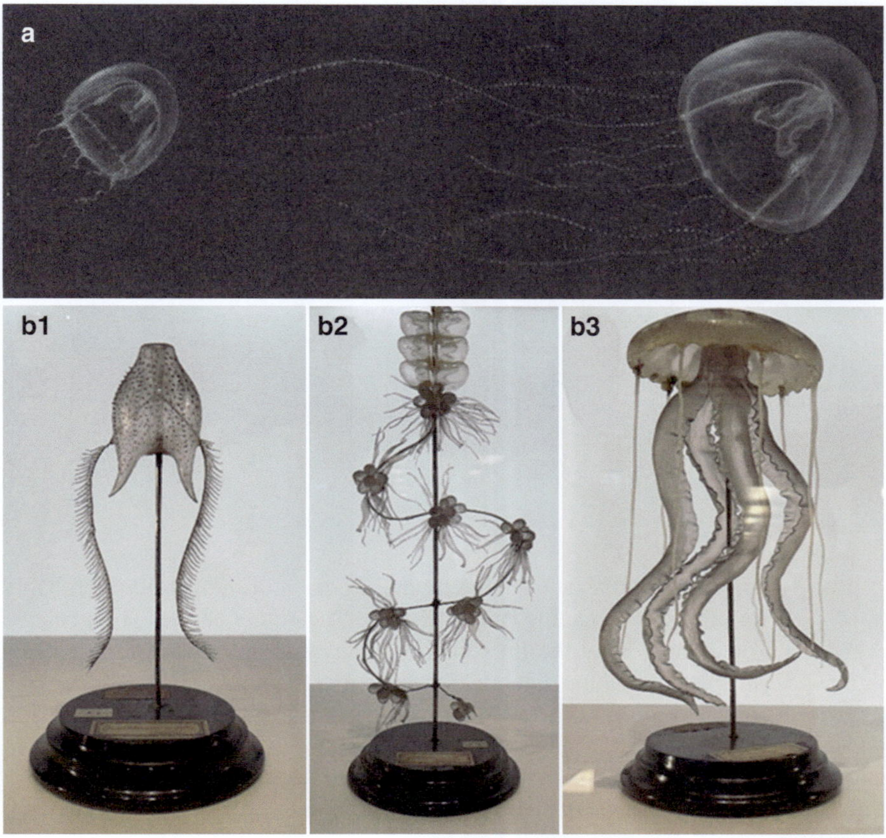

Fig. 2.25 Top (**a**), Hydrozoa medusa *Fialella zappai* (early and mature stage) illustrated by A. Gennari. Bottom (**b**), jelly plankton realized in glass by Leopold and Rudolf Blascha (Natural History Museum of Vienna, Austria). (**b1**) *Callianira* (Ctenophora); (**b2**) *Apulemia* (Cnidaria Hydrozoa Siphonophora); (**b3**) *Chrysaora* (Cnidaria Scyphozoa) (photographs, G. Belmonte)

water column. In the case of Annelida, however, the progressive transformation of the small first larval stage (**trochophore**) in larger specimens, passes through the appearance of devices (**chaetae**, **parapodia**), which somehow help the prolongation of a pelagic life (this advanced larval stage can receive the name of ***nectochaeta***).

The development of a flat body, with extended parapodia and long chaetae, is the final solution for holoplanktonic annelid species that, as in the families **Alcyopidae**, **Lopadorrhynchidae**, and **Tomopteridae**, complete their whole life cycle in the pelagos (Fig. 2.26, atlas Plate 17).

Such holoplankton species are mainly active predators that aggress their prey by a rapid eversion of the pharynx (few species of Iospilidae are, however, filter feeders). The search for prey is assisted by telescopic eyes.

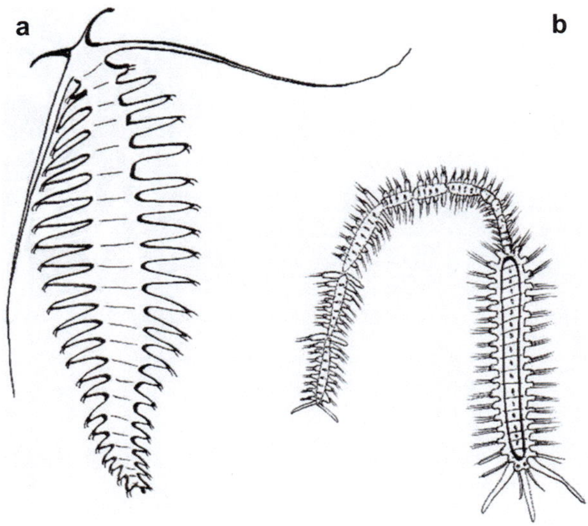

Fig. 2.26 Pelagic Polychaeta. (**a**) Scheme of a Tomopteridae. Parapodia elongation is the most evident adaptation to the pelagic lifestyle. (**b**) Scheme of an epitoch (with a chain of asexually generated individuals), the pelagic form of an otherwise benthic worm. (Courtesy G. Belmonte)

Long cirriform appendages of the second segment in **Tomopteridae** probably have a sensorial role in detecting prey presence. Regardless of their diet, pelagic **Annelida Polychaeta** show the highest metabolic rates among the bathypelagic organisms, suggesting that the diet must be energetically rich.

Evolution has interested the eyes of **Alcyopidae,** which appear large, complex, and telescopic. Such eyes have lenses controlled by muscles that can turn the axis in any direction. The eyes can perceive the intensity and direction of light, other than the outline and size of objects. Incredibly similar to the eyes of **Cephalopoda** and **Vertebrata**, these eyes are indicated as examples of convergent evolution.

The predaceous habitus obliges pelagic Annelida to follow the vertical migrations of their prey (zooplankton) inside the epipelagos or between the epi- and meso-pelagos. This notwithstanding, some species are typical of deep pelagos and do not migrate.

Actino = rod *Cephalo* = head; *Podos* = foot *Chaeta* = needle, bristle *Chorda* (L) = axile structure, flexible (rope) *Cteno* = comb *Foro* (L) = bring *Nekto* = swimming; *Calyx* = cup *Para* = close to *Pneumato* = air filled, lung *Podos* = foot *Ptera* = wing, flattened appendage *Rinco* = proboscis *Tunica* (L) = traditional dress *Vertebra* = from (L) *vertere* = to turn

Guglielmo et al. (2019), in a detailed study on pelagic Polychaeta in the South Adriatic, found that *Pedinosoma curtum* (1 species on a total of the 25 listed pelagic Polychaeta in that study) was never present in samples collected at depths above 400 m.

Flatworms **Nemertea** have trochophore-like ***pilidium*** larvae (Fig. 2.27, atlas Plate 14) and are generally benthic species after metamorphosis. This

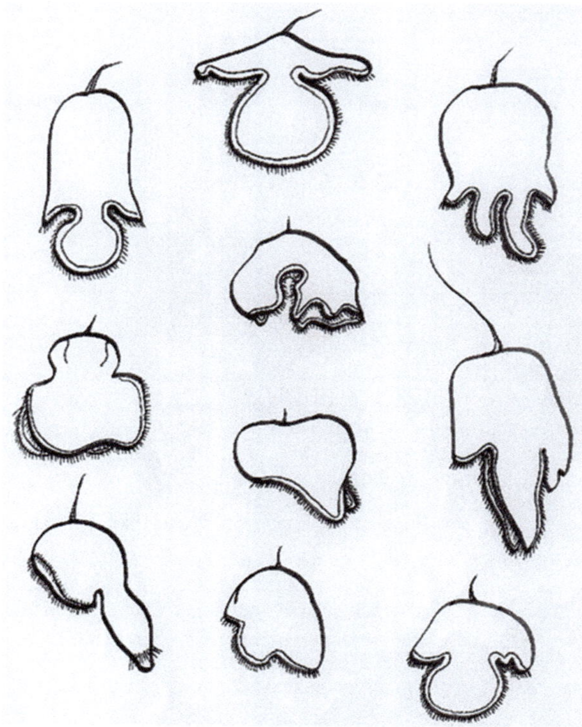

Fig. 2.27 Pilidia of Nemertea. Each pilidium received a name from the first scientists, who considered them independent plankton species (and not larvae of already existing benthic species). (Courtesy G. Belmonte)

notwithstanding, approximately 100 species are defined as "**bathypelagic**", being linked to the deep and dark pelagos, where they exert the ecological role of predators on other zooplankton.

Such Nemertea are probably the largest organisms able to move without specialized appendages, and they use the flat, leaf-like body to avoid fast sinking and to swim, producing body waves (Table 2.1).

Ctenophora, more than pilidium, are the largest organisms among those relying on cilia to swim. In the case of Ctenophora, however, cilia are extremely elongated and associated together in a series of paddling appendages (***cteni***) similar to combs, aligned along eight lines (meridians) stemming from the apical organ and going towards the inferior part of the jelly body.

Alcyo = kingfisher *Lopado* = naked, unarmed *Meta* = after, successive; *chrono* = time (a metechronal movement is that where some elements move before others thus giving the effect of a wave) *Nema* = wire *Pilidium* = helmet of Roma soldiers *Poly* = many; *Chaeta* = bristle, hair *Pteros* = wing; *Podos* = foot; *Tomo* = septate, cut *Trocho* = wheel; *Foro* = bearing *Trachaea* = rough arteria *Taenio* = ribbon; *Glossa* = tongue

Table. 2.1 Size range (linear dimension) among marine ciliated larvae (of 116 different species). The etymology of names is reported in the text when each larva is presented (Fig. 2.28)

Larva name	Taxon	Size range, μm	Number of taxa considered
Planula	Cnidaria Hydrozoa	200–900	13
Actinula	Cnidaria Hydrozoa Trachilindae	300–1000	2
Planula	Cnidaria Scyphozoa	200–450	6
Planula	Cnidaria Anthozoa	450–1500	4
Cerinula	Cnidaria Anthozoa Ceriantidae	2000	1
Zoanthina	Cnidaria Anthozoa Zoantidea	2000	1
Cidippus	Ctenophora	500–4000	2
Pilidium	Nemertea	200–500	10
Muller's	Platyhelminthes	100–300	3
Trocophora	Annelida Polychaeta	90–900	19
Trocophora	Mollusca	150–400	4
Veliger	Mollusca Gastropoda	200–500	7
Veliger	Mollusca Bivalvia	35–300	10
Cyphonautes	Bryozoa	375	1
Actinotrocha	Foronidea	750–2000	5
Glottidium	Brachiopoda	280–400	2
Pelagosphaera	Sipunculida	500–1250	3
Doliolaria	Echinodermata Crinoidea	400–2500	2
Pluteus	Echinodermata Ophiuroidea	400–2000	5
Bipinnarial	Echinodermata Asteroidea	400–600	2
Brachiolaria	Echinodermata Holothuroidea	600–6000	6
Doliolaria	Echinodermata Holothuroidea	400	1
Pluteus	Echinodermata Echinoidea	300–700	5
Tornaria	Enteropneusta	700	2

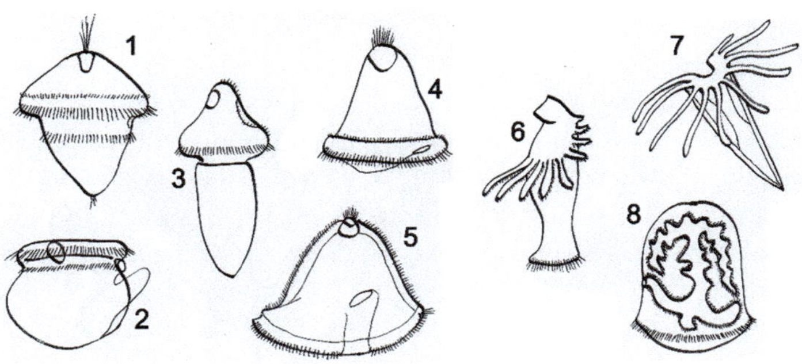

Fig. 2.28 Schematic representation of some larvae (only ciliary belts and apical organs are stressed). (1) Polychaeta trochophore; (2) Bivalvia veliger; (3) Sipuncula trochophore; (4) Entoprocta larva; (5) Bryozoa cyphonautes; (6) Phoronida actinotrocha; (7) Brachiopoda glottidium; (8) Enteropneusta tornaria. (Courtesy G. Belmonte)

The movement of *cteni* along the same meridian is **metachronal** and produces waves running along each meridian where the different orientation of *cteni* reflects differently the light, thus producing evident and multicolour effects.

Holoplanktonic Mollusca pass from the ciliated larva (***veliger***) to pelagic adults adopting other solutions: the body is generally small and transparent (e.g., **Gastropoda Heteropoda**), and more regularly, the shell is reduced (**Pteropoda Techosoma**) or completely absent (**Pteropoda Gymnosoma**). All pelagic Heteropoda are carnivores predators; they use image-forming eyes to detect preys and protrusible ***radula*** to capture them. The foot is laterally compressed into a fin-like ventral structure.

Normally, Heteropoda swim with their ventral side (and the fin-like foot) upwards; consequently, the dorsal shell or the visceral mass hangs below the body. At the posterior ventral edge of the fin, a sucker is present. As in other pelagic organisms, the eyes are relatively large compared with those of taxonomically related benthic species. The ***radula*** (type **taenioglossa**) hosts enlarged lateral and marginal hook-shaped teeth with elongated shafts.

Bodies and shells are mostly transparent, giving more evidence to the dark visceral mass and eyes. Sex is separate, and fertilization is internal.

Eggs are protected by mothers, and the first free individual is a **veliger** with a multiple lobed velum (six lobes in **Atlantidae** and **Carinariidae**; four lobes in **Pterotrachaeidae**).

The first shell (**protoconch**) of a veliger evolves into the final shell of an adult. This can be large enough to host the entire mollusc when retracted (as in *Atlanta*), a small protection for the only visceral mass (as in *Carinaria*), or absent at all (as in *Pterotrachaea*). Another group of Mollusca Gastropoda adapted to the pelagic lifestyle is the heterogeneous (not *taxon*) group of **Opistobranchia Pteropoda** (informal name alternative to **Prosobranchia** Heteropoda), containing the two orders **Thecosomata** and **Gymnosomata**.

Thecosomata are shelled and contribute significantly to the Carbon cycle in oceans because they are **microphagous** (filter feeders) and locally very abundant and efficacious in acting as Carbon segregators because they are used in papyraceous shells. **Thecosomata** have a very light shell (**teleoconch** in adults) and a bipartite foot forming two "wings" used to swim (Fig. 2.29, atlas Plates 15 and 16).

Some of them show the character of **hermaphroditism** (common among Opisthobranches), in some cases also the direct development (without the larval stage, as in the genus *Limacina*), and examples of asexual reproduction (as in the genus *Clio*).

Antarctic waters are particularly rich in **Thecosomata,** whose abundance is linked to the seasonality of phytoplankton production. They collect microplankton (mostly unicellular) with a mucous web that can reach enormous sizes (even 100 times the animal size). When the mucous web is replenished with planktonic organisms, it is ingested by the mollusc, which starts to produce another mucous web as soon as possible.

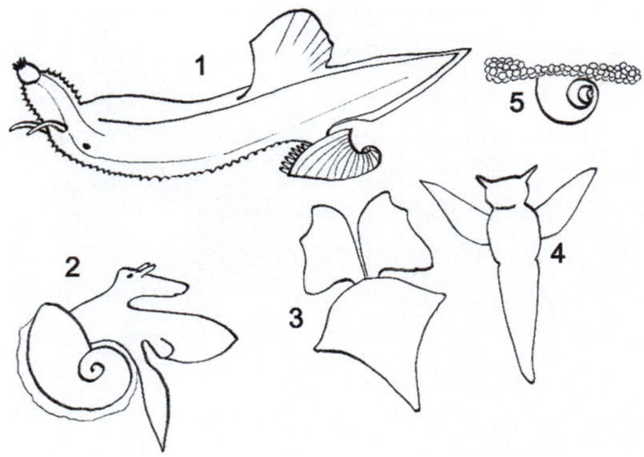

Fig. 2.29 Pelagic Mollusca Gastropoda. Sketches of (1) Carinariidae, (2) Atlantidae (Pterotrachaeidae), (3) Cavoliniidae, (4) Clionidae, (5) Janthinidae. (Courtesy G. Belmonte)

Due to such feeding behaviour, **Thecosomata** are strictly related to **epipelagos** (where **phytoplankton** are abundant), and only horizontal partitioning can be recognized among species.

Clio piramidata, a cosmopolitan species, appears to be influenced by climate conditions and is present in different sub-species in different geographic areas.

Atlantis = mythic kingdom in the ocean *Carina* (L) = keel *Ein* = internal; *Demos* = house *Gastro* = intestine, ventral area; *Podos* = foot *Gymno* = naked; *Soma* = body *Hermes* = Greek God; *Aphrodites* = Greek Goddess *Heteros* = different *Micro* = small; *Phagos* = I eat *Poly* = many; *Chaeta* = bristle, hair *Pteros* = wing; *Podos* = foot; *Trachaea* = rough arteria *Radula* = from (L) *radere* = to scrape *Taenio* = ribbon; *Glossa* = tongue *Velum* = veil

Endemism can be considered rare, but *Cavolinia tridentata* (forma *tridentata*) seems endemic to the Mediterranean Sea, and *Diacavolinia flexipes* seems endemic to the Red Sea.

Thecosomata perform diel vertical migrations from a few meters to 1500 m of extension, reaching the surface layers during the night for feeding. Few species are mesopelagic or bathypelagic and thus never enter the epipelagos.

In **Gymnosomata,** the larval shell (**protoconch**) is lost during metamorphosis, and a completely naked adult swims in the pelagos. Gymnosomata are solitary because hunters and carnivores whose prey could be, among all, **Thecosomata**. Mesopelagic and bathypelagic species are more common among Gymonsomata, which "wait" for their prey in the deeper part of their daily vertical migration range.

Final examples of pelagic Gastropoda are **Janthinidae**, a family with only 8 species (genera *Janthina* and *Recluzia*), macroscopic (up to 3–4 cm) with a well-developed shell. They produce a raft of mucous bubbles at the sea surface (the **Neuston** habitat). The mollusc lives upside-down attached to such a raft and preys on neustonic Cnidaria (Siphonophora and Velellidae).

During its larval condition (veliger), *Janthina* has an ***operculum*** that could be used to protect the larva inside the **protoconch**, but in reality, it is used to hang the

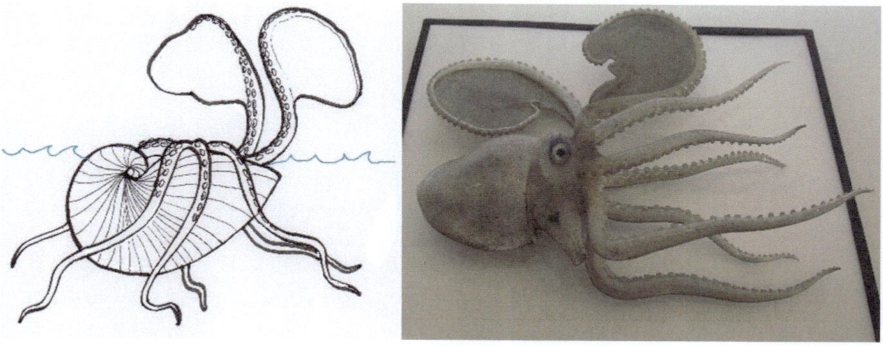

Fig. 2.30 *Argonauta*, female. Left, a sketch of the position at the sea surface to keep wind with the two modified arms. (Courtesy G. Belmonte). Right, a glass reproduction by Blaschka artisans (Natural History Museum, Vienna; photograph G. Belmonte)

veliger to the surface of the sea (from below), thus also linking the larval stage to neustonic life.

In *Recluzia,* the bubble raft is exclusive to the female, thus suggesting an original function linked to egg protection (an egg sac).

A group of Mollusca with holopelagic representatives (not holoplanktonic because they are capable of fast and long-distance movements, typical of the **Nekton**) are the **Cephalopoda**. Apart from the typically nektonic species (which will be considered in the chapter on nekton), we have to consider that all Cephalopoda show a precocious stage (generally termed "larva"), which is of plankton pertinence. They are never dominant (in terms of the number of specimens) in plankton, also because they are able to escape capture based on the use of standard plankton nets. In any case, they are characteristically abundant in some geographic areas, such as the Bering and Okhotsk Seas.

The **Octopoda Argonautidae** (genus *Argonauta*) are not nektonic as adults but neustonic. When the female is still a few mm in size, she starts to build a papyrus shell with the first couple of arms (longer than the other three couples). A shell produced by arms (which derive from the foot in the scheme of Mollusca ancestor) does not correspond to the standard shell of Mollusca (which is produced by the mantel), and for this reason, such a structure receives the name of **niche**.

The female can reach a total size of 150–200 mm and passes her life in the neuston, where she moves around, abandoning herself to the wind. To tell the truth, she can even produce two membranes extended in each of the other two arms, thus realizing seals to be offered to winds and improve navigation at the water surface (Fig. 2.30).

Appendices = appendix *Argo* = mythic Greek ship; *Nauta* = mariner *Calan* = from the Irish caelan, slender *Chaeto* = needle, spine; *Gnathos* = jaws *Colony* = from (L) *colere* = cultivate, and (L) *colonia* = farm community *Cope* = oar; *Podos* = foot *Deuteros* = secondary, successive; *Stoma* = mouth *Jan* = violet; *Anthos* = flower *Niche* = from (L) *nidus* = nest, and (F) *nicher* = make a nest *Octo* = eight; *Podos* = foot *Operculum* = from (L) *operire* = to cover *Opistos* = behind; *branchia* = gills *Proso* = ahead; *branchia* = gills *Proto* = primitive; *conch* = shell *Techo* = case, covering; *Soma* = body *Teleo* = distant; *Conch* = shell *Thalia* = mythic Greek muse (comedy and poetry)

In the case of **Cephalopoda Nautiloidea**, the mollusc is conspicuous (up to 20 cm of shell maximum dimension) and able to adopt a fast movement to capture prey or escape. The shell is rigid, external, and concamerated, a complex structure that allows the suspension in the water at the desired depth, with or without cameras filled with gas. Wide vertical migrations are performed by nautili, but in any case, the maximum depth (approximately 500 m) is where the shell structure can resist the external water pressure.

Another Cephalopoda group shows concamerated shells and mesopelagic lifestyles. They are species of the genus *Spirula*, a type of small squid (not more than 3–4 cm) equipped with a spiralized and concamerated shell positioned inside the mantle cavity. The position of the shell (posterior) and its gas content (distributed in different cameras) oblige the mollusc to stay quasi-vertical, with the head below. Additionally, in the case of *Spirula*, the fragility of the shell does not allow adults to stay more than 500 m below the sea surface. Very small individuals, however, with a more compact shell have been collected at more than 1000 m, suggesting that birth areas are deeper than those where adults thrive.

Many species of holoplankton can be classified as "carnivores", feeding on smaller plankton organisms or on those not able to escape. The presence of large eyes, although functional for other purposes, can be related to feeding behaviour in **Cephalopoda**, **Polychaeta**, **Mollusca Gymnosomata**, and fish larvae under twilight conditions. Such a generalization can also be useful in the reverse sense, i.e., to recognize, with not-well developed eyes, those species that could be classified as filter feeders or primary consumers. This is the case for **Mollusca Thecosomata**, but more obviously, such a feeding habit is common to many other species (typically abundant because of access to the immense food resource represented by primary producers).

The absence of large eyes, however, does not classify **Chaetognatha** among filter feeders. These "arrow worms", in fact, are the maximum adaptation to the predaceous existence and the most effective hunters/predators of mesozooplankton.

Many organisms of the benthos perform a lifestyle when in the larval condition in the water column as filter feeders mainly on micro- and nanoplankton, because this happens during their early postembryonic development, when they are tiny, just before settling on a substrate and growing in size until adulthood.

Notwithstanding the typical life cycle of benthic adults and pelagic larvae, **Chordata Tunicata** shows approximately 120 species living their entire existence in the pelagos. They are **Thaliacea** (50 species) and **Appendicularia** (70 species).

Commonly considered part of the so-called jelly plankton, they are Metazoa of independent ancestral derivation from jelly medusae (contrary to the majority of invertebrate marine fauna, they are **Deuterostomia**), and they share with them simply the jelly appearance.

In contrast to Cnidaria medusae, which are predators, Pelagic Tunicata can be considered primary consumers, being filter feeders on micro- and nanoplankton (Alldredge and Madin 1982). The filtering activity is of two types: internal and based on mucous-ciliary apparatus (**Thaliacea**), or external and based on the production of mucous sieves where the water is forced to pass through by the movement of a muscled tail (**Appendicularia**). In **Thaliacea,** the sieve is represented by the gill pharynx, a wall perforated by micropores equipped with cilia that move the water. The gill-pharynx wall divides the animal into anterior and posterior parts.

The water pushed by cilia backwards is deprived of microparticles during its passage throughout the pharynx. These particles are embedded in a mucous matrix and transferred by other cilia to the mouth and the intestine.

The whole individual moves slowly forward, while the water flows through its sieve apparatus. The clearance power (capacity to subtract particles from the water) of these animals is high, and when they appear in high numbers in the environment, they can subtract food for other primary consumers (e.g., crustacean zooplankton), thus shifting the supply from one trophic chain to another.

Swarms of **Thaliacea** are not common but are evident due to the size of specimens, which are also associated with long chains of individuals (**colonies**) (Fig. 2.31).

Appendicularia represent a more constant component of meso-zooplankton but are known only to specialists, due to their small size. Their presence in neritic plankton is important because of their role as primary consumers. They, however, adopt a completely original filtering style. Each individual produces **a mucous "house"** divided into afferent and efferent spaces with the interposition of a sieve that catches microparticles.

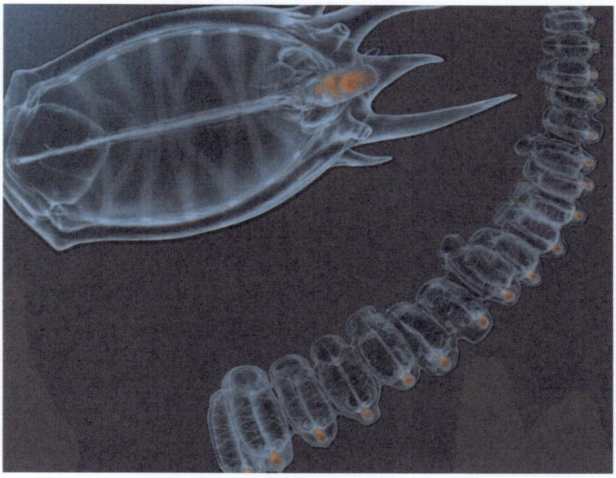

Fig. 2.31 *Salpa* sp., solitary and colonial form (Drawing of A. Gennari)

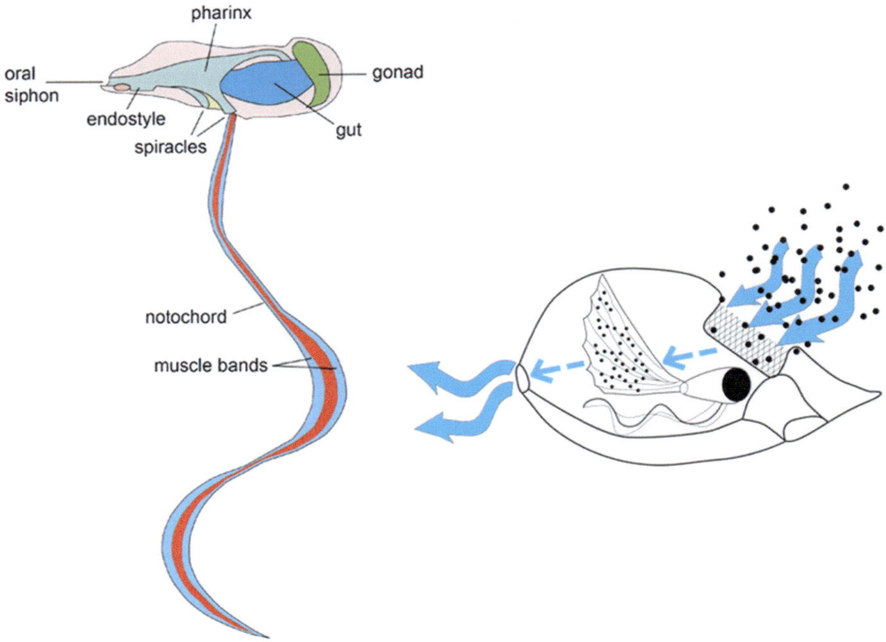

Fig. 2.32 Scheme of Appendicularia (left) and of the functioning of its mucous house (right) (graphic composition of A. M. Miglietta)

The water is forced to pass through the sieve by the oscillatory movement of the muscled tail of the individual. When the sieve is clogged with microfood particles, the specimen simply ingests this food concentrator, destroying the whole mucous house, whose remains sink to the sea bottom and represent one of the most important components of the "**marine snow**" (Fig. 2.32).

The cycle of house-building and abandonment can last even a few hours in areas rich in microfood particles, and **Appendicularia** spend their entire lives in successive cycles of building and abandonment.

Micro- and nano-plankton are food for many Metazoa *taxa*, together grouped in the category of primary consumers and generally filter feeders. Apart from the holoplanktonic **Thaliacea** and **Appendicularia**, many larval stages (**meroplankton**) are tiny and ciliated and thus adapted to profit of small particles and organisms captured between cilia.

A large group of primary consumers, however, do not use cilia to collect microfood particles. This large, even dominant, group is that of **Crustacea Copepoda**, holo-planktonic Metazoa deprived of cilia, obliged to use bristles and setae in catching micro- and nano-plankton. **Copepoda** are generally small in size. Among **Calanoida** (the most abundant Copepoda of the zooplankton and, consequently, the most abundant animals of the planet), the water flow is generated by a movement of second antennae; the unification of mouth parts (the metameric appendages: mandible, maxillula, maxilla, and maxilliped) realizes a basket-sieve where food particles are collected (Fig. 3.23, Chap. 3).

The possibility of independent movement of the single components of such a basket sieve allows the individual to perform even a selection of desired food particles. Most of the food particles, in fact, cannot be suitable for the copepod diet, so simply avoid their capture. When the copepod individuates, the presence of suitable food particles immediately opposes the basket sieve to the water main flow and captures them.

This filtering behaviour is voluntary and used ad hoc; hence, we can situate copepods among "hunters" of selected prey more than among "grazers" of abundant food crops.

The hunter-predator attitude is even enhanced in the early developmental stages of **Copepoda** (larvae **nauplius**), which are smaller than adults and have only mandibles as filtering appendages around the mouth. Their feeding behaviour can classify them as small hunters preying on unicellular moving vegetal organisms because their size is very close to that of unicellular prey.

2.4.2 Pelagic Life Cycles, and the Trochaea Theory

Cnidaria are among the best-known pelagic organisms, and their life cycle is very complex. The **medusa** is the sexually mature stage (hence, it is the adult) and produces gametes destined for fertilization and the production of new individuals with re-elaborated genetic patrimony (derived from the unification of paternal and maternal DNA). The product of embryonic development is a free ciliated gastrula, which has the name **planula**. The planula generally settles on a substrate and generates a **sessile polyp**.

The **polyp** is a sexually immature stage, but it is able to reproduce asexually by **lateral budding**, with the formation of many genetically identical individuals (a so-called **clone**) that generally remain physically connected in the formation of a **colony** where each polyp can be devoted to a different role (**cooperation**).

The alternation of sexual and asexual generations in Cnidaria is called **metagenesis** and represents the realization of alternative genetic programs in each individual genome (Fig. 2.33).

Thousands of Cnidaria species, however, show a high number of variations in the general scheme. One category with the most divergent life histories is that of holoplanktonic species, which sometimes renounce the benthic stage.

Box 2.4 Life Cycles and Life Histories

Life cycles describe the condition of organisms in each, and sometimes distinct, part of their lives. Generally, the different conditions experienced during the life span of organisms are called **stages**. When these are distinguishable and different from each other, we understand that the organism lives in different conditions before arriving at reproduction, which represents the final biological goal of its existence. In each life cycle, we can distinguish gametes, embryos, larvae, juveniles, and adults. **Gametes** are destined to fertilization (they fuse together). **The embryo** is the precocious multicellular stage, which is not free in the environment but protected inside the egg shell or the mother body. **Larva** is the first free stage that is able to move in the environment. The larval environment, however, is generally different from that of juveniles or adults, and for this reason, the larva is morphologically and functionally distinguishable and has to perform a drastic change (**metamorphosis**) to reach the juvenile and/or adult condition. **Juveniles** live in the same environment as adults, being smaller in size and having adult external features, although they are still sexually immature. **Adults** are sexually mature individuals who provide for the generation of new individuals. Each condition or stage can be interested in more than one **age** or **instar**, which is useful for growing until the size proper for passage in the successive stage. Life cycle studies are a matter for multicellular organisms, but for unicellular organisms, life cycle stages can also be distinguished (although with different terms) if they show a kind of sexual reproduction. Asexual reproduction can be performed by unicellulars, embryos, larvae, juveniles, or adults, thus declaring that it is not involved in the compartmentation of life cycles.

Life histories, on the other hand, are the chronicles describing (numerically) the life of each organism. They communicate the number of eggs produced by each adult female (**fertility**), in unique or multiple events (**spawning**), the percentage of eggs fertilized and their hatching (**fecundity**), the time length of each stage (in time units), and the life length (days, years).

In terrestrial habitats, only juveniles and adults are generally evident and actively interact with the environment. However, in marine environments, the free circulation of larvae is a rule (Fig. 2.34). The possibility of being free in the environment is also present at the stage of embryos (which, for this reason, are larvae) and of gametes. Such tiny organisms perform fertilization and growth of developmental stages as free in the water column, thus belonging to the pelagos world at least in a brief, precocious stage of their life. (For a better distinction between life cycles and life histories, see Giangrande et al. 1994).

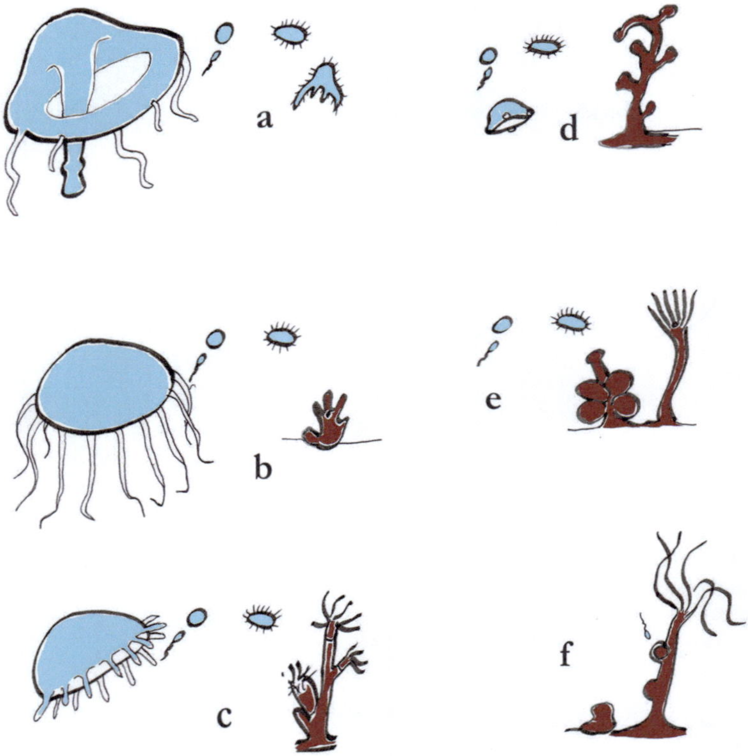

Fig. 2.33 Metagenesis diversity in Hydrozoa (Cnidaria). (**a**) *Liriope*; (**b**) *Gonionemus*; (**c**) *Obelia*; (**d**) *Millepora*; (**e**) *Hydractinia*; (**f**) *Hydra*. From (**a**) to (**f**) diminishes the pelagic part (light blue) in favour of the benthic part (brown) of the life cycle. (Courtesy G. Belmonte)

Among Hydrozoa, in **Narcomedusae** and **Trachymedusae,** the free ciliated planula generates a pelagic **actinula**, thus avoiding the benthic stage of a hydropolyp and the associated colony.

A **holoplanktonic** ***taxon*** **of Cnidaria** is the order **Siphonophora**, which also thrives at the sea surface in a particular system called **Neuston**. Siphonophora comprises approximately 150 species, holopelagic, and some colonial. Colonies, technically, derive from the asexual reproduction (budding) of a juvenile stage (the polyp). Hence, in this case, the polyp colony is pelagic and not benthic. Among the Siphonophora colonies, the cooperation between different polyps and, consequently, their morphological and functional specialization is extreme, and some scientists describe them as **superorganisms** more than colonies of individuals (Fig. 2.35).

An axile, long **coenosarc** (common basis) connects all polyps. Generally, the coenosarc enlarges at one extremity in an organism that plays the role of a buoy (**pneumatophore**), which produces a gas inside itself and ensures the floatation of the whole colony. Below the pneumatophore (of different size and shape according

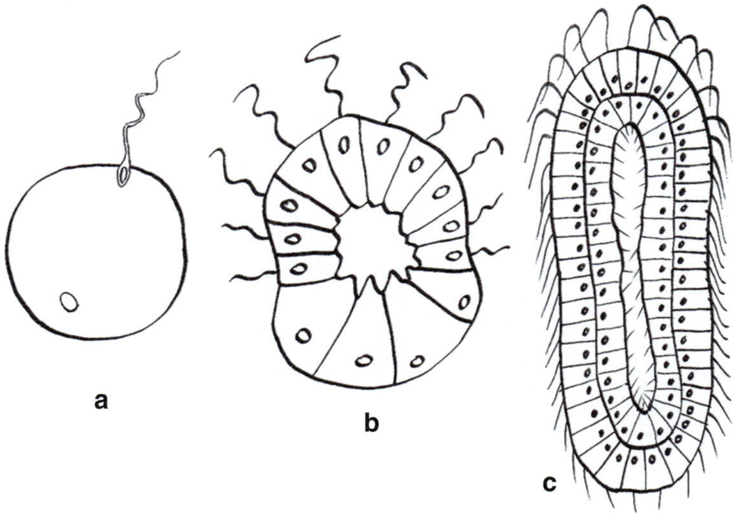

Fig. 2.34 Precocious stages of marine organisms, free living, and pelagic as life styles. (**a**) Egg (large) and sperm (small with flagellum). (**b**) A precocious developmental stage of Porifera corresponding to a blastula (only one cell layer). (**c**) A precocious stage of Cnidaria corresponding to a Gastrula (two cell layers). (Courtesy G. Belmonte)

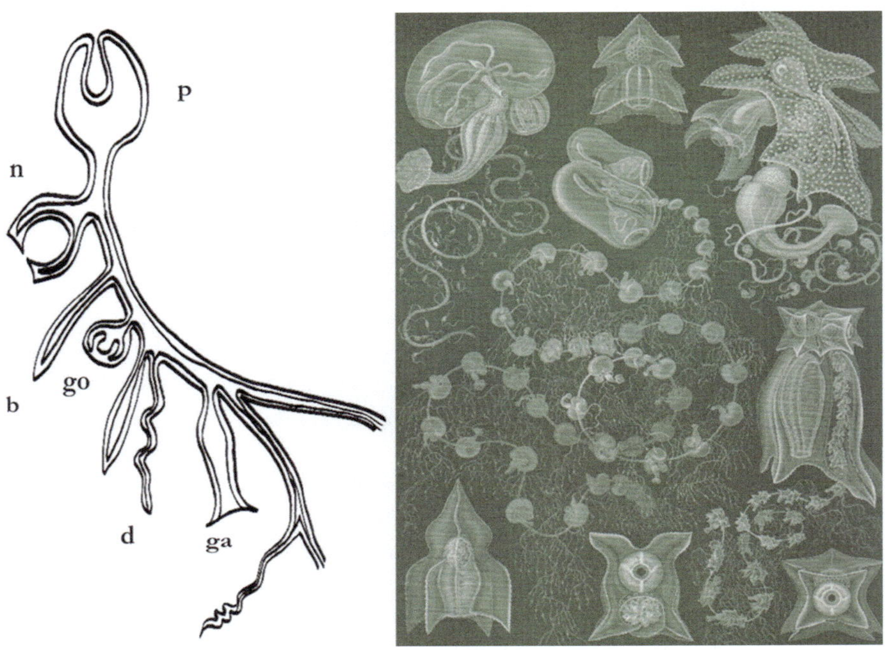

Fig. 2.35 Left, Organization scheme of a siphonophore. p = pneumatophore; n = nektocalyx; b = phyllozoid; d = dactilozoid; go = gonozoid; ga = gastrozoid. (Courtesy G. Belmonte). Right, Siphonophora in the Haeckel's (1904) atlas

to the species), the long coenosarc brings modified polyps (**zoids**) with different functions:

1. **nectocalyx**, with a swimming role based on jet propulsion determined by pulsating contraction of an umbrella (the medusa swimming style);
2. **dactilozoid**, with a defensive role, being rich in nematocysts;
3. **gastrozoid**, with an alimentary role, and an intestinal lumen common to all the other individuals throughout the coenosarc;
4. **gonozoid**, with the role of producing sexual medusas (mature males and females);
5. **phyllozoid**, with the role of protection (over all of the gonozoids).

Small preys are directly ingested by gastrozoids, but larger preys (even some fish) are paralyzed by the injection of the poisonous content of extruded nematocysts and entrapped by a net of tentacles that allow gastrozoids to digest outside the food and absorb it in small aliquots.

In *Physalia physalis* (the largest of Siphonophora), gonozoids directly produce gametes (the same colony can produce either male or female gametes), and for this reason, some scientists recognize them as medusa stages never detached from the mother colony. After fertilization (external), early embryonic development starts, but no one knows where it happens or what shape has the first developmental stages. The growing polyp can only sink during its growth and/or the asexual budding of other polyps. Only at the appearance of the pneumatophore and/or nectocalyx specimens does the young colony go up to the surface, where the final large colony lives.

Actinula = larva having rods *Clone* = twig, a-sexually deriving from a mother body *Coeno* = common; *Sarcos* = flesh *Narco* = numbness/stiffness, probably consequent to the stinging of the medusa *Siphono* = tube, pipeline; *phora* = bringing *Anthos* = flower; *a-* = depriving prefix; *techa* = case, box *Chryso* = golden; *Mitra* (Sanskrit) = friend; protector *Conus* (L) = conus *Intro* = internal; *verton* = from (L) vertere: to turn *Porpe* = brooch, clasp *Rata* = from (Polynesian) chestnut or (Māori) timber tree for oars

A large pneumatophore can pass 30 cm and protrude up to 15 cm out of the sea surface.

This allows the structure to catch the wind and transport the whole colony to the seas without consuming energy for swimming. The *Physalia* colony/superorganism is a predator of other smaller animals, but it is in turn the prey of other specialized predators, immune to the poison of nematocysts (Fig. 2.36).

Among Cnidaria Hydrozoa, another group of colonial organisms adopted the pelagic lifestyle and, in some cases, the neustonic lifestyle. Species of the order **Anthoathecata**, family **Porpitidae**, show a modified lifestyle very similar to that of *Physalia*.

The most known of these pelagic colonies is the superorganism, which has the name *Velella velella*. The pneumatophore is discoid and of chitinous consistence.

Fig. 2.36 Details of the *Physalia* (Siphonophora) organization in a drawing of Haeckel (1904)

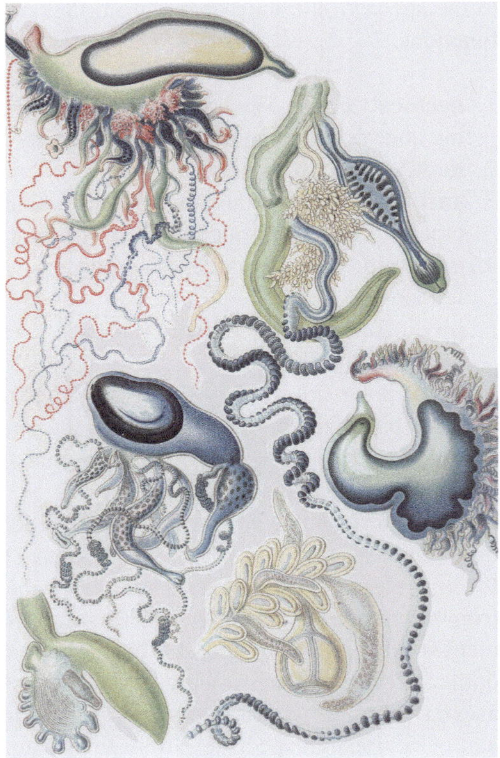

Obliquely on the elliptic surface of the pneumatophore, a triangular sail protrudes outside of the sea surface to catch the wind and move the whole colony. Below the pneumatophore disk, many stinging polyps (gastro- and dactylo-zooids) extend into the water, waiting for prey to capture. A large central polyp plays the role of the mouth of the colony and ingests the largest food particles.

Velella is a colony of only polyp organisms. In fact, some polyps (gonozoids) bud small medusae (**chrisomitra**), which sink to depths of 600–1000 m, where they produce gametes whose fertilization gives origin to an early larval stage, the **planula** (practically a **gastrula**), which successively originates a sphere (the **conaria**) with a conical **introverton** of embryonic material.

From the conaria, a **rataria** will develop, which practically corresponds to the polyp stage. In fact, the rataria "balloons" up to the surface, where it grows in size and buds other polyps, thus forming the final *Velella* colony (Fig. 2.37).

Another polyp colony similar to *Velella* and of the same family is *Porpita porpita*. This is a disk colony living entrapped in the film of the water surface (pleuston), where it feeds on neuston organisms (copepoda and decapoda larvae) (Fig. 2.38).

Porpita, similar to other neuston/pleuston species (e.g., *Physalia*, *Velella*), does not perform vertical migration and never abandons the very surface layer. This

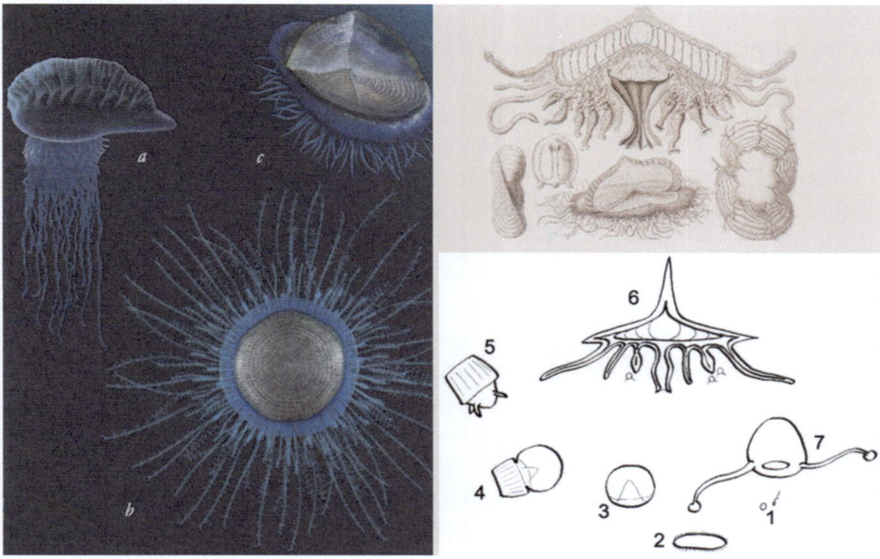

Fig. 2.37 Left, Hydrozoa Siphonophora of the neuston. (**a**) *Physalia*, (**b**) *Porpita*, (**c**) *Velella* (drawings of A. Gennari). Right top, details of *Velella* colony organization from the drawing of Haeckel (1904). Right bottom, Metagenesis of *Velella.* (1) Gametes (egg and sperm) emitted by small medusae (7); (2) planula; (3) larva conaria; (4) upside-down conaria; (5) rataria (with germs of tentacles); (6) multi-individual colony; (7) small medusae generated by budding. (Courtesy G. Belmonte)

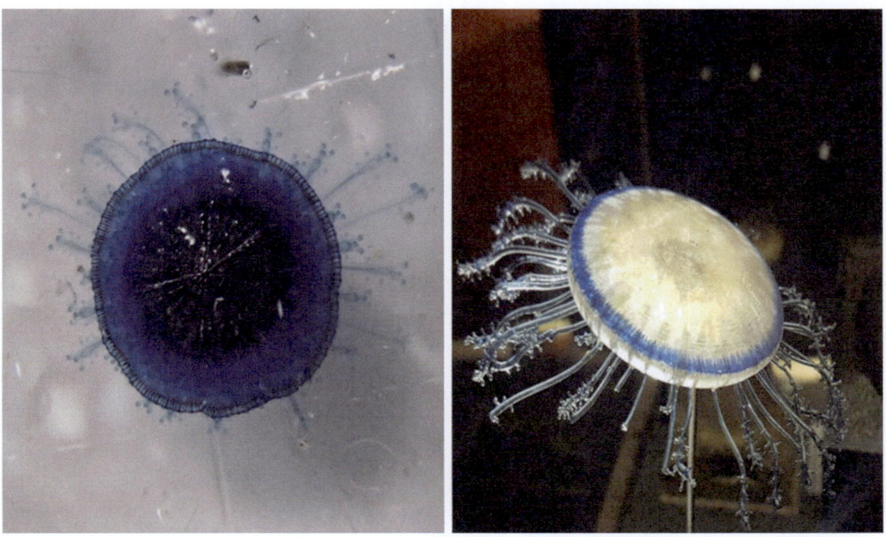

Fig. 2.38 Left, the blue button *Porpita porpita* (Hydrozoa) collected in the Neuston of central Mediterrtanean Sea. Right, the same colony represented in glass (Blaschka collection, Natural History Museum of Liege, Belgium) (photographs G. Belmonte)

position exposes it to UV rays and/or predation from birds, thus inducing the evolutionary adaptation of blue colouration, which is mimetic and (probably) protective against UV radiation.

Such a blue colouration is also typical of their predators (*Glaucus* and *Jantina*, two Gastropoda of the neuston environment).

Among Cnidaria, the pelagic lifestyle is only exceptionally linked to the juvenile stage (the polyp), but it is more common as a condition of the adult (the **medusa**).

All the other phyla of marine organisms, however, show the possibility of a pelagic lifestyle in the larval stage. Generally, such larvae are ciliated and, obviously, small in size, thus representing an important component of primary consumers feeding on planktonic primary producers.

Early larval stages are morphologically very simple **planulae** (a sort of gastrula) among Cnidaria, but larvae in other phyla are more complex and morphologically distinguishable. The ectoderm of planulae is a monolayer of columnar, ciliated cells. Cells of the ectoderm are mainly supportive epithelio-muscular, and glandular, ganglionic, sensorial, and defensive (cnidocyte) types are also present. The basal extremity of ectodermal cells is imbedded in a thin **mesoglea**. Opposed to the **ectodermal** layer cells, a layer of **endodermal** cells is present to separate the mesoglea from the virtual lumen of the intestine. The monolayer of the endoderm is composed of columnar supportive cells with a digestive role. These columnar endoderm cells are somehow associated with interstitial cells, and both types are able to differentiate into all cellular types necessary for organisms to function.

Specialized cells derived from endodermal interstitial cells also migrate into ectoderm to play their role. In some Hydrozoa (**Trachymedusae**), the planula originates a pelagic polyp, **actinula**, with eight arms (tentacles), which successively originates the benthic polyp or even directly a medusa, thus completing the whole life cycle in the pelagos (Fig. 2.39).

Among Cnidaria **Scyphozoa,** the planula settles on a substrate and gives origin to a benthic polyp (the **scyphistoma**) that produces a series of swimming elements, the **ephyrae**, due to transversal budding.

Ephyrae are the prelude to a well-formed medusa, but they show a typically lobed bell with eight arms that are well distinguishable and live from weeks to years before the completion of the final medusa stage.

The planulae of Cubozoa and Anthozoa can be simply larger in size, but they maintain their simple morphology. The size of planulae, in Anthozoa, can reach a length of some mm, and in general, the mouth extremity is forwardly directed during swimming.

Among Anthozoa, planulae are the only pelagic stage, but many sessile adults are able to migrate for long distances if settled on floating objects (rafting).

Cteno = comb; *phora* = bringing *Trachi* = rough skinned

Among **Ctenophora,** the larvae are not well known, but the stage ***cidippus*** is common. Ctenophora, contrary to Cnidaria, are mostly holoplanktonic and someway more specialized, having anal pores around an apical organ and lacking the

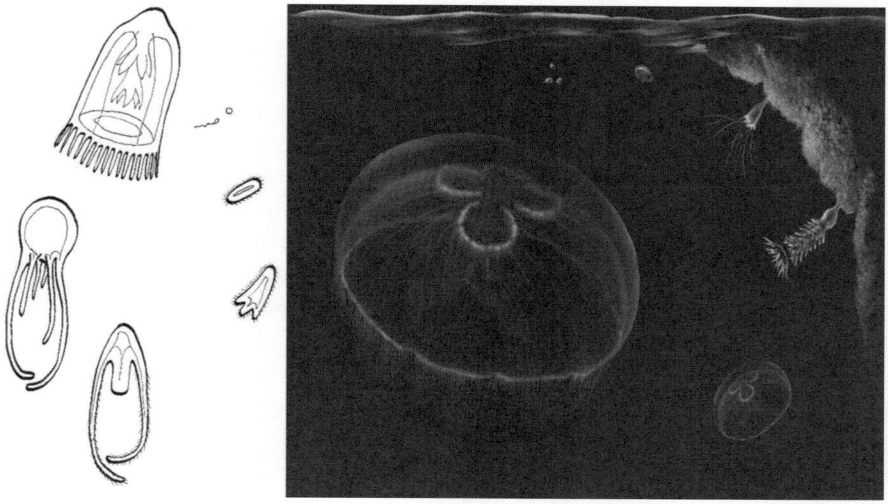

Fig. 2.39 Left, scheme of the Trachymedusae (Hydrozoa) life cycle. From the medusa (on the top), the development goes clockwise with gametes (egg and sperm), **planula**, and **actinula** (with tentacle buds). (Courtesy G. Belmonte). Right, representation of the Scyphomedusae (Scyphozoa) life cycle, with the planula originating a benthic **scyphistoma** budding **ephyrae**, before the medusa stage. (Drawings of A. Gennari)

possibility of asexual reproduction (this is probably proof of an evolutionary specialization of cells, not flexible in their final destination, as in Cnidaria).

Sexual reproduction proposes the gametogenesis of both sexes in the same individual (**hermaphroditism**), and in some species, sexual maturity is reached twice, during the *Cidippus* (juvenile) condition and during the final-adult condition (**dissogony**).

Even if predators and superpredators are present in some cases, Ctenophora populations locally can reach high concentrations of individuals due to their fast reproductive cycle and adaptability to variable diet regimes.

One of the first, and more utilized, subdivisions for the pelagic inhabitants regards their link to the water column for the whole life (**holo-plankton**) or only for a part of it (**mero-plankton**). Swimming ability and/or floatability are developed in a large number of species, representing the majority of animal *phyla* and algae. The small size generally favours floatability, but macro-organisms evolved devices to avoid sinking even without active swimming.

Pilidium is the typical planktonic larva of many **Nemertea**. It shows ciliary bands for movement and food collection and an **apical tuft** of probable sensorial role. Additionally, the larvae of **Kamptozoa** show the typical conic shape with a ciliary apical tuft. This common scheme is also repeated in other and more specialized *phyla*, thus suggesting a probable common ciliated ancestor. In Mollusca and Annelida, such a type of larva is called a ***trochophore*** (Fig. 2.25) and represents an interesting morphological scheme useful for understanding successive development in these phyla.

In Mollusca, the trochophore is free and present in only archaic *taxa*; all the other groups show a more typical (and successive) larval stage (the ***veliger***), which is ciliated and completely planktonic, notwithstanding the presence of a small shell.

In Annelida, the trochophore is well known and used to represent the anatomical scheme of all trochophore-like larvae. An apical organ equipped with a tuft of long cilia is always present together with a ciliated band (**protrochus**) anterior to the mouth. This ciliary band circumscribes the larva, passing just above the mouth, and divides the larva into two parts: **the episphere and the iposphere**.

The feeding trochophores also show a **metatrochus** posterior to the mouth that collects food particles and transfers them to a ventral ciliated field (**gastrotrochus**), whose cilia drive the particles upwards directly to the mouth.

Cidippe = Greek mythic beautiful priestess of Arthemide *Disso* = double; *gonia* = generation power, reproduction *Ephira* = Greek ancient city (in Epyrus) *Epi* = up *Gastro* = ventral *Ipo* = down *Kampto* = bent, curved; *zoa* = animals *Medusa* = mythic personage with serpentine hair *Meta* = successive, final *Proto* = anterior *Scyphos* = cup *Stoma* = mouth *Trochus* = weel

In the iposphere, an elongation process of the larva occurs before the abandonment of the pelagic lifestyle, and a series of growing metameric segments is added, ornamented with **chaetae** and sometimes with additional ciliary bands (Fig. 2.40).

Post trochophore larvae (those with chaetae) also start to show **parapodia** (metameric appendages), which help the swimming behaviour. Before the appearance of

Fig. 2.40 Larval development in Annelida Polychaeta, with the trochophore finally equipped with chaete. (Courtesy G. Belmonte)

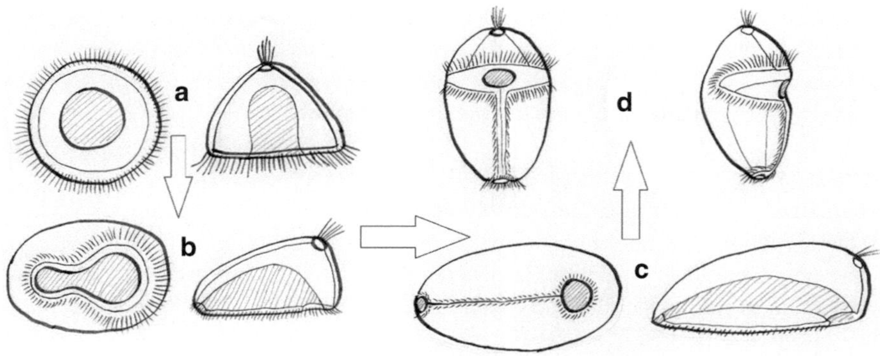

Fig. 2.41 Scheme of the trochaea affirmation. (**a**) The planktonic trochaea swims in the ancestral ocean (it has only the archaeotrocus and the apical organ); (**b**) it is already well equipped to settle and creep on the bottom, where it evolves an elongate body; (**c**) the elongation gives origin to bilateria and worms; (**d**) worm-like organisms on the bottom produce swimming replicates properly "trochophoral" (with a three-divided trochus). (From Nielsen 1995, modified by G. Belmonte)

parapodia, cilia were the only mobility devices in organisms having a body size ranging from 150 to 900 μm.

The link between life cycles and the water column, even in the case of only temporary life forms, suggests that an evolutionary reason has to exist.

In addition, early stages are generally small (even unicellular, as fertilized free eggs) and not affected by gravity force in the water column. Additionally, for these reasons, Nielsen (1995) proposed a planktonic origin, in the ancestral sea, of the macroorganisms that today live on the planet.

Nielsen started with the hypothesis of von Baer (1828) and Haeckel (1866) on the possible repetition of evolutionary steps in the embryonic development of animals.

The embryologic passage from a fertilized egg to successive stages with more cells (**morula**, **blastula**, **gastrula**) was considered the necessary repetition in each embryonic development of what happened in evolutionary time because it was contained in a "genome memory" and represented the successive appearance of microorganisms of growing complexity (**citea, morea, blastea, gastrea**).

What Nielsen proposed was the addition of a further step to the evolutionary correspondence of the growth process. In fact, he noted that the anatomical organization of very young larvae could also be read as shared among many *phyla* and that the adult body architectures can all be derived from a fundamental larva, the trochophore, corresponding to a hypothetic common ancestor, the **trochaea** (Fig. 2.41).

The trochophore is a tiny, ciliated larva, typical of the development of Annelida Polychaeta, but with similar representatives among Bryozoa, Kamptozoa, Nemertea, and primitive Mollusca.

It is characterized by a conic shape, with an evident ciliary tuft on the apical extremity strictly associated with a concentration of neuronal cells in the so-called apical organ. Such a concentration of neuronal cells (a sort of small brain) is the starting site of neuronal connectives commanding a long and sinuous ciliary band that is devoted to the motility and capture of food particles.

The ancient, hypothetic, trochaea was not more than a specialized free gastrula, with an external layer of cells (**ectoderm**) and a primitive intestine (**archenteron**) delimited by an internal layer of cells (**endoderm**) and communicating with the external environment throughout the **blastopore**.

In the water column, the blastopore is supposed to be oriented downwards, and surrounded by a band of cilia (**archaeotrochus**) longer than the others present elsewhere on the larval surface, and densely associated. The ciliary beat of such a fringe of cilia, commanded by the apical organ, produces the swimming of the larva and, contemporaneously, the collection of small food particles and microscopic organisms. In an evolutionary scenario, if such a swimming larva was obliged to sink to the bottom, it was already adapted to creep on the bottom and collect food particles with cilia around the blastopore (which is oriented towards the bottom).

Arche = ancient, primitive; *enteron* = intestine *Blaste* = blaze, burst, origin of something *Citus* = cell *Ecto, Endo* = external, internal *Gastro* = stomach, intestine *Mora* from Armenian *mor* = blackberry *Poro* (L) = hole, mouth *Trochus* = weel (referred to the curved line designed by special cilia on the larva surface)

This benthic lifestyle (in a 2D world) probably obliged the conical larva to creep in preferential directions, thus identifying an anterior and a posterior part of its anatomy.

The anterior lip of the blastopore was the most interesting because it is the first to contact the substratum and the collection of food particles, in comparison with the posterior part. At the posterior extremity of the blastopore, the ingestion activity was substituted by the expulsion of indigested material. The same opening (the blastopore) was used for oppositely directed material flow but separated by preferential borders (anterior and posterior, respectively).

The next step of evolution was the better separation of the two functional parts of the blastopore with the junction of the ventral-lateral lips. Such a medial junction leaves an anterior opening (the **mouth**) and a posterior opening (the **anus**). The archaeotrochus (primitive ciliary band), in this new arrangement, surrounds not only the mouth but also from the mouth to the anus in a ventral double band (**gastrothrochus**) and surrounds the anus.

Due to the double function (swimming and food collecting), the ciliary band around the mouth, where both functions are needed, became stronger and more evident, separating its design into a portion preceding the mouth opening (**prototrochus**), one immediately subsequent to the mouth (**metatrochus**), and a final smaller portion around the anus (**telotrochus**).

The new antero-posterior organization of the body plan gave birth to **bilateral** animals and forced the apical organ to be in the anterior position. Such a microorganism creeping on a solid bottom does not need any additional devices to enter the water column.

As a trochophore, in fact, it is simply a bilaterial larva swimming in a 3D environment.

All animals derived from this basic anatomic scheme are called **Protostomia** because the mouth derives from the primitive opening (the blastopore) of the early larval stage. This anatomical design is opposite to that of **Deuterostomia,** where the mouth opens secondarily and is not derived from the blastopore.

Such an alternative type of larva is that belonging to the **tornaria** group, a ciliated larva typical of Echinodermata, Chordata, Enteropneusta, and Pogonophora, with some similarities also among Foronida and Brachiopoda (Fig. 2.25).

In the tornaria scheme (which should represent the ancestral **Tornaea** in body architecture), some small canals connect secondarily the inner part of the archenteron with the external environment, very far from the blastopore.

These canals (hypothetically four) produce a series of four pores on the surface of the larva, just around the apical organ.

Bis = two; *Latus* (L) = side *Blaste* = blaze, burst *Cado* = bulb, bulbaceous; *Phore* = bring *Coelum* (L) = void space, sky *Deuteros* = secondary, successive *Dolium* = barrel *Monos* = singular; *Phylum* = group, lineage *Neo* = new *Oo, Ovo* = egg *Peri* = around; *os oris* (L) = mouth *Protos* = primitive, first of all *Pyro* = fire *Soma* = body *Stoma* = mouth *Strobila* = plug of lint shaped as a pine cone; body stemming apical copies of itself *Trochus* = weel (curved line of cilia) *Throphe* = feeding; *Zoos* = animal *Tunica* (L) = undergarment, membrane

One of these pores has been proposed as giving origin to the mouth, thus individuating the ventral side; the two adjacent pores become respiration pathways, and the dorsal opening remains as a **coelomatic** hydropore. The new mouth is not equipped with a preexisting ciliary band and uses the cilia of the surface ectodermal epithelium as an alternative to the archeotrochus.

Such a new band of cilia, similar to the archeotrochus in function and position around the mouth but not derived from it, receives the name **neotrochus**.

The archaeotrochus, consequently, in tornaea, is restricted to the only circumscription of the anus opening. The neotrochus is generally well developed and sinuous around the mouth opening, producing typical designs in the larvae of each Deuterostome *taxon*.

To confirm that neotrochus is not a part of the ancient ciliary band, the beating direction of the neotrochus cilia is exactly opposite to that of the archaeotrochus cilia. In Deuterostomia larvae, the neotrochus cilia push the water far from the mouth, thus collecting food particles from the oral side of the ciliary band. The food collected is then transported by the cilia of the peri-oral area towards the mouth, where ingestion occurs (Fig. 2.42).

In contrast, water is pushed towards the mouth by the cilia of Protostomia **peri-oral** trochus, and food particles pass throughout the ciliary band to be transferred, by the cilia of the peri-oral surface, to the mouth for the final ingestion.

The Nielsen theory of **Trochaea** elegantly demonstrates how the two main groups of animals (**Protostomia** and **Deuterostomia**) can be derived from a common ancestor (the Trochaea), thus belonging to a unique evolutionary lineage (the **monophyletic** interpretation of Metazoa evolution). All the Metazoa should derive from a unicellular ciliate ancestor joining many other individual cells (aggregations) and being able to divide among them roles, thus acquiring the advantage of a best survival.

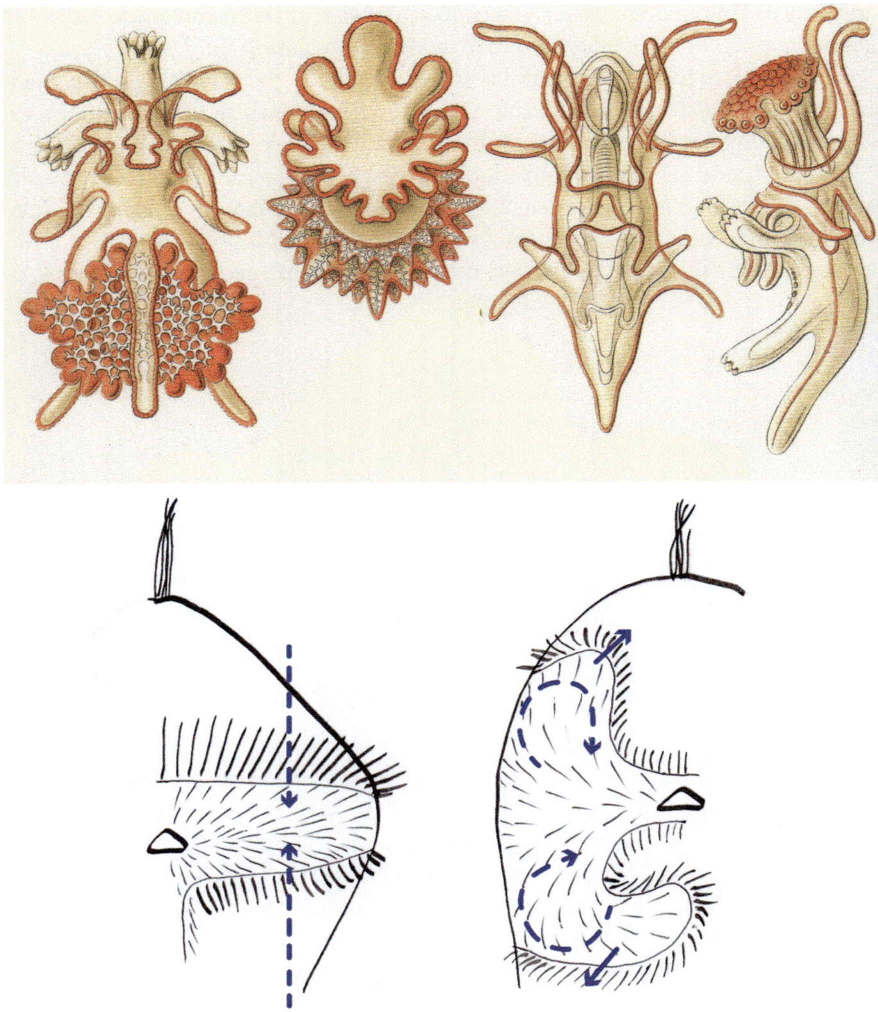

Fig. 2.42 Tornaea-type larvae. Top, Echinodermata larvae from Haeckel's atlas (1904); bottom, the scheme of food particle circulation in trochaea-type larvae (left) and in tornaea-type ones (right). (Courtesy G. Belmonte)

The amelioration of the stay together passed from an elimination of the solitary lifestyle with time. The precocious assemblage at the end was perfect if descendants of each unicellular individual remained connected at the moment of their formation under asexual divisions of mother cells.

This style of coexistence (between individuals having the same genome), which does not derive from the aggregation of initially separated cells but from the lack of separation of daughter cells coming from the same cellular division of the "mother", has the name **colony** and represents the maximum level of cooperation, up to the loss of individuality.

When such an incomplete division also occurs in multicellular organisms, the process receives the name "**budding**", with sons not completely divided from the stemming mother.

The **colonial lifestyle** derives directly from the possibility of individuals multiplying themselves asexually. This possibility seems linked to a low evolutionary level due to the necessary presence of stocks in the organisms of undetermined cells (able to form all the cell types in a new individual). These cells with a multiple destiny (not specialized) are not common only in early Metazoa (Porifera, Cnidaria) but also in Protostomia (Annelida) and in Deuterostomia (Chordata). The presence of asexual reproduction and asexual generations complicates the life cycles, and **Thaliacea** (pelagic Chordata Tunicata) represent a typical expression of such complexity among Deuterostomia.

Each thaliacean organism is able to produce many clones of itself (**blastozoids**) thanks to an asexual reproduction of specialized parts and cells (Fig. 2.43). Among **Pyrosomidae** (eight species), each pelagic individual (**oozoid**) produces a colony of tubular shape (a blind tube closed at one extremity) where single and independent blastozoids are immersed in a common matrix (the **tunic**).

Each blastozoid is positioned with the inhalant part of the body on the external side and the exhalant extremities at the inner side of the tube.

The colonies range in size from a few cm (*Pyrosoma aerniosum*) up to 20 m (*Pyrostremma spinosum* has an internal section of approximately 2 m in diameter). Their name derives from "fire" because they are able to produce bioluminescence. **Doliolidae** are 21 Thaliacea species, mostly epipelagic, with oozoids never passing 5 cm in length. Oozoids are generated by the development of a **tadpole larva** similar to that of Ascidiacea. When individuals reach a certain size, a process of transversal budding (**strobilation**) starts from a region (**stolon**) of the internal ventral

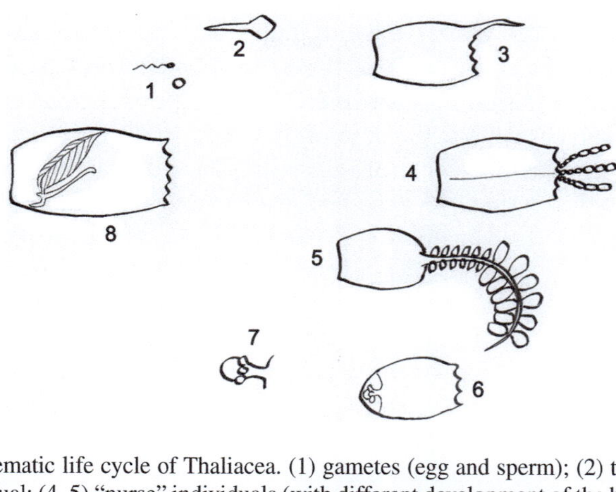

Fig. 2.43 Schematic life cycle of Thaliacea. (1) gametes (egg and sperm); (2) tadpole larva; (3) oozoide individual; (4, 5) "nurse" individuals (with different development of the budding cadofore with gastrozoids); (6) phorozooid individual budding gonozoids; (7) detail of the gonozoid budding element; (8) free gonozoid (hermaphrodite). (From Alldredge and Madin 1982, modified by G. Belmonte)

side and produces a series of buds (blastozoids) that are carried by specialized ameboid cells up to a postero-dorsal filiform process (the **cadophore**). The first buds that arrived on the cadofore are positioned in two rows at the two sides of the process and will remain there for the rest of their lives, originating a floating chain-like colony (two chains of individuals interconnected, stemming from a nurse oozoid) and assuming the role of feeding individuals (**trophozoids**), each one of 1–2 mm size. A central row of blastozoids finally formed by the ventral stolon, each of which bears a ventral peduncle from which ulterior strobilation produces a chain of buds.

These buds bearing blastozoids are called **forozoids** and form daughter colonies that abandon the nurse oozoid. Individuals on the peduncle of forozoids are called gonozoids; if they detach from the colony and become independent, they can produce gametes from mature gonads. The embryo generated after fertilization produces, at this point, a new **tadpole larva,** and the cycle restarts.

Salpidae are 41 pelagic species of Thaliacea that are typically absent only in cold oceans. Their body shows muscular ribbons that are not complete (they are not connected on the ventral side), thus differentiating the group from **Doliolidae**, where muscular ribbons are circular and complete.

Among Salpidae, blastozoids and oozoids can be very different from a morphological point of view. The embryo develops in the nurse-individual body, arriving to occupy all its volume. Once independent, the oozoide starts with the strobilation process, producing chains of buds where individuals of the same age (cohorts) are easily distinguishable from younger or older individuals.

The filter-feeding lifestyle of Salpidae is one of the most efficacious processes for using energy from food, and for this reason, they are the Metazoa with the maximum growth rate.

The colonial lifestyle is of great interest to evolutionists because they recognize in such an organization the stem of multicellularity.

According to Nielsen (1995), the unicellular ancestor of the colonies of future Metazoa is a flagellated organism of the group **Choanozoa Choanoflagellatea**. Such unicellular organisms are **Protozoa** characterized by the presence of crested mitochondria, one centriole at the base of the unique flagellum, and the flagellum having a basis inserted in a collar of microvilli. These characters, not shared with other unicellular Eukaryota, are sometimes evident (even if not together) in at least some cells of the Metazoa anatomy (for example, in the spermatozoon).

The evolutionary pathway that connects unicellular Choanoflagellatea with Trochaea is, however, probably not unique. The evolutionary branch of sponges (not true Metazoa but Parazoa due to the lack of true tissues) took care of choanoflagellated cells without producing Trochaea. In addition, Platyhelminthes never adopted the anus in their anatomy and seem to derive from Protozoa Ciliophora, more than Choanozoa (Fig. 2.44).

Among Ctenophora, the presence of four anal pores around the apical organ in adults seems to be a relict of what the Gastrea originated along the evolutionary path of Deuterostomia.

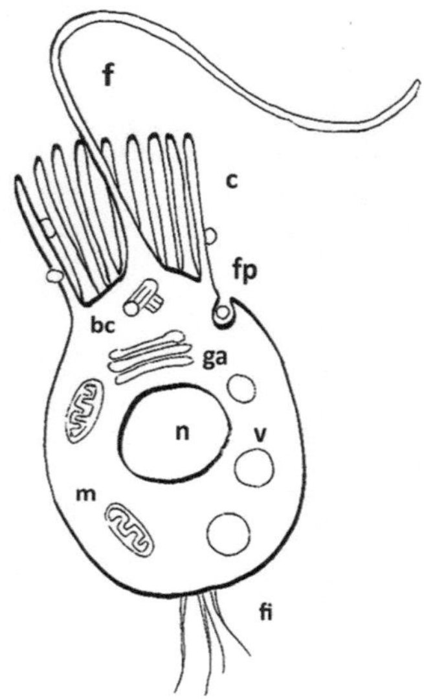

Fig. 2.44 Choanoflagellata scheme. bc, basal corp; c, microvilli collar; f, flagellum; fi, filopodia; fp, food particles; ga, Golgi apparatus; m, mitochondria; n, nucleus; v, food vacuoles. (From Hoffmeyer and Burkhardt 2016; modified by G. Belmonte)

The Nielsen theory of Trochaea proposes a feeding larva (due to the double use of ciliary bands for swimming and feeding) in the plankton as the ancestor of most Metazoa. Today, this kind of larva is not widespread in metazoan life cycles because it starts to be active at a very precocious postembryonic stage and has to feed on plankton microorganisms (it is a **plankto-trophic** larva) to grow and arrive at the final formation of the adult. This long-life history is dangerous for the survival of the species because it enhances the period of risk for a larva to become food for other planktotrophic organisms. Actually, the **plankto-trophic larval lifestyle** is very expensive in terms of mortality rate and is counterbalanced by a high maternal investment in egg production. This, however, is the current situation, with many predators feeding on larvae of any type. When the story of Metazoa started, approximately 700 million years ago, the environment did not host such a predator risk, and first, Metazoa (tiny and ciliated) had the possibility to grow in size and to evolve complex coexistence strategies without any predaceous pressure.

Most likely, the extra-large ciliate organisms of this first group (resulting from the competition escape towards a large size) sank to the bottom, where they became benthic organisms.

The water column remained the adaptation environment for tiny and small ciliated organisms. Only the adoption of new structures and/or solutions allowed such proto-planktonic ciliate organisms to reach large sizes. Oil droplets and gas bladders resolved the buoyancy problems.

Paddled appendages and fluctuating portions of the body substituted the ciliary movement and resolved problems of swimming for large organisms.

As in the best adherence of ontogenesis to phylogenesis, the advanced larval stages, or the post- metamorphosis individuals, also depict the transition between

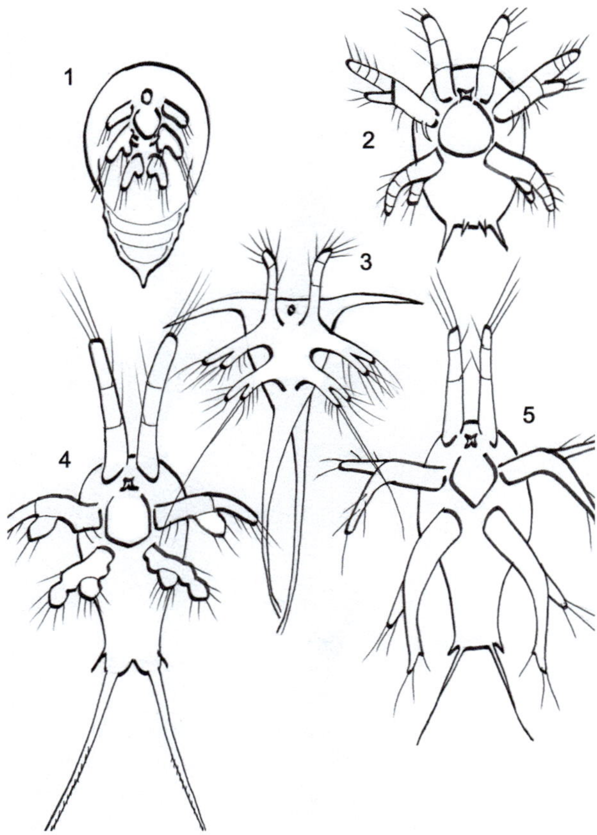

Fig. 2.45 Different morphologies of Crustacea nauplius. (1) Facetotecta; (2) Copepoda Cyclopoida; (3) Cirripedia Lepadomorfa; (4) Copepoda Calanoida (Temoridae); (5) Malacostraca Decapoda (Penaeidae). (Courtesy G. Belmonte)

cilia and other devices useful to remain in the water column among the **holoplankton** as adults.

The most representative model of a nonciliate plankton organism using bristles to swim and catch food is undoubtedly the **nauplius**, the typical larva of Crustacea (Fig. 2.45).

Gono = germinal *Nauplius* = Greek mythic driver of the ship Argo
Phoro, *Foro* (L) = pore *Salpa* (L) = fish type

References

Albrecht C, Wilkie T (2008) Ancient Lake Ohrid: biodiversity and evolution. Hydrobiologia 615:103–140

Alldredge AL, Madin LP (1982) Pelagic tunicates: unique herbivores in the marine plankton. Bioscience 32(8):655–663

Arrow Aladin NV (1995) Ecological state of the fauna of the Aral Sea during the last 30 years. GeoJournal 35:29–32

Azam F, Fenchel T, Gray JG, Meyer-Reil LA, Thingstad T (1983) The ecological role of water column microbes in the sea. Mar Ecol Prog Ser 10:257–263

Belmonte G (2012) Species richness in isolated environments. The disregarded role of time. J Biodivers 3(4):273–280

Briggs JC (1995) Global biodiversity. Elsevier, Amsterdam, 452 pp

Connell JH (1978) Diversity in tropical rainforest and coral reefs. Science 199:1302–1310

Dodson SI (1991) Species richness of crustacean zooplankton in European lakes of different sizes. Verh Internat Verein Limnol 24:1223–1229

Dodson SI (1992) Predicting crustacean species richness. Limnol Oceanogr 37:848–856. Ver Theor Angew Limnol Vehr 24:1223–1229

Ebert TA, Balko ML (1987) Temporary pools as islands in space and time: the biota of vernal pools in San Diego, Southern California, USA. Arch Hydrobiol 110:101–123

Fisher RA, Corbet AS, Williams CB (1943) The relation between the number of species and the number of individuals in a random sample of an animal population. J Anim Ecol 12(1):42–58

Gause GF (1934) The struggle for existence. Williams and Wilkins, Baltimore, p 163

Giangrande A, Geraci S, Belmonte G (1994) Life-cycle and life-history diversity in marine invertebrates and the implications in community dynamics. Oceanogr Mar Biol Annu Rev 32:305–333

Gould SJ (1994) The evolution of life on the earth. Sci Am 271:84–91

Grassle JF, Maciolek NJ (1992) Deep-Sea species richness: regional and local diversity estimates from quantitative bottom samples. Am Nat 139(2):313–341

Guglielmo R, Bergamasco A, Minutoli R, Patti FP, Belmonte G, Spanò N, Zagami G, Bonanzinga V, Guglielmo L, Granata A (2019) The Otranto Channel (South Adriatic Sea), a hot-spot area of plankton biodiversity: pelagic polychaetes. Sci Rep 9:19490

Haeckel E (1866) Generelle Morphologie der Organismen. Verlag von G. Reimer, Berlin, p 455

Haeckel E (1904) Kunstformen der nature. Bibliograpisches Institut, Leipzig

Hoffmeyer TT, Burkhardt P (2016) Choanoflagellate models—*Monosiga brevicollis* and *Salpingoeca rosetta*. Curr Opin Genet Dev 39:42–47

Howarth FG (1990) Hawaiian terrestrial arthropods, an overview. Bishop Mus Occas Pap 30:4–26

Hubbell SP (2001) The unified neutral theory of biodiversity and biogeography. Princeton University Press, Princeton, 448 pp

Hutchinson GE (1959) Homage to Santa Rosalia or why are there so many kinds of animals? Am Nat 93:145–159

Hutchinson GE (1961) The paradox of the plankton. Am Nat 95:37–145

Lynch JD (1988) Refugia. In: Myers AA, Giller PS (eds) Analytical biogeography. Chapman & Hall, London, pp 311–342

MacArthur RH, Wilson OE (1967) The theory of Island biogeography. Princeton University Press, Princeton, p 224

Nielsen C (1995) Animal evolution. Interrelationships of the living phyla. Oxford University Press, Oxford, 467 pp

Norris RD (2000) Pelagic species diversity, biogeography, and evolution. Palaeobiology 26(sp 4):236–258

Preston FW (1948) The commonness, and rarity, of species. Ecology 29(3):254–283

Rosenzweig ML (1995) Species diversity in space and time. Cambridge University Press, Cambridge, 436 pp
Sepkowsky JJ Jr (1984) A kinetic model of Phanerozoic taxonomic diversity. III. Post-Palaeozoic families and mass extinctions. Palaeobiology 10:246–267
Simberloff DS, Wilson EO (1970) Experimental zoogeography of Island. A two year record of colonization. Ecology 51:934–937
Sturmbauer C (1998) Explosive speciation in cichlid fishes of the African Great Lakes: a dynamic model of adaptive radiation. J Fish Biol 53(Suppl A):18–36
Terborgh J (1973) On the notion of favorableness in plant ecology. Am Nat 107:481–501
Timoshkin OA (2001) Lake Baikal: fauna diversity, problems of its "immiscibility" and origin, ecology and "exotic" communities. In: Timoshkin OA (ed) Index of animal species inhabiting Lake Baikal and its catchment area. Nauka Publishers, Novosibirsk, pp 16–73
Von Baer KE (1828) Uber Entwickelungsgeschichte der Thiere. Beobachtung und Reflexion, vol I, Konigsberg: Borntfiiger, 230 pp
Whittaker RH (1972) Evolution and measurement of species diversity. Taxon 21(2/3):213–251
Williamson M, Gaston KJ (2005) The lognormal distribution is not an appropriate null hypothesis for the species–abundance distribution. J Anim Ecol 74:409–422

Mediterranean Sea Biodiversity

3

3.1 Knowledge of Mediterranean Sea Biodiversity

An analysis of the literature demonstrated that only 364 (26.4%) of 1375 surface units in the whole Mediterranean Sea have been interested in at least one collection of zooplankton samples, and only 104 (7.4%) of the same geographic units have been interested in multiple collections (timely repeated, thus considering the seasonality) in the past 200 years (Belmonte and Potenza 2001) (Fig. 3.1).

This situation is also the rule in other parts of the world, and it is the only available knowledge that we can use in generalizing the biological description of pelagic communities.

As in all the marine habitats of the world, Mediterranean zooplankton are dominated by Copepoda (overall **Calanoida** but also **Cyclopoida**, **Harpacticoida**, and **Monstrilloida**, in order of species richness), and **Acartiidae** (Calanoida), **Oncaeidae,** and **Oithonidae** (Cyclopoida) are the most representative families of the group in neritic and coastal waters (those mostly interesting for human studies). The knowledge on Acartiidae is continuously updated, and the number of species in a restricted geographic area as the Mediterranean passed from 7 (beginning of the twentieth century) to 18 in recent times (Belmonte 2021).

A = without; *Cartos* = comb action; *Acartos* = not combed *Calan* = from Irish caelan, slender *Cyclos* = rounded; *Ops* = eye; *Cycl-ops* = mythic giant with a lone eye in front *Harpacticon* = rapacious predator *Monstrilla* = small monster *Oithona* = Gaelic name meaning "virgin of the wave" *Oncaea* = swelling, tumor

G. Belmonte, *Elements of Pelagos Biology*,
https://doi.org/10.1007/978-3-031-74280-4_3

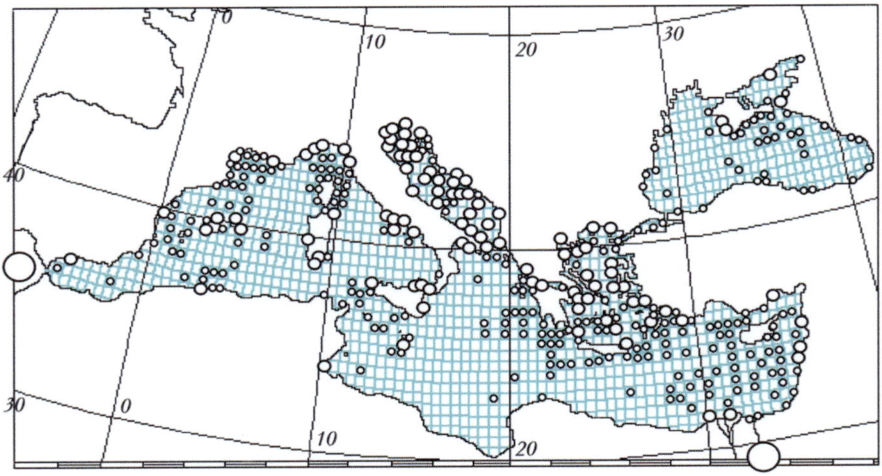

Fig. 3.1 Map of the Pontic-Mediterranean basin, subdivided into 1375 rectangular areas (each, approximately 55.6 × 42.7 km); 535 areas (39% of the total) contain a coastline trait, and 840 (61%) are completely pelagic. On the basis of >200 scientific publications on Copepoda from 1873, small circles indicate 363 areas (26.4% of the total) interested by single sample collection; large circles indicate 104 areas (7.6%) interested by repeated sample collections. Extra-large circles in the Atlantic Ocean and in the Red Sea indicate the presence of the taxon (Copepoda) also out of the Mediterranean Sea. (From Belmonte and Potenza 2001)

Such an increase in species richness is due to the growth of samples collected (and of new sites) and analysed, the better distinguishability of species, and the historically recent arrival of **nonindigenous species** (NIS) from elsewhere.

The growth of sites interested in studies allowed the discovery of *Pteriacartia josephinae*, a species widespread in coastal areas but not present outside the Mediterranean Sea (it is **endemic**). The improvement of investigation techniques allowed the recognition of *Acartia margalefi* in the second half of the past century as a new species, although it was surely also present in ancient times and probably corresponded to a dwarf form of *Acartia clausi* reported by scientists as typical of confined coastal waters. The arrival of NISs is the reason for the addition of *Paracartia grani* and *Acartia tonsa* to the Mediterranean list of Acartiidae species.

The generalized human ignorance about the marine environment is obviously due to the "habitat" typical of the researchers, because humans are linked to land and not living in the sea. This condition provoked a historical retardation of knowledge of marine habitats compared with freshwater and land habitats.

The land-based mind approach of humans is evident even on the geographic maps where the political borders of states do not take into consideration the inclusion of territorial seas and limits are drawn (erroneously) as those at the level of the coastline.

The unbalanced perception of the sea environment by the Mediterranean (land) scientific and political community is finally evident in the institution of **marine protected areas**, mostly along the coasts. In the whole of Europe, marine protected areas manage a coastal marine environment that never passes the **isobath** line of −50 m, thus confirming the tight relationship of marine protection by human society with the very coastal sites (a very limited reality of the marine habitat).

The Italian Seas (587,000 km^2), in the centre of the Mediterranean Sea, well represent the Mediterranean Sea from a biological point of view. The Italian Marine Biological Association censused **12,306 species in the Italian Seas** (mainly benthic, 9307 faunal, 2957 floral, and 42 fungal) in 2008 (Relini 2008), representing 73% of the **16,848 total** species reported from the whole **Mediterranean Sea** in 2010. The Mediterranean pelagos is less known than the coastal environments, but it is the elective habitat for more than 50% of Mediterranean floristic species (**1484 autotrophic species**, mainly unicellular algae of phytoplankton).

The heterotrophic compartment is very diverse, containing representatives of all the phyla, both holoplankton and meroplankton, from unicellular Protozoa to the largest animal of the planet (the blue whale).

Some *phyla* (e.g., Ctenophora, Chaetognatha) are planktonic, and others dominate the pelagic communities, even in deep layers of the sea, such as Crustacea Copepoda Calanoida (with more than 270 species only in the Italian Seas), or pelagic fish.

The **endemism** level of the Mediterranean Sea is very high, with a total of **3350 species** (approximately **20.2%** of the total species richness) as endemic (absent in any other part of the world) and highly representative, mostly of benthic communities (e.g., *Posidonia oceanica*, *Pinna nobilis*, and *Corallium rubrum*) (Ben Rais Lasram et al. 2010). Some endemism also exists in the pelagos, but in this case, it is represented by a smaller percentage of species (as the already-cited *Pteriacartia josephinae*) (Fig. 3.2).

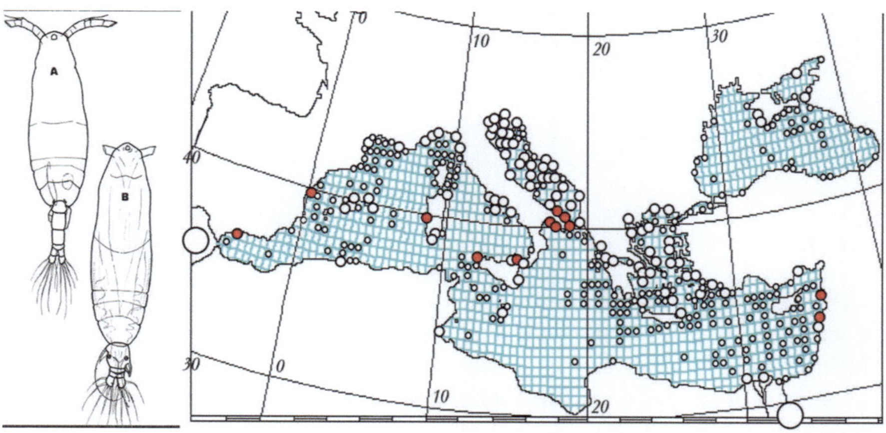

Fig. 3.2 Left, *Pteriacartia josephinae* (Copepoda) dorsal view; an endemic representative of Mediterranean fauna; (**A**) male; (**B**) female. Antennulae are not drawn. Right, map of *Pteriacartia josephinae* geographic distribution in the Mediterranean Sea

Among fish (Osteichthyes and Chondrichthyes), for example, Mediterranean-Pontic endemics account for 11.4% of the global species richness, and most of them are reported as "benthic" fish.

The endemism of over 20% of the global species richness, however, authorizes the indication of the area as a **biogeographic province** typical and distinguishable from the neighbouring seas.

The percentage of endemic species over the total species of each taxon varies with the considered *phylum*, being maximum among Porifera (48%), Crustacea Mysida (36%), and Chordata Ascidiacea (35%), and minimum among Chordata Osteichthyes (12%), Mollusca Cephalopoda (10%), and Crustacea Decapoda (10%).

The number of described **Protozoa** and **Chromista** species in the Mediterranean is approximately **4400**, representing either autotrophic or heterotrophic organisms. However, many **morpho**-species should include a number of cryptic or pseudocryptic variants, which should increase the total number of species.

At least for Prokaryota, in general, in the coastal northwestern Mediterranean, season is more important than geographic position in determining community structure. **Microbial communities tend to be similar on the horizontal space and much more variable on the vertical profile**.

In the last decade, optical and electron microscopy, often in combination with molecular and culturing techniques, have considerably increased the taxonomic knowledge of many of these groups and highlighted the potential existence of extended cryptic or unknown diversity.

As an example of taxon representativeness, in the northwestern Mediterranean Sea, diatoms (**Bacillariophyta**) are present with **736** species. Dinoflagellates (**Dinophyta**) are present with **673** species. **Coccolithophoridae** is the third most important *taxon* of unicellular autotrophs (Phytoplankton), with **166** species (Coll et al. 2010).

Most unicellular organisms, however, are heterotrophic and more dispersed along the water column than phytoplankton, which are linked to surface waters because they need light to perform photosynthesis.

One of the most diversified taxa of heterotrophs is **Ciliophora**.

Autos = by itself, independently; *Trophic* = linked to food *Bacillus* (L) = rod; *phytos* = plant *Bio* = life; *Gea* = earth; *Graphos* = drawing *Chroma* = colour *Dino* = power for movement; *phytos* = plant *ein* = inside; *deimos* = house *Coccos* = rounded; *lithos* = stone; *Phero* = to bring *Iso* = equal; *Bathos* = depth *Morphe* = shape *Protos* = primitive; *Zoos* = animal

They are roughly divided into loricate and aloricate according to the presence of an external **lorica** protecting the cell. Among the marine loricate Ciliophora, **Tintinnina** are a very diversified group.

Numbers of Tintinnid species ranging from **40 to 68** were recorded in surveys of various Mediterranean sites.

Other groups, such as **Foraminifera**, which have Calcium carbonate tests, and **Radiolaria**, which produce siliceous or Strontium sulfate cytoskeletons, have been the subject of many stratigraphic and palaeoceanographic studies (Fig. 3.3).

Radiolaria also have the most complex and rich architecture of a unicellular organism. Notwithstanding, life cycle and biodiversity studies on living Foraminifera and Radiolaria in the Mediterranean remain scarce.

A total of **650 fish species** occur in the whole Mediterranean Sea, and an updated list reveals that the Mediterranean currently contains **116 exotic** (NIS) species.

Approximately 80 fish species are **Chondrichthyes**, although the status of some is uncertain because of their infrequency, and 9 of them do not breed in the Mediterranean Sea.

Nine species of marine **Mammalia** are encountered regularly in the Mediterranean: eight Cetacea and one Pinnipedia.

Of these species, five are Delphinidae, and the others are Ziphiidae, Physeteridae, Balaenopteridae, and Phocidae. Additional 14 Cetacea species are sporadically sighted throughout the basin and are considered "visitors" or "non-residents".

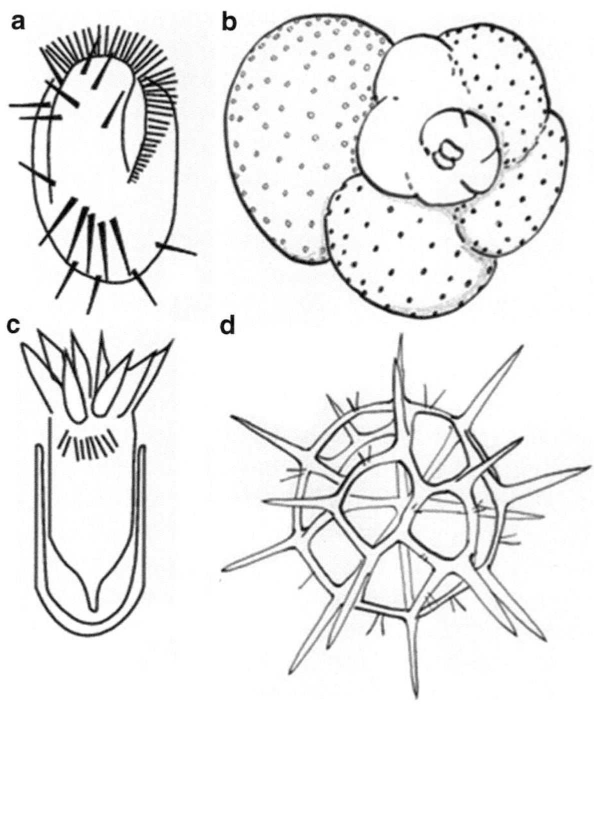

Fig. 3.3 Scheme of some representatives of unicellular Eukaryota of the zooplankton: (**a**) *Euplotes* (Ciliophora), (**b**) *Globigerina* (Foraminifera), (**c**) *Tintinnopsis* (Ciliophora), (**d**) Acantharia

Of the seven existing species of **sea turtles** (Reptilia), **two** (the green and the loggerhead *Chelonia mydas* and *Caretta caretta*, Cheloniidae) commonly occur and nest in coast of Mediterranean Sea, and one (the leatherback turtle *Dermochelys coriacea*, Dermochelyidae) is regularly sighted but without any evidence of nesting sites.

Seabirds (Aves) from the Mediterranean are **15 species**, and their population densities are small, consistent with a relatively low-productivity ecosystem compared with other marine areas of the planet, particularly upwelling regions. Ten of the Mediterranean species are gulls and terns (Charadriiformes), four are shearwaters and storm petrels (Procellariiformes), and one is a shag (Pelecaniformes).

The Mediterranean has few areas deeper than 3000 m, and bathyal or abyssal taxonomic groups are not abundant. This notwithstanding, it is a suitable model for investigating deep-sea biodiversity patterns along longitudinal gradients across its different sub-basins.

The deep-water fauna of the Mediterranean has a lower degree of endemism than that of the Atlantic at similar depths. As a consequence of this situation, while the Mediterranean Sea is recognized as one of the most diverse regions on the planet, its deep part contains a much lower diversity than the deep-sea regions of the Atlantic and Pacific oceans.

This result is easily attributable to the small extension of the Mediterranean Basin and to the even smaller extension of deep areas inside it.

Finally, we never have to disregard the story of the Mediterranean, which is a relatively "young" sea, overall in consideration of the last Miocene crisis (7–5 million years ago), when an enormous loss of water produced a sea level drop of approximately 2000 m and the formation of separate hyperhaline lakes in the deepest parts.

This condition determined the reduction to zero of the preexisting fauna (that of Tethys origin), and when a connection reopened at its western extremity (more or less at the level of Gibraltar), only the surface fauna of the Atlantic was allowed to enter due to the existence of a threshold that impeded the communication of the deepest parts of the newly connected worlds.

The biodiversity of the Mediterranean has been definitively influenced by the introduction or spontaneous arrival of species from other geographic areas. Since the first review of exotic species in the Mediterranean, studies in this field have intensified. Currently, **more than 600 metazoan** species have been recorded as **alien** (3.3% of the total species richness). However, this estimate is continuously increasing and may be as high as **1000 species** if unicellular organisms are included.

Most of these **non-indigenous species** (NIS) are littoral and sublittoral benthic or demersal species (or their symbionts).

The species most likely to be introduced by the main routes and activities (the Suez Canal, vessels, and aquaculture) are shallow-water benthic species.

A taxonomic classification of the NIS showed that Mollusca (33% of NISs) was the most frequently reported, followed by Arthropoda (mainly Crustacea) (18%), Chordata (17%), Rhodophyta (11%), and Annelida (8%) (Galil et al. 2018). The data are presumably most accurate for large and conspicuous species that are easily distinguishable from the native biota, for species that occur along a frequently sampled or fished coast, or for those geographic areas where taxonomic expertise is available.

Chondros = cartilage; *Ichthyos* = fish *Cilia* (L) = eyelash; *Phora* = bringing *Coccus* = small sphere; *Lithos* = stone *Dino* = moving ("dynamic") *Mammalia* (L) = having the brest *Metazoan* = final animal

Data are entirely absent for many of the small members of the invertebrate *phyla*. Thus, the total number of NISs is certainly downward biased.

The provenance area of the Mediterranean NISs is most commonly (69%) the Indo-West Pacific with Indo Pacific (41% of total NISs), followed by Indian Ocean species (16%) and the Red Sea species (12%). Some NISs have a **pantropical** or **circum-tropical** distribution (19%). NISs exclusively of Atlantic derivation have a small importance (12%).

As far as can be deduced, the majority of NISs entered through the **Suez Canal** (**Lessepsian migrants**) (53%), after its opening in 1869. Vessel-based introductions from other parts of the world account for 22% of NISs, and aquaculture accounts for 10%. The methods of introduction differ greatly among the *phyla*: whereas the NIS **macrophytes** were introduced through aquaculture and vessels (41% and 25% of the total NIS, respectively), the majority of Crustacea, Mollusca, and Pisces NISs are Lessepsian migrants (59%, 64%, and 86%, respectively), and those introduced by aquaculture practices are few (4%, 5%, and 4%, respectively).

The number of NISs that have been recorded during the twentieth century has increased in recent decades.

The increasing role of the Mediterranean as a hub of international commercial shipping, a surge in the development of marine shellfish farming over the last 35 years, and the continued enlargement of the Suez Canal have contributed to the increase in introductions since the 1950s (Fig. 3.4).

A comparison of the NISs recorded along the Mediterranean coasts of Spain and France and an equivalent length of coast in the Levantine Sea (from Port Said, Egypt, to Marmaris, Turkey) showed marked differences in their numbers, origin, and methods of introduction.

There are nearly four times as many **NISs** (of any provenance) **along the Levantine coast** (**456 species**) as along the **northwestern coast of the Mediterranean** (**111 species**). The majority of NISs in the Eastern Mediterranean entered using the Suez Canal (68% spontaneously), whereas aquaculture (42%), vessels (38%), or both (5%) are the main vectors of introduction in the Western Mediterranean (Galil et al. 2018) (Fig. 3.5).

Climate change favours the affirmation of Red Sea species in the southeastern Mediterranean and their rapid spreading northwards and westwards. It similarly favours species coming from the African Atlantic coasts to enter the Western basin via the Gibraltar Strait.

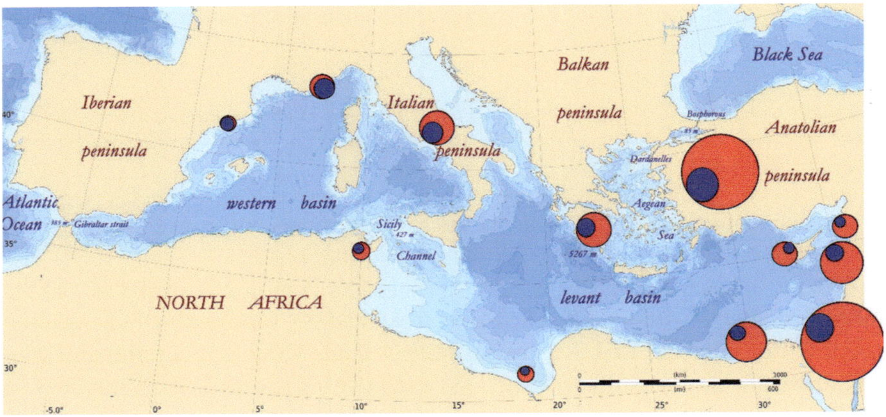

Fig. 3.4 Map of Lessepsian species distribution in the Mediterranean Sea. Disk diameter represents the total number of non-indigenous species (NIS) in each area. The red colour of each disk indicates the percentage of Lessepsian NISs (from the Red Sea). The blue colour indicates species from the Atlantic Ocean or elsewhere. (From Galil et al. 2018, modified)

Fig. 3.5 Lessepsian migrants: left, *Rhopilema nomadica* (Cnidaria, Scyphozoa) (drawing of A. Gennari); right, *Pterois miles* (aquarium of San Sebastian—Spain; photograph G. Belmonte)

In the Mediterranean Sea, a high number of invertebrate taxa show a North-West-to-South-East **decrement in species richness**, with a highly heterogeneous distribution. The reverse abundance of NISs, whose number decreases from SE to NW, is simply the consequence of the intense arrivals from the Red Sea and Suez Canal.

A few exceptions exist to this distribution rule, and some pelagic taxa are particularly involved in such an opposite trend. For example, while there are the same number of *Euphausia* species in the western and central basins, several other invertebrate groups are richer in the Levantine Sea than in the West Mediterranean. This is the case for Copepoda, which is enumerated in a total of 83 species in the Balearic Sea (West Mediterranean) (Fernandez de Puelles et al. 2003) and 192 species in the Turkish Levantine Sea (East Mediterranean) (Siokou-Frangou et al. 2010).

Additional information from the literature on spatial distribution patterns of Mediterranean marine diversity suggests that the measurement of local ***α*-diversity** is not enough to draw a clear picture for the whole Mediterranean basin.

The pattern of a generally decreasing diversity with increasing depth is obvious for photosynthetic organisms, which do not find light for photosynthesis at increasing depths below sea level. The same frame, however, was also documented for animal (invertebrate and fish) species.

Diversity appears concentrated along coastal areas and on continental shelves, mainly in the **euphotic** layer (above 200 m depth). However, the **patterns did not necessarily show a monotonic decrease with depth**. Larger numbers of **Cumacea** (Fig. 3.6) species have been found either in shallow waters of **0–99 m depth (48 species)** or between 200 and 1400 m depth.

Other studies carried out on the depth-related distribution of marine biodiversity in the deep Mediterranean suggest a uniform scheme of species diversity distribution, the highest values of which are observed at intermediate depths (approximately 2000 m) and lower values at upper bathyal and abyssal plains.

Box 3.1 Diversity Indexes

***α*-diversity** is the number of species found in one sample or within a single habitat,

***β*-diversity** depicts the extent of species replacement along environmental and/or climate gradients, also termed "turnover diversity",

***γ*-diversity** represents the diversity of an entire geographic area, including more habitats.

***S* species richness.** The number of species in a sample or a habitat, without any indication of their importance.

***D* diversity index of Simpson.** The number of species *S* is considered, as is the relative importance *P* of each species in terms of abundance. $D = 1/\Sigma(P_i)^2$

The *D* value depends on the number of species *S* and on the level of uniformity (evenness) of the distribution of individuals among the species. The *D* value grows if the uniform distribution of specimens is high. The diversity *D* of a situation with many species *S* can be lower than a situation with a low *S* if the uniformity-evenness is low.

***E* evenness/equitability.** It can be expressed as D/D_{max}, where D_{max} represents the maximum possible uniformity in species importance. D_{max} corresponds to the number of species *S* (when all are uniformly represented); hence, $E = D/S$, and it is between 1 and 0.

***H*, *J*, and Shannon diversity/equitability indexes**. The formula contains logarithmic values (basis 10, *e*, or 2), and this presence lowers the final score of *H* but increases that of *J* in comparison with indexes *D*/*E*.

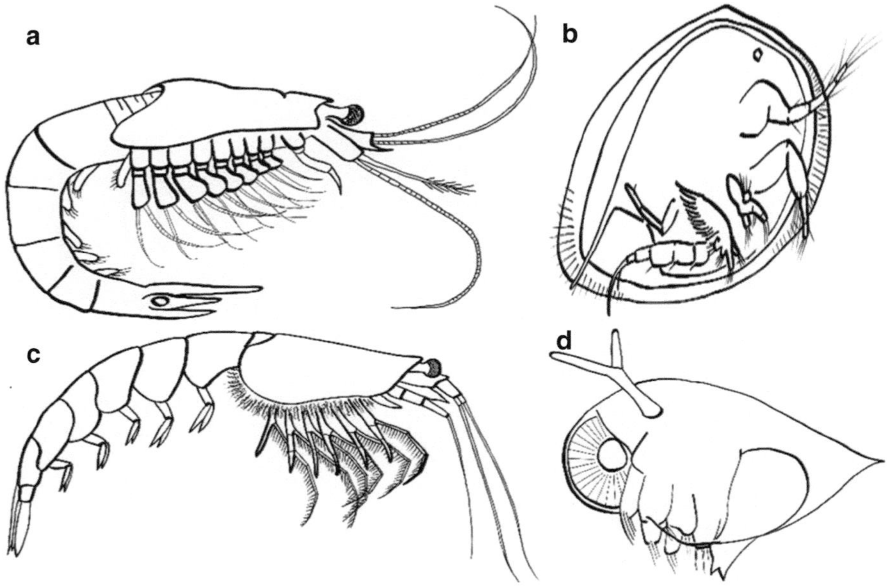

Fig. 3.6 Sketches of planktonic Crustacea other than Copepoda: (**a**) Mysida, (**b**) Ostracoda, (**c**) Euphausiacea, (**d**) Cladocera (original)

Always waiting for a clarification of knowledge on the vertical distribution of pelagic biodiversity, we have to register recent studies on benthic *taxa*, which demonstrated that the rule of the highest value at 2000 m is not always recognizable. In open-slope systems, bathymetric gradients of species diversity have been widely documented. In the Mediterranean Sea bottoms, although **Nematoda** diversity decreases with depth, ample ranges of biodiversity are observed at the same depth.

These data suggest that the supposed wide vertical distribution of species (**eurybathy**) in the Mediterranean fauna could be lower than previously thought, and a parallel with pelagic fauna is waiting for a demonstration.

3.2 Temporal Trends

Environmental factors have led to profound changes in the abundance, distribution, and composition of Mediterranean marine species in the distant past.

For example, during the **Cretaceous (145-65 million years** ago), the geographic area that today hosts the Mediterranean Sea (at those times a part of the **Tethys** Ocean) was connected with a "young" Atlantic Ocean on the West, whose opening started approximately 190 million years ago. The two oceans (Atlantic and Tethys) contributed very different faunas to the proto-Mediterranean area, and the Tethys predominated in this role. During the **Miocene (23-5 million years** ago), the proto-Mediterranean corresponded to the modern Sea with some small differences; apart from the communication at the West extremity with the Atlantic Ocean, it also communicated on the East extremity with the "new" Indian Ocean, whose opening counterbalanced the progressive closure of the ancient Tethys. During the final Miocene (**7-5 million years** ago, the so-called **Messinian** stage), the proto-Mediterranean was isolated from the Indian Ocean, and the connection with the Atlantic Ocean was also closed (Fig. 1.12). During this Messinian isolation, the Mediterranean underwent severe desiccation (and the derived **salinity crisis**) that led most species to extinction (Gautier et al. 1994).

Chondros = cartilage *Cuma* = ancient city, close to Naples, site of the Sybill *Eury* = good performant; *Bathos* = depth *Ichthyos* = fish *Eu* = good, *Photo* = light *Mysida* = (mysis) closure of eyes *Nema* = row *Synthesis* = formation

Although some shallow sea lakes remained on the two sides of the dried Sicily–Tunisian connection, **allopatric speciation** events were probably contained in the poor fauna adapted to hyperhaline conditions, and the reopening of communication with the Atlantic Ocean realized a restocking of Mediterranean biodiversity based on Atlantic fauna and flora. Until the nineteenth century AD, the only oceanic connection to the Mediterranean was that one.

At the end of the last **ice age** (approximately **30,000 years** ago), there were notable climate-driven fluctuations but also human-induced changes due to the long periods of exploration and exploitation.

The heaviest impact, however, is the recent (opening, 1869) connection with the Indian Ocean through the Suez Canal and the Red Sea, which added its effects to the globalization of commerce and trade, the increasing pollution and eutrophication of densely inhabited coastal areas, and coastal habitat modification and loss, all inserted into a general frame of global climate change.

Early evidence of **human interaction** with marine fauna in the Mediterranean Sea comes from the **Palaeolithic** and continues through the **Mesolithic** and **Neolithic** periods (approximately 20,000–4000 BC). **Zooarchaeological remains** are found in Franchthi Cave in the southern Argolid, Greece; Las Cuevas de Nerja in southern Spain; Athlit Yam, south of Haifa Bay in Israel; Cape Andreas Kastros in Cyprus; and the Strait of Gibraltar. Fish bones of large Tunnidae, Sparidae, and Mugilidae were found. Zooarchaeological remains in Spain include 20 *taxa* and show a reduction in mean fish size and range over time that has been considered an indication of overfishing (Tortosa et al. 2016).

In Cyprus and in Athilit Yam, 90% of the remains are grey trigger fish (*Balistes capriscus*), and this fact is interpreted as intensive fishing regardless of size. In Gibraltar, remains of Mediterranean monk seals and Mollusca consumed by humans were found. Stable isotope analyses of human bones show that between 10,000 and 8000 BC, the main Mediterranean coastal populations did not rely significantly on marine food.

This allows us to consider the start of marine resource exploitation by humans a very recent episode in the frame of human history.

The exploitation of marine resources is clearly reported by Aristotle in his zoological works dating to the fourth century BC, where he focused his scientific interest on fish and invertebrates exploited by humans in various ways. Fisheries in the Aegean communities by that period were characterized by variability both in the nature and abundance of the exploited fish and in the manner of their exploitation. Mollusca and other invertebrates were part of the diet of the ancient Greeks, and their consumption was connected with the treatment or prevention of various health problems and diseases.

Commercial fishing and fish processing activities play an important role in the economy of the Pontic area (Black Sea). The export of fish and fish products needed pelagic habitat exploitation for the production of salt fish (*tarichos*) and fish sauce (*garum*), mainly from anchovies, which became common in the Roman market.

The occupation of islands became common, and this was also the beginning of the introduction of alien species into ecosystems biogeographically separated from continents. Some of these introductions (rats, carnivores) triggered the extirpation of seabird colonies on islands.

Seafood became increasingly popular toward the end of the Roman age (fourth to fifth centuries), probably because of the proximity of and access to marine resources. There is historical evidence of overfishing in some parts of the West Mediterranean in the early Roman Imperial period. Even then, certain fishing techniques (such as fishing by torchlights at night) were prohibited to manage or counteract the decline in fish stocks, and efforts were made to boost natural availability with introduced fish and shellfish stocks.

There are also pictorial remains that show fishing gear and a large variety of targeted species during Roman times. Gastropoda Muricidae, the red coral *Corallium rubrum*, and several species of Porifera were exploited on an industrial scale in those periods.

Fishing, fish processing, industrial exploitation of several marine species, and the development of improved fishing gear continued during the Byzantine period (fifth to fifteenth centuries).

Various literature sources point out that targeted species, among them the currently overfished tuna, were conspicuous in the early Byzantine period.

Zooarchaeological material from the Israeli coastline dating from the Byzantine through the Moslem Crusader and Mamluk periods (tenth to fourteenth centuries) points to a high consumption of marine and freshwater species that are still fished in Israel today, such as the thin-lipped grey mullet (*Liza ramada*), Sparidae, and the parrot fish Scaridae.

There was noticeable fishing activity dating from the Byzantine, Moslem (tenth century), and later Norman periods (eleventh to thirteenth centuries) in southern Italy and in Sicily, where blue fin tuna was the main target species exploited by drifting traps (*tonnara*).

Another example of human exploitation of marine resources from historical times is the hunting of seabirds on islands, particularly the shearwater Procellariidae, which probably constituted the only source of protein in periods of scarcity on small islands.

Allaio = stranger, born elsewhere; *Patria* (L) = from *Pater* (phater), familiar site, local *Creta* (L) = chalk. Abundant at the end of the geological period *Ichthyos* = fish *Eury* = good performant; *Bathos* = depth *Lithic* = stony *Meso* = intermediate (as age) *Mio* = (meion) less; *coenos* = (kainos) recent, closely related *Nema* = row *Neo* = new *Palaeo* = ancient; *lithos* = stone *Photo* = light *Sipunculus* (L) = small tube *Synthesis* = formation *Tethys* = Greek nymph of the sea, wife of Peleus, and mother of Achilles

In places such as Formentera (Balearic Islands), humans contributed to the depletion and partial extinction of the Balearic shearwater (*Puffinus mauretanicus*) (Fig. 3.7), with consequences at the level of the marine trophic web.

Human impacts on marine biodiversity were prosecuted with increasing intensity as Mediterranean cities and ports continued to enlarge and advance in technology.

It is assumed that since the fourteenth century, the adoption of new fishing methods (such as the *tonnara*) in the West Mediterranean, their spread to South Italy, and their introduction to the Adriatic in the seventeenth century increased fishing catches.

Fishing catches increased to the extent that even the early fishermen's organizations (sixteenth century), such as *Cofradias* in Catalonia and the *Prud'hommes* in Provence, were concerned about possible negative effects on exploited stocks. Such effects were further intensified in the nineteenth century, with an increase in the efficiency of existing fishing gear (e.g., otter trawl) and the introduction of new ones (such as midwater pelagic trawls, hydraulic dredges, and iron-toothed dredges). Industrialized fishing has increased impacts on species, habitats, and ecosystems. Several studies have also shown historical changes in fish communities in different parts of the basin. These findings pointed to **a general, severe depletion of top predators** in the Mediterranean basin, including blue-fin tuna, which is considered

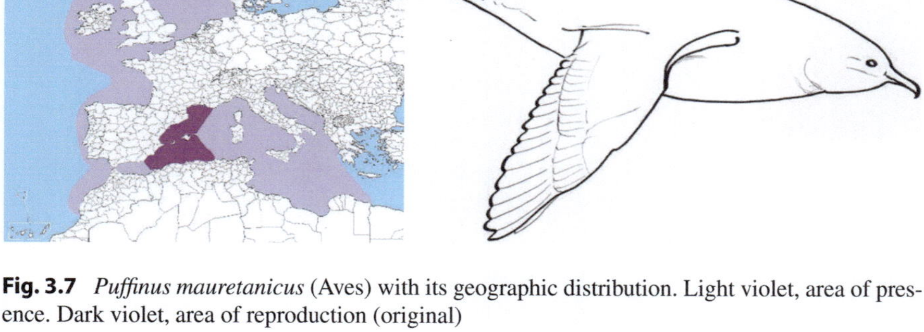

Fig. 3.7 *Puffinus mauretanicus* (Aves) with its geographic distribution. Light violet, area of presence. Dark violet, area of reproduction (original)

critically endangered according to the declining trend observed in the second half of the twentieth century.

Historical fluctuations in the abundance of this species have been described on the basis of a centuries-long series of tuna trap catches, starting in the seventeenth century, and have suggested a link to climate fluctuations.

3.3 Estimates and Patterns of Marine Diversity in the Mediterranean Sea

A recent estimation (Coll et al. 2010) of approximately **17,000 species** for marine species in the Mediterranean Sea (benthic and pelagic) updated and exceeded previous values, which were on the order of 8000 (1992)–12,000 (2000) species. In comparison with the 1992 estimation, the total number of recorded species has substantially doubled in 25 years. Although new species arrivals and introductions are continuously recorded, the number of species has grown principally due to the continuous study and exploration of the environment. Estimates of invertebrates and protists, in particular, have undergone an upward revision in recent years as a result of recent efforts and improvements in analytical methods and instruments. Current estimates of species for Porifera, Cnidaria, Polychaeta, Mollusca, Arthropoda, Echinodermata, and Ascidiacea all sensibly exceed those dating back to the early 1990s.

However, since most microbial diversity is basically unknown, total numbers and their modifications are uncertain (Table 3.1).

Table 3.1 Number of species per taxon in the Mediterranean Sea (from Coll et al. 2010)

Taxon	Number of species
Archaea?	
Bacteria (including Cyanobacteria)	165
Bacillariophyta	736
Dinophyta	673
Coccolithophoridae	166
Retaria (Foraminifera, Radiolaria)	600
Other Protozoa	277
Chlorophyta	190
Rhodophyta	657
Magnoliophyta	7
Porifera	681
Cnidaria	757
Platyhelminthes	1000
Mollusca	2113
Annelida	1172
Arthropoda Crustacea	2239
Bryozoa	388
Echinodermata	154
Chordata Tunicata	229
Chordata Vertebrata	693
Other invertebrates	2168

The **West Mediterranean** shows the highest **species richness**, likely owing to the long-term injection of Atlantic species and the wide range of physicochemical conditions.

The central Mediterranean, Adriatic, and Aegean Seas are areas with the second highest species richness, although with exceptions. The Adriatic Sea shows lower species numbers, probably because of the restriction of water mass (and species) exchanges with the remaining part of the Mediterranean basin and a generalized confined situation (decreasing water column towards the north, excessive presence of fresh water at the surface, larger amplitude of temperature variations, smaller geographic extension). However, this sub-basin shows a large number of endemics, possibly due to its higher geographic confinement.

The Levantine Basin and southeast part generally have the lowest species richness, which suggests the presence of unfavourable conditions for species richness, although we must also consider this datum as a byproduct of a less intense sampling effort. In fact, a lack of data is evident in several east and south portions of the Mediterranean basin. This may have strongly influenced some of the results regarding spatial patterns; hence, generalizations in this sense have to be proposed carefully.

Therefore, as new species are assessed in the eastern and southern areas, patterns may be modified. In addition, the biodiversity in the east extremity of the Mediterranean Sea is most influenced by new species introductions. The **Suez Canal, opened in 1869** (see Box 1.5 and Fig. 1.15), has restored the connection between the Mediterranean and the Indian Ocean, and in recent years, we have witnessed an exponential increase in the number of Indo-Pacific species recorded in the Eastern Mediterranean. This trend will continue to influence the biodiversity of the Mediterranean Sea.

Explanations for the observed heterogeneity of species richness in the Mediterranean Sea include the presence of **a threshold in the Sicily–Tunisian Strait** that divides the Mediterranean into two sub-basins, other than the palaeobiogeographical history of the different parts of the Mediterranean Sea.

The western basin shows more biological similarity with the Atlantic Ocean, hosting a higher number of cold-temperate species, while the eastern basin shows more biological similarities with the Indo-Pacific and hosts a larger number of warm-temperate (even subtropical) species.

The Sicily–Tunisian Strait sill partially acts as a barrier to the exchange of many species between the two basins and constitutes the contact area of their geographic distributions.

Diversity distributions in the Mediterranean are also associated with a productivity gradient. Higher productivity areas show higher diversity, partially because they are important feeding and reproductive sites for several taxa. Most of these areas occur in the West Mediterranean and the North Adriatic. A large productive area is evident between Tunisia and Libya, but high biodiversity is not reported. Most likely, in this case, is the absence of studies and their continuity, which affect the final result. The area with the highest productivity, however, is the northwestern Black Sea. Additionally, in this case, the datum is not associated with high biodiversity, and here, the impoverishment of biodiversity can probably be imputed to the general confinement that characterizes the basin.

The main categories (**drivers**) explaining the **biodiversity** in the Mediterranean are (a) **bathymetric gradients**, which are associated with increasing pressure and decreasing food availability in deeper bottoms; (b) **geographical and physico-chemical features**, which are responsible for the NW–SE gradient of trophic conditions and someway coincident with confinement opposition to a high biodiversity; (c) **environmental heterogeneity** (e.g., grain size distribution, habitat complexity, distribution of food inputs); and (d) **different distribution of research institutions and of studies** (overall in the southern part).

3.4 Threats to Mediterranean Biodiversity

Normally, geological and physical changes are responsible for the most dramatic changes in biodiversity in the Mediterranean Sea over a geological time frame. Today, human activities are essential elements to be considered as well.

The most important threats to marine biodiversity in the Mediterranean region are **habitat loss, degradation and pollution**, **overexploitation of marine resources**, and the **arrival and affirmation of NISs**, in addition to **global climate change** (Fig. 3.8).

Fig. 3.8 Evolution of ocean bio-communities from a fish dominion towards a jellyfish dominion, driven by the fishery pressure on carnivorous fish species. The future would possibly be oriented towards a community rich in jellyfish predators. The size of some species (such as the sunfish or the turtle) grows from left to right if species are jellyfish eaters (Drawn by A. Gennari)

Habitat loss and degradation are currently the most common threats. Human interventions, such as coastal area modification, that can be traced back to before the Roman age have important consequences for diversity. The growth of the human population around the Mediterranean basin is an element to be considered in the process of habitat loss, degradation, and pollution. **Human pressure on coastal areas** (defined as **anthropization**) provokes sediment loading and water pollution, reduces the extent of important habitats, such as seagrass meadows, oyster reefs, maerl bottoms, and macroalgal beds, and increases their negative consequences on Mediterranean neritic ecosystems, with a consistent increase in the last two centuries. Most species depend strongly on their habitats; hence, habitat loss and degradation have the highest consequences for marine diversity.

Cultural eutrophication, particularly in confined basins such as the Adriatic Sea, can also be traced back for centuries.

Human pressure on coastal Mediterranean habitats reached its peak in the 1980s and may have been the cause of jellyfish outbreaks, red tide occurrences, bottom anoxia events, benthic mass mortalities, and mucilage events, which have been recorded in the recent history of the Adriatic Sea.

Direct and indirect pollution can be generated directly from the coast or through river transport from internal continent areas.

The oldest and one of the most important human activities that has become a threat to diversity is the exploitation of marine resources. The high demand for marine resources has recently also been attributed to mineral extraction, but the fishery is undoubtedly the main human activity, and recently, the growth of the planet population has multiplied the demand for food coming from fishing. Several fish resources are highly exploited or overexploited.

A few Mediterranean invasive NISs have drawn the attention of scientists, managers, and media for their conspicuous impacts on the native biota.

Tropical species have entered the Mediterranean through either the Suez Canal (**Lessepsian migration** (Gautier et al. 1994)) or the Strait of Gibraltar during recent decades, mainly by ship transportation.

The Mediterranean Sea is highly susceptible to ship-vehiculated biological invaders: in 2006, just to analyse 1 year, 13,000 merchant vessels made 252,000 calls at Mediterranean ports, and an additional 10,000 vessels passed through the basin without stopping at one of his ports. The increase in shipping-related invasions may be attributed to the increase in shipping volume throughout the region, changing trade patterns that result in new shipping routes, improved water quality in port environments, augmented opportunities for overlap with other introduction vectors, and increasing awareness and research effort (Fig. 3.9).

The swarms of vessel-transported American comb jelly (*Mnemiopsis leidyi*) that spread across the Mediterranean from Israel to Spain in 2009 raise great concern because of their notorious impacts on the ecosystem and fisheries (as certified during the crisis of the Black Sea in the second half of the twentieth century) (Fig. 1.17) (Tortosa et al. 2016).

Moreover, with the development of large-scale marine aquaculture (mariculture) in the late twentieth century, the commercially important NIS shellfish *Crassostrea*

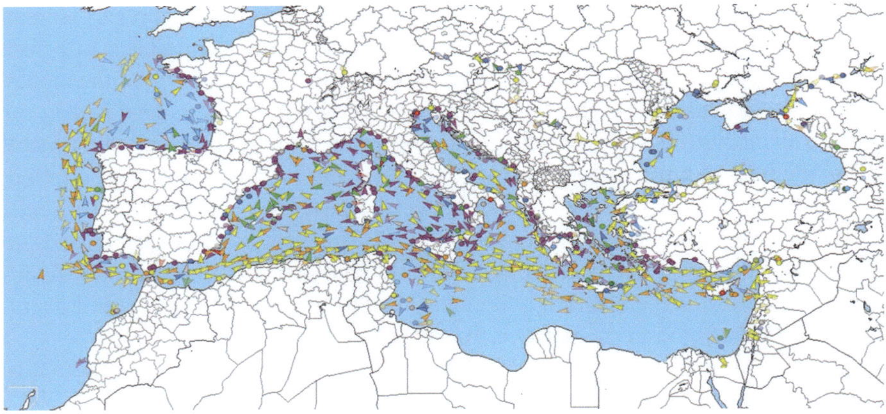

Fig. 3.9 Map of vessels present at the sea surface during just 1 day of the year, at noon, in the Mediterranean basin

gigas and *Ruditapes philippinarum* were intentionally introduced to the Mediterranean. The high permeability of aquaculture facilities, transport, and intentional transplantation of these species have resulted in many unintentional introductions; e.g., oyster farms have become veritable gateways into Mediterranean coastal waters for NIS macrophytes.

The massive "official" and "unofficial" importation of foreign spat (young bivalves both before and after settlement) in the 1970s and 1980s coincided with a marked increase in NISs around oyster farms, and the aliens were considered to have unintentionally profited from the imported oysters as vectors. Segments of the industry may still resort to illegal importation: neither the Turkish authorities nor the UN Food and Agricultural Organization were aware of the importation of the bilaterally ablated female banana prawn (*Fenneropenaeus merguiensis*) that was found in the Bay of Iskenderun, Turkey.

The disequilibrium provoked by affirmed NISs is responsible for reducing the population of some native species. This notwithstanding, many NISs are voluntarily introduced due to their market value, and other NISs involuntarily introduced have become locally valuable fishery resources. Some Lessepsian migrants (sensu (Por 1978)) were exploited commercially almost as soon as they entered the Levantine Sea, and their economic importance was quickly acknowledged. Levantine fisheries statistics record the growing prominence of the Lessepsian migrants: the Eritrean prawns are highly prized, and beginning in the 1970s, a shrimp fishery developed in the Levantine Mediterranean Sea. Nearly half of the trawl catches along the Levantine coast consist of Eritrean fish, but the commercially exploitable species were accompanied each summer by swarms of the undesired Scyphozoa jellyfish *Rhopilema nomadica* (Fig. 3.5) and washed ashore along the Levantine coast. The shoals of jellyfish adversely affect tourism, fisheries, and coastal installations, and severe jellyfish envenomation requires hospitalization. The recent spread of

Fig. 3.10 Lessepsian migrant *Lagocephalus sceleratus*, a species dangerous for life if ingested without caution (Drawing of A. Gennari)

silver-striped blasoppie fish (*Lagocephalus sceleratus*) (Fig. 3.10) and striped catfish (*Plotosus lineatus*) poses severe health hazards.

Pronounced temperature fluctuations and a significant increase in the average temperature of Mediterranean waters during the past two decades have coincided with an enlarged pool of warm-water NISs that have become established and expanded their distributions. These **thermophilous** NISs have a distinct advantage over native Mediterranean species. Although no extinction of a native species is yet attributable to the invasion of NISs, sudden declines in abundance, concurrent with proliferation of NISs, have been recorded.

The increase in Mediterranean water temperature alters jellyfish population dynamics and may, in addition, indirectly impact fishing. The seawater temperature of the Mediterranean Sea has increased substantially since at least the 1970s. Rising temperatures enlarge the pool of NISs that could settle, enable the warm-water species present in the sea to expand beyond their present distributions, and provide thermophilic NISs with an advantage over native Mediterranean species. The appearance of NISs of tropical origin is leading to what is called the "**tropicalization**" of the Mediterranean Sea. Although tropical invaders have been recorded in the northern sectors of the Mediterranean, tropicalization is especially obvious in the southern sectors, where species of tropical origin now form a significant portion of the entire biota.

Tropical species that entered the Mediterranean Sea via the Suez Canal (Lessepsian migration) or the Strait of Gibraltar for decades used to remain in proximity of their entrance doors in the East or West Basins, respectively. In the last two decades, however, the number of subtropical species that have also spread through the entire Mediterranean Sea has grown. Examples of **Lessepsian migrants** that, coming from the Levantine basin, crossed the Strait of Sicily and arrived in the Western basin include algae, many invertebrates, and fish.

The Strait of Sicily is currently a crossroad for species of distinct tropical origins (Atlantic and Indo-Pacific), expanding their geographic distribution in the longitudinal sense within the Mediterranean.

If the southern sectors of the Mediterranean are going to be "tropicalized" (higher occurrence of tropical NISs), the northern sectors are going to be "**meridionalized**" (increased proportion of Mediterranean indigenous thermophilic species). The destiny of species of boreo-Atlantic affinity is uncertain. These entered the Mediterranean during glacial periods and have been established in the northern and colder areas of the basin. Because they cannot move farther northward, due to the interposition of the European continent, their populations may dramatically decrease or even extinguish. Although the total extinction of flora and fauna from a basin as wide as the Mediterranean may be unrealistic, the signs of increased rarity or even disappearance of cold-water species deserve further attention. An example is the deep-water white coral, *Lophelia pertusa*, whose reefs of which have become rare in the Mediterranean. The coldest parts of the Mediterranean (Gulf of Lions, North Adriatic) could act as a refuge for cold-temperate species, but if the ongoing warming intensifies, those areas may act as traps without any neighbouring cool water where species can escape in the future (Fink and Wienberg 2015).

However, there may also be habitat fragmentation and local extinction of species unable to undertake migrations. The lack of evidence of species extinctions, coupled with the establishment of NISs, is apparently leading to an increased species richness in the Mediterranean, which represents a debated issue.

Species richness is increasing at the whole-basin scale (**γ-diversity**) as a consequence of the addition of studies and sites analysed.

Climate warming, moving physiological barriers, and inducing spatial overlap between NISs and indigenous species causes biotic homogenization and, hence, a depression in **β-diversity**.

Thus, the relationship between tropicalization, and biodiversity is not straightforward.

Seawater acidification may also be a threat to Mediterranean marine biodiversity.

The most obvious consequence of the increased concentration of atmospheric CO_2 (in turn responsible for the greenhouse effect) is the warming of ocean surface layers, which corresponds to a diminution of the dilution power of CO_2 in seawater. This decrease in CO_2 availability in seawater reduces the formation of calcified structures in marine organisms (based on $CaCO_3^-$ content). This could affect both planktonic and benthic communities (Suwa et al. 2014).

Anthropos = human *Meridian* = relative to the midday. Half of the sky where the sun is at noon *Thermo* = temperature; *phylein* = appreciator

Calcareous phytoplankton (**Coccolithophoridae**), living in the upper layer of the oceans, the layer more interested in warming and CO_2 depletion, play a significant role in the primary productivity of the sea, whereas many benthic habitats of the neritic areas are engineered by sessile organisms that lay down carbonate crusts.

The effects of global climate warming are also confused by the low level of knowledge of Mediterranean biodiversity. The study of Mediterranean marine diversity over many years has produced a reliable amount of information. However, this information remains incomplete with the discovery and description of new species, especially smaller ones that are less conspicuous or belong to cryptic biota. The biodiversity in the Mediterranean Sea could be much higher than is currently known.

Most of the small fauna of the Mediterranean is typical of current scientific knowledge: in one of the best-known geographic areas of the world, there are many regions and habitats that remain insufficiently studied, and several taxonomic groups in deep-sea areas and portions of the southern region are still poorly known.

The description of **new species for Cumacea, Mysida, Polychaeta,** and **Ascidiacea** (described or first recorded) shows that the scientific community is still far from an estimation of the actual existing species. No asymptote has been reached yet in such a "species accumulation curve", and there has been no slowing in the rate of discovery for less conspicuous species in the Mediterranean Sea.

The shortage of taxonomists for many taxa is a particularly serious problem worldwide, and it also applies to Mediterranean countries. Several of the main invertebrate specialists have retired or are close to retirement, and a few are being replaced. Many samples are not properly analysed for species identification, which leads to an underestimation of biodiversity.

The current spread of NISs requires serious taxonomic attention. Many, if not most, taxa are subject to anthropogenic threats in one way or another, and researchers must work against time to avoid losing valuable biological information.

Sampling biases are another source of uncertainty in the estimation of marine biodiversity. In particular, the three-dimensional reality of marine pelagic ecosystems requires many more studies but is poorly understood. Light intensity decreases with water depth, and species perform extensive migrations within the water column or along the seabed.

Marine species react to light intensity cycles, which may include movements in and out of our windows of sampling time. Information gathered without attention to such rhythms will affect perceived population distribution, biomass, and biodiversity estimation.

The list of the known species in the Mediterranean Sea is not only informative on the total number but also illustrates what high-level *taxa* are the most biodiverse. From the list, it appears evident that Mollusca and Crustacea are the most represented *phyla*, followed by Vertebrata, Porifera, and Polychaeta. As a consequence of this indication, in each monitoring program of biodiversity (at the regional, environmental, or habitat scale), we must confirm such a rule.

A clear classification of biologically valuable areas, to avoid a time-consuming assessment of total species richness, can be directed at a few well-known and identifiable *taxa* with the assumption that they should represent the whole community. For the species-abundance curves, in fact, the two values (number of species and abundance of individuals) stay under the same rules of the species richness distribution in a habitat (Fig. 2.14), and high-level taxa should contain a few very abundant and represented species and many rare ones (Hubbel 2001). Today, the identification of biodiversity hot spots will be focused on updated data on macroalgae and seagrasses, selected invertebrate *taxa*, and seabirds. Only larger geographic areas will also consider pelagic life forms, in terms of large fishes and mammals, which facilitate monitoring because they have a longer life duration and a larger body size, making them easily identifiable.

European laws protect a category of marine species, the **seabird species** (with the exception of some large gulls), due to their small or declining populations or the small number of breeding sites. Nine species are included in Annex II of the EU list of endangered or threatened species. The **Balearic shearwater** (Fig. 3.7) is critically endangered, and the monitored colonies of Cory's and Mediterranean shearwaters are slowly declining. Although information is incomplete for macroalgae and invertebrates, a total of 11 macroalgae, 3 flowering plants, 9 Porifera, 3 Cnidaria, 17 Mollusca, 2 Crustacea, and 3 Echinodermata are now listed as endangered or threatened in Annex II of the Barcelona Convention for the Protection of the Marine Environment and the Coastal Region of the Mediterranean (1995), with no one holo-pelagic species included in the list.

The identification of endangered species as "benthic" does not mean the absence of pelagic species from EU attention. Although holo-pelagic species are not considered, we cannot forget that benthic species need a pelagic stage to complete their life cycle or, in any case, use the pelagos to send propagules around to expand their geographic distribution.

An attempt to integrate the spatial data and temporal trends visualizes macroecological patterns at the Mediterranean scale. These results depict a **situation of high diversity and heterogeneity** for the Mediterranean Sea, but they also evidence the need for further study of geographical areas that are largely unexplored, mainly along the southern border of the basin, along the North African coasts, and together with the deep sea.

The Mediterranean Sea is a complex biogeographic province where ecology and human activities meet and strongly interact, representing a large and growing potential impact on marine biodiversity.

Although much is known about individual threats, knowledge is very limited about how multiple impacts will interact. Therefore, there is a need to develop a comprehensive analysis of conservation and management initiatives to preserve Mediterranean biodiversity.

3.5 Pelagic Organisms of the Mediterranean Sea

Regarding the only pelagic world, all studies to date concur in identifying members of the SAR11 group as the most abundant Mediterranean Bacteria, followed in importance by other **Alphaproteobacteria** more common in coastal regions and during algal blooms. **Cyanobacteria** (***Prochlorococcus*** and ***Synechococcus***, Fig. 3.11) form the rest of the Prokaryota diversity, together with many cultivable (**Alteromonadales**) and uncultivable Gammaproteobacteria and Bacteroidetes, with some differences with depth and distance from emerged continents.

The deep Mediterranean hosts several extremely peculiar and interesting ecosystems, such as the deep **hypersaline anoxic "lakes"** on the bottom of the Ionian Sea, which are reported to include several new and little-known microbial lineages.

Although not well studied, many prokaryotic archaic organisms thrive in submerged volcano injections of water rich in reduced molecules. Such a set of reduced

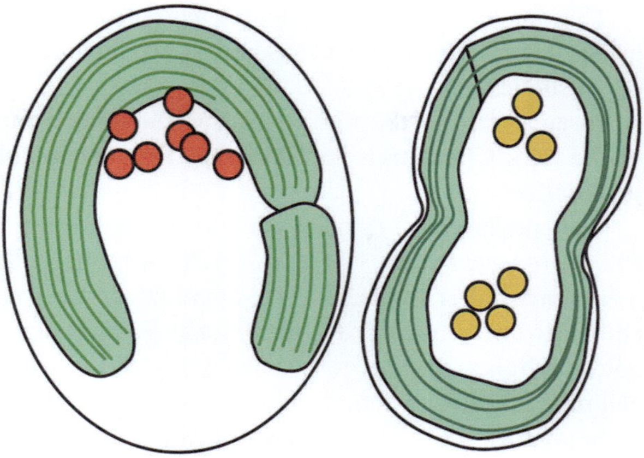

Fig. 3.11 Sketches for a schematic representation of: left, *Prochlorococcus*, and right, *Synechococcus*. The two Prokaryota organisms are very simple. The most evident character of their cell is the chloroplast (green). The DNA is organized in a chromosome dispersed in the cytoplasm (original)

molecules (mainly sulfuric) sustains complex bacterial assemblages, where energy coming from chemical oxidation can be used by Prokaryota in the synthesis (**chemiosynthesis**) of organic molecules supplying trophic chains, not based on the use of sunlight energy (Cavanaugh et al. 2006). Thermal fields (where warm, anoxic, and sulfuric waters sustain chemiosynthesis) are generally deep and close to middle oceanic ridges where volcanic activity is intense and continuous. In the Mediterranean, such a situation of middle-ocean ridge continuous volcanic activity is present together with volcanism related to subsidence arcs (explosive and infrequent activities), and the scientific community has started to explore the existence of life in such very ancient conditions.

Marine bacterial diversity peaks in tropical latitudes, and at Mediterranean latitudes, the number of detectable taxa, in terms of DNA characteristics (the "operational taxonomic units", **OTUs**), is between **100 and 360** for every **surface water sample.**

Archaea diversity is known to be lower, but new technique data from the northwestern Mediterranean indicate that the numbers should be raised to approximately **1000 "bacterial species" per water sample**.

In the Mediterranean Sea, excluding the volcanic spots, the bacterial concentration is quite stable, and their production is low.

This general rule must be considered in the context of a certain variability: (a) the existence of **a west–east gradient** of decreasing **bacterial production** and (b) the existence of enhanced metabolic activities and production related to specific hydrologic discontinuities, such as currents, eddies, and frontal areas. Following the general pattern of east-ward increasing oligotrophy, bacterial production is several times lower in the southeast than in the northwestern basin.

As in most oligotrophic and subtropical oceanographic regions, low biomass values in the Mediterranean Sea are generally associated with the dominance of Cyanobacteria, prochlorophytes, and tiny flagellates.

This smallest fraction of phytoplankton (**picoplankton**) can only be quantified using special techniques and has largely been ignored in classical light microscopy-based studies.

As an average of the whole basin, Mediterranean picoplankton account for **59% of the total Chl *a* presence, and they are responsible for 65% of the primary production**.

The prokaryotic fraction of the picoplankton, e.g., *Synechococcus* and *Prochlorococcus* (Fig. 3.11), can reach abundance values of up to 10^4 cells mL^{-1}. *Synechococcus* is dominant in the upper layers during stratification periods (different water layers overlap due to seasonal reasons), and despite the pronounced oligotrophy, it is apparently responsible for maximum photosynthetic efficiency values and probably able to cope with low nutrient conditions.

In contrast, as in other oceans, prochlorophytes are most often found in deeper layers in stratified conditions (summer and early autumn), with a sharp peak near the inferior limit of the euphotic zone (approximately 200 m), being abundant at the surface during the non-stratified period (winter—early spring).

However, two distinct ecotypes of *Prochlorococcus* exist, which show preferences for high- and low-light conditions. Both types, substituting each other along the water column and accordingly adjusting the light intensity existing at different depths, have been identified, for example, in the Strait of Sicily.

Chemio = chemical; *synthesis* = creation

In addition to Prokaryota, quite a high diversity of **Eukaryota** may be found within the picoplanktonic fraction (**Prasinophyta**, **Pelagophyta**, **Prymnesiophyta**, and **Chrysophyta**; Fig. 3.12).

Based on epifluorescence counts, tiny (<3 µm) autotrophic and heterotrophic organisms were dominant (in the order of 10^3–10^4 cells mL^{-1}, approximately 75% of the <10 µm size fraction) across the Mediterranean Sea in June. Several non-colonial picoplanktonic **Bacillariophyta** (e.g., *Chaetoceros*, *Thalassiosira*, and *Skeletonema* species, Fig. 3.13, atlas Plates 1–4) have also been found to be abundant in some cases, although their small size may prevent their identification even at the class level.

Ostreococcus is the smallest eukaryote known to science (first described from a coastal lagoon of Mediterranean France). It is a cell with only one chloroplast, one mitochondrion, and one sac for the Golgi apparatus. Chrysophyta is a taxon also containing diatoms (Bacillariophyta).

The small **Eukaryota** of the plankton (the **nanoplankton**) are photosynthetic, heterotrophic, or mixotrophic organisms with **low morphological variability**. Thus, we must rely on molecular techniques to unveil their diversity. Molecular work has allowed the discovery of new groups of eukaryotes present in such a small class.

The size components smaller than 20 µm, commonly defined as **nanoplankton**, are mainly composed of small flagellates, mostly naked, in association with

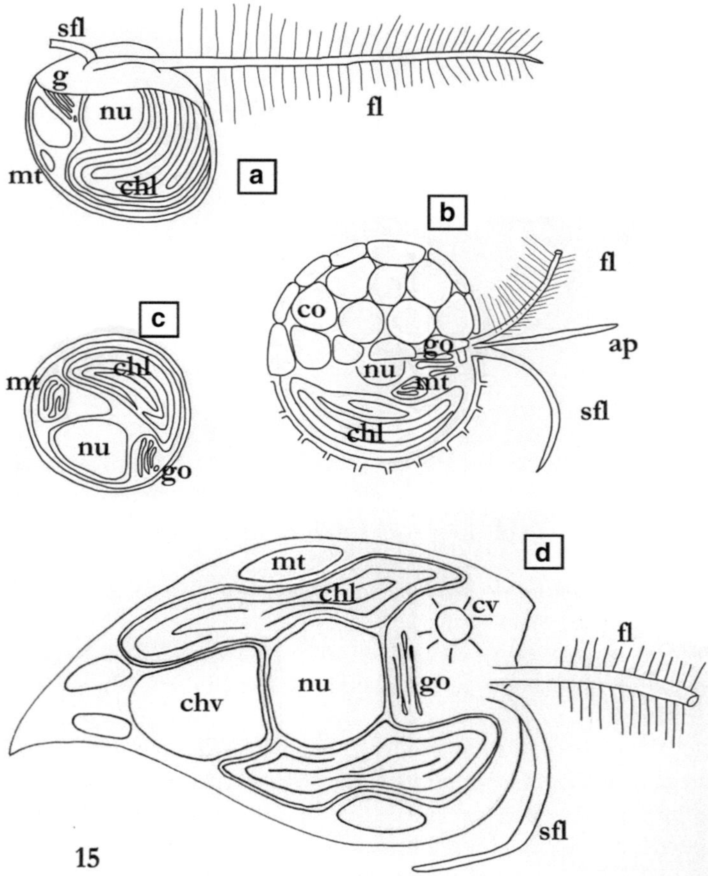

Fig. 3.12 Schematic representation of picoplankton. (**a**) Chrysophyta (*Pelagomonas*), (**b**) Prymnesiophyta (*Emiliania*), (**c**) Prasinophyta (*Ostreococcus*), (**d**) Scheme of Chrysophyta. chl, chloroplast; chv, chrisolaminarin vesicle; co, coccolites; cv, contractile vacuole; fl, long flagellum; go, Golgi apparatus; mt, mitochondrion; nu, nucleus; sfl, short flagellum. Scale bar indicates μm (original)

encapsulated forms such as **Coccolithophorida** (Prymnesiophyta) and with a limited number of small solitary diatom species (**Bacillariophyta**).

Nanoflagellates (Dinophyta of small size) are the dominant group, in terms of cell numbers, most of the year in open Mediterranean waters.

Similar to the pico-eukaryotic fraction, a longstanding lack of taxonomic resolution for these heterogeneous nanoplankton species has led to the view that they do not vary significantly in quality and quantity over time and space, probably because they are controlled by equally fast-growing predators. However, there are several indications that nanoflagellates vary in space and time and may also contribute significantly to blooms.

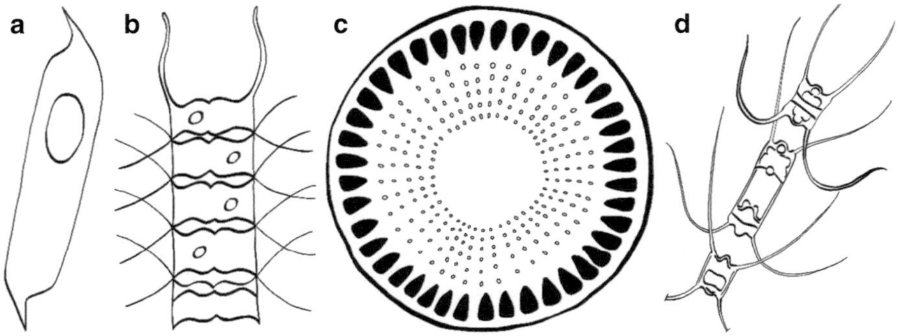

Fig. 3.13 Bacillariophyta diversity. (**a**) *Proboscia*, (**b**) *Skeletonema*, (**c**) *Thalassiosira*, (**d**) *Chaetoceros* (original)

Based on pigment presence, **Prymnesiophyta** represent the largest fraction of nanoflagellates most of the year and at several localities. Among them, **Coccolithophorida** deserve special mention, as they show high diversity in the Mediterranean.

The widespread ***Emiliania huxleyi*** is generally the most frequent and dominant species, although it does not form spectacular blooms.

Coccolithophorida have been found to constitute a large part of the nanoplankton in autumn and winter in many parts of the Mediterranean Sea, with a preference for the Southeast basin.

Much less known is the diversity and distribution of noncalcifying Prymnesiophyta (e.g., ***Imantonia*** and ***Chrysochromulina*** species), which can only be identified with special techniques (electron microscopy or molecular analyses). Among the few species mentioned in this group are those of the genus ***Phaeocystis***, which may form large, recognizable colonies but are not easily detected when they are in the flagellate solitary stage (Fig. 3.14).

Cryptophyta, often only detected by their marker pigment **alloxanthin**, are generally more abundant in the same season as Bacillariophyta, e.g., in winter and spring. ***Plagioselmis prolonga*** is one of the most frequently mentioned species in this taxon. However, the species list of Cryptophyta obtained with light microscopy investigations should all be reconsidered since most species can only be recognized in live samples or using electron microscopy.

For small flagellates, Cryptophyta mainly include naked autotrophic and heterotrophic species that are poorly known and are not identifiable under light microscopy.

In addition, their pigment track may overlap with that of other flagellate groups.

In the southeast basin, **Dinophyta** were reported to be dominant in different seasons and especially in conditions of water stratification. Some small thecate species, such as those of the genera ***Prorocentrum***, ***Heterocapsa***, or ***Scrippsiella*** (Fig. 3.15, atlas Plates 5 and 6), are also part of the nanoplankton, but they are generally not abundant in offshore waters.

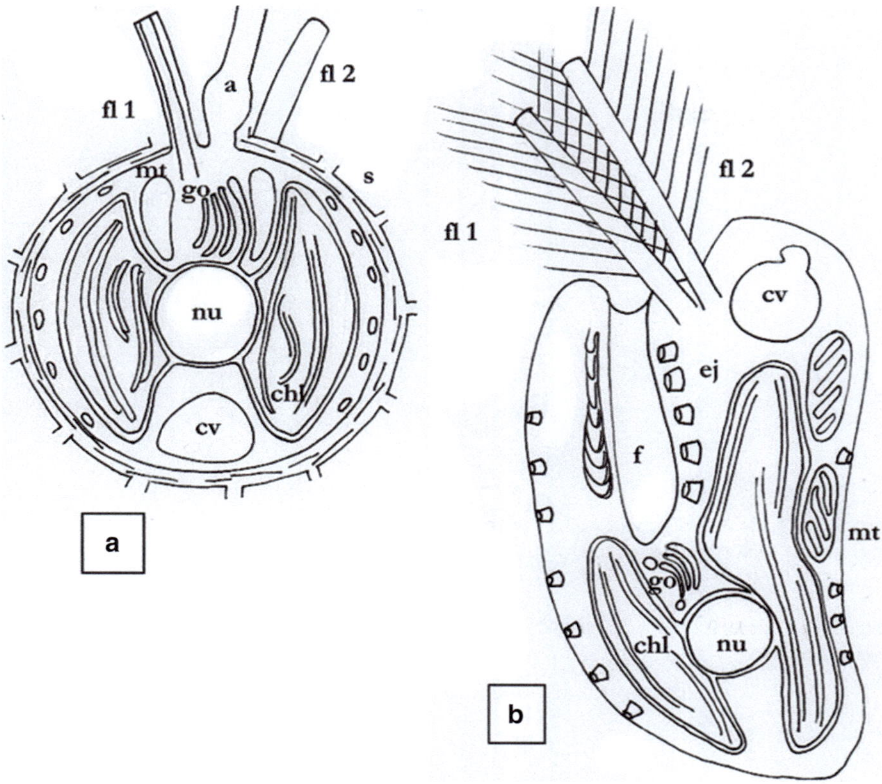

Fig. 3.14 Schemes of (**a**) Prymnesiophyta (*Chrysochromulina*) and (**b**) Cryptophyta (*Plagioselmis*). a, haptonema; chl, chloroplast; cv, contractile vacuole; ej, ejectosome; fl 1 and fl 2, flagella; go, Golgi apparatus; mt, mitochondrion; nu, nucleus. In Prymnesiophyta (a), the two flagella are of different lengths, and a haptonema is present as an additional filament. In Cryptophyta (b), the two flagella are of similar length (original)

Alpha = first letter of the Greek alphabet *Altero* = tall and fierce *Chrysos* = gold *Cyano* = blue *Eu* = good; *Karyon* = nucleus *Monade* = fundamental unit *Pelagos* = open sea *Pico* = point; small quantity *Prasino* (L) = green *Pro* = primitive; *Chloro* = green; *Coccus* = round body *Proteo* = Greek God of transformation *Prymnesio* = cable for mooring *Synecho* = hold together, compress

The contribution of picoplankton and nanoplankton decreases along with the increase in Chl *a* concentration. Where this occurs, colonial and microplanktonic Bacillariophyta belonging to several genera (***Asterionellopsis***, ***Chaetoceros***, ***Pseudonitzschia***, ***Thalassionema***, and ***Thalassiosira*** Figs. 3.13 and 3.16) become more important.

An increase in the abundance of Bacillariophyta is evident at many sites in the Mediterranean Sea during the winter (February–March). However, these events are ephemeral and, hence, not always detected in offshore waters. Notably, massive sedimentation events associated with Bacillariophyta blooms during the winter

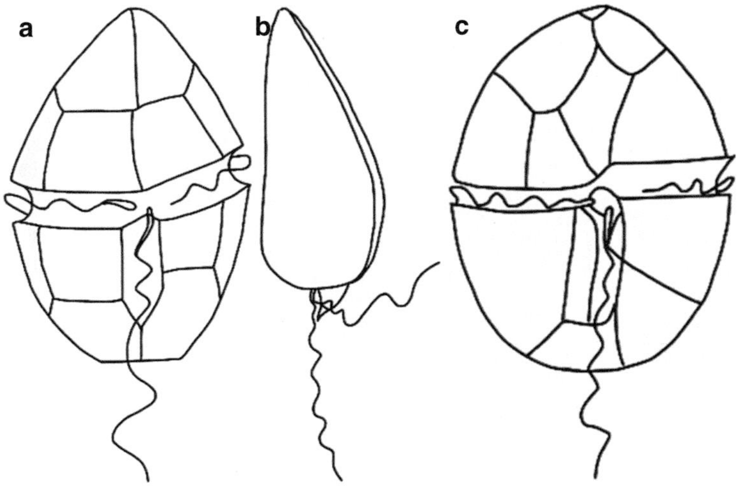

Fig. 3.15 Sketches of Dinophyta. (**a**) *Heterocapsa*, (**b**) *Prorocentrum*, (**c**) *Scrippsiella*. The two flagella are orthogonally oriented (original)

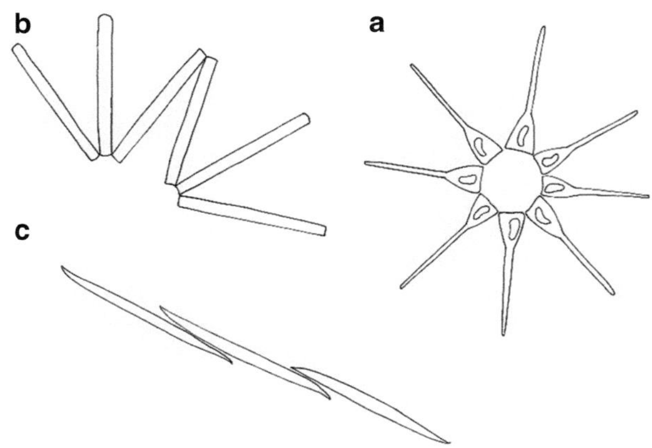

Fig. 3.16 Colonial associations in Bacillariophyta. (**a**) *Asterionellopsis*, (**b**) *Pseudonitzschia*, (**c**) *Thalassionema* (original)

suggest that these phytoplankton are scarcely exploited by zooplankton. In contrast, their sinking represents a resource for the zoobenthos of the underlying bottom.

In addition to the coastal winter bloom, Bacillariophyta also dominate, and for longer periods, in deep convection areas. Colonial species belonging to the genera *Pseudonitzschia* and *Chaetoceros* dominate during spring in the deep convection areas of the North Balearic Sea, and in the Otranto Channel, high concentrations of healthy *Chaetoceros* were found as deep as 500 m.

Colonial, bloom-forming Bacillariophyta belonging to the genera *Chaetoceros*, *Thalassiosira*, ***Proboscia***, ***Rhizosolenia***, and ***Leptocylindrus*** are generally the main contributors to high Chl *a* patches in waterfront areas and pelagic gyres. These patch structures have been defined as the "oases" of the Mediterranean desert (in terms of productivity). The biological phenomena that they drive strictly depend on water mass dynamics and hence are spatially heterogeneous and show a very high temporal dynamic, as well as marked interannual variability.

Bacillariophyta-dominated Chl *a* peaks are often found in subsurface waters, as in the case of a monospecific bloom of a *Thalassiosira* forming gelatinous colonies (10^7 cells L^{-1} and 23 µg Chl *a* L^{-1}), which was detected at 54 m depth in the Almeria-Oran front area (West Mediterranean).

Bacillariophyta show intermittent and probably mostly underestimated abundance in deep waters. This variability might explain the mismatch between the relatively few reports of Bacillariophyta in phytoplankton samples and the high amount of biogenic silica (mostly of Bacillariophyta origin) found in surface sediments and sediment traps.

Bacillariophyta and other phytoplankton species seem to distribute themselves in distinct layers of the water column, suggesting the existence of a tightly structured system, within which the distinct phytoplankton components may have different ecological roles.

Colonial *Chaetoceros* species are a rather constant feature of diatom-dominated plankton, but the accompanying assemblages seem to vary from area to area. For example, ***Rhizosolenia*** and *Thalassiosira* were abundant genera in the Catalan-Balearic Sea, while *Leptocylindrus danicus*, *Pseudonitzschia delicatissima*, and ***Thalassionema*** *nitzschioides* were abundant in the Southern Adriatic Sea. To the east, *Bacteriastrum*, ***Hemiaulus***, and *Thalassionema* were found south of Crete in July, whereas *P. delicatissima*, ***Dactylosolen*** *fragilissimus*, and *Thalassionema frauenfeldii* were found north of Crete in June. While the above-mentioned colonial species often appear in relatively high concentrations, other large Bacillariophyta are found at much lower concentrations in offshore waters. In fact, these "large" Bacillariophyta have been reported as responsible for a substantial but underestimated fraction of primary production in oligotrophic waters characterized by a strong seasonal thermocline and nutricline.

However, they have a patchy or sparse distribution and are generally not sampled properly.

Asterios = star shaped *Chaeto* = bristle; *ceros* = horn *Cocco* = round body; *lithos* = stone; *phoro* = bearing *Chrysos* = golden; *Chromo* = colour *Dactylo* = finger; *Solen* = razor, knife *Dino* = moving (dynamic) *Imantonia* = dedicated to Irene Manton *Hemi* = half, semi; *Haulus* = tibia *Heteros* = different; *Capse* (L) = case, capsule *Leptos* = delicate; *Cylindrus* = rolled, wallowed *Nano* = small, dwarf *Phaeo* = brown; *Cystis* = capsule *Plagio* = not clear, not linear; *Selmis* = with spiny leaves *Proboscia* (L, G) = Elephas' trunk; device for collecting food *Proro* (L) = prow; *Centrum* (L) = centre *Pseudo* = false; *Nitzschia* = dedicated to Nitzsch *Rhizo* = rooth; *Solen* = razor *Scripps* = dedicated to Scripps laboratories *Thalassos* = one of the Greek names of the sea; *nema* = reel, rope; *sira* = brilliant

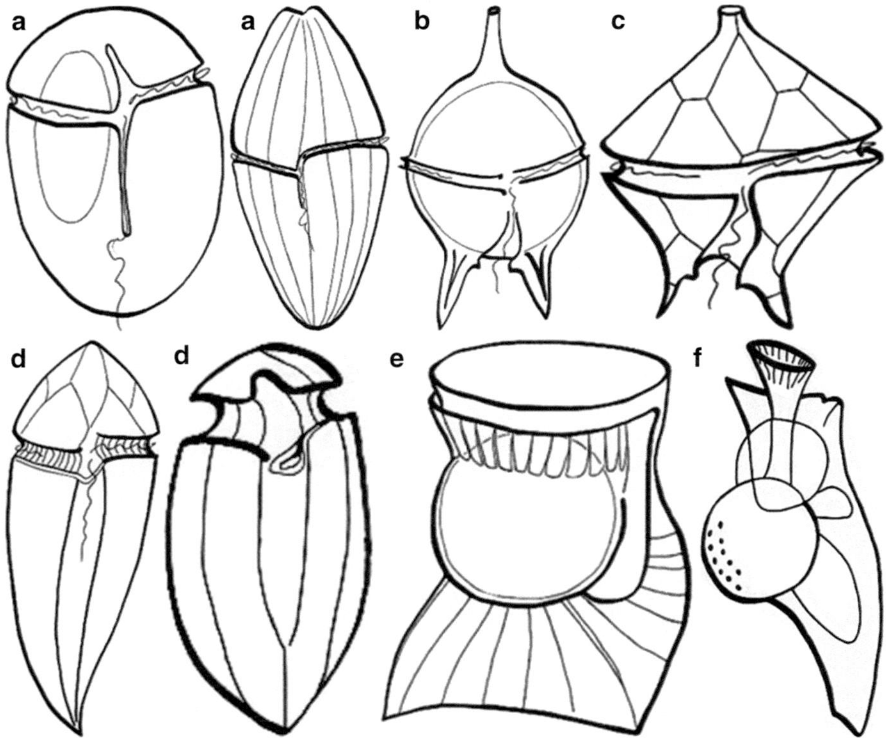

Fig. 3.17 Sketches of facultatively or obliged heterotrophs among Dynophyta. (**a**) *Gymnodinium*; (**b**) *Peridinium*; (**c**) *Protoperidinium*; (**d**) *Oxytoxum;* (**e**) *Ornithocercus*; (**f**) *Histioneis* (original)

Among them are some of the large *Rhizosolenia* species, which may form migrating mats in oceans and, along with species of the genus *Hemiaulus*, may host the **diazotrophic** cyanobacteria ***Richelia intracellularis***, thus playing a role in nitrogen fluxes in pelagic ecosystems. With the exception of the high productivity events mentioned above, Dinophyta are generally more abundant than Bacillariophyta in the cell size fraction >20 μm. The species most commonly reported belong to the genera ***Gymnodinium***, ***Gyrodinium***, ***Neoceratium***, ***Protoperidinium***, and ***Oxytoxum*** and are generally associated with warm and stratified waters. Rarely, the percentage contribution of single microplanktonic Dinophyta is high (***Oxytoxum*** spp. reaches 12% of the total cell counts in the Alboran Sea). Indeed, similar to Bacillariophyta, larger Dinophyta are not present in the samples most of the time, but their role can be significant despite their low abundances. Some of them, e.g., ***Ornithocercus***, ***Histioneis***, and ***Citharistes*** (Fig. 3.17), can host endosymbiotic Cyanobacteria that allow them to survive even under conditions of severe Nitrogen limitation.

Box 3.2 N_2 Fixation

Marine **primary production** is often limited by **Nitrogen** (N). In the open ocean, the biological reduction of di-Nitrogen gas (N_2) to bioavailable N (a process called **biological N2 fixation, BNF**) is a critical source of new N for the ecosystem and controls the uptake and sequestration of CO_2 over long periods.

In the upper ocean, under sunlight, it was traditionally thought that BNF was largely restricted to the subtropical and tropical gyres and mediated by a few groups of larger-sized N_2-fixing (**diazotrophic**) Cyanobacteria and **holobionts** (>10 μm): the colony-forming *Trichodesmium* spp. (filamentous cyanobacteria) and cyanobacterial symbionts (*Richelia intracellularis* and *Calothrix rhizosoleniae*) of Bacillariophyta (**diatom–diazotroph associations; DDAs**). More recently, solitary Cyanobacteria have been detected outside tropical belts.

Some solitary (unicellular) Cyanobacteria, **UCYN-A**, live in symbiosis with small single-celled eukaryotes (Haptophyta) (Fig. 3.18)

Other **UCYNs** (**B**) live singly, colonially, or in symbioses with the large chain-forming Bacillariophyta *Climacodium frauenfeldianum*.

Finally, noncyanobacterial diazotrophs (NCDs) also exist, including Archaea and Bacteria lineages.

The first global ocean database of diazotrophs regarded those that can be identified by microscopy (*Trichodesmium*, *Richelia*, and *Calothrix* (Suikanen et al. 2021)).

Several oceanographic regions displayed high abundances (hotspots) of diazotrophs that had not been previously sampled. For example, in the Mozambique Channel between Madagascar and Africa, diazotrophs constituted up to 30–40% of the large bacterioplankton.

Many regions contain only a small abundance of diazotrophs. For example, the percentages of diazotrophs in the Atlantic Ocean, the Southern Ocean, and the Mediterranean Sea reach maximum values of only 0.4%, 1%, and 4% of the whole bacterioplankton, respectively.

The distribution of diazotrophs appears to depict a patchy biogeographical pattern.

Cyanobacterial diazotrophs are mainly found as assemblies of abundant groups.

With a few exceptions where ***Trichodesmium*** was the main component, there was a trend of co-occurrence for several cyanobacterial diazotrophs. For example, in the Red Sea, both larger (*Trichodesmium*, *Richelia*) and smaller (***Crocosphaera***) diazotrophs were detected. In contrast, only small diazotrophs codominated in the South Atlantic Ocean.

Ultrasmall diazotrophs consist of Proteobacteria and **are abundant in the Arctic Ocean**. Ultrasmall prokaryotes are unusual due to their reduced cell volume (these cells can pass through 0.22 μm filters, a size usually expected to exclude most microorganisms), and thus, they are thought to have reduced genomes and to lack the proteins needed to carry out more complex metabolic processes (Fig. 3.10).

However, there is recent evidence that they do indeed participate in complex metabolism.

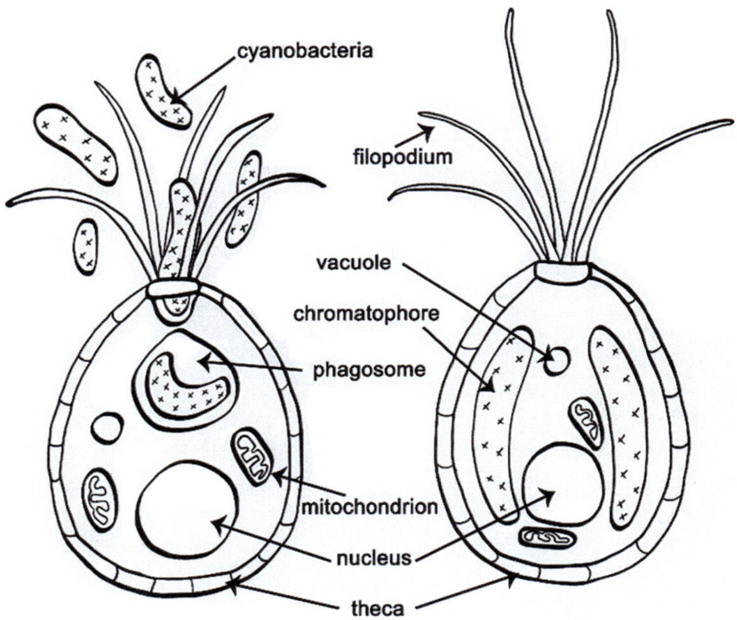

Fig. 3.18 Schematic representation of unicellular organisms establishing a tight association (endosymbiosis) with a cyanobacterium (original)

Species of the widespread genus ***Neoceratium*** may be mixotrophic. They distribute themselves in separate layers of the water column.

Finally, *Protoperidinium* spp. and several **athecate** Dinophyta in the genera *Gymnodinium*, *Gyrodinium,* and ***Lessardia*** are completely **phagotrophic** and may be considered part of the microzooplankton, but their importance in the offshore Mediterranean Sea has rarely been assessed.

A = (privative); *Techa* = box *Ceratos* = horn *Chitaris* = music instrument (citara) *Chloro* = green; *Phytos* = plant *Di* = two; *Azo* = obsolete reference to Azote (Nitrogen), it means "without life"; *Trophe* = to eat *Di* = two; *Stephanos* = crown *Dictyo* = network *Gymnos* = naked; *Dinium* = power *Gyros* = turning *Halo* = saline; *Sphaera* = sphere, ball *Histio* = web, tissue *Lessardia* = dedicated to E. Lessard *Neo* = new; *Ornitos* = bird; *Cercus* = appendage *Oxy* = containing Oxygen; *Toxum* = seed *Phago* = swallow; *Trophe* = to eat *Prasino* = light green; *Phytos* = plant *Proto* = primitive; *Peri* = around; *Dinium* = energetic power *Richelia* = energy reserve, helper *Tintinnulum* (L) = small bell; *Trichos* = hair; *Desmos* = link *Viridis* (L) = green

Among other unicellular groups, **Silicoflagellata *Dictyocha*** and ***Distephanus*** are also constant, although scarce, components of offshore Mediterranean microplankton, and their abundance reaches the highest values in surface waters in winter and in deeper waters in spring and summer.

In addition, a few other flagellates that can form large colonies are also part of the offshore microplankton, at least in some phases of their life cycle. One of these is the key species, ***Phaeocystis*** cf. *globosa*, which can form spherical colonies reaching some millimetres in diameter.

Another interesting species is ***Halosphaera viridis*** (**Prasinophyta** or **Chlorophyta**), which is found up to depths of 100 m in autumn–winter but then rises to a closer distance from the surface in spring.

Such extensive vertical migrations could account for considerable upward recycling of Carbon and nutrients. Unfortunately, as in the case of large Dinophyta and Bacillariophyta, there are few data on the distribution of these interesting microplankton *taxa* in offshore waters.

Each phytoplankton organism needs sunlight to perform photosynthesis; thus, its presence will be more certain in the first 100 m than in the deepest layers.

The light intensity and wavelength useful for phytoplankton growth and reproduction are not those at the actual sea surface, where the maximum sunlight energy is available, but those at a certain depth (in the South Adriatic, between **40 and 80 m**), where only the green and blue wavelengths penetrate. This layer is called the **Depth Chlorophyll Maximum** (**DCM**).

Regarding the heterotrophic components of plankton, more importance has been given in the last 50 years to heterotrophic microorganisms and their role in recycling organic matter (the so-called **microbial loop** (Azam et al. 1983)). In the Mediterranean Sea, strong primary production based on photo-autotrophy is not common, and the microbial loop is an important alternative source of organic molecules for the whole trophic web. Many Prokaryota and unicellular Eukaryota are normally mixotrophic, showing the possibility of shifting from auto to heterotrophy if necessary. In unicellular Eukaryota, such a possibility is allowed by the presence of autotrophic Prokaryota inside the cells (**endosymbiosis**).

Box 3.3 Symbiosis

The evolution has favoured associations of individuals when such a cohesive behaviour gave advantages in energy capture and its management (to stay alive and reproduce their own genes). This was probably the birth story of Eukaryota (supposed to be the result of symbiosis between large and small primordial cells). The same driving force probably acted on the association of individuals of the same species, thus originating multicellularity and coloniality.

The alliance between very different organisms is justified by the necessity of completing the biochemical budget of each organism to reach better life conditions.

Each species shares its life with many other species in each ecological community. The types of interaction, for this reason, are numerous and variable with time. Mutualistic symbiosis, however, is an obligatory collaboration that does not allow single collaborators to stay alone and survive. Practically no species exists without the presence of "collaborators", which resolve some metabolic problems. The tightest symbiosis involves organisms from different realms, where single participants are still recognizable but not able to conduct an independent life. In the marine environment, the most famous symbiosis is probably that of zooxantellae (Dinophyta) inside the polips of reef-building corals (Cnidaria). Zooxantellae *Symbiodinium* provide polips with products derived from **photosynthesis** (carbohydrates, glycerin, and amino acids). In addition, they remove excreted material. Corals, in addition, enhance their calcification, which is essential to coral skeletal growth. This symbiosis involves two eukaryotes (one unicellular and one multicellular), and the genome of the symbiont is three times larger than the genome of its host. Other Prokaryota and Viruses are also associated with corals and *Symbiodinium*, but their roles are mostly uncharacterized.

Another case of mutualistic symbiosis regards almost exclusively pelagic organisms and is **bioluminescence**, the product of a symbiosis with Bacteria (at least 11 species, mostly belonging to the genus *Aliivibrio*), hosted in defined body districts or organs (of at least 76 animal species), and used to produce light.

The term **holobiont** (Margulis 1990) was coined to indicate that evolution has worked mainly through symbiosis-driven leaps that merged organisms into new forms ("holobionts"). Rohwer et al. (2002) used the word holobiont to describe a unit of selection *sensu* Margulisfor corals (polyp + zooxanthellae + various ectosymbionts + viruses).

Aquatic habitats favour holobiont affirmation because higher chemical connectivity and signaling favour present between macro- and microorganisms and because the phylogenetic diversity at broad taxonomic scales is higher compared to land.

Ciliophora are unicellular Eukaryota able to perform autotrophy when equipped with internal autotrophic prokaryotes. Patterns of taxonomic diversity in the Mediterranean Sea have been investigated, with particular attention on Ciliophora **Tintinnina** (Fig. 2.20, atlas Plate 8). Along a West–East Mediterranean transect sampled at the start of summer (June), the concentration of such Ciliophora varied little, but the number of **species and genera as well as their diversity indexes increased eastwards** (as for Copepoda, and contrary to the normal trend of diversity diminution along the same geographic axis).

The diversity indexes of heterotrophs were positively correlated with the DCM depths of the sampled area but negatively correlated with the Chl *a* concentration. The west–east variation in Tintinnina diversity was parallel to shifts in the Chl *a* size-diversity estimate.

The body size of Ciliophora sometimes overlaps with that of smaller multicellular zooplankton (Metazoa) and is generally included in the category of meso-zooplankton. The meso-zooplankton distribution in Mediterranean epipelagos is characterized by a generalized scarcity, with high values only in a few areas. Total abundance and biomass values are comparable to those measured at the same latitudes in the North-East Atlantic. A west-to-east decrease in population abundance emerged from the surveys across the basin, with a mean value at midday in the 0–200 m layer double in the west than in the east basin of the Mediterranean.

Zooplankton distribution patterns show high local variability, with notable spatial variation even during the same season. Additionally, the vertical distribution is not homogeneous in the first 200 m below sea level.

Sampling with finer mesh nets (<200 μm) or with large bottles revealed that microplankton biomass and abundance can increase by 2 to 20-fold when smaller Metazoa (50–200 μm) are considered in addition to proto-zooplankton.

In the open Mediterranean Sea, the bulk of epipelagic meso-zooplankton is concentrated in the upper 100 m layer and sharply decreases below this depth. In this upper layer, meso-zooplankton play a major role in biological processes based on their trophic connection with phyto- and micro-zooplankton in the euphotic zone. During the daytime, the progressive decrease in meso-zooplankton abundance with depth is interrupted by a small-scale increase in the level of the maximum Chl *a* concentration (DCM), where the highest abundance of Copepoda larvae (nauplii) and juveniles (copepodids) is reported.

Similar to the world ocean, the Mediterranean epipelagos are enriched during the night by the **diel migrants** that ascend from the mesopelagos below.

Calo = from Irish caelan, slender *Centropages* = centre-hill *Clauso* = from Claus, an eminent Austrian scientist *Coryfeus* = summit, leader *Cteno* = comb *Euterpe* = Greek muse of music *Farran* = dedicated to Farran *Halo* = saline *Micro* = small *Oithona* = Celtic virgin of the wave *Oncaea*= hooked *Sub* = inferior; *Eu* = good *Temora* (Hebrew) = tall

In the Catalan Sea (western Mediterranean), almost half of the zooplankton standing stock residing at the DCM at depths of 50–90 m during the day moves to the 0–50 m layer during the night in summer.

Over the annual cycle, meso-zooplankton abundance in offshore waters oscillates within a narrower range and reveals lower seasonal variability than in coastal waters.

Peaks of abundance occur between the end of winter and the beginning of spring in the western Mediterranean.

Notwithstanding differences in amplitude, the timing of the zooplankton annual cycle along coastal-offshore gradients is synchronous. A time series conducted on a monthly basis between 1994 and 1999 off Mallorca (Spain) well depicted the meso-zooplankton composition and dynamics in the open Mediterranean Sea (Fernández de Puelles et al. 2003).

This effort highlighted the influence of wide-scale climatic factors (e.g., the North Atlantic Oscillation), proposing a warming climate (e.g., a higher water temperature) as responsible for a flexion in zooplankton abundance.

Epipelagic meso-zooplankton assemblages in the open Mediterranean Sea are highly diversified in terms of taxonomic composition, and Copepoda is the most represented taxon in terms of species richness, abundance, and biomass. The dominance of small Copepoda species (mostly <1 mm in total length of adults) in terms of both numbers and biomass represents the major feature of the structure of meso-zooplankton at a basin scale.

Plankton individuals of 0.5–1 mm size contribute 45–58% to the total meso-zooplankton abundance in the open East Mediterranean Sea, even if the collection is executed with "coarse" mesh nets (>300 μm). The importance of such small Copepoda has also been highlighted in other Mediterranean coastal waters.

Among hundreds of species, a few small-sized and species-rich genera of **Calanoida** (e.g., ***Acartia***, ***Paracalanus***, ***Clausocalanus***, and ***Calocalanus***, also comprising ***Ctenocalanus vanus***) and **Cyclopoida** (***Oithona***, ***Oncaea***, and ***Corycaeus***) account for the bulk of copepod abundance and biomass in the epipelagic layers of the Mediterranean Sea.

The contribution of *Oithona* and *Oncaea* to species abundance and diversity becomes extremely significant in samples collected by fine mesh nets (200 μm), where they compete for numerical dominance with Harpacticoida ***Microsetella*** and ***Euterpina*** (Fig. 3.19, atlas Plates 20 and 21).

Several species of the above genera display distinct distribution patterns along the water column and/or over the seasons, suggesting differences in their ecological traits, as observed for other species of the Atlantic Ocean.

Although their populations largely overlap, the peaks of cogeneric *Clausocalanus paululus*, *C. pergens*, *C. arcuicornis* and *C. furcatus* succeed each other from winter to autumn in the open Tyrrhenian and Ionian Seas and in the Strait of Sicily.

In the Ionian and South Aegean Seas, the autumnal dominant *Clausocalanus furcatus* and *Oithona plumifera* replaced the smaller-sized *C. paululus* and *O. similis* in the spring.

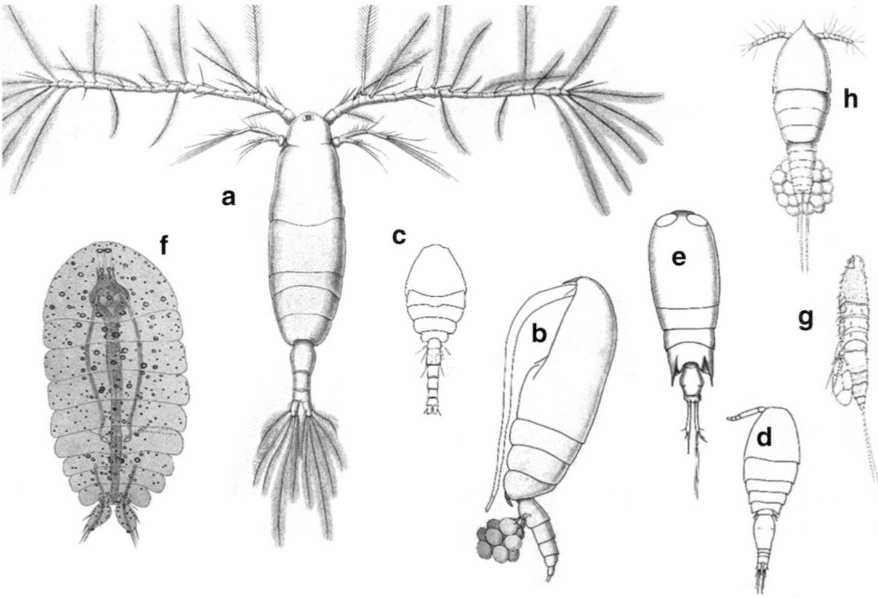

Fig. 3.19 Schematic representations of Calanoida (**a**, *Acartia*; **b**, *Clausocalanus*); Cyclopoida (**c**, *Oithona*; **d**, *Oncaea*; **e**, *Corycaeus*; **f**, *Sapphirina*), and Harpacticoida (**g**, *Microsetella*; **h**, *Euterpina*) typical of neritic zooplankton. (From Giesbrecht 1892; rearranged)

In the eastern Mediterranean and in autumn, *O. plumifera* is abundant in the 0–50 m layer, whereas *O. setigera* dominates in the 50–100 m layer.

Similar **abundance partitioning** is also perceivable in coastal waters for the species of the families Acartiidae and Paracalanidae.

West-to-east differences in the percentage contribution of some important species of the whole copepod assemblage might reflect differences in species biogeography, but they also indicate different structural and functional features of these systems. For example, ***Centropages typicus*** and ***Temora stylifera***, which are very common in neritic and coastal waters, are dominant species only on the northern side (Western Mediterranean, Adriatic Sea, and North Aegean Sea). In contrast, ***Calocalanus*** spp. (e.g., *C. pavo*, *C. pavoninus*), ***Haloptilus longicornis***, ***Oncaea*** spp. (e.g., *Oncaea "media"* group, *O. mediterranea)*, and **Corycaeidae** (e.g., ***Farranula rostrata***) heavily contribute to total copepod abundance in the southern part (practically all the Cenrtral/East-Mediterranean) (Fig. 3.20, atlas Plates 20 and 22).

The occurrence of large (>1 mm) Calanoida, such as ***Calanus helgolandicus,*** is much less important in the open Mediterranean than in the North Atlantic. This species mainly inhabits the intermediate and deep layers of the northern parts of the Mediterranean and ascends to epipelagic waters in late winter–spring. Its presence is considered extremely rare in the Eastern Mediterranean Sea.

Seasonal and vertical patterns similar to those of *C. helgolandicus* are reported for the large ***Subeucalanus monachus*** in the Alboran, Ionian, and Levantine Seas.

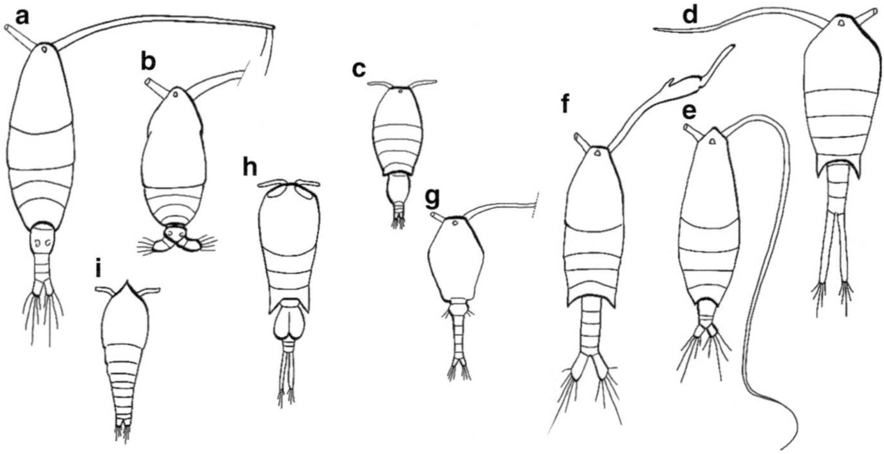

Fig. 3.20 Sketches of common genera of plankton Copepoda present in the Mediterranean Sea (with stress of characters useful for identification). (**a**) *Calanus* (long setae at the extremity of the antennula); (**b**) *Calocalanus* (divergent rami of the terminal furca); (**c**) *Oncaea* (short antennulae); (**d**) *Temora* (long rami of the terminal furca); (**e**) *Haloptilus* (extralong antennula); (**f**) *Centropages* (adult male, with right antennula modified for clasping the female); (**g**) *Oithona* (second part of the body armed with short appendages), (**h**) *Corycaeus* (two eyes); (**i**) *Euterpina* (pointed front) (original)

C. helgolandicus was found in highly dense patches in the frontal zone of the open Ligurian Sea in association with high phytoplankton concentrations. Here, the particular environmental conditions allow the proliferation of large Copepoda and Euphausuacea, which are at the bases of large Cetacea feeding.

Subeucalanus monachus was very abundant in the Rhodes Gyre during the spring of 1992, when an upwelling of waters rich in nutrients sustained a high phytoplankton biomass dominated by large Bacillariophyta. The distribution of the large Calanoida suggests that they are vicariant species that peak in separate areas of the Mediterranean.

Copepoda, reported as the strongest vertical migrants in the Mediterranean Sea, i.e., ***Pleuromamma gracilis***, *P. abdominalis*, and ***Euchaeta acuta***, enter the epipelagos only during their nocturnal ascent from deeper waters.

Whatever the species succession (seasonal or along a geographic gradient), what could be proposed as a rule is the passage from small to large species, even of the same genus, passing from neritic to open areas or from summer to winter seasons.

The other meso-zooplankton groups that contribute to community diversity and biomass in the open Mediterranean are less abundant than Copepoda but are locally and temporarily important (Siokou-Frangou et al. 2010). **Cladocera** are very abundant in epipelagic and coastal waters and expand their distribution beyond the continental slope only in narrow platform areas. Among Cladocera, ***Evadne spinifera*** contributes up to 6% to meso-zooplankton abundance (Fig. 3.6, atlas Plate 19).

Ostracoda, which are not numerous in the meso-zooplankton at temperate latitudes, present a remarkably consistent distribution in different Mediterranean sectors. Their contribution to total zooplankton numbers increases gradually with depth and varies from <2% in the upper 50 m to 11% in the 200–300 m layer. The highest Ostracoda abundance was recorded in the winter period in neritic waters, probably in relation to temperature conditions and/or the scarcity of potential predators.

Gelatinous zooplankton represent a heterogeneous and important group of organisms that play different and significant roles in pelagic communities, either as efficient filter-feeders or voracious predators. However, they are generally underestimated because standard sampling devices used for meso-zooplankton damage or destroy their fragile bodies and are therefore inappropriate for their quantitative estimation.

Pelagic filter-feeding **Tunicata**, especially **Thaliacea**, are known to occur periodically in dense swarms, sometimes with outbreaks lasting for days or weeks. **Doliolidae** and **Salpidae** together account for up to 9% of the total zooplankton in restricted sectors and/or seasons. **Appendicularia** (Fig. 3.21, atlas Plate 32) represent a more constant component in open-water zooplankton, and their high population growth rate under favourable conditions seems to depend on the selected

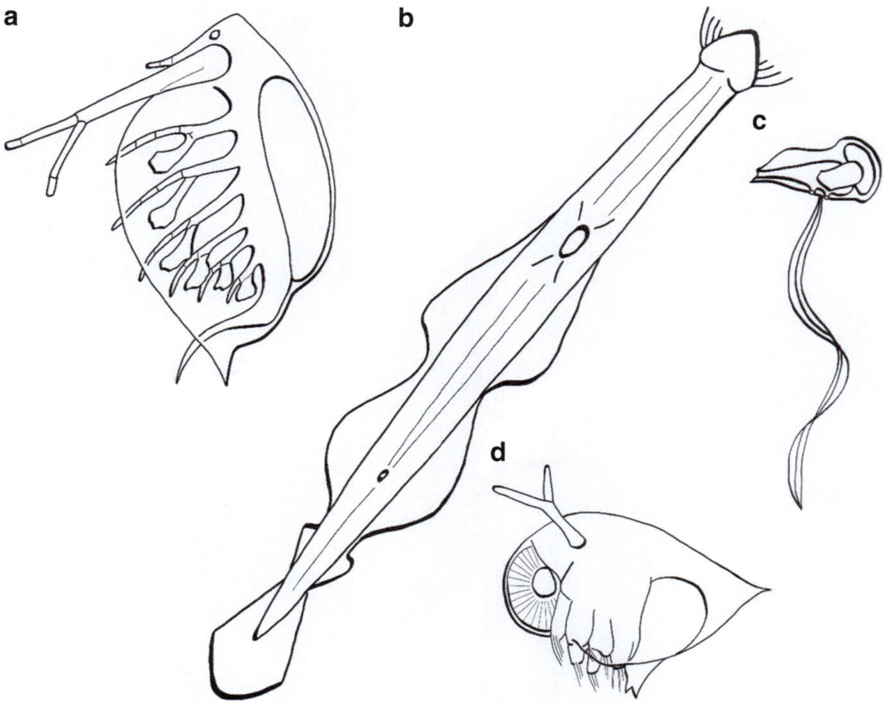

Fig. 3.21 Common non-Copepoda species in the zooplankton of the Mediterranean Sea. (**a**) *Penilia* (Crustacea Cladocera); *Sagitta* (Chaetognatha); (**c**) Appendicularia (Chordata Tunicata); (**d**) *Evadne* (Crustacea Cladocera) (original)

sampling area and time. They accounted for 1–23% of the total spring zooplankton abundance in some open sectors.

The contribution of Appendicularia to zooplankton abundance can be very high (38%) in the Ligurian Sea in winter, with a dominance of ***Fritillaria* spp**. Some Copepoda (family Oncaeidae) are adapted to the profit of Appendicularia products (the mucous nests), and their number appears noticeable in association with Appendicularia in the water column.

Among carnivorous zooplankton, **Chaetognatha** are more abundant than **Siphonophora (Cnidaria)**, probably because they are specifically adapted to prey on Copepoda. In both the West and East Mediterranean during autumn, the most abundant Chaetognatha (Fig. 3.21, atlas Plate 28) are species belonging to the genus ***Sagitta***. During the same season, Siphonophora ***Muggiaea***, ***Abylopsis***, ***Lensia***, ***Eudoxoides***, and the hydrozoan medusae ***Rhopalonema***, ***Liriope***, ***Aglaura*** are common in the western Mediterranean (although the role of carnivores limits their abundance to an ecologically sustainable presence) (Fig. 3.22).

Meso-zooplankton abundance and biomass display patterns at the sub-basin scale that broadly follow hydrological features, similar to the distribution of primary producers.

In the **Alboran Sea**, the highly energetic dynamics favour a plankton biomass and diversity higher than on the Atlantic side close to the Strait of Gibraltar.

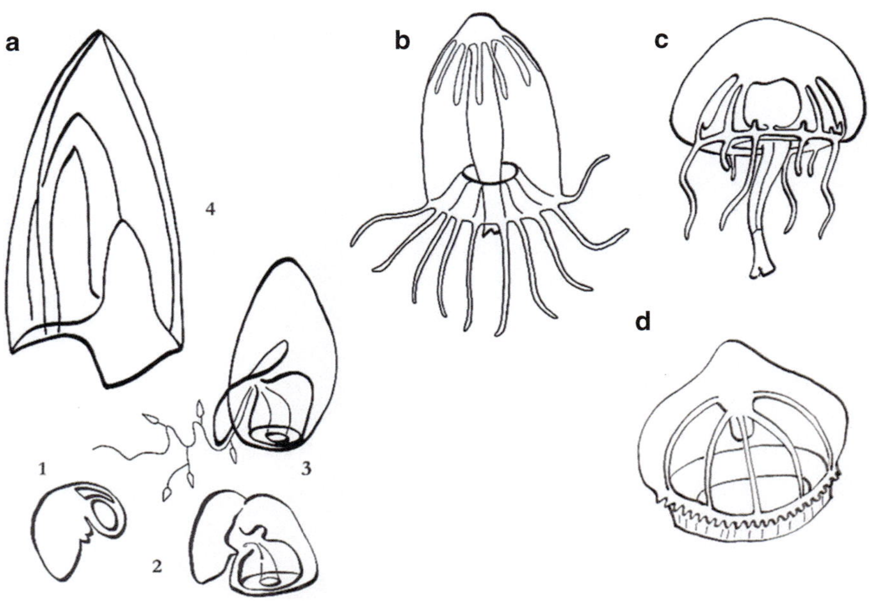

Fig. 3.22 Planktonic Hydrozoa. (**a**) Holoplanktonic *Muggiaea* (Siphonophora), with growing stages of development (1–4); meroplanktonic medusae (**b**) *Aglaura*; (**c**) *Liriope*; (**d**) *Rhopalonema* (original)

In the same area, the sustained productivity caused by processes linked to the Atlantic water inflow results in high zooplankton abundance (dry mass, 18 mg m^{-3}; copepod abundance, 5000 ind. m^{-3}) in the upper 200 m.

This enrichment is probably responsible for the demographic structure among Chaetognatha and Siphonophora, which are found to have a high proportion of juveniles and **eudoxids**, respectively. A great spatial variability of biomass values (5.5–25 mg m^{-3}) is possible in this sector in sites positioned within different water masses and hydrological features at distances of 15–20 NM.

Increased meso-zooplankton standing stock values are associated with **marine water fronts** in the Balearic, Catalan, and Ligurian Seas.

The zooplankton abundance in the **Strait of Sicily** seems to be enhanced by intermittent **upwelling**.

In the northeastern Aegean Sea, the strong **thermohaline front** between the inflowing modified Black Sea water and the Aegean Sea water harbours the most abundant meso-zooplankton recorded in the epipelagos (0–100 m) of the very oligotrophic eastern Mediterranean (up to 3875 ind. m^{-3} and 26.73 mg m^{-3} dry mass).

The permanent or semipermanent **cyclonic gyres** of the East Mediterranean (e.g., the Rhodes Gyre and that south–west of Crete) revealed higher meso-zooplankton abundance than the neighboring **anticyclonic gyres**.

Mesoscale water circulation and hydrodynamic features affect not only the abundance but also the composition and structure of meso-zooplankton. In the Alboran Sea, the Copepoda *Centropages typicus* and *Clausocalanus furcatus* revealed a preference for the frontal area.

An interesting aspect regarding the influence of mesoscale features is revealed when studying cyclonic and anticyclonic eddies synoptically.

In the **Algerian basin**, the eastern edge of an anticyclonic eddy seems to represent favourable conditions for Calanoida (*Paracalanus*, *Clausocalanus*, *Calocalanus*, and *Calanus*) due to the extension of the Chl *a* distribution down to a depth of 200 m.

Aglaura = splendor, brightness *Clado* = branch division; *cheiros* = horn, branch *Eu* = good; *Chaeta* = bristle *Eudoxia* = good thought *Evadne* = mythologic girl, eu-adne: true holy; *Doxoides* = involved in glory, honour *Lensia* = involved in reliability and/or generosity *Lirio* = flower; *Ops* = appearance *Muggia* = city close to Trieste (North Adriatic) *Ostracos* = hard covering, shell *Pleuro* = lateral; *Mamma* = from the first name *omma*, eye, button, present on the side of some copepods *Rhopalo* = club like; *Nema* = wire *Sagitta* (L) = arrow

Chaetognatha were more abundant in the centre of the above cyclonic structure. In the neighbouring cyclonic eddy, the highest abundance of filter feeders (Ostracoda, Cladocera, Tunicata Doliolidae and Salpidae) was attributed to enhanced trophic conditions.

Dissimilarities in copepod assemblages between cyclonic and anticyclonic gyres in the eastern Mediterranean are recorded only in the subsurface layer (50–100 m). The cyclonic gyres were characterized by the Copepoda *Clausocalanus pergens* and

Ctenocalanus vanus, while the anticyclonic gyres were dominated by *Clausocalanus paululus*, ***Mecynocera*** *clausi* and ***Lucicutia*** *flavicornis*; differences were attributed to the higher Chl *a* concentrations of the cyclonic gyres compared with the anticyclonic gyres.

In the hydrodynamically very active area of the **Ligurian Sea**, zooplankton assemblages show distinct patterns at small spatial scales due to both the physical environment and animal behaviour. The Copepoda ***Calanus*** *helgolandicus*, ***Centropages*** *typicus*, ***Oithona*** spp., and ***Oncaea*** spp. were **associated with the frontal zone**, while ***Clausocalanus*** and ***Paracalanus*** appear not to be affected by this spatial variability and do not show a clear distribution pattern. Thus, an ecological distinction can be proposed in the Liguria Sea, at least for species thriving in the frontal zone.

Data on meso-zooplankton production in the open epipelagic Mediterranean are restricted in space and time. In the **northern sectors** (Gulf of Lion, the Catalan Sea, and the northeastern Aegean Sea), **Copepoda production values** oscillate from **19 to 58 mgC m^{-2} d^{-1}** according to the season. The values are much lower in the South and North–West Aegean Sea, in accordance with the recognized oligotrophy of these areas.

Most studies on meso-zooplankton production in the Mediterranean are limited to coastal species and sites and are based mainly on egg production estimates; their results are hardly useful for open waters dominated by Copepoda species with different reproductive biology (still to be deeply studied).

A few studies were carried out in the open Mediterranean for the egg-carrying Copepoda species of Oithonidae and Oncaeidae (Cyclopoida). As the egg-carrying strategy implies lower egg production but also lower egg mortality in comparison to a broadcast egg-laying strategy, these species can maintain quantitatively limited but persistent populations in a wide range of trophic conditions. In fact, the cosmopolitan and abundant *Oithona similis* has very low and similar egg production rates in the North Aegean Sea and in the more eutrophic North Atlantic Ocean, without significant seasonal differences between sites.

In contrast, for the broadcast-spawners *Centropages typicus*, *Temora stylifera*, and *Clausocalanus lividus*, egg production rates recorded in the Catalan Sea are lower than rates reported for the same species in oceanic waters. This indicates that their egg production can be limited by the oligotrophic conditions of the Catalan Sea.

The dominant copepod species in the Mediterranean are extremely diverse, not only in terms of taxonomy and morphology but also in life history traits and behaviour, greatly affecting their modes of interacting with prey and predators. For example, ***Calocalanus*, *Ctenocalanus*,** *and* ***Paracalanus*** cruise slowly and **create feeding currents**, likely collecting more successfully the non-swimming phytoplankton cells. In contrast, ***Clausocalanus*** species move continuously **without creating feeding currents** and **capture cells that enter a restricted volume** just in front of their mouths. **Oithonidae** stand motionless for most of the time, **perceiving hydromechanical signals** from prey with their rich array of setae and hairs before aggressing them. **Oncaeidae** and **Corycaeidae** swim primarily with a **jerky forward motion** and have peculiar **mouth appendages** that allow them to **scrape food items** from particle aggregates, such as discharged Appendicularia houses and marine snow. The comparison of individual activity and motion behaviour between *Clausocalanus furcatus* and *Oithona plumifera* has revealed substantial differences in their sensory and feeding performances, which apparently allow them to coexist in the autumn period.

All these distinct behaviours correspond to different functional roles of the species in the epipelagos and justify their coexistence with the occupation of **distinct niches even in the apparently homogeneous habitat** represented by a 3D water world. The overall picture emerging from the diversified characters of the small copepods prevailing in the open waters of the Mediterranean Sea indicates that they may efficiently exploit the whole spectrum of resources available, even in oligotrophic conditions.

Although their natural diet and feeding performances have been measured only rarely in the open Mediterranean Sea, many species of **Copepoda** are likely to **prefer Ciliophora** (microzooplankton) over autotrophic unicellulars (phytoplankton). Indeed, in the northeastern Aegean Sea, clearance rates of some Copepoda (*Calanus helgolandicus*, *Centropages typicus*, *Paracalanus parvus*, *Oithona similis*, and *Oncaea* spp.) were one order of magnitude higher on Ciliophora than on Chl *a*-containing unicellulars. Moreover, Copepoda seemed to consume almost the entire Ciliophora production but only part of the available primary production, thus suggesting that **not all autotrophs provide adequate food supply** in terms of quality and/or size.

Rare measurements of feeding rates in the open Mediterranean Sea, together with laboratory results and data coming from coastal areas, agree in establishing that the ingestion rates depend on food availability (quantity and quality).

During a spring bloom in the **Alboran Sea**, Copepoda ingestion rates on natural particle mixtures varied between 0.5 and 5.8×10^6 μm^3 particles mg^{-1} zooplankton dry weight h^{-1}, and the highest value was recorded in the DCM layer (Chl *a* maximum concentration). Most of the in situ studies have provided evidence that mesozooplankton feeding varies in relation to the prevailing type of food and that **feeding can be selective** even when a homogeneous food assemblage is available (Fig. 3.23).

For example, at a precise site, filtration rates rose from 0.54 to 1.89 mL $copepod^{-1}$ h^{-1} when the diet switched from mixotrophic to heterotrophic nanoflagellates.

In offshore waters of the northwestern Mediterranean Sea, zooplankton are dominated by four genera of Copepoda (*Clausocalanus*, *Paracalanus*, *Oithona*, and *Centropages*), feeding on phytoplankton in June, when cells >10 μm occur, and relying on microzooplankton or detritus in October, when phytoplankton are of smaller size (cells <10 μm), thus vanishing the capture ability of Copepoda, and the same shift has also been observed under strong oligotrophic conditions (scarcity of phytoplankton cells).

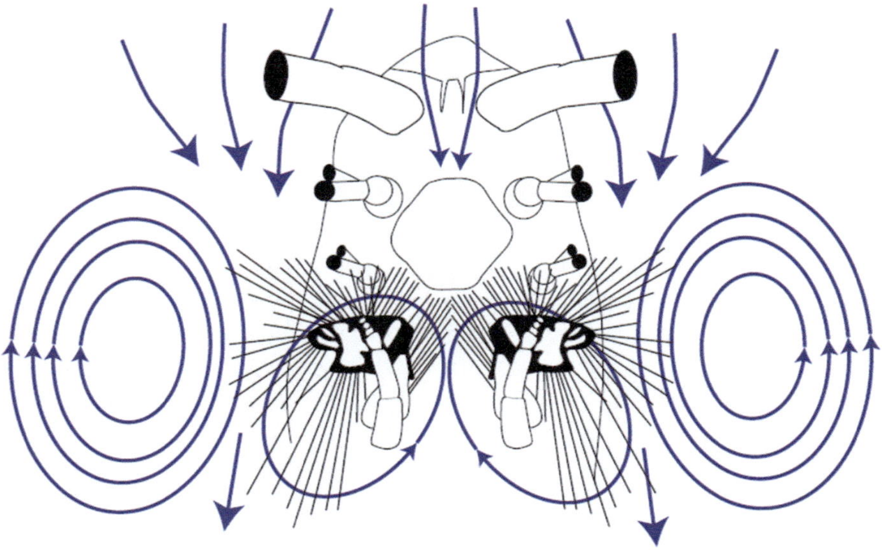

Fig. 3.23 Scheme of feeding currents around the cephalic part of a Calanoida. (From Sverdrup 1942, re-drawn)

A similar seasonal shift was observed in the Catalan Sea, where Copepoda were strongly coupled with the autotroph biomass during phytoplankton blooms in March and with heterotrophic organisms in late spring and early summer, when phytoplankton abundance was lower.

Switches in feeding preferences and performances might result from real group/genus adaptability in response to different food environments. However, it is also possible that this apparent flexibility corresponds to the prevalence of congeneric species that are very similar morphologically but show different food requests. This second case can be hypothesized for ***Clausocalanus pergens*** and ***C. paululus*** by observing their distribution in different sectors of the Mediterranean Sea and the Atlantic Ocean. Both species are widespread in the epipelagos of the open Mediterranean Sea in late winter–spring, but the former prevails in the presence of phytoplankton blooms and the latter in oligotrophic sectors (e.g., the Ionian Sea), suggesting a separation of their trophic niches.

Given the general oligotrophic status of the Mediterranean Sea, food availability can affect meso-zooplankton dynamics. **Food limitations** may lead to competition among meso-zooplankters in the open sea. However, the **ability to exploit different food sources** may overcome this problem and **allow the co-occurrence of various and numerous, even cognate, *taxa*** (Fig. 3.24).

Finally, what must be carefully considered is also the **size variation** of most holoplankton during life and the obliged shift from the diet of early-born larvae (nauplii) to that of adults within the same species.

The prey selection of Copepoda could significantly affect prey abundance (in the case of Ciliophora), exerting a strong top-down control on the Protozoa populations.

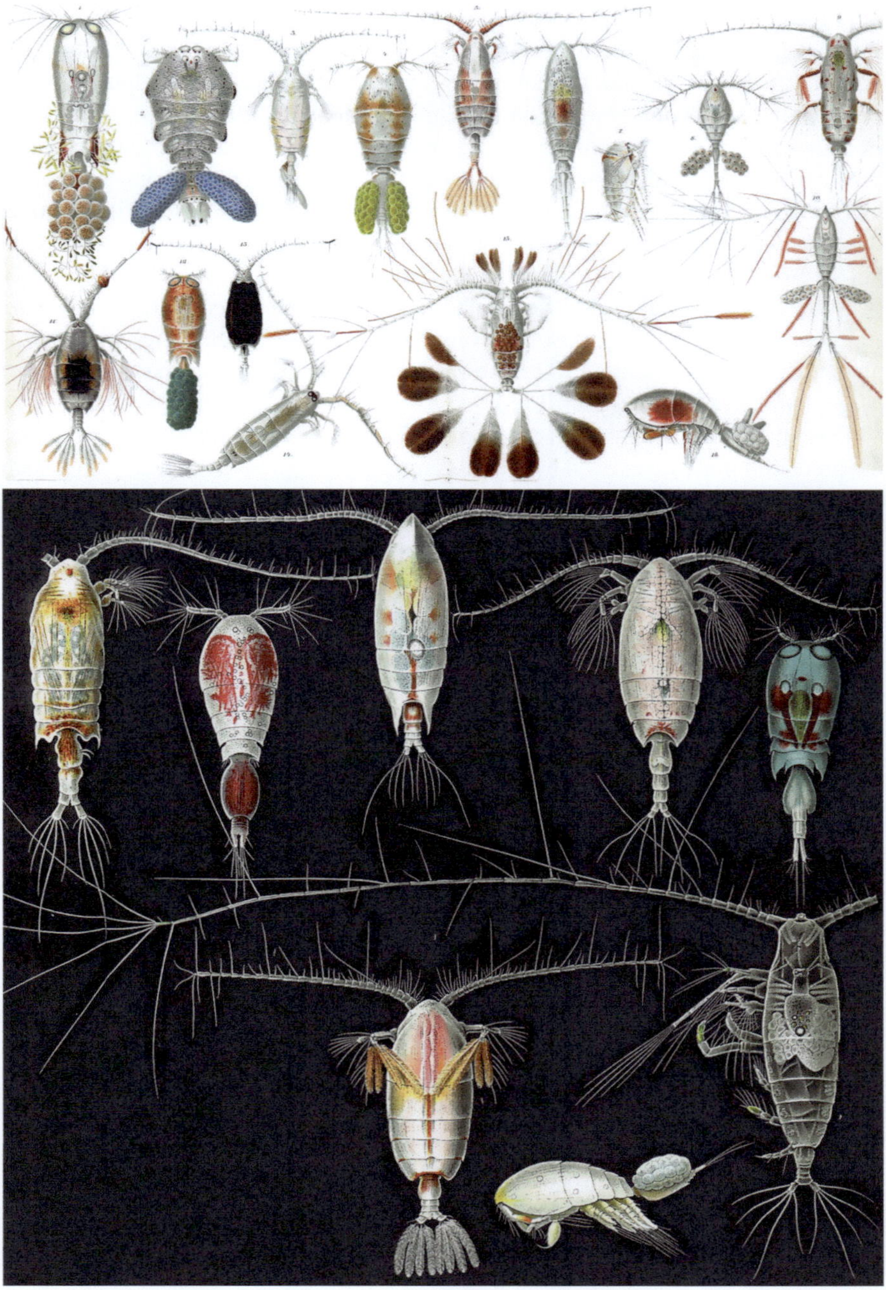

Fig. 3.24 Copepoda are the most represented and abundant category of meso-zooplankton. Here is proposed a selection of those illustrated by Giesbrecht (1892) in a famous atlas used by generations of plankton students to identify different species of Copepoda in the Mediterranean Sea

This control has been hypothesized to bed the major factor for the low standing stock of Ciliophora across the entire Mediterranean Sea. In addition, in situ measurements indicate that the impact of meso-zooplankton grazing on phytoplankton can also be significant. Namely, in the **Gulf of Lions**, the percentage of **primary production removed by** zooplankton **grazing was estimated to be important** in both winter (47%) and spring (50%). Previous estimations in the area suggested that half of the phytoplankton loss (in terms of biomass) from March to April should be due to zooplankton grazing. In the very oligotrophic South Aegean Sea, Copepoda grazes 14–35% of the primary production from cells >3 μm.

In the northeastern Aegean Sea, small Copepoda (*Oncaea* spp., *Clausocalanus* spp., *Paracalanus parvus*) showed a considerably higher grazing impact on phytoplankton production (almost 100% during September) than larger species (*Calanus helgolandicus*, *Centropages typicus*).

Few studies are available on **the feeding impact of carnivorous zooplankton**; the predation pressure exerted by **Chaetognatha on Copepoda** appears, from such scant information, to be overall negligible in the Catalan Sea (<1%), whereas it **reaches 7.8% in some sectors of the east Mediterranean Sea**.

A good coupling between meso-zooplankton and their prey is suggested by the horizontal patterns of zooplankton distribution in the open sea, which match those of autotrophic biomass and production (e.g., the west-to-east decrease trend) and, to a lesser extent, those of micro-heterotrophs. However, occasionally, the sectors of the maximum zooplankton abundance do not coincide with those of the highest phytoplankton concentration, and this contrast might be attributed to factors other than nutrition, such as zooplankton survival due to lack of predation.

Indeed, in the very West Mediterranean Sea (Algerian basin, the Catalan Sea, and the Gulf of Lion), **larvae** of the fish ***Engraulis encrasicolus*** feed mostly on Copepoda (*Centropages typicus*, *Temora stylifera*, *Microsetella rosea*, Clausocalanidae, Paracalanidae) and, to a lesser extent, on Mollusca, Cladocera, and Appendicularia. In particular, Copepoda nauplii and juveniles are the major prey items of anchovy and sardine larvae not only in the West Mediterranean but **also in the Adriatic Sea**. Here, sardine adults and larvae seem to also feed on phytoplankton, while sprats feed on Copepoda, Decapoda larvae, Cladocera and Chaetognatha. The principal prey of all size classes of *E. encrasicolus* juveniles in the NW Adriatic Sea are small Copepoda (0.2–0.6 mm in **prosome** length), such as *Euterpina acutifrons* and *Oncaea* spp. The observed preference of anchovies for these few copepod species might be related to their abundance but also to species-specific behavioural (e.g., swimming patterns, patchy distribution) and/or physical characteristics (e.g., colour, bioluminescence) of the prey.

The constant and important presence of Copepoda in the diet of anchovies and Clupeidae reported above and the relevant portion (20%) of total zooplankton production estimated to be consumed by adult anchovies in the Catalan Sea definitely indicate that **meso-zooplankton play a crucial role as the major link between plankton and small pelagic fish production**.

After the first reports of a food web dominated by small plankton in the Liguria Sea, several studies showed that such a situation (indicative of a **"microbial loop"**) **dominates in large parts of the Mediterranean Sea**.

The widespread Phosphorus deficit in Mediterranean waters leads to the investigation of how P limitation could shape the microbial food web and influence carbon flow in the very oligotrophic Levantine Sea. The fast transfer of the added P to the particulate form, along with an unexpected slight decrease in Chl *a*, led to two possible scenarios:

(a) **Bacteria** utilize dissolved organic N, thus **outcompeting** the N-limited **autotrophers of phytoplankton**. The heterotrophic biomass would then have been quickly available to larger consumers, including meso-zooplankton, with a sharp increase in Copepoda egg production (**bypass hypothesis**) independent of phytoplankton.
(b) An enhanced accumulation of P in both bacteria and pico-phytoplankton (presumably less in the latter) forms a P-enriched diet for grazers, stimulating the observed increase in Copepoda egg production (**tunnelling hypothesis**).

The leading role of heterotrophs in the Mediterranean Sea, as it emerges from a mainly heterotrophic plankton standing stock dominated by the **pico-**size group, is the recurrent situation in the basin.

Heterotrophic/autotrophic biomass ratios vary from 0.5 to 3.0 in the northwestern Mediterranean Sea and from 0.9 to 3.9 in the Aegean Sea, with higher values more frequently found in oligotrophic areas and during the season of water stratification. The spatial trend is consistent with this pattern, with ratios increasing along a longitudinal transect spanning from the Balearic Sea to the Levantine basin.

The above outlook suggests two different scenarios for the Mediterranean epipelagic food web:

1. The system is mainly heterotrophic, dominated by **Bacteria and unicellular Eukaryota**. In this scenario, Bacteria outcompete autotrophs of phytoplankton in Phosphorous uptake, relying on allochthonous Carbon, which would be used more or less efficiently according, e.g., to the season or the area.
2. The system is in balance between production and consumption, and **the situation reflects either seasonally biased sampling or higher turnover rates in autotrophs than in heterotrophs**.

Despite the dominance of heterotrophic microbes and pico-autotrophs in the offshore Mediterranean, the **prevalence of nano- and micro-autotrophs** can occur, after intermittent nutrient pulses.

These shifts are typically associated with highly dynamic mesoscale physical structures that favour deep vertical mixing and nutrient upwelling in frontal areas, as well as in areas close to extended and highly productive coastal systems enriched by large river outflows (Gulf of Lions, Adriatic, and North Aegean Seas).

Pro = anterior; *Soma* = body

Such **pulses** are **spatially limited and concentrated**, in contrast to the nutrient inputs from the atmosphere, which are spread over large areas and diluted. They determine a considerable variability of **the food web structure** along trophic gradients, which varies not only in space but also in time, spanning from the **recycling systems** to **new production systems**.

Areas characterized by a high level of primary and secondary productivity sustained by upwelling are often driven by physical processes and provide favourable reproductive habitats for fish.

Indeed, the Alboran Sea, the Gulf of Lions and the nearby Catalan Sea, the Adriatic Sea and the North–East Aegean Sea are known as successful spawning grounds and areas of high yield of small pelagic fish, mainly anchovy and sardines.

In these areas, the larvae of small pelagic fish feed mainly on Copepoda, but unicellulars (Ciliophora and small flagellates) have also been found to contribute to their diet.

Biomass production also occurs at the DCM (Deep Chlorophyll Maximum), with frequent dominance of diatoms (Bacillariophyta) close to the **nutricline** over broad periods. Whether the DCM hosts a characteristic plankton is still an open question. A first approximation may then be that **the microbial food web is the prevailing structure in the offshore waters** of the Mediterranean Sea, with few exceptions where larger, bloom-forming phytoplankton might fuel the "classical" food web.

When estimated with normal counts, the heterotrophic Bacteria included groups capable of diversified metabolic strategies, and the Mediterranean Sea should not be different from the global ocean in this respect.

Some naked flagellates, Ciliophora, and Dinophyta are mixotrophic, and their contributions are significant in the Eastern Mediterranean. Metazoans also display a wide range of feeding strartergies and food preferences.

Among Copepoda, the species most abundant in the Mediterranean Sea are known to exploit a large variety of food resources, including faecal pellets (e.g., *Oithona*) and marine snow (e.g., Oncaeidae).

Appendicularia, which are capable of feeding on pico- and nanoplankton, constitute a bypass from the lower trophic levels to fish, contributing to enhance the efficiency of the food chain.

The variable grazing impact on primary producers >5 μm by meso-zooplankton, despite the prevalence of Ciliophora in their diet during both mixing and stratified seasons, indicates a flexible connection between autotrophs and micro-heterotrophs on one side, and with higher trophic levels on the other side.

Nutri (L) = that gives food; *Cline* = passage, transition   *Oligo* = few; *Trophe* = food

This frame suggests that the Mediterranean Sea is characterized by a "**multivorous food web**", including a *continuum* of trophic pathways spanning from the herbivorous food web to the microbial loop and dynamically expanding or contracting with seasonality along the year, areas, and transient processes.

The high diversity in species, feeding style, reproduction modes, and consequently in functional roles, might support a more efficient energy transfer to the higher trophic levels, a common feature among oligotrophic systems.

Systems can be modeled according to two possible alternatives:

1. The **low abundance of autotrophs** is the consequence of **low availability of nutrients,** which, as typical for **oligotrophic conditions**, determines **low abundances of intermediate and top predators** (bottom-up control).
2. The **low abundance of autotrophs** is the consequence of an effective **top-down control** that propagates along the food web, ultimately affecting the top predators.

The two alternatives are not in contrast, although the former is driven by geochemical constraints, while the latter emphasizes the ecosystem dynamics as a whole. Indeed, bottom-up control does exist and determines the carrying capacity of the system.

This is relatively low because of the moderate nutrient fluxes and the continuous loss of the internal nutrient pool into the Atlantic Ocean.

On the other hand, the view of a top-down control is supported by the structure of the planktonic food web discussed above as well as by the evidence that fisheries in the Mediterranean are richer than expected on the basis of measured Chl *a* and nutrient concentrations (the so-called "**Mediterranean paradox**" (Estrada 1996)).

We can thus hypothesize that intricate and very flexible food webs contribute to minimizing Carbon loss in deeper layers. Predators, in this framework, can profit from Carbon produced and transformed within the system, thus being the ultimate controllers of plankton abundance in the Mediterranean Sea. The paradox becomes even less paradoxical if one takes into account the significant role of external organic matter inputs (that coming from continents), which corroborates the view of the **Mediterranean Sea as a "coastal ocean**".

References

Azam F, Fenchel T, Gray JG, Meyer-Reil LA, Thingstad T (1983) The ecological role of water column microbes in the sea. Mar Ecol Prog Ser 10:257–263

Belmonte G (2021) Acartiidae Sars G. O, 1903. ICES identification leaflets for plankton no. 194. 29 pp. https://doi.org/10.17895/ices.pub.7680

Belmonte G, Potenza D (2001) Biogeography of the family Acartiidae (Calanoida) in the Ponto-Mediterranean Province. Hydrobiologia 453/454:171–176

Ben Rais Lasram F, Guilhaumon F, Albouy C, Somot S, Thuiller W, Mouillot D (2010) The Mediterranean Sea as a "cul-de-sac" for endemic fishes facing climate change. Glob Change Biol 16:3233–3245

Cavanaugh CM, Mckiness ZP, Newton ILG, Stewart FJ (2006) Marine chemosynthetic symbioses. Prokaryotes 1:475–507

Coll M, Piroddi C, Steenbeek J, Kaschner K, Ben Rais Lasram F, Aguzzi J, Ballesteros E, Bianchi CN, Corbera J, Dailianis T et al (2010) The biodiversity of the Mediterranean Sea: estimates, patterns, and threats. PLoS One 5:e11842

Estrada M (1996) Primary production in the northwestern Mediterranean. Sci Mar 60(2):55–64

Fernández de Puelles M, Pinot J-M, Valencia J (2003) Seasonal and interannual variability of zooplankton community in waters off Mallorca Island (Balearic Sea, Western Mediterranean): 1994–1999. Oceanol Acta 26:673–686

Fink HG, Wienberg C, De Pol-Holz R, Hebbeln D (2015) Spatio-temporal distribution patterns of Mediterranean cold-water corals (*Lophelia pertusa* and *Madrepora oculata*) during the past 14,000 years. Deep Sea Res Part I Oceanogr Res Pap 103:37–48

Galil B, Marchini A, Occhipinti-Ambrogi A (2018) East is east and west is west? Management of marine bioinvasions in the Mediterranean Sea. Estuar Coast Shelf Sci 201:7–16

Gautier F, Clauzon G, Suc JP, Cravatte J, Violanti D (1994) Age and duration of the Messinian salinity crisis. C R Acad Sci Paris (IIA) 318:1103–1109

Giesbrecht W (1892) Pelagische Copepoden. Sistematik und Faunistik. In: Fauna und Flora des Golfes von Neapel. Monographie XIX. Verlag, Berlin. 881 pp, 52 tables

Hubbel SP (2001) The unified neutral theory of biodiversity and biogeography. In: Monographs in population biology, vol 32. Princeton University Press, Princeton. 392 pp

Margulis L (1990) Words as battle cries—symbiogenesis and the new field of endocytobiology. BioScience 40(9):673–677

Por FD (1978) Lessepsian migration: the influx of Red Sea biota into the Mediterranean by way of the Suez Canal. Ecological studies, vol 23. Springer-Verlag, Berlin. 288 pp

Relini G (2008) Checklist della Flora e della Fauna dei mari Italiani. Introduzione. Biologia Marina Mediterranea 15(Suppl 1):V–XI

Rohwer F, Seguritan V, Azam F, Knowlton N (2002) Diversity and distribution of coral-associated bacteria. Mar Ecol Prog Ser 243:1–10

Siokou-Frangou I, Christaki U, Mazzocchi MG, Montresor M, Ribera d'Alcalà M, Vaquè D, Zingone A (2010) Plankton in the open Mediterranean Sea: a review. Biogeosciences 7:1543–1586

Suikanen S, Uusitalo L, Lehtinen S, Lehtiniemi M, Kauppila P, Makinen K, Kuosa H (2021) Diazotrophic bacteria in planktonic food webs. Food Webs 28:e00202

Suwa R, Hatta M, Ichikawa K (2014) Proton-transfer reaction dynamics and energetics in calcification and decalcification. Chem Eur J 20:13656–13661

Sverdrup (1942) The oceans, their physics, chemistry, and general biology. Prentice-Hall, New York, 1061 pp. http://ark.cdlib.org/ark:/13030/kt167nb66r/

Tortosa EA, Jordá Pardo JF, Álvarez-Fernández E, Pérez Ripoll M, Avezuela Aristu B, Morales-Pérez JV, Rodrigo García MJ, Marlasca R, Alcover JA, Jardón P, Pérez Herrero CI, Pardo Gordó S, Maestro A, Villalba Currás MP, Salazar-García DC (2016) Palaeolithic–Epipalaeolithic Seapeople of the Southern Iberian coast (Spain): an overview. Actes de la séance de la Société préhistorique française de Rennes, publiés sous la direction de Catherine Dupont et Gregor Marchand Paris, Société préhistorique française, pp 69–92

Nekton

4

4.1 Identikit of Nekton Organisms

Large organisms generally show a swimming power that allows them to be independent from the main water currents. Their swimming power is enough even to allow them to progress counter current.

This community of organisms with their completely pelagic lifestyle have the name **nekton**.

Large organisms subtract themselves from water currents better than smaller organisms. The appurtenance to such a community, however, is not simply a matter of size. Smaller organisms certainly cannot escape the main currents, but many large organisms (Scyphozoa medusae, Siphonophora colonies, and Thaliacea colonies) are not included in the nekton and are classified as Plankton because their swimming power does not allow them to escape the main water currents.

Aside from such a distinction, we must additionally consider that fast swimming species cannot be considered nekton for their entire life cycles. In fact, eggs and larvae cannot escape water currents, and for this reason, their species belong to plankton (**Ichthyopklankton**) for at least this early part of their life history.

The swimming power of the nekton organisms, generally of large size, cannot be produced by cilia or their derivates (as *cteni* of Ctenophora), and it is based on the presence of osteo-muscular apparatuses (**tails** of fish and Cetacea; **arms** of seals, penguins, and turtles) or muscular apparatuses (the **mantel–siphon** system of Cephalopoda). Arm-based swimming (seals, penguins, and turtles) is sometimes linked to breathing air organisms (Mammalia, Aves, and Reptilia) that, also due to their link with land (at least for reproduction), are sometimes not considered "true" nekton. The use of arms is not common among fish, and it derives from the evolutionary development of such structures during the land occupation by the ancestors of reptiles, birds, and mammals.

Many nekton organisms are born from **pelagic eggs**, and they are so small at their early existence that they are considered plankton. Such a "temporary" plankton (the ichthyoplankton, if derived from fish) is a delicate stage of the life history

G. Belmonte, *Elements of Pelagos Biology*,
https://doi.org/10.1007/978-3-031-74280-4_4

of each species, and it is under the attention of many scientists and fishery managers due to its economic importance in generating and/or preserving future fish stocks.

The nekton species, either fish or squid, tend to constitute large aggregations of individuals, which are ruled by body size. Individuals similar in body size are generally of the same age (if they belong to the same species), and for this reason, such associations are also called "**schools**" (Fig. 4.1) (Stocker 1999).

Even very large species, such as tuna (*Thunnus thynnus*), can group together with very small species, such as anchovies (the common name of many species in the family Engraulidae), simply during the development age, which corresponds in size to that of the smaller species (**size-based fish aggregation**). Such aggregations are typical of the smallest pelagic species, and monospecific assemblages are generally rare. In fact, although one species is dominant in each assemblage, some other species can participate in the group, with individuals of the same body size, even if of different ages.

Ichthyos = fish; *Plankton* = going around *Nekton* = swimming *School* = group of individuals of the same age

Size assemblages show coordination in movement execution during feeding activity and defensive reactions.

The growth of size obliges large species to abandon interspecies relationships and gregarism with small pelagic fish. When isolated in monospecific aggregations of large individuals, these mostly show migration performances or cooperative aggression on smaller prey.

The large size, although somehow favoured by the water medium that pushes up on each immerged body, accordingly with the water volume occupied (the physicist Archimede's principle), obliges individuals to actively avoid sinking. This can be resolved by **associating** horizontal **swimming with a fin apparatus** (generally the pectoral fins) that directs upwards each horizontal movement. More than 60% of

Fig. 4.1 Left, a school of fish in a large tank of the San Sebastian aquarium (Spain) (photograph G. Belmonte). Knowledge of fish behaviour allows biologists the management of large numbers of specimens in restricted volumes, also mixing different species (with a similar body size). Right, the grouping of fish is also related to a defensive strategy against large predators (Drawing A. Gennari)

living fish, however, and all lunged marine animals are equipped with an internal **gas tank** that concentrates gas or eliminates it, allowing them to remain suspended at a desired depth. This device is the **gas bladder** of bony fish, and it is generally less important in strong swimmers and more important in sedentary species (Wittemberg 1958) (Fig. 4.2). The gas bladder can develop as a vesicle in the dorsal wall of the oesophagus, and the connection with the alimentary canal can remain for all life or can be reabsorbed. In the first case, fish are called **physostomes**; in the second case, they are called **physoclisti**.

In any case, the epithelium of the anterior wall of the gas bladder is specialized in absorbing and/or producing gas from or to the bladder and obtaining, as a consequence, the regulation of the suspension at desired depths.

Physos = open passage; *Stoma* = mouth, opening; *Klistos* = closed, shut

Lunged animals (Cetacea, Pinnipedia, Chelonia, and Spheniscidae) have **lungs** for breathing that are secondarily used for such buoyancy regulation, maintaining the air inside when at the surface and expelling the air during diving.

Apart from approximately 1000 cartilaginous fish species (**Chondrichthyes**) that lack the gas bladder, most of the bathypelagic and abyssopelagic bony fish (**Osteichthyes**) do not have the gas bladder because its inflation would require an impossible effort under the high external pressures existing where such species thrive in the very deep ocean.

The gas bladder aiding buoyancy is also present in many Cephalopoda (Mollusca) in the form of a specialized and multichambered shell, among **Nautiloidea** and **Coleoidea**, which were very abundant in primitive oceans and today reduced to no more than 750 species. All Mollusca equipped with such a device, however, are mostly "planktonic" (not able to make fast and/or long horizontal movements), and only a small portion of species are typically nektonic. The true nektonic Cephalopoda belongs to Decapoda **Teuthoidea** (squid) (Denton et al. 1969) (Fig. 4.3). Unlike

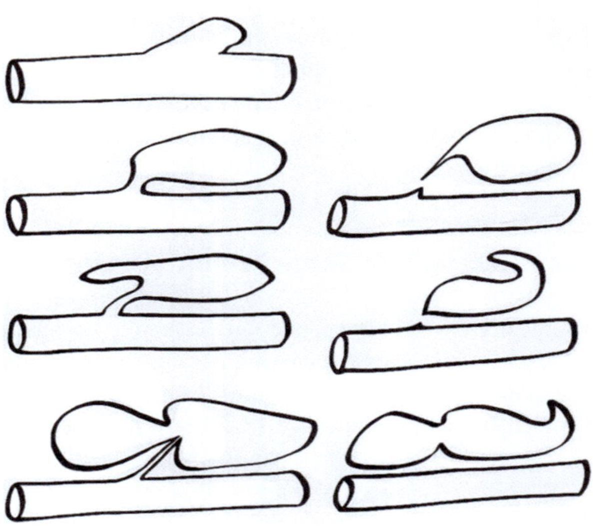

Fig. 4.2 Scheme of the development of the two types of gas bladder. In the column on the left, the realization of a physostome condition; on the right, a physoclist condition. In both sequences, the realization proceeds from top to the bottom. The tube below the gas bladder represents the oesophagus. (Courtesy G. Belmonte)

Fig. 4.3 Left, a sketch of a Cranchiidae, the glass squid (generally distinguishable for the short arms, apart two, other than for transparency of the body); right, a section of the *Nautilus* shell, with evidence of successive chambers. (Courtesy G. Belmonte)

other Cephalopoda, many squids (half of the species) have a fluid-filled receptacle in the **coelom,** or connective tissue.

This liquid reservoir acts as a **buoyancy chamber** based on chemistry, replacing the heavy metallic cations (Na) derived from seawater with low-molecular-weight ammonium (NH_4) ions derived from the excretive metabolism. The small difference in body density provides a small contribution to buoyancy per unit volume, so the mechanism requires a large buoyancy chamber to be effective. Since the chamber is replenished with liquid, it has the advantage over a swim bladder in that it does not change significantly in volume with an increase in pressure. The so-called glass squids of the family **Cranchiidae** have an enormous transparent coelom containing ammonium ions and occupying approximately two-thirds the volume of the animal, allowing it to float at the desired depth.

4.2 Details on Some Pelagic Species

Practically all fish species are nekton simply because they swim. Very rare specimens can walk or others can fly, but this happens when they need to temporarily abandon the swimming habitus. Among the **28,000 different species** of **Osteichthyes**, approximately **18,000 are marine,** and most of them are linked to marine bottom habitats. However, when they move in their habitat, they do so by swimming, and for this reason, at least during these short times, they are nekton.

The largest bony fish is the sunfish (***Mola mola***), which can pass 2000 kg, or the oarfish (genus ***Agrostichthys***, ***Regalecus***), which can reach 10 m in length (Fig. 4.4). In the Jurassic, however, ***Leedsichthys problematicus*** passed both of these limits.

The largest fish, however, is not bony but cartilaginous and belongs to the class **Chondrichthyes**, which today lists approximately **1000** marine species. Apart from mythological "giant sharks" whose existence is sure for the past (***Megalodon megalodon*** of the Miocene, probably reaching 20 m in length) and still to be proven for the current fauna, the largest fish of the ocean is *Rhincodon typus* (**whale shark**), which reaches 18 m.

4.2.1 The Small "Oil Fish"

The fish species with the most abundant populations belong to the taxonomic group of Clupeiformes (anchovies, herrings, and sardines, of the families Engraulidae and Clupeidae), which are completely pelagic but small in size.

They are **physostome** species (the gas bladder is connected with the oesophagus) of archaic morphology (having the pelvic fins in posterior position), with a **homocercal tail** fin and **cycloid scales** in the skin.

Among the 140 species that stay under the common name "**anchovies**", a total of 9 **species** are objects of intense fishing on the whole planet. Their populations are immense, and unlike other fish, the average size of specimens seems not to be negatively affected by crowding. Consequently, schools of anchovies reach impressive

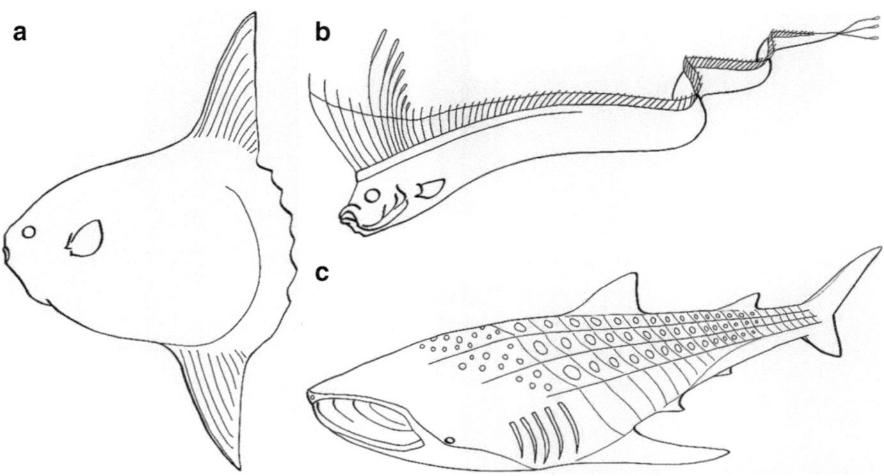

Fig. 4.4 (**a**) Sunfish (*Mola mola*); (**b**) oarfish (*Regalecus* sp.); (**c**) the whale shark (*Rhincodon typus*). (Courtesy G. Belmonte)

numbers and biomass, arriving to address the economy and policy of ocean coastal countries.

During early life stages, anchovies are visual predators of small zooplankton (the small anchovy is only slightly larger than small Copepoda), but when they grow, their feeding habit switches to a lower **trophic level** (between 3.0 and 2.5), and adults are filter feeders, including large amounts of phytoplankton (Dinophyta and Bacillariophyta) in their diet. This feeding behaviour (switch from a less abundant resource to a more abundant one) probably sustains their impressive numbers and the apparent independence of demography from the crowding effect.

Box 4.1 Trophic Levels

Each species has its own position/role in the ecosystem, and it can be assigned to a **trophic level** to numerically describe this fact (Pauly et al. 1998). The way that each species finds energy for life gives us the possibility to organize species of the same ecosystem in a network of roles, or, better, a web where each species is more or less linked to many others. The trophic level of an organism is the number of steps it takes from the initial point of the chain, where energy is captured and made available for the entire community. Such an initial point (level 1) is occupied by primary producers: species able to build energetic molecules using non-living sources of energy. When the first level is occupied by the microalgae of phytoplankton (performing photosynthesis), the second level is that of organisms that feed on them (fitoplanktivorous) and are called primary consumers or secondary producers. Species feeding on phytoplanktivores are carnivores (or predators) of level 3. Larger carnivores (typically in marine trophic chains) of levels 4 and 5 (sometimes called super predators) are those species that eat carnivores.

Organisms, however, often gain energy at more than one trophic level. For example, some carnivores also eat microalgae, and some microalgae can shift from autotrophic to heterotrophic roles. This is particularly true for species of the marine environment, which start their lives as tiny larvae and arrive as large and voracious adults. The system is further accomplished by the possibility of each organism eating its own conspecifics (cannibalism).

The fisheries experts set the values of trophic levels to 1.0 in plants and detritus, 2.0 in herbivores and detritivores (primary consumers), 3.0 in secondary consumers, and so on. The definition of the trophic level, **TL**, for any consumer is:

$$\mathrm{TL}_i = 1 + \sum\nolimits_j \left(\mathrm{TL}_j \cdot \mathrm{DC}_{ji} \right)$$

where TL_j is the fractional trophic level of prey j, and DC_{ji} represents the fraction of j in the diet of i. That is, the trophic level for each species is 1.0 plus the average of how much different trophic levels contribute to its alimentation.

In the case of marine ecosystems, the trophic level of most fish and other marine consumers takes a value between 2.0 and 5.0. The upper value, 5.0, is unusual, even for large fish, although it can be attributed to apex predators of marine mammals, such as polar bears and killer whales (Pauly et al. 1998).

In addition to studies of animal behaviour and the quantification of animal stomach contents, TL can be quantified through stable isotope analysis of animal tissues. This is because there is a consistent increase in the Nitrogen isotopic composition at each trophic level caused by fractionations that occur with the synthesis of biomolecules; the magnitude of the increase in Nitrogen isotopic composition at each trophic step is approximately 3–4‰.

Fishery biologists noted that abundance peaks of anchovies (with the only exception of Peruvian *anchoveta*) are reported from the same years between populations of different geographic areas or even different species, thus depicting worldwide (or at least oceanwide) **synchronous dynamics**.

The adults of European anchovy (*Engraulis encrasicolus*) go in neritic coastal areas during spring-summer and stay at the inferior limit of the epipelagic layer (approximately −200 m) during the remaining part of the year (in the Black Sea, such a vertical transition is limited at 60–70 m below sea level).

The seasonal vertical transition is integrated with a vertical daily migration during spring-summer. During the day, anchovies stay close to the bottom (in neritic areas), and during the night, they distribute widely along the whole water column.

The sexual maturation (and the time for the first reproduction) is reached in 1 year at a size (mouth–tail length) of 9 cm.

Each female broadcasts in the water up to 40,000 eggs, typically elliptic and pelagic.

Eggs are fertilized in the water, and juveniles (white small fish) are particularly searched for their taste and flavour (such a fishery, in any case, is rigorously controlled and ruled). **European anchovies** live up to **5 years,** reaching a final size of a maximum of **20 cm** (more commonly 13–14 cm).

The European anchovy, however, represents **8.3% of the total fished anchovies** (6.2 million tons per year).

Japanese anchovy (*Engraulis japonicus*) represents **19.5% of the world** catch, and **Peruvian *anchoveta* (*Engraulis ringens*)** represents **68.5%**.

As in the case of anchovies, herrings (***Clupea arengus***, in the Atlantic Ocean) and sardines (*Sardina pilchardus*) also show impressive population sizes and sustain the fishery economy of many coastal human communities. The **North Atlantic herring** reaches a **size of 45 cm** (commonly 30 cm) and lives for approximately **22 years**. The **European pilchard (*Sardina pilchardus*)** reaches **27 cm and** can attain an age of **15 years**. These categories, however, occupy a trophic level (between 3.5 and 3.0) higher than that of anchovies (i.e., they are "more carnivorous" than anchovies).

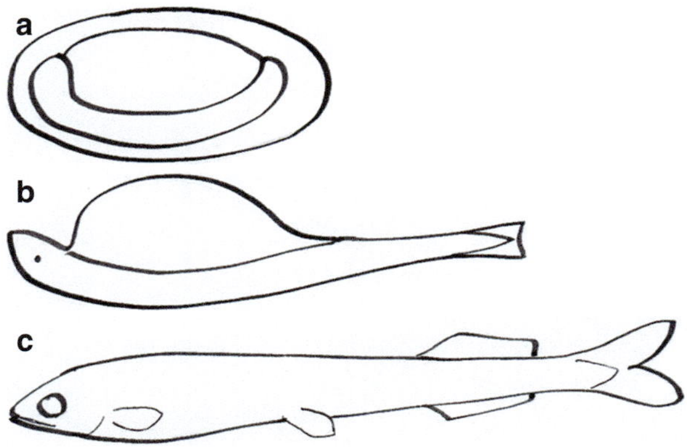

Fig. 4.5 Sketch of the life cycle of *Engraulis*. (**a**) Egg; (**b**) larva with the yolk sac; (**c**) juvenile. (Courtesy G. Belmonte)

Biologists noted that **large amounts of larvae and juveniles** of such "oil fish", particularly those **of anchovies**, which should represent the life stages closest to the reproductive moment of the preceding generation, **often stay in very low-productive areas** of the oceans (Bakun 1996). If the reproductive areas can be indicated as those where eggs and larvae stay and grow, it is not clear why the species choose such a poor environmental situation as a nursery, but we have to consider that this could be an attempt to avoid areas rich in potential predators (Fig. 4.5).

Coleos = turned on itself *Chondros* = cartilaginous; *Ichthyos* = fish *Mola* (L) = millstone *Nauplius* = the pilot of the Greek mythic ship Argo *Osteos* = bony; *Ichthyos* = fish

4.2.2 The Large "Oil Fish"

Tuna are large and voracious predators, typically pelagic, belonging to the **family Scombridae**. They show a moon-shaped tail fin and a typical peduncle with finlets on the opposite dorsal and ventral sides. Pelvic fins stay in a pectoral position. Under the jargon name tuna, **at least 15 species** can be considered, from the smallest bullet tuna (***Auxis rochei***) of 50 cm and 2 kg to the largest blue fin tuna (***Thunnus thynnus***) of 4.5 m and 680 kg. The **trophic level** of all the species ranges **from 3.7 to 4.5**, which allows us to consider them plentiful carnivores.

Juveniles live in surface waters and are already predators equipped with teeth and large eyes (Fig. 4.6). A blue-fin tuna can live for **50 years**.

The Atlantic blue fin tuna reaches sexual maturity at the age of 5 years and typically migrates, forming high concentrations of individuals for mating.

Approximately 21 species in the family **Scombridae** are commonly called **mackerel**. The Atlantic mackerel (*Scombrus scombrus*) is smaller (maximum 65 cm) and has shorter lives (18 years) than tuna fish, which are also members of the

Fig. 4.6 From top to bottom, tuna larva (with yolk sac), two juveniles, small adult. (Courtesy G. Belmonte)

Fig. 4.7 Mackerel (*Scombrus*), with dark dorsal stripes useful to communicate the position of each individual inside the school (Drawing, A. Gennari)

same family. Their **trophic level ranges between 3 and 4**; thus, they can feed on anchovies, but they can be prey of tuna.

The evident colour stripes on the dorsal side of mackerels seemingly do not provide camouflage against broken backgrounds because mackerel live in midwater pelagic environments that have no background (Denton and Rowe 1998). Scombridae have an **optokinetic reflex** in their eye–brain system that seems to be sensitive to moving stripes. For fish to school efficiently, they need feedback impulses that help them align with adjacent fish and match their speed. The stripes of adjacent fish provide "**schooling marks**", which signal changes in relative position (Fig. 4.7).

Platelets present on the stripes produce additional information to other fish about how a given fish moves. As the orientation of the fish changes relative to another fish, the amount of light reflected by this layer also changes. Such a cluster of visual information gives the mackerel considerable advantages in being able to react quickly while schooling and feeding.

Marlins are approximately ten species in the family **Istiophoridae**. The **Atlantic blue marlin** (***Makaira nigricans***) is the largest species, and it can reach 5 m in length and 818 kg in weight. They are popular sport fish in tropical areas, and the overfishing pressure endangers their existence. Marlins are fast swimmers and can pass, during short bursts, **100 km/h**.

Agrostis = related to a grass field; *ichthyos* = fish *Clupeus* (L) =allis shad *Cyclos* = rounded, quasi circular *Engraulis* = anchovy *Homo* = similar; *Cercus* = branch *Leeds* = city of England *Problematicus* (L) = rising problems *Sardina* = from the Mediterranean island Sardinia

The Mediterranean version of Istiophoridae marlins is the spearfish ***Tetrapturus belone***, with a short rostrum and a maximum size of 2 m in length and 60 kg in weight.

The Mediterranean Sea, however, is characterized by the presence of another type of fish with a long rostrum, the "sword fish" ***Xiphias gladius*** (family **Xiphiidae**).

Such a pelagic species is widespread in all the oceans and is particularly known in the Mediterranean Sea, where locally, it is subjected to a traditional fishery (Fig. 4.8).

Adults are large, highly migratory, predators with a long, flat, pointed rostrum (the "sword"). Swordfish are elongated, round-bodied, and lose all teeth and scales as adults.

Fig. 4.8 Messina (Italy). *Feluca* with *antenna* (vertical) and *passerella* (horizontal) to hit swordfish with harpoon. Traditional boat for the hunting of swordfish in the Strait of Sicily (Italy) (photograph G. Belmonte)

These fish can typically be found from the surface to a depth of 550 m and even down to 2000 m. They reach a maximum of 4.5 m in length and 650 kg in weight.

4.2.3 Common Species Without Market Importance

Pelagic fish are not only surface or epipelagic fish. Less known, but suspected to be very abundant, are mesopelagic species of the genus ***Cyclothone***.

All species in the genus *Cyclothone* live in the mesopelagic zone at 300–1500 m depth. There, the light is virtually absent (<1% of the sunlight at the surface), and organisms should not rely on their eyes to catch prey, avoid predators, or find mates.

Cyclothone fishes also cannot tell us how they live because they cannot survive when brought to the surface and therefore cannot be reared in a laboratory setting for observation of their lifestyle.

In 1912, a "**Lilliputian fauna**", mainly composed of *Cyclothone signata*, was reported from the deep sea, formed by high quantities of small fish that were not easily fished due to their size, deep living, and ability to escape nets. Eyring et al. (1948) reported the existence of layer scattering sounds in all the oceans (with the exception of polar ones) between 300 and 800 m in depth (the **deep scattering layer, DSL**) which could be represented by large schools of such mesopelagic fish.

Johnson (1948) discovered the diurnal vertical movement of such deep scattering layers, identifying it with "zooplankton masses". The scattering, however, is due to the presence of air/gas bubbles normally absent from zooplankton and only existing (as gas bladders) among fish whose existence was largely ignored until that moment. The fish fauna responsible for the DSL was identified as composed by small species belonging to **Gonostomatidae**, **Sternoptychidae**, and **Myctophidae** (Fig. 4.9).

Parallel net trawls allowed us to catch specimens and establish the presence of a **gas bladder** in 37 of 40 species. The absence of DSL in polar oceans is also accompanied by the absence of deep fish fauna, thus confirming the DSL-fish correspondence and eliminating any suspicion of a possible role of other large zooplankton, such as Euphausiacea (very abundant in polar seas).

The difficulty of catching fish and/or precisely estimating such fish schools (probably the largest populations of vertebrates on the planet) is also associated with a sort of metamorphosis that interests all the specimens.

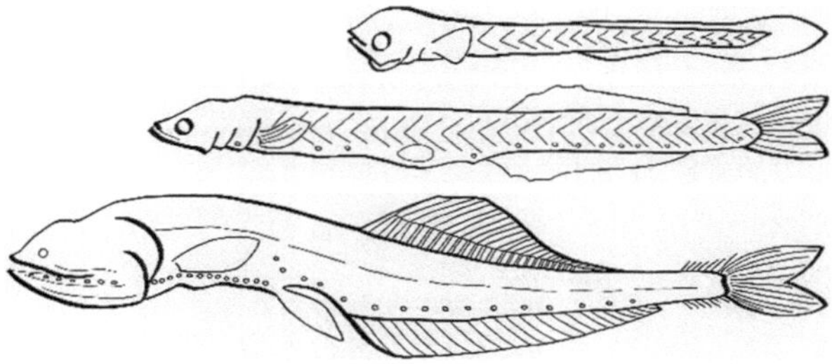

Fig. 4.9 *Cyclotone signata*. Different ages of the developmental process. Small specimens have a gas bladder. The older specimen lose it. (Courtesy G. Belmonte)

Belone = needle *Chondros* = cartilage *Cyclos* = rounded, circular *Gladius* (L) = Roman short sword *Ichthyos* = fish *Istios* = sail; *Fero* = bring *Gonios* = angle *Lilliput* = a mythic place in the Gulliver's travel *Makaira* = sword *Micteo* = nose; *Ophis* = snake *Othos* = ear *Sternos* = breast bone *Opto* = eye related; *Kinetic* = movement *Stoma* = mouth *Tetraptos* = four-strong; *Ura* = tail *Thunnus* = round bodied *Xiphias* = sword

In fact, only young stages (2–8 cm) have a small gas bladder, and this is replenished by oil or fat with age; thus, larger specimens do not form any DSL, disappearing from ultrasound studies.

The ecological importance of such mesopelagic schools is enormous, although it is still not completely perceived. The abundance of mesopelagic fish is the basis of the diving behaviour of some lunged predators (seals, for example), which seem to have evolved to profit from such a trophic resource that is evidently easy to harvest, notwithstanding the positioning at more than 200 m in depth.

Sharks (**Chondrichthyes**) are common predators of deep pelagos, but they do not form large monospecific schools, and consequently, their entrance into the fish market is occasional and linked to the capture of large and heterogeneous assemblages of fish.

The filter feeding lifestyle of some shark species (***Cetorhinus***, ***Rhincodon***) and the absence of any gas bladder oblige the pelagic sharks to conduct a never-stopping (continuously swimming) existence to avoid sinking.

Manta rays (***Mobula birostris***) are enormous ray fish, adopting filter feeding for approximately 30% of their diet and feeding for the remaining 70% on small mesopelagic fish.

Generally, Chondrichthyes stay in subsurface waters just for feeding purposes and generally inhabit deep and dark waters for the remainder of their lifetime.

The same could be affirmed for large filter feeders such as **Cetacea Mysticeti**, which stay in the epipelagos for feeding reasons, but contrary to the other large animals of the sea, they cannot abandon the very surface ocean, needing to breathe air time by time.

Although they are targets of specialized fisheries, they are generally protected and not hunted in many parts of the world today. In contrast to other Mammalia (such as seals and Pinnipedia in general) or other marine-lunged animals (such as turtles and penguins), Cetacea evolved complete independence from the land and reproduced in the open sea.

They comprehend **the largest animals** even lived on the planet, either among the filter feeders (the blue whale) or among predators (the sperm whale). Their existence is completely linked to the pelagos and geographically extended water areas, also due to their migratory behaviour.

4.3 Invertebrate Nekton

Most Cephalopoda (Mollusca) remain close to the bottom and/or are typical of neritic/coastal areas. This notwithstanding, a certain number of species are pelagic for at least a period of their life cycles or of the day. These species (mostly belonging to the genus ***Illex***) have great importance for the fishery economy.

Bi = double; *Rostrum* = part anterior to the mouth *Calipto* = covering *Ceto* = large fish; *Rhinos* = nose *Eu* = true, good; *Phausia* = shining, light emitting *Foro* = bring *Furca* (L) = fork *Illex* (L) = false *Krill* = (Norwegian) name for pelagic shrimps *Meta* = posterior, at the end; *Nauplius* = pilot of the Greek mythical ship Argo *Rhincos* = corner, angle; *Odon* = tooth; *Pereion* = middle part of a body, thorax *Sperma* = seed (male); *Thoracos* = torax; *pos-podos* = foot

In fact, ***Illex argentinus*** (the Argentine squid) is **the second most fished species worldwide**, and its large populations do not appear to be conditioned by fishery subtraction because the life cycle is short (approximately 1 year) and the life histories of single specimens are asynchronous (Brunetti and Ivanovic 1992).

The eggs are deposited on the seafloor (even at a depth of 1000 m), but they hatch asynchronously, thus producing a **continuous injection of recruits** in the active populations.

Large swarms of Argentine squids are attracted to light sources used by fishermen during the night and thus collected with large nets at the surface in the south western Atlantic Ocean.

The presence of large populations of such carnivores requires large amounts of prey availability.

Illex species generally vary their prey target with age (and size), ranging from crustaceans to fish and small Cephalopoda (sometimes even being cannibalistic on small specimens of their own species).

Their extreme abundance, on the other hand, makes them the diet basis for many predators (Pisces, Mammalia, Aves).

Such **complex trophic nets** are the rule in ocean environments and are not suitable for modelling because they are associated with a migratory behaviour that moves them for hundreds of kilometres in the ocean (up to 1500 km in the case of the North Atlantic *Illex*).

Although they are among the fastest swimmers of the sea, Cephalopoda are not completely pelagic because a part of their life cycle, or of their daytime, is spent on or close to the seafloor, and part of their food (and of their predators) is composed of benthic or demersal animals.

Other groups of very abundant animals, completely pelagic, can be considered not completely nekton due to their size. This is the case for **Euphausiacea** (Crustacea), whose adults reach conspicuous sizes (4–10 cm, according to the species) in the same size class as the mesopelagic "Lilliputian Fauna", which generates the deep scattering layer in the mesopelagos.

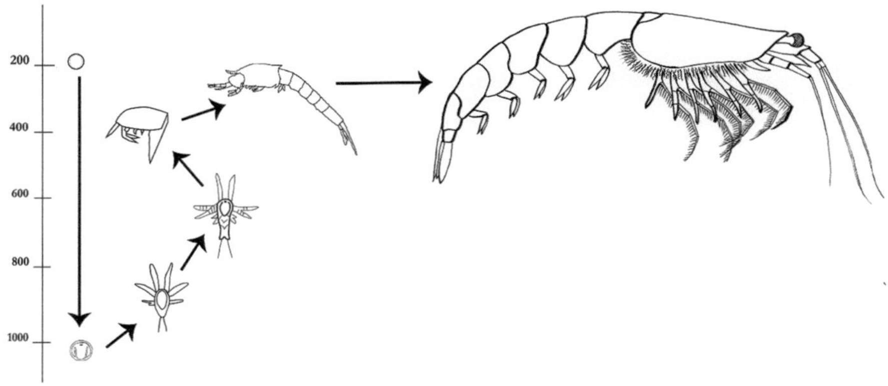

Fig. 4.10 Life cycle of *Euphausia superba* with positioning of its developmental stages in the water column. (Courtesy G. Belmonte; graphical composition of A. M. Miglietta)

Euphausiacea, however, are considered more abundant than *Cyclothone* (the DSL fish), and consequently, they comprise the most abundant schools of the planet, commonly termed **krill** (Fig. 4.10).

The *Euphausia superba* of the Antarctic region is a filter feeder that feeds on small phytoplankton during swimming thanks to a set of bristles equipping each thoracic leg. They use, in this way, an abundant food resource, positioning themselves around the **value of 2.0** in the **trophic level**. As Crustacea (rich in nutritive contents) and so abundant, Euphausiacea are food for many animal species and, overall, for the largest animals of the planet (the whales).

After the transfer of a **spermatophore** from males to females, internal fertilization involves 6000–10,000 eggs, which are broadcast in the water. During embryonic development, each 0.6 mm egg sinks to the bottom or down to 1000–3000 m in oceanic areas. The egg hatches a **nauplius**; once this has moulted into a **metanauplius** (after two moults), the organism starts migrating towards the surface in a migration known as **developmental ascent (Siegel** 2000).

Nauplius and metanauplius do not eat but are nourished by the energetic storage resources derived from the egg. After 3 weeks, the young krill finish the ascent.

They can appear in enormous numbers, counting 2000 per m^3, in 60 m water layers. Larger, additional larval stages follow metanauplius (second and third **calyptopis**, first to sixth **furcilia**), characterized by increasing development of the additional legs and the formation of compound eyes. At a size of 15 mm, the juvenile krill already have the habits of adults. Krill reach maturity after 2–3 years. Like all Crustacea, krill have to moult in order to grow. Approximately every 13–20 days, krill shed their chitinous exoskeleton and abandon it behind as exuvia.

In aquaria, if they are not fed, krill **shrink in size** after moulting, which is exceptional for animals of this size.

It is likely that this is an adaptation to the seasonality of their food supply, which is limited or even not available in the dark winter months under Antarctic sea ice.

A well-nourished krill would have eyes proportional to body length, if compared to a starving krill that would have eyes larger.

Modified thoracic legs (**thoracopodia** or **pereiopodia**) form the feeding basket of the filter apparatus.

Euphausia superba directly ingests minute phytoplankton cells, which no other animal of krill size can do.

Carcino = crab; *Phago* = eat *Hydro* = water; *Ergo* = work *Lepto* = subtle; *Onycos* = claw, nail *Lobo* = lobed; *Odon* = tooth

This is accomplished through filter feeding, using the krill's highly developed front legs, which form an efficient filtering apparatus: the six **thoracopodia** (or **pereiopoda**) create a "feeding basket" moved through the water to collect phytoplankton cells and transfer them into the mouth.

Antarctic krill can scrape off the green lawn of ice-algae from the underside of pack ice. Krill have developed special rows of rake-like setae at the tips of their thoracopodia and graze the ice in a zig-zag fashion. One krill can clear an area of one square foot in approximately 10 min.

Recent discoveries have found that ice algae films are well developed over vast ice surfaces, often containing much more carbon than the whole water column below. Krill components find an extensive energy source here, especially in the spring after food sources have been limited during the winter months.

Euphausia superba has a circumpolar distribution. The Southern Ocean covers 32 million km^2 (10 times the size of the Mediterranean Sea). In the winter season, more than three-fourths of this area becomes covered by ice, whereas 24,000,000 km^2 becomes ice-free in the summer. The surface water temperature fluctuates between −1.3 and −3 °C.

The waters of the Southern Ocean form a system of currents, the surface *stratum*, which travels around Antarctica in an easterly direction. Near the Antarctic continent, the East Wind Drift runs counterclockwise. At the front between both, large eddies develop, for example, in the Weddell Sea. The krill swarms stay inside these water masses to establish one single stock all around Antarctica, with gene exchange over the whole area.

Currently, there is little knowledge of the precise migration patterns since individual krill cannot yet be tagged to track their movements. The largest shoals are visible from space and can be tracked by satellite. One of these recorded swarms covered an area of 450 km^2 of ocean, with a thickness of 200 m, and was estimated to contain over 2 million tons of krill.

Recent studies suggest that krill do not simply drift passively in these currents but actually modify their positioning. By moving vertically through the ocean on a 12-h cycle, swarms play a major part in mixing deeper, nutrient-rich water with nutrient-poor water at the surface.

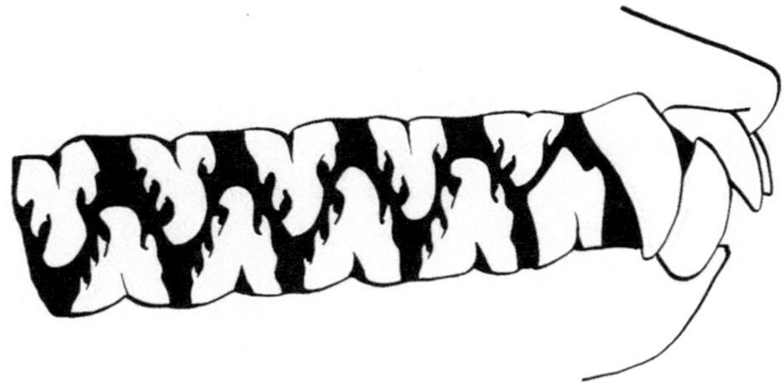

Fig. 4.11 Crabeater seal *Lobodon carcinophaga*, sketch of the specialized teeth in a lateral view of the jaws. (Courtesy G. Belmonte)

Crabeater seals (***Lobodon carcinophaga***) have even developed special teeth as an adaptation to catch abundant food sources: their unusual **multilobed teeth** enable this species to sieve krill from the water (Fig. 4.11). Its dentition looks like a perfect strainer, but how it operates in detail is still unknown.

Crab eaters are **the most abundant seals** in the world; 98% of their diet is made up of *E. superba*. These seals consume over 63 million tons of krill each year.

Leopard seals (***Hydruga leptonyx***) have developed similar teeth (45% krill in diet), but they are also predators of crabeater puppies.

All seals consume 63–130 million tons of krill per year, more than all whales (34–43 million tons). Marine birds consume 15–20 million tons, squids 30–100 million tons, and fish 10–20 million tons, reaching a total of approximately **313 million tons of krill consumption per year**.

The **biomass of Antarctic krill was estimated** in 2009 **to be** 0.05 giga tons of Carbon (Gt C), **close to the total biomass of humans** (0.06 Gt C).

The reason Antarctic krill are able to build up such a high biomass and production is that the waters around the icy Antarctic continent harbour one of the largest plankton assemblages in the world, possibly *the* largest.

The ocean surface is filled with phytoplankton; as the water rises from depth to the euphotic epipelagos, nutrients produced by the degradation of organic matter from all of the world's oceans rise into the photic zone where they are once again available to living organisms.

The primary production—the basis of the food chain—deriving from pelagic phytoplankton has an annual C fixation of **1–2 g/m^2 in the open ocean**. Close to the ice, it **reaches 30–50 g/m^2**. These values are not outstandingly high in comparison with very productive areas such as upwelling regions, but the area over which upwelling takes place is enormous, even if compared with other large primary producers such as rainforests. In addition, during the Austral summer, there are many hours of daylight to fuel the process. All of these factors make plankton and krill a critical part of the entire planet's eco-cycle.

4.4 Species Readapted to Marine Life (Reptilia, Aves, Mammalia)

There are currently 100 recognized species of marine Reptilia, 555 Aves (17–20 **Sphenisciformes**, 125 **Procellariiformes**, 60 **Suliiformes**, 350 **Charadriiformes**), and 129 Mammalia. The total of these species still show adaptations to life in aerial conditions, such as the necessity to breathe air, and most aquatic organisms need a land for reproduction and parental care.

In the evolutionary past, Reptilia showed the most diverse group of marine species, dominating the planet scenario from the **Permian** to the end of the **Cretaceous** (300 to 65 million years ago).

During the Mesozoic era, many Reptilia *taxa* adopted a marine lifestyle, including **ichthyosaurs**, **plesiosaurs**, **nothosaurs**, **placodonts**, **sea turtles, thalattosaurs,** and **thalattosuchians** (Fig. 4.12).

The class suffered the great extinction of the Cretaceous (65 million years ago), but some types of gavialid crocodilians remained abundant and widespread until the end of the **Miocene**.

Some marine Reptilia, such as ichthyosaurs, plesiosaurs, **metriorhynchid** thalattosuchians, and **mosasaurs,** became so well adapted to a marine lifestyle that they were incapable of venturing onto land and giving birth in the water. Others, such as sea turtles and saltwater crocodiles, return to shore and emerge from the water to lay their eggs. Some marine reptiles also occasionally rest and bask on land.

The Reptilia best adapted to ocean lifestyles today are marine turtles that have developed hydrodynamic profiles and finned legs. They, however, are still linked to the land habitat (sand beaches) for egg laying, embryo development, and birth.

The reason for the many species of birds included in the definition of **"marine birds"** is their dependence on marine food (Fig. 4.13).

Fig. 4.12 Left, a representation of ancient oceans (dominated by Reptilia) (drawing A. Gennari). Right, Silhouettes of marine Reptilia from ancient oceans the human size is represented by the red scuba diver (graphical composition of A.M. Miglietta)

Fig. 4.13 Silhouettes of main marine birds and their behaviour in fishing activity. Over the water surface, dipping species; below the water surface, diving species. (Courtesy G. Belmonte)

The best marine birds are **Sphoenisciformes** (penguins), which lack the ability to fly and have evolved physiological adaptations for long and deep dives, although they maintain the obligation to stay on land for reproduction.

Additionally, among Mammalia, many species are more or less linked to aerial existence, but the whole group of **Cetacea** remains in the pelagos for reproduction.

Cetacea (81 species) are the largest group of marine mammals, consisting of dolphins, porpoises, and whales. Among the *taxa* of terrestrial origin that readapted to an aquatic lifestyle, Cetacea have made the most complete transition and spend their entire lives in the water.

They breathe air through the lungs and have a nostril (blowhole) (single, or double) positioned on top of their skulls to allow them to stay submerged during air breathing. Typically, they lack posterior legs (although they maintain bone rudiments of such appendages inside the body and close to the posterior part of the vertebral column) and develop an evident horizontal caudal fin.

Their skin is naked or has very few hairs, probably to facilitate the hydrodynamic power of the body, which is streamlined and remarkably fish-like.

In the suborder **Odontoceti**, toothed species are grouped. They have teeth on both the superior and inferior jaws, such as dolphins and killer whales, or at least in the inferior jaw (mandible), such as sperm whales.

Odontoceti echolocate (they send sounds through water and receive echoes from small cavities containing gas) principally to individuate prey (fish equipped with the gas bladder) in the dark layers of the epipelagos.

The **melon**, an ovoid-shaped fatty organ found in the forehead of all toothed whales, is believed to be used in **echolocation**. Dolphins are able to generate sounds in the form of clicks within nasal sacs at the top of the skull, just behind the melon, which probably plays an amplificatory role in the produced sound. When the generated sound waves strike an object, some of them are reflected back and perceived by the ear apparatuses of cetaceans.

Among Odontoceti, dolphins can be distinguished for the following anatomical characters (Fig. 4.14):
An elongated snout (also, not correctly called rostrum).

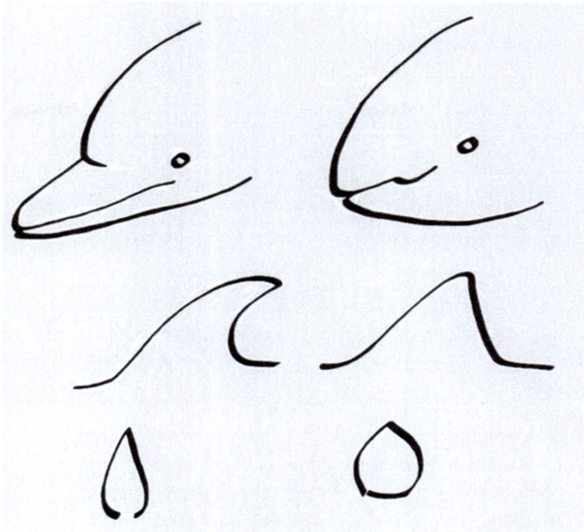

Fig. 4.14 Anatomical differences between dolphins (**a**) and porpoises (**b**) for their identification in the wild. From the top to the bottom, snout, dorsal fin, tooth. (Courtesy G. Belmonte)

A falcate dorsal fin.
Conical-shaped teeth.
In contrast, porpoises show:
A blunt snout (sometimes not correctly indicated as a rostrum).
A triangle-shaped dorsal fin.
Spade-shaped teeth.

In the suborder **Mysticeti** (baleen whales), teeth are absent, and the alimentary structures equipping the mouth are parallel rows of keratinous plates (baleen) hanging from the mouth ceiling (palate). The baleen structure is used as a strainer to collect zooplankton, krill, and small fish after filling the mouth with a large volume of water and squeezing it out through the baleen due to the pushing action of the tongue.

Cetus = monster (large) fish *Charadrius* = plover *Creta* = Greek Island *Ichthyos* = fish; *Sauros* = lizard *Noto* = dorsal; *Sauros* = lizard *Perm* = city of East European Russia *Placo* = plate; *Odon* = tooth *Plesio* = close to (resembling); *Sauros* = lizard *Procella* (L) = marine tempest *Spheniscus* = wedge *Sula* (L) = little she-bear, (L) nickname of Ursula *Thalatto* = marine; *Sauros* = lizard; *Suchus* = crocodile

Box 4.2 Cetacea Evolution

Whales evolved from dog-sized land mammals approximately 52.5 million years ago (**Eocene**), just after the great extinction at the end of the Cretaceous (65 Million years ago), thus participating in the evolutionary adaptation of Mammalia in the substitution of large Reptilia, which have dominated the planet to date. ***Pakicetus***, one of the oldest known cetacean ancestors, had a long-snouted skull with primitive carnivorous teeth lining its jaws. ***Pakicetus inachus*** is a cetacean ancestor from the Early Eocene of Pakistan with nose openings and eyes oriented upwards in the dorsal part of the skull. The species is believed to have been semiaquatic because its dense cortical limb bones, instead of spongy bones, are considered to reduce buoyancy in the water. Extant hippopotamuses have the same bone structure as their cetacean ancestors.

Indohyus mayor is the most suspected species, as it causes the descent to successively produce Cetacea (Thewissen et al. 2007). The animal was an **artiodactyl**, even-toed herbivore and lived in the Middle Eocene in a geographic region today occupied by Kashmir.

The modern artiodactyl species morphologically most similar to *Indohyus* is probably the African mousedeer ***Hyemoschus aquaticus,*** which lives near streams and feeds on land (Fig. 4.15). *Indohyus* had more pronounced aquatic specializations than *Hyemoschus* does, and it probably spent a considerably greater amount of time in the water either for protection or when feeding. *Indohyus* either came onshore to feed on vegetation (as the modern ***Hippopotamus*** does) or foraged on invertebrates or aquatic vegetation in the same way that the modern muskrat *Ondatra* does. Such an ancient artiodactyl belongs to a *taxon* (family **Raoellidae**) that is considered the sister group to Cetacea, and this implies that aquatic lifestyle originated before the Order Cetacea and in freshwater.

The great evolutionary change that occurred at the origin of Cetacea is thus driven by dietary specialization. This is consistent with the cranial and dental **synapomorphies** identified. Molars of *Indohyus* are markedly different from those of pakicetids, and it is widely assumed that pakicetids, differently ftrom *Indohyus*, ate aquatic prey. The compared morphology of *Indohyus* and *Hippopotamus* towards Cetacea shows a progressive migration of nostrils from the anterior extremity of the skull to the blowhole position at the top of the skull (Nummela et al. 2006). Further adaptation is perceivable in the progressive lack of importance of the posterior legs, which no longer protrude from the body in modern Cetacea. These last, however, clearly show hind limbs in their embryonic early stages.

In 2001, the first archaeocete **astragalus** (a short bone at the level of the foot–leg articulation) was found. The double-pulley astragalus is a **synapomorphy** present in Cetacea and Artiodactyla (Thewissen et al. 2007) (Fig. 4.15b).

The Sulaiman Mountain Range (Pakistan) was once the ocean floor of the Tethys Ocean. This water body was a nursery for the budding archaeocete lineage.

Another important deposit of archaeocete fossils is the Wadi Al-Hitan (Arabic for "Valley of the Whales"), Fayoum, western desert of Egypt, a territory submerged by sea during the Eocene. Among archaeoceti, ***Ambulocetus natans*** (48 million years ago) was still partially terrestrial and partially aquatic.

Apart from the above-described ancestors, the first archaeocete discovered by humans was ***Basilosaurus*** (40 million years ago) in 1830.

The species was named erroneously because the *-saurus* suffix implies this animal was a reptilian (lizard) like the much longer-extinct plesio-saurs! *Basilosaurus* was the largest creature of its time, marine or terrestrial.

Basilosaurus contrasted with the primordial *Ambulocetus* skeletons due to its long body with negligible hind legs.

In the **Miocene** (23-5 My ago), the land–sea transition of archaeocetes was completed, and many Cetacea appeared to anticipate the arrival of modern species. In the **MAUS** collection (University of Salento), 60 Cetacea specimens (or parts of them) are present, 12 classified, from 12 to 10 million years ago (Miocene).

Three of them are holotypes (***Messapicetus longirostris*, *Zygophyseter varolai*, *Archaescrictius ruggieroi***) (Fig. 4.16).

The Italian territory, in addition to North Egypt and South Pakistan, is an elected site for the search for prehistoric Cetacea. This is due to the geographic position of these territories, once a time submerged by the Miocenic seas (the western part of the Tethys Ocean).

Pisa, in Italy, hosts one of the most important and beautiful collections of modern Cetacea (19 complete skeletons in a gallery of the Natural History Museum of the Calci Certosa) (Fig. 4.17).

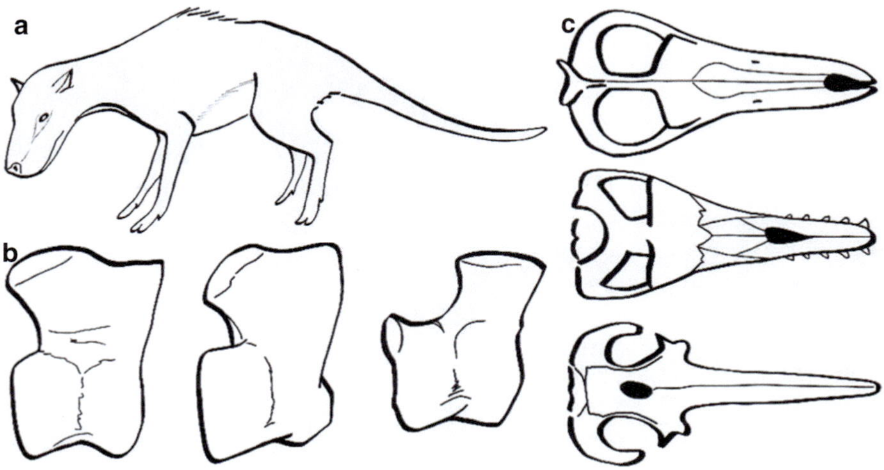

Fig. 4.15 (**a**) An interpretation of the *Indohyus* based on anatomical characters derived by fossils; (**b**) astragalus comparison between *Pakicetus* (on the left) and modern Arctyodactila (pig and deer); (**c**) nostril migration in the skull from *Ambulocetus* (on the top) to modern dolphins (on the bottom). (Courtesy G. Belmonte)

Fig. 4.16 The skull of a large cetacean coming from the "Lecce limestone" (Miocene), *Zygophyseter varolai*, from the MAUS Museum of the Environment, University of Salento (Italy) (photograph G. Belmonte)

Ambulocetus, Ambulo = walk; *Cetus* = large fish *Aquaticus* (L) = water related *Arche* = ancient; *Scrictius* = written *Artio* = articulated; *Dactilus* = finger *Astragalus* = ankle bone *Basilosaurus*, *Basileus* = emperor; *Saurus* = lizard *Echo* = sound phenomenon; location = site *Inachus* = a Greek god of rivers. The first king of Argo *Indo* = from India; *hyus* = pig *Odontos* = teeth; *Cetus* = large fish *Paki* = from Pakistan *Hippopotamus*, *Hippo* = horse; *Potamus* = river *Hyemoschus*, *Hyus* = pig; *Moschus* = musk deer *Longus* = long; *Rostrum* = snout *Messapicus* = belonging to Messapia (ancient name of Salento Peninsula, Southern Apulia); *Cetus* = large fish *Natans* = swimming *Synapomorphy*, *Syn* = together/share; *Apo* = original; *Morphe* = form/anatomic detail *Zygophyseter varolai*, *Zygos* = fused/joined; *Physeter* = blowpipe, blowhole *Varolai* = dedicated to A. Varola

Fig. 4.17 Top, the Calci *Certosa* Gallery of Cetacea (Museum of Natural History, University of Pisa, Italy) (photograph G. Belmonte). Bottom, interpretation of *Zygophyseter* from the Lecce Miocene (Drawing A. Gennari)

4.5 Mediterranean Common Nekton Traits

The investigation of patterns and relationships between anatomical traits of organisms is especially informative for geographic areas that host a high biodiversity concentration, such as the Mediterranean Sea. In addition, the Mediterranean Sea has a long history of human presence that interfered with the natural populations of marine species, both directly (by fishing) and indirectly (by pollution).

An extensive dataset of **23 traits** relative to **235 nektonic species** (197 Osteichthyes, 20 Chondrichthyes, 10 Cephalopoda, 8 Crustacea) of the Mediterranean Sea was considered by Tsanatos et al. (2020) to depict the main traits of the Mediterranean Nekton.

Most (89.6%) of the species examined were found to have a **life span ranging from** 1.1 to 118.0 years (**average, 14.4 years**).

Regarding size, Mediterranean Nekton species showed maximum lengths ranging from 5.0 to 800 cm (**average, 63.7 cm**).

Fecundity showed a narrow variability, with extreme values represented by only a few species.

In particular, the values of **low fecundity** are a characteristic **typical of Chondrichthyes** (the 20 lowest fecundity values). The highest values of fecundity were obtained by a heterogeneous group of Osteichthyes: *Thunnus thynnus, Labrus merula, Psetta maxima, Scophthalmus rhombus, Polyprion americanus.*

Regarding the **trophic level**, **primary consumers** and/or herbivores were rare, and a **general positioning of the lower level species** at trophic level **3.0** (predators of herbivores) have been found.

Resuming the dominant trait categories of Mediterranean Nekton, we can affirm that a **typical** species is a **bony fish** (83% of considered species), **gonochoric** (86%), with a **subtropical geographic distribution** (64%), **summer spawner** (61%), with **benthopelagic lifestyle** (63%), **carnivore** (60%), **eurytherm** (66%), **stenohaline** (61%), and **solitary** (57%).

The relationships between continuous traits indicated **eight statistically significant correlations** among traits. Long-living organisms are larger, occupy a high trophic level, dwell deeper and have a lower age at maturity (calculated as a percentage of their life span).

Finally, in three cases, nonlinear relationships provided better residual fit than linear ones: **fecundity** was **higher in species with low and high life spans** ($R^2 = 0.286$, $p < 0.001$), and the same trait showed the highest values in intermediate water depths ($R^2 = 0.055$, $p = 0.004$).

The life span is maximal in **flat-shaped species**, in **ambushing and active predators**, and in **pelagic spawners**.

Fecundity is maximal in **fusiform and pelagic** species.

The statistical analysis (hierarchical clustering) cut at a high value of dissimilarity (e.g., 70%), distinguishes the three major groups of **pelagic**, **benthopelagic**, and **benthic species**.

The rarity (even below 5% of the species total) of autumn spawning (but also winter and all year spawning) as well as that of herbivory was a surprising result.

The **planktivorous diet** may be considered **rare** if **related to the number of species,** but **very common if related to the specimen abundance** (such as that of the small pelagic or benthopelagic species that possess this trait). Nevertheless, the fact that it is shared by only a handful of species may be a risk for ecosystem functioning, especially taking into account the fact that these species are known to have interannual abundance fluctuations.

Some of these species, such as the European anchovy ***Engraulis encrasicholus*** and the European pilchard ***Sardina pilchardus,*** are under intense fishing pressure and have been shown to be affected by climate variation.

The **positive relationship between size and trophic level** is evident because most predators are typically larger than their prey. Additionally, a relationship was found between **maximum body length and average depth** of species distribution.

The traits of the family Sparidae, such as compressed body shape, hermaphroditism (e.g., *Sparus aurata*) and grazing feeding (e.g., *Sarpa salpa*), were found to have a relatively high number of positive associations with other traits.

Regarding **fisheries-derived effects**, **size** (also for the various significant relationships existing with other traits and its implications for management) is a crucial trait.

The study of nektonic trait distribution and combinations can be useful to elucidate trait interactions significant for indirect alterations of ecosystem functioning, especially today with a multitude of anthropogenic stressors acting on marine environments and on specific traits of marine biodiversity.

The documentation of **rare traits** (such as **winter and autumn spawning**, **herbivory**, very low or high **size or fecundity**, hard substrate type and **ambushing predation**) in species together with the appraisal of the significance of traits at various scales indicates aspects of crucial importance that need to be preserved.

Engraulis = anchovy; *Encrasicholus* = related to anchovies *Gone* = related to sexual gland; *Chorus* = together *Eury* = good, easy; *Termos* = related to warmth *Labrus* = leap, reem *Maxima* = the largest *Merula* = blackbird *Mi* = close/at short distance; *Ops* = look; *Coenos* = stay together *Pilchard* = without value *Planktos* = vaguing; *Voros* = eater *Poly* = multiple; *Prion* = saw *Psetta* = flatfish *Rhombus* = spinning top, of bicone shape *Ruggiero* = Ruggiero, Livio First Director of MAUS of University of Salento *Sardina* = from the Mediterranean island Sardinia *Steno* = restricted, difficult; *Halos* = related to salinity *Scophthalmus* = from *skopelos* (a lantern fish) and *ophthalmos* (eyes) *Tethys* = nymph of Greek mythology, mother of Achilles

In any case, the documentation of a multitude of relationships between functional and ecological traits indicates how the **environment**, through the delimitation of species distribution in space and time depending on their traits, **filters functional traits**, while the validation of functional trait associations hints at functional interdependencies determined by evolution.

References

Bakun A (1996) Patterns in the ocean. Ocean processes and marine population dynamics. California Sea Grant College System, and Centro de Investigaciones Biologicas del Noroeste, La Paz, 323 pp

Brunetti NE, Ivanovic ML (1992) Distribution and abundance of early life stages of squid (*Illex argentinus*) in the south-west Atlantic. ICES J Mar Sci 49:175–183

Denton LEJ, Rowe DM (1998) Bands against stripes on the backs of mackerel, *Scomber scombrus*. Proc R Soc Lond B 265:1051–1058

Denton EJ, Gilpin-Brown JB, Shaw TI (1969) A buoyancy mechanism found in cranchid squid. Proc R Soc Lond Ser B Biol Sci 174(1036):271–279

Eyring CF, Christensen RJ, Raitt RW (1948) Reverberation in the sea. J Acoust Soc Am 20:462. https://doi.org/10.1121/1.1906399

Johnson MW (1948) Sound as a tool in marine ecology, from data on biological noises and the deep scattering layer. Contributions from the Scripps Institution of Oceanography, vol 380, pp 443–458

Nummela S, Hussain ST, Thewissen JGM (2006) Cranial anatomy of Pakicetidae (Cetacea, Mammalia). J Vertebr Palaeontol 26:746–759

Pauly D, Trites AW, Capuli E, Christensen V (1998) Diet composition and trophic levels of marine mammals. ICES J Mar Sci 55:467–481

Siegel V (2000) Krill (Euphausiacea) life history and aspects of population dynamics

Stocker S (1999) Models for tuna school formation. Math Biosci 156:167–190

Thewissen J, Cooper L, Clementz M, Bajpai S, Tiwari BN (2007) Whales originated from aquatic artiodactyls in the Eocene epoch of India. Nature 450:1190–1194

Tzanatos E, Moukas C, Koutsidi M (2020) Mediterranean nekton traits: distribution, relationships and significance for marine ecology monitoring and management. Peer J 8:e8494

Wittemberg JB (1958) The secretion of inert gas into the swim-bladder of fish. J Gen Physiol 41(4):783–804

Pelagic Lifestyles in Confined Environments

5

Box 5.1 The Paralic System

In the paralic system scheme proposed by Guelorget and Perthuisot (1992) (Fig. 5.1), six levels were recognized with a growing (from I to VI) distance from marine conditions. Level 1 (I) corresponds to the marine domain; level 6th corresponds to an environmental evolution towards a complete freshwater condition (estuaries) or a hyperhaline condition (salt ponds) where diversity, biomass, and productivity, if compared with those of the central paralic system, are somehow reduced.

In the paralic system schematization, level 2 (II) is a transitional situation between marine domain (level 1) and level 3 (III) which is characterized by a maximum of benthos biomass and production.

The benthos importance diminishes at level 4 (IV) where contemporaneously the importance of plankton biomass and production grows. These last are the maximum at level 5 (V).

This general scheme implies also a transition in body size of organisms from those of the benthos (larger) to those of the plankton (smaller) and, after level V, to the size of Bacteria.

More in detail, microphytobenthos dominate level II, macrophytobenthos dominate level III, phytoplankton dominate levels IV–V, and Bacteria dominate levels V–VI.

The reduced presence of large organisms, from level IV to VI, corresponds to a lack of consumption of the primary production realized by phytoplankton and Bacteria. The lack of consumption determines a net balance in favour of organic matter accumulation/sedimentation and possible Oxygen depletion due to degradative processes at levels V and VI.

G. Belmonte, *Elements of Pelagos Biology*,
https://doi.org/10.1007/978-3-031-74280-4_5

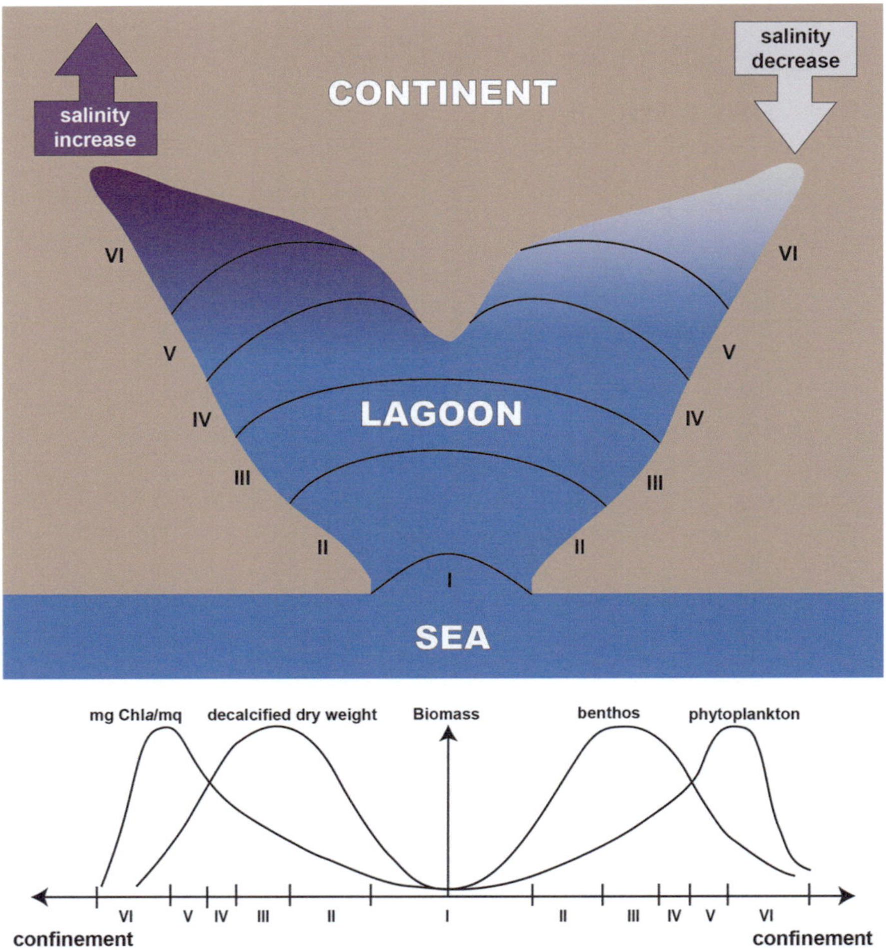

Fig. 5.1 The paralic system of Guelorget and Perthuisot (1992) (redrawn by A.M. Miglietta). On the top, the spatial representation of transition from the sea to the continental waters. On the bottom, a scheme of biomass abundance under different levels (I–VI) of confinement

Such a confinement scheme could be even exported on a larger geographic scale to verify if the ecological rules of decreasing body size of community-components, reduction of species richness, and growth of productivity, can be recognized in large seas, if considered as confined parts of oceans.

An attempt of zonization (according to the six levels of the Paralic System scheme) was proposed for the Baltic Sea, only weakly connected with the Northern Sea (part of the North-east Atlantic Ocean).

Here the paralic level 1 could correspond to the North Sea, level 2 describes the situation of Skagerrak passage, and the whole Baltic Sea should belong to levels 3 and superior. Level 3 of confinement starts at the southern Danish islands and terminates at the level of the Sweden–Poland conjunction. All the remaining Baltic Sea is labelled as belonging to levels 4 and 5. The deep part of the Baltic, between Sweden and the Gulf of Finland, is so isolated and without remixing of the water column that level 6 has been proposed (Fig. 5.2).

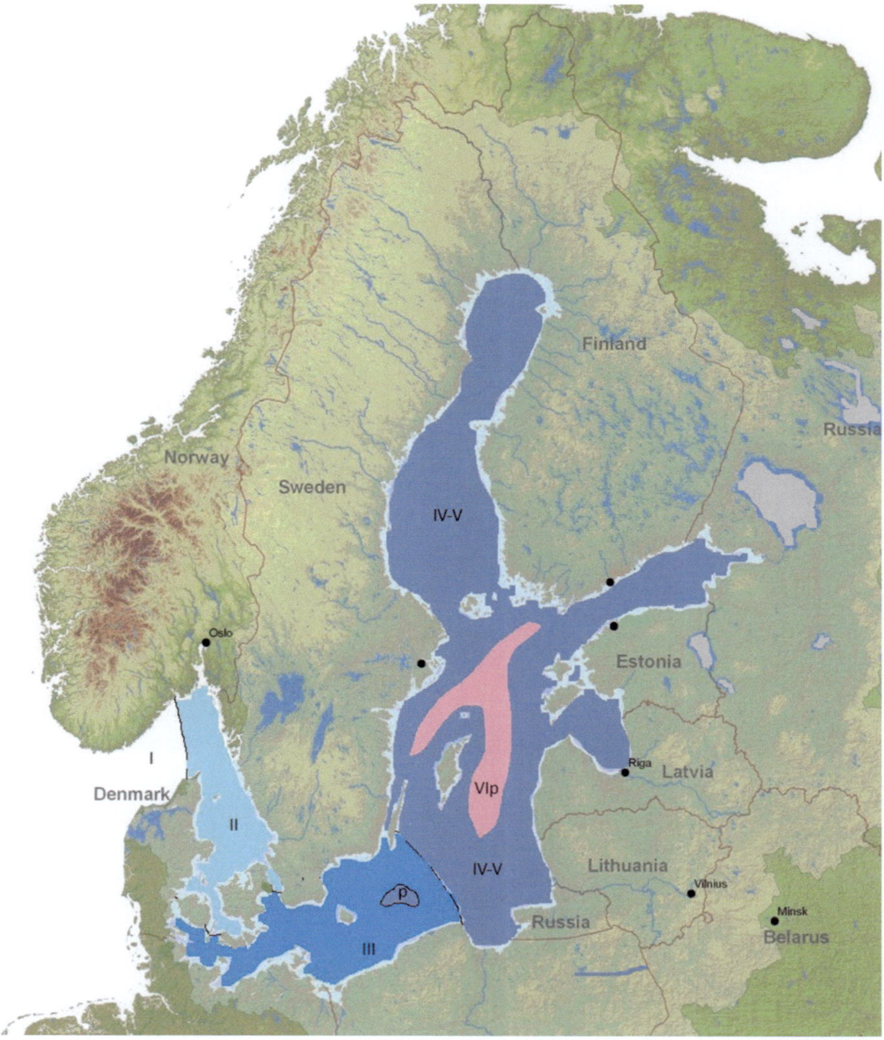

Fig. 5.2 The Baltic Sea as a paralic system. I–VI, levels of confinement; p, deep parts of the basin (depth confinement) (original)

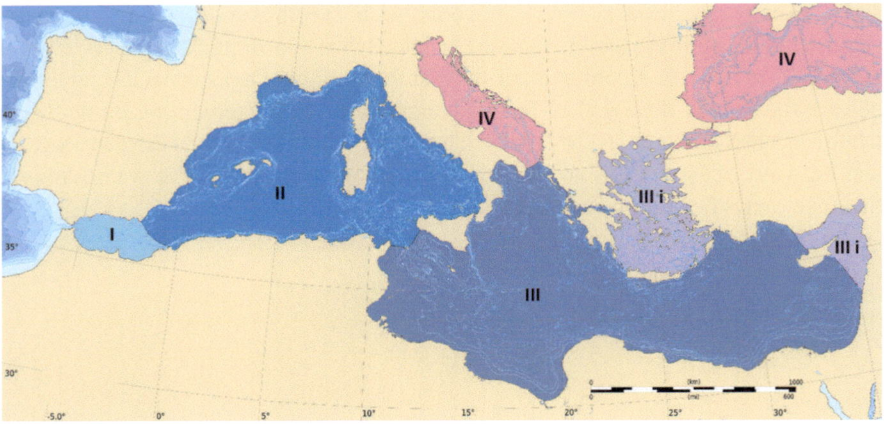

Fig. 5.3 The Mediterranean Sea as a paralic system. Confinement levels (I–IV) also contain an extension of level III, as "internal" (III i). The Azov Sea, not represented in the map, receives level V (original)

Such a confinement classification could be proposed for every internal-coastal sea, and the Mediterranean Sea could represent a confined part of the Atlantic Ocean, in terms of species richness, productivity, and body size diminution of organisms, in practise a lagoon if compared with the neighbouring ocean.

Inside the same Mediterranean Sea, the Levant Basin could be considered a part more confined than the Western part, and a gradient of confinement could be identified along the transition W part–E part–Black–Azov Seas (Fig. 5.3).

Macro = large; *Micro* = small; *Phyto* = vegetal; *Benthos* = bottom system *Para* = close to; *halos* = saline *Phytos* = vegetal; *Plankton* = going around

5.1 Basic Features of Confined Environments' Pelagos

To obtain information on the pelagic communities of the confined environments, some "laws" of the confined systems have to be considered:

(a) Communities in confined environments are composed of small species, with short life cycles in comparison with open seas.
(b) Stress conditions are more intense than in the open sea.
(c) Variations can occur on a time and/or space scale smaller than in the open sea.

The higher variability of environmental conditions both in space and in time in confined environments, coupled with the short life cycles of plankton, is the main reason for the adoption of a common lifestyle strategy: the presence of species only under favourable conditions. This also means that species are somewhere else when the conditions are hostile, and this is of fundamental importance in organizing sample collections devoted to understanding the biodiversity content and the functioning of confined ecosystems.

5.2 Techniques and Methods

The enhanced variability of abiotic conditions on a year-scale, the general diminution of body sizes, and the consequent shortening of life cycles of Metazoa plankton, coupled with the shallowness and brackish condition of the water, depict a complex frame where sample collections have to be strategically planned and carried out, paying more attention than in open sea waters.

The variability of abiotic conditions imposes the adoption of a seasonal collection rhythm (four times per year at temperate latitudes) at least. The variability of abiotic conditions, however, is not limited to the year period; it also characterizes the space, due to the influence of locally delimited coastal currents, the inflow of freshwater from rivers, the proximity of human settlements and activities.

The collection plan for samples must consider such elements of diversification to obtain reliable results on the actual ecological conditions of the coastal system under study.

The plankton net adopted should be equipped with a mesh size smaller (e.g., 80 μm) than that usually used in the open sea to collect plankton with body sizes generally smaller and/or characterized by larvae coming from benthic adults. The narrower mesh size of plankton nets, requested by a collection plan in confined environments, cannot ensure the same capture efficacy of meso-zooplankton of the largest size, which can be obtained with a coarse mesh size. To obtain comparable collections in areas with a different confinement degree, both mesh sizes (e.g., 80 and 200 μm) should be used.

The sampling stations should be numerous enough to describe the plankton diversification in the space, and the sampling should be frequent enough to correctly describe the plankton seasonality.

For the robustness of the statistics that are applied to the final results, each station should be interested in a certain number of repetitive collections (replicates) just to circumscribe the casual variability. If more mesh sizes of nets are used, the replications have to be interested in each mesh size adopted.

Finally, the analysis of the samples must avoid a "simplification" in light of their heterogeneity in terms of community composition. In fact, larvae are sometimes abundant, and their identification is not possible at the level of species, contrary to the adults of the holoplankton. This leads to a focus of attention on the only identifiable species, with a restriction of the analysis to a part of the whole existing biodiversity (Fig. 5.4).

For Phytoplankton, the use of different mesh sizes does not matter because the sampling is performed with the collection of volumes of water (1–5 L). Also in this case, however, the multiple collection of samples from the same station, the

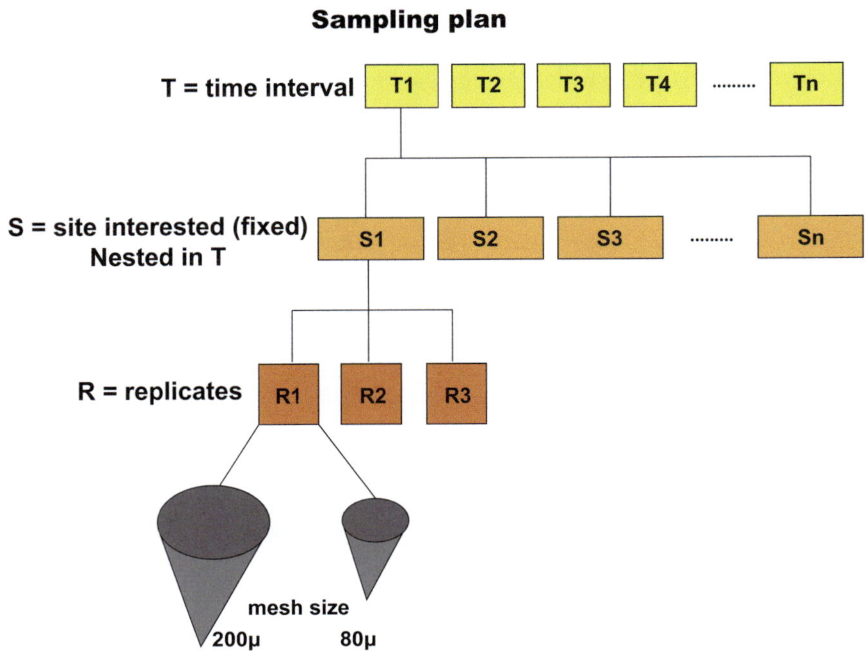

Fig. 5.4 Sampling plan scheme to be considered for plankton collection in confined environments. The employment of plankton nets (here with mesh sizes of 200 and 80 µm) has to be enriched by water collected with bottles for the recognition of the smallest body size fraction components (graphic composition, A. M. Miglietta)

adoption of more stations to better describe the variability of community with space, and a shorter period dividing two successive sampling dates are necessary.

For phytoplankton and micro-zooplankton of confined areas, it also should be useful to collect sediments to analyse the presence of resting stages periodically produced by holoplankton.

5.3 Life Cycle Strategies

The description of ecological systems relies upon the individuation of ecological compartments (production, consummation, decomposition) where to collocate living organisms as functional units that help the student in reading the flow materials and energy for the understanding of ecosystem functioning.

Such compartments are useful for modelling, but they do not consider the concrete parts (species), each somehow different from the others, in the adoption of life strategies to solve the same environmental problems.

Plankton, in particular, has been always considered as a compartment, or a category with few functional compartments (phyto-, zooplankton) of living beings. This accepted simplification led to the ignorance of what are interspecies differences and where the species are, when they are not in the plankton.

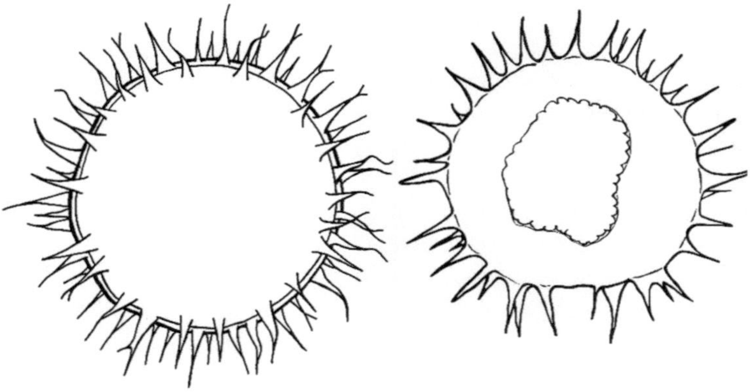

Fig. 5.5 Left, Resting Egg for resting Metazoa embryos that do not encyst themselves but, in turn, could be considered "encysted" (by accessory cells of the mother). Right, cyst of unicellular organisms which encyst themselves, generally after the sexual reproduction. (From Belmonte and Rubino 2019 re-drawn)

With or without each species the plankton continues to exist (because single species succeed each other in performing their compartmental roles) and to offer a contribution to the functioning of the ecosystem.

Thus, the Phyto-plankton suffers from the absence of salt nutrients in the water, and when remixed conditions introduce nutrients into the system, the phytoplankton blooms, offering available energy (in terms of food) to the whole trophic net. Phytoplankton, however, is not a unity but an assemblage of dozen (sometimes hundreds) different species, and when the rarest disappear at the lack of nutrients, the simple renewal of the nutrient availability (e.g., in upwelling areas) cannot be responsible for the re-appearance of disappeared species. Each species, in fact, derives only from representatives of the same species, and (as Redi demonstrated in the seventeenth century), individuals cannot be "spontaneously" produced by inorganic material (nutrients).

An easy answer to the problem could be that species are elsewhere or that they are composed of so few individuals that our sampling methods are not able to detect them. Although this last is a possibility, we have to face the embarrassing consideration that "planktonologists work badly" because they do not find species which are part of the system or, more embarrassingly, that the evolutionary rule of genetic drift does not function on the evolution of plankton organisms. In fact, the survival of a few rare individuals each year would sexually refuel the total population without suffering the principle of genetic drift, because this, with time, should be responsible for the appearance of new species.

Recently, however, the scientific community of marine biologists has started to understand what limnologists have known for a long time: species of plankton can produce resting stages that allow them to pass unfavourable periods without abandoning the geographic site. Such a suspension of the life cycle is now reported from plankton species, mostly coastal, of Diatomeae, Dinophyta, Ciliophora, Rotifera, Cladocera, Copepoda (Belmonte and Rubino 2019) (Fig. 5.5).

Among unicellular organisms (Protozoa and Chromista) the single cell encysts itself, possibly after a sexual reproduction act. Among Metazoa, however, we consider that the resting stage (an embryo inside the egg envelope) has been reduced in the encysted condition by a contour of dedicated gonadic cells before abandoning the mother body (thus the embryo does not produce the cyst that envelopes it).

For marine plankton, the existence of resting stages has been ascertained for the first time by Wall and Dale (1968) among Dinophyta, for unicellular organisms, and by Sazhina (1968) among Copepoda, for Metazoa.

At present, more than 500 species of coastal marine plankton have been listed among those producing cysts, and this number seems still far from reality.

Cilium (L) = eyelash; *Phora* = bringing *Clados* = branch; *Cheiros* = horn *Cope* = oar; *Podos* = feet *Dia* = space, interval; *Tome* = separated *Dia* = space, interval; *Pausa* = stop, suspension *Phytos* = plant; *Plankton* = going around *Rota* (L) = weel; *Phora* = bringing *Zoos* = living organism with a soul; animal

The rest for plankton has the same wide classification and terminology used for all the other living organisms. All the suspensions of life, in fact, can be grouped under the term "dormancy" (or lethargy, for vertebrates).

A first subdivision regards the dormancy induced by environmental factors and a second subdivision regards the genetic programation of individuals. In the first case, environment-induced dormancy, the correct and distinctive term is quiescence.

It occurs as a consequence of unpredictable stress, which somehow produces a physiological response useful to avoid the interaction of the individual with such an unpredicted situation. In the second case (genetically programmed dormancy), the distinctive term is diapause. It occurs as an evolutionary adaptation (thus a message content in the DNA of organisms), which anticipates a predictable stress (as the seasonally recurrent ones).

Quiescence acts only after an external solicitation on normal developmental stages that are obliged, at that point, to subtract themselves from the stressful situation by suspending their lives. Such a condition of suspended life can be re-produced if the stress remains, thus obliging the sleeping organism to a long absence of activity (Fig. 5.6).

Diapause acts under genetic control and can produce a sleeping status that is even different from the corresponding "normal" one of the life cycle.

The sleeping condition is deeper than in the quiescence, at least in terms of duration. This, in fact, either long or short and insensitive to external cues, produces a life suspension that is more intense and persistent, resolved only by the deadline of the refractory period. The possible absence of suitable conditions at the end of the refractory period re-dives the developmental stage into a deep sleeping cycle, which can be re-proposed many times (Giangrande et al. 1994).

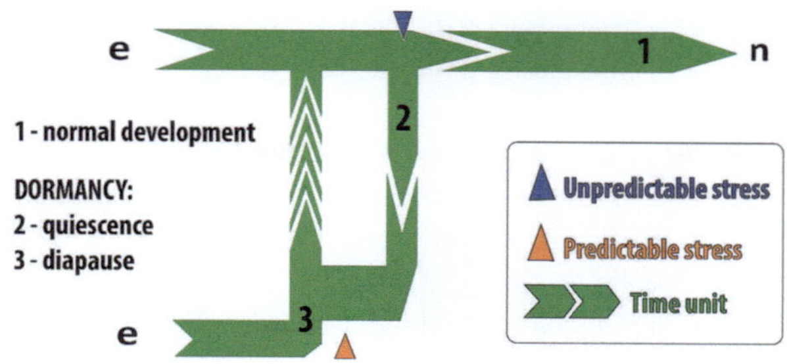

Fig. 5.6 Scheme of alternative paths of the embryonic development in Copepoda Calanoida. (From Giangrande et al. 1994, modified by A.M. Miglietta). 1—normal development (two time units = 2 days) from e, egg laying, to n, egg hatching and nauplius birth. 2—occurrence of an unpredictable stress that induces embryo dormancy inside the egg envelope (quiescence). Development is obliged to pass through a certain number of time units (days) before returning to the hatch pathway. If the stress condition persists, however, the hatch does not occur and the rest is prolonged. 3—The e, egg, is produced with the aim of resting and has not possibilities to hatch a nauplius (n) before a certain period (refractory phase). This dormancy (diapause) is genetically programmed and executed in prevision of a recurrent stress (before its occurrence)

Several environmental factors are hypothesized to be responsible for the onset of diapause. For populations of the freshwater Cyclopoida *Cyclops strenuus*, it was hypothesized that diapause is primarily cued by the length of day (photoperiod).

The marine Calanoida *Labidocera aestiva* reared at 15 °C produce subitaneous eggs under a "summer" photoperiod of 18hL to 6hD, whereas females exposed to a "winter" photoperiod of 8hL to 16hD produce mainly diapause eggs.

In Cyclopoida the rest involves the copepodid stage, and in *Cyclops vicinus*, a response to environmental stimuli is sexually differentiated, with males being more involved than females.

Annual seasonality (variation of photoperiod and temperature) is the most common environmental factor driving diapause, but other cycles are possible, such as the lunar cycle or its associated tidal cycle.

Copepoda are well equipped with mechanical and chemical receptors that are likely to detect environmental information, for example, the presence of predators, changes in phytoplankton abundance, or population crowding. Such a set of stimuli could directly affect the induction and termination of individual dormancy.

The timing of diapause is possibly coupled with still unknown internal triggers. In *Pseudocalanus* sp., the nutritional state (i.e., lipid content) of the copepodid V acts on physiological events that induce diapause.

An abundant granular secretion in neurosecretory cells of the cerebral region of *Calanus finmarchicus* and *C. helgolandicus* (Calanoida) was found to be at the basis of the control of diapause, moulting, and overwintering metabolism and behaviour (Maps et al. 2012). In addition, there is widespread mention in the literature of a long-term endogenous timer (i.e., an internal or biological clock) cueing the arousal and especially the termination of diapause among Copepoda.

The type of eggs (subitaneous or diapause) produced by *Eurytemora affinis* is determined by the environmental conditions experienced during the naupliar stage of the mother.

Calanus finmarchicus also undergoes dormancy to be successful in the highly seasonal boreal Atlantic. However, it does not produce resting eggs but enters the lethargic stage as a copepodid, according to mechanisms that are still not completely understood.

This lack of understanding hampers our ability to model and forecast the response of *C. finmarchicus* population dynamics to a rapidly changing marine environment. Dormancy in *C. finmarchicus* is an extended period spent at depth below the surface mixed layer, during which lethargic stages do not feed. Dormant individuals rely on lipid (wax esters) storage to fuel their lowered metabolic demand (during a long period of starvation) and to support gonad maturation and the ontogenetic vertical upward migration occurring at the wake-up after moulting of the sleeping copepodid V into the adult (copepodid VI). The lipid accumulation window (LAW) hypothesis proposes that there is a window of favourable environmental conditions during which *C. finmarchicus* copepodids are able to accumulate lipid reserves sufficient to meet the dormancy period and the maturation process upon dormancy arousal. Venturing into dormancy with low lipid levels would enhance the mortality risk because of either starvation at depth or a mistimed migration to an unfavourable surface environment (Fig. 5.7).

An evolutionary stable life-cycle strategy for *C. finmarchicus* requires dormancy to be initiated while the environment remains favourable for reproduction, growth,

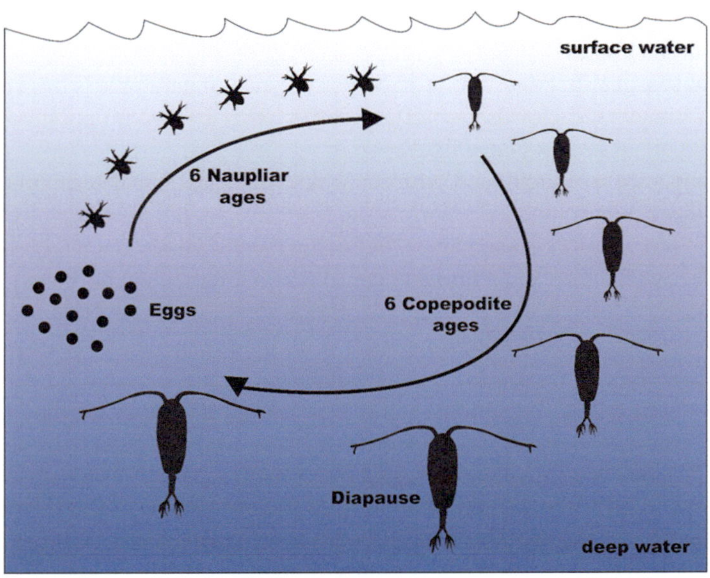

Fig. 5.7 Dormancy activity cycle in *Calanus finmarchicus*. The cycle involves different depth layers of the ocean (graphic composition, A. M. Miglietta)

and survival, and that the lipid reserves of the individuals should be destined for the dormancy period.

The timing of initiation and termination of dormancy varies by several weeks for the same species on the western North Atlantic continental shelf. In the Gulf of St. Lawrence and the Gulf of Maine, local populations of *C. finmarchicus* produce stocks of dormant individuals, profiting from the presence of local topographic refuges at depths of a few hundred meters. However, these two neighbouring areas offer contrasting temperatures and food conditions for different populations of the same species. The Gulf of St. Lawrence is cold, with winter sea-ice cover and a primary production period extending from spring to fall. It is characterized by a cold intermediate layer (at the core of which temperatures are <0 °C), which isolates surface waters from the deep ones. Inter-annual variability in temperature is low within the deep water mass that shelters the dormant *C. finmarchicus*.

Dia = separation; *Pausa* (L) = stop, pause *Dormant* (L) = sleeping *Lethe* = forgetful; *Argos* = idle, inert *Quiescentia* (L) = calm, rest

The Gulf of Maine, in contrast, is much warmer than the Gulf of St. Lawrence, with primary production peaking twice in separate spring and fall blooms. Its intermediate-depth layer never falls below 5 °C, and its deep basin is filled with a mixture of cold slope Labrador water and southern warm slope water, whose relative proportions determine the actual temperature of the deep water mass. The presence of abundant populations of *C. finmarchicus* is significant for marine food webs in both regions.

This species arguably plays a key role as a lipid-rich food resource for fish, including herring, sand lance, mackerel, and capelin, as well as larger species such as the northern right whale. Hence, understanding the mechanisms controlling the population dynamics of *C. finmarchicus* and how populations in both regions may react to environmental variation has broader ecological and economic applications. This is of particular interest for the Gulf of Maine, located at the southern margin of the species' biogeographic range, where local water temperatures are likely at the upper edge of tolerance for the species seasonally.

Based on predictions of warming sea surface temperature, a model predicts the disappearance of *C. finmarchicus* from the southern margin of the North Western Atlantic over the next few years.

C. finmarchicus (probably the most abundant animal on the planet) is a crucial component of the planktonic ecosystems. The Gulf of Maine is at the southern edge of its biogeographic range, and here the species appears more vulnerable to the impacts of global warming.

The interruption of the life cycle is common among the plankton of seasonally variable regions. The lethargic condition, however, is not easily recognizable if it occurs without evident modifications of the interested individuals, as it occurs in *C. finmarchicus*.

The species that do not rest as sleeping individuals, produce rsting eggs and cysts. A morphological variation is recognizable in the case of cysts from

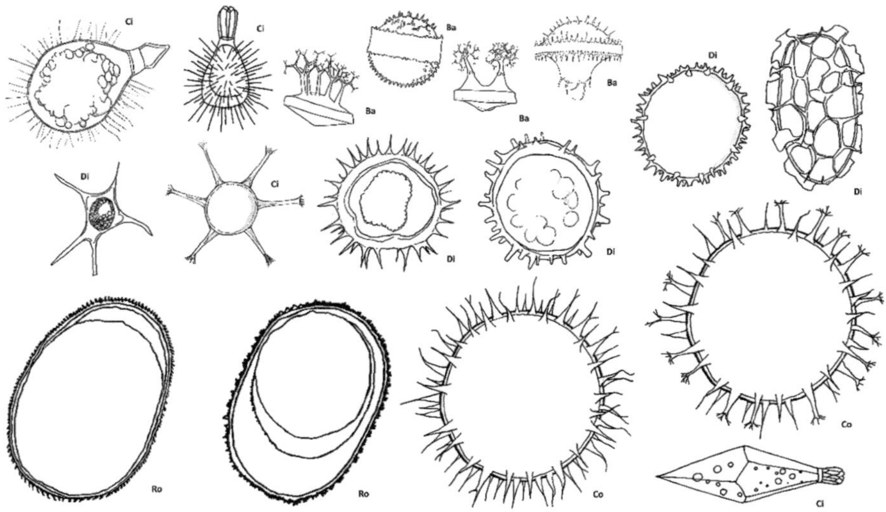

Fig. 5.8 Spiny resting stages. Ba, Bacillariophyta; Ci, Ciliophora; Co, Copepoda; Di, Dinophyta; Ro, Rotifera. (From Belmonte and Rubino 2019 re-arranged)

unicellular plankton or resting eggs of planktonic Metazoa. In these cases, resting stages differ from the corresponding active stages, and their morphology is unexpectedly diverse.

About 80% of marine cysts appear typically spiny or sculptured at their surface (Belmonte et al. 1997) (Fig. 5.8). The presence of spines at the cyst surface has been considered to be responsible for either an acceleration of sinking rate to the bottom or, on the contrary, a sinking retardation device. In the first case, acceleration is the consequence of the clinging of many cysts together (as in the case of Bacillariophyta) to form a larger and heavier body mass. In the second case, sinking can be retarded if spines allow the clinging with floating and/or moving materials and organisms.

In the sediments, the spiny covering of cysts could be useful for other functions. They enlarge the contact surface of the cysts with the environment, thus allowing a better perception of environmental variations and the hatching time. In addition, but probably the most important role, spines represent a passive defence against scraping by sediment grains due to pressure.

In fact, viable cysts have been found down to 20 cm from the sediment surface, where they can stay an unpredictable length of time (and have to support a certain pressure from the overlying sediment).

In general, the enlargement of diameter realized by the spine's presence could make spiny cysts larger than the mouth opening of some predators.

What is less immediate, but proved to be real, is the advantage the spines represent for the embryo inside each spiny egg among Copepoda.

The spiny surface, in fact, could help the egg to retard the sinking to sediment bottoms that, in confined shallow areas, are periodically under hypoxia.

Such a condition kills early embryonic stages of Copepoda; thus, the retarded sink offered by spines somehow allows embryos to complete their sensible first development steps and, once ultimate, to enter a resting stage insensible to external hypoxic conditions.

For some reasons (abundance of producers, absence of predators, sinking, and accumulation favoured by low hydrodynamism), the coastal confined areas are the preferred sites of cyst accumulation in the so-called marine cyst banks.

What emerged from studies in the Mar Piccolo of Taranto, however, was that the distribution of encysted *taxa* is not homogeneous in the whole extension of the marine area, and in addition, it is also not uniform along the vertical sediment column (Fig. 5.9). In some cases, in fact, different species showed different maxima of cyst accumulation at different depths, suggesting that both the horizontal and the vertical distribution of cysts in sediments are not due to a common external driver but in some way have to be conditioned by single species characteristics.

Hydro = water; *Dyna* = movement   *Hypo* = low, minimum; *Oxy* = oxygen

The fact that the vertical distribution of cysts varies without a common pattern in the sediment suggests that the abundance of each cyst type depends on the demographic history of the corresponding species in the water column. Under this aspect, the cyst abundance in a particular layer corresponds to the end of the abundance of the species in the water column in a certain period, and, like in the water column, also in the sediment, a succession of species can be read or perceived if the layering of sediments is not "disturbed" or haltered by re-mixing events.

In a study of a well-layered sediment core from the north Adriatic Sea, the radio-dating of some isotopes allowed to establish that the history could be read (about one century of years) (Fig. 5.10). What accomplishes such studies is, however, the large number of undetermined variables that exist. Apart from the age of sediment (determinable only in well-layered situations), what lacks is the knowledge of cysts to propose a complete history of the plankton (cysts are morphologically different from active stages). What comes to help the study is the possibility of using molecular tools to recognize species or an intrinsic element of cysts; they are resting stages hence, in some way, we can hope to observe their germination (thereby discovering the corresponding active stage) even after long periods of rest.

At present, nobody knows what the resting capability of cysts is. But from the sediment core of the north Adriatic cysts, germinations occurred along the whole sediment core (about 10 cm) with an active swimming dinoflagellate derived from the hatching of a cyst coming from a sediment position corresponding to 80 years ago (Fig. 5.10).

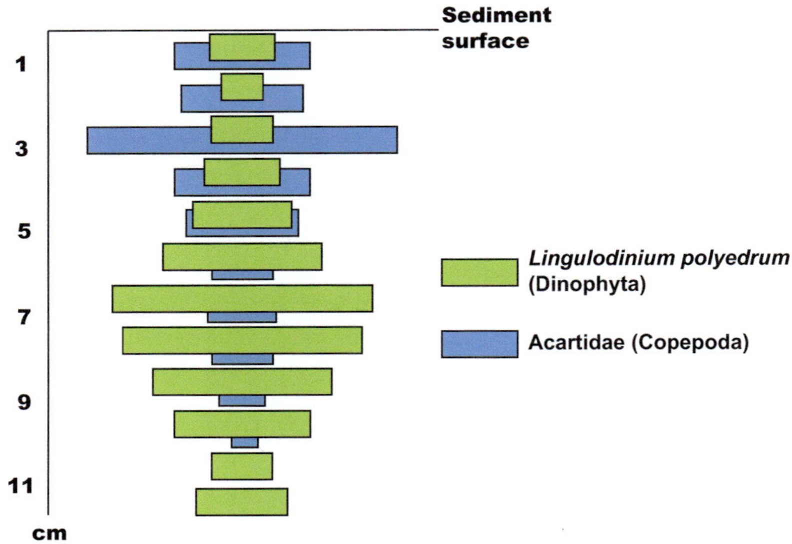

Fig. 5.9 Vertical distribution of cysts in the sediment (Brindisi harbour, Italy). The maximum of *Lingulodinium polyedrum* (Dinophyta) cyst abundance was at 7 cm below the surface. The maximum of Acartiidae (Copepoda Calanoida) resting eggs (in the same site) was at 3 cm. (From Belmonte et al. 1995, modified by A. M. Miglietta)

In marine sediments, ages of 40 years are common for resting eggs of Copepoda, but the same animal group, in freshwater, showed the hatching of an egg coming from a sediment layer dating more than 300 years (Hairston et al. 1995).

The presence of plankton species in a resting form in ecological compartments different from the water column and yet sympatric with their own active stages is fundamental to understanding the dynamics of single populations and that of the whole plankton.

The presence of a resting stage in the life cycle of a species is devoted to avoiding adverse environmental conditions, but something suggests that it could be a "standard" trait even under more stable environmental conditions to avoid co-existence and competition in crowded communities. In fact, the search for resting stages in bottom sediments gave results also in coastal seas, out of confined and variable lagoons. In studies carried out with a multisampling approach (water column and bottom sediments), the existence of a marine cyst bank also far from the coastline was ascertained, and below water columns relatively high (about 200 m). In an investigation in the South Adriatic, Rubino et al. (2009) collected microplankton in the water column and cysts from sediments from 27 different coastal stations. The total number of *taxa* recorded in the water column was 120, and 81 were the cyst morphotypes collected from the bottom sediments. Among these, only 16 *taxa* were found in both compartments, and the remaining 65 cyst types in sediments represented an unexpressed part (about 35%) of plankton diversity (Fig. 5.11).

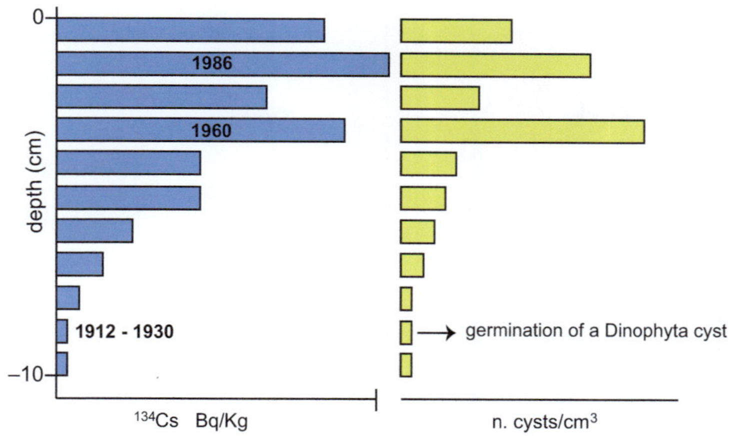

Fig. 5.10 A sediment core from an anoxic basin in the North Adriatic. Ordinated layering of the sediment was established with measures of radioactive Caesium. The radio dating allowed the recognition of the two radioactivity peaks of 1960 (when nuclear weapons were experimented in the atmosphere) and 1984 (the Chernobyl Nuclear Power Plant accident). Cysts from each level were put in conditions to germinate. The last germination (Dinophyta) occurred from a layer of 80 years old (graphic composition of A. M. Miglietta)

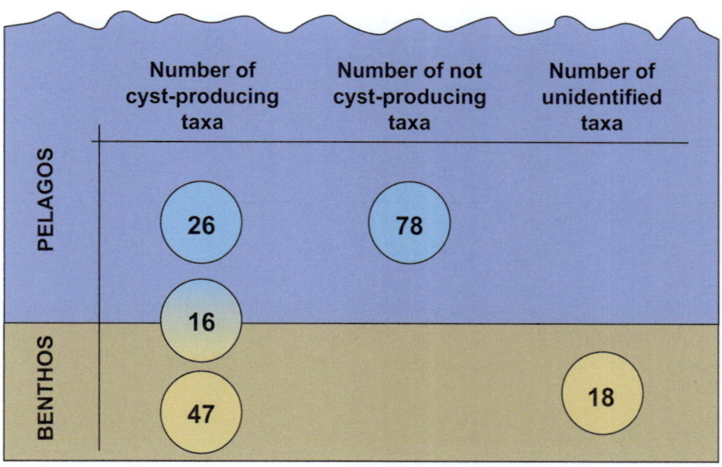

Fig. 5.11 Results of an investigation on the microplankton distribution in the South Adriatic Sea (Rubino et al. 2009). The number of cyst-producing taxa in the water column (26 + 16 found also in sediments) was lower than the number of cyst morphotypes in the sediment (16 found also in water column + 47 not found in the water column + 18 unidentified). Although the cyst production is a characteristic of only a part of the community, the information content of sediment (81 cyst morphotypes) is an important fraction (about 35%) for knowledge of community composition and its dynamics (graphic composition of A. M. Miglietta)

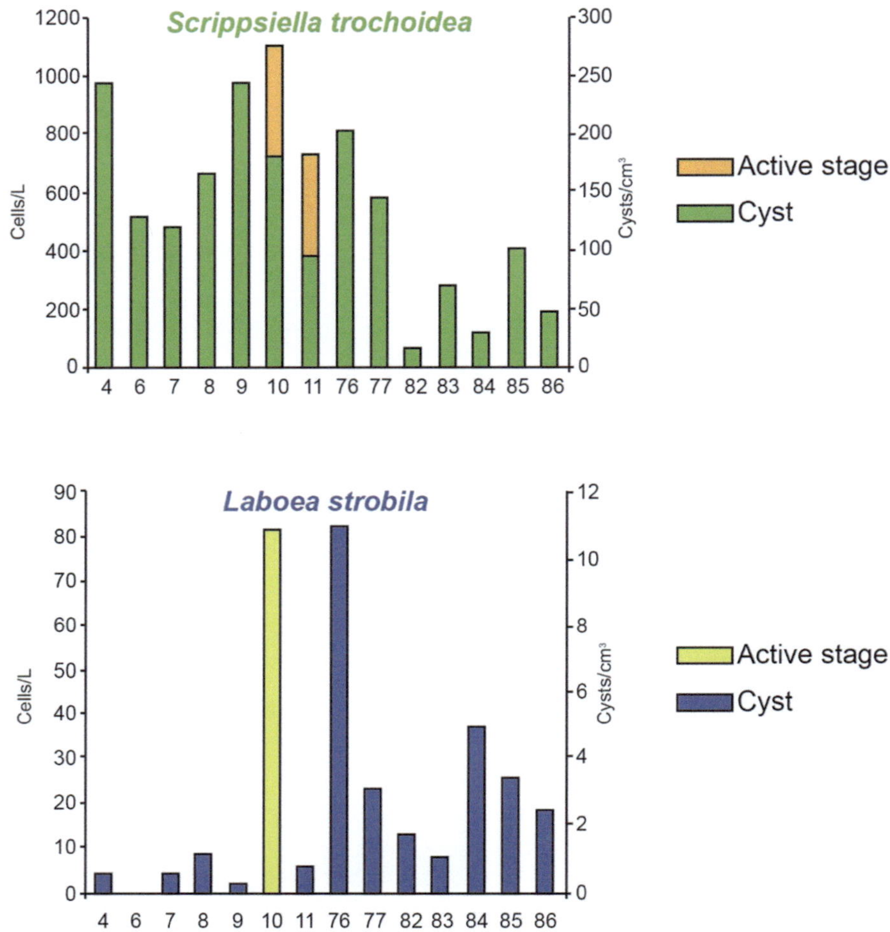

Fig. 5.12 *Scrippsiella* sp. (Dinophyta) and *Laboea* sp. (Ciliophora) examples of distribution in both water column and sediments. In the case of *Laboea* sp. its sediment cyst bank is present just and only where the active cell is absent from the water column. (From Rubino et al. 2009, modified by A. M. Miglietta)

In detail, *Scrippsiella trochoidea* (Dinophyta) was present in the plankton of only 2 stations, but it was found in sediments of the whole geographic area (all the 27 stations) as cyst. In the case of *Laboea strobila* (Ciliophora) swimming cells were present in the plankton of the whole area, with the only exception of one station. The sediments of the whole area do not have cysts of this ciliate, with the only exception of the station where the species was absent from the water column. The alternate presence in the water and in the sediments, in this case, has been of paramount evidence (Fig. 5.12).

If the defence of each individual is one of the problems of its existence, it is not clear how a "dormant" stage can defend itself when, by definition, it is not active.

The external envelope (generally spiny) of cysts probably plays an important role in defending an inactive organism. The inactivity, however, should be defined because some functions should be necessary to avoid aggression from Bacteria, Fungi, or Protista typical of the degradation of organic matter.

In cysts of *Artemia franciscana* (Crustacea Anostraca) scientific instruments were not able to determine any metabolic activity.

This non-living condition, however, is not enough to make this cyst available for degradative processes. A primitive immunologic (antibacterial) defence in the shell of *A. franciscana* cysts was described, which probably plays a role in avoiding bacterial aggression (Stabili et al. 1999). The cyst envelope, in itself, is a barrier against the entrance of microbes.

And, finally, the environment where a long-lasting rest occurs (e.g., an anoxic and/or chemically reducing bottom mud) could be prohibited to life in general. Altogether, these reasons cannot explain the long rest (even of hundreds of years) of some organisms, and we are obliged to add the capability, internal to the cyst, of living materials and metabolic processes to be secured against alteration during the rest time (Fig. 5.13).

Hydro = water; *Dyna* = movement   *Syn* = together; *Patria* = birthplace

Resting stages of Crustacea, Rotifera, Ciliophora, and Dinophyta have also been recognized as resistant to chemicals and disinfectants (normally used to destroy microbes).

Formalin, NaOCl, $KMnO_4$, $CuSO_4$, and organic Iodine, did not significantly modify the hatching rate of resting stages either in acute or chronic expositions (Pati and Belmonte 2003).

This high rest capacity and the widespread presence of the phenomenon among living organisms suggest that the recovery of life forms, after global disasters and mass extinctions, could be attributed to the existence of a long rest capacity (many years for some species) in waiting for the return of favourable conditions. The hatching of resting cysts, after an acute environmental problem, could also repopulate the planet after the mass extinction of 65 million years ago at the transition between the tertiary and quaternary periods.

Bottom sediments seem to be the obvious site for the accumulation of cysts from the overlying water column, but the sea surface is also a border where positive buoying material can accumulate. In cold seas, cysts floating at the surface could also be trapped in the sympagic habitat of the surface ice (Fig. 5.14).

Both the benthic and sympagic compartments are the object of studies due to the existence of complex autochthonous communities (meiobenthos within sediments, sympagic biota within ice), although the non-active biological components have been disregarded in the past. In the last 20 years, scientists have suggested that ice habitats also provide a biological reservoir. The trapped plankton organisms, both as a planktonic stage and cysts, may remain dormant in the ice during the adverse season (glaciated and dark), and they repopulate the water column when the ice melts. Although scientists admit the data need confirmation, they consider the

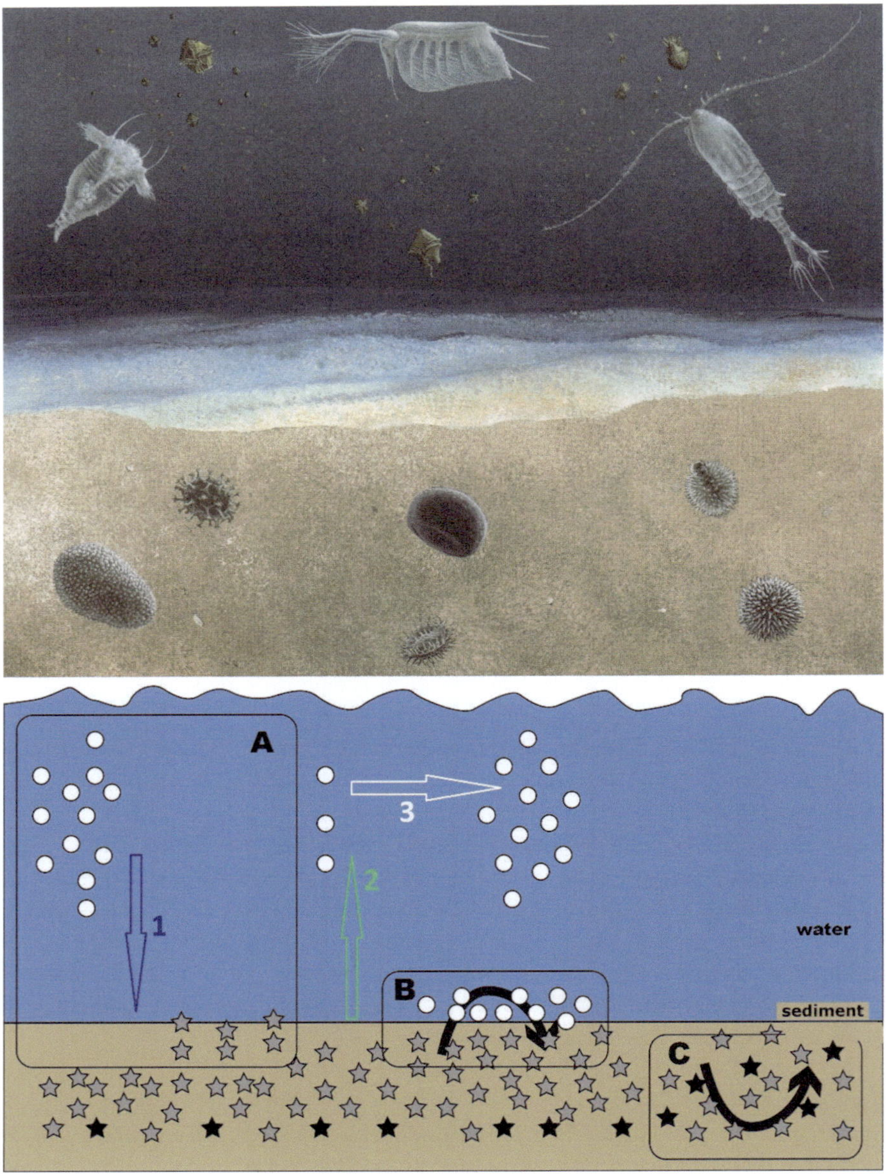

Fig. 5.13 On the top, representation of resting stages (in the sediment), and corresponding plankton forms (in the water column) (drawing of A. Gennari). On the bottom, schematic representation of the plankton dynamics when the cyst sediment bank is considered. A, plankto-benthic exchange. B, hyperbenthos–benthic exchange. C, intra-sediment moving (due to reworking). 1, cyst production and sinking. 2, cyst germination. 3, reproduction of active stages. (From Belmonte and Rubino 2019, modified by A. M. Miglietta)

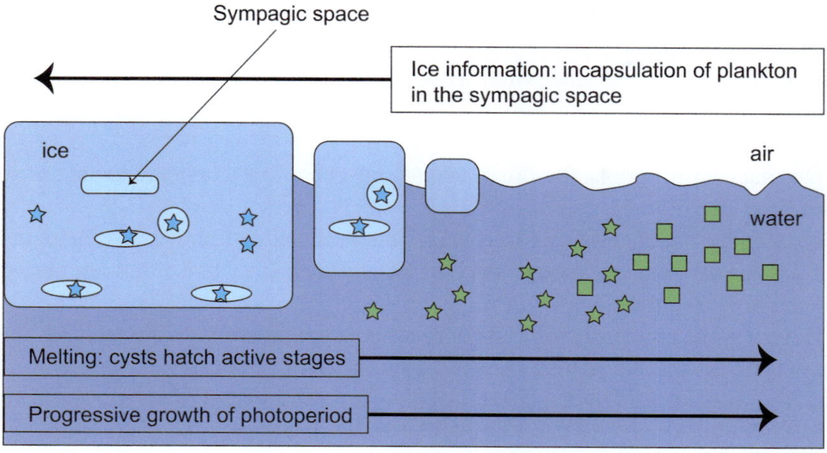

Fig. 5.14 Hypothesis of plankton cycle in the ocean surface at polar latitudes. The ice forming at surface entraps plankton cells. These survive the long dark period of polar seas by resting as cysts in the sympagic spaces. At the melt, cysts are set free in the water and germinate active stages. 1, cysts in the sympagos; 2, cysts in the plankton; 3, active cells generated by cyst germination. (From Belmonte and Rubino 2019, modified by A. M. Miglietta)

biodiversity found sufficient to hypothesize that cysts provide an injection to initiate plankton blooms at the start of each ice-melting season.

In some ice-cores, cysts of unknown origin, which were not attributed to either diatoms or flagellates, contributed a high proportion of the algal abundance.

The overwintering of plankton species within ice during the dark polar season is an admitted possibility, and the rich number of cysts collected by sediment traps close to the bottom could derive (at least partly) from the melt-mediated release of sea ice-dweller cysts to the water column.

However, further research is needed to clarify where the rich assemblages of sympagic benthic species originate, when the sympagic assemblages appear in the ice formation, and where the organisms go when the ice melts.

Taxa potentially able to produce cysts are well represented in the sea-ice biota, and many cysts have been reported from that habitat, some of them being undetermined.

Other studies reported planktonic *taxa* in the sea ice biota, thus fuelling the suspicion that plankton can be entrapped by the ice during its formation and/or that resting specimens, and cysts wake up and originate active organisms when managed by scientists for the analysis. The suspicion is that specimens do not compose an active community in the ice habitat, but simply they sleep and wait for the ice melting.

The extreme conditions present within the ice during the long-lasting polar winters (low temperature, low light or complete darkness, and high salt concentration in liquid water segregated in ice crevices) suggest the suitability of a resting strategy for the survival of at least some species.

The sea-ice biota is a peculiar compartment where planktonic species, trapped during the autumn in the forming ice, may have evolved the possibility of producing cysts to survive the extreme winter conditions. In any case, the presence of cysts could allow the immediate restoration of planktonic communities when favourable conditions return (Fig. 5.14).

Dormant biotic pools should instead represent a link between distant activity periods.

In both the meiobenthos and sympagic biota, the attention of researchers in the past has mostly focused on exclusive or typical *taxa* (i.e., those that are perfectly adapted to local conditions).

However, the abundance of cysts in sediment may be even greater than the active components of the meiofauna, and the term *Inactive Temporary Meiobenthos* has been proposed to refer to this fraction (Pati et al. 1999). The provocative term "resurrection" was adopted to describe the hatching and/or the re-activation of resting forms. The idea was completely adopted in the more impactful *resurrection ecology,* proposed by Kerfoot and Weider (2004).

Confined coastal areas are able to capture and accumulate large numbers of cysts in their sediments. Cyst production and sinking to the bottom are evident during the abundance peaks of plankton populations.

Cyst production is consequent to the nutrient depletion caused by rapid phytoplankton growth. Peak abundances of cysts in the water and sediment, however, are always separated by a time lapse, with cyst peaks delayed with respect to the active stage.

It is not entirely clear whether the richness of confined coastal areas derives from the abundance of cyst producers that bloom in these areas, from an accumulation effect of small particles due to the low hydrodynamism, or even from the lack of benthic predators. In open neritic habitats, cyst-producing species are present, but there the sediment tends to be coarser (i.e., the area is subjected to high hydrodynamism) and to host a lower number of cysts. In contrast, confined coastal areas, although with a shallower water column and less oxygenated sediments (due to their limited hydrodynamism), show high concentrations of cysts in sediment. This links the cyst abundance to the confined environments, and to the size of sediment particles.

In addition, the abundance of cysts in confined environments could be enhanced by the absence of potential predators, which are poorly adapted to the periodical hypoxia present in such sediments.

It should be pointed out that some spatial situations that are not properly identifiable as confined (and thus not typically hypoxic), such as the German Bight and Kiel Bay, show rich assemblages of cysts in sediment (Nehring 1997).

Studies carried out in the Baltic Sea are the only ones that shed light on the complex processes involving plankto-benthic interactions in terms of food and energy flow. *Monoporeia affinis* (Crustacea, Amphipoda) specimens rework sulphide sediments and re-oxygenate them down to a depth of 10 mm, thereby playing an indirect role (through oxygenation and bio-reworking) in the realization of suitable conditions for the germination of cysts. In this way, *M. affinis* possibly affects the recruitment of active stages into the water column because its burrowing activity can re-suspend buried cysts or drive cysts deeper into the sediment (Viitasalo 2007).

Hyperbenthic Crustacea (Mysida, *Mysis mixta*) feed selectively only on certain cyst types, thus possibly affecting plankton composition by differentiated subtraction of propagules (in the specific case of the Cladocera *Bosmina longirostris maritima*). Interestingly, the indicated Mysida are planktivorous and probably exploit cyst banks when the corresponding active stages of their prey are absent from the water column. Also, Mollusca Bivalvia *Macoma baltica* is indicated as a consumer of *Bosmina* cysts and Amphipoda *M. affinis*, as responsible of a deeper burying of cysts into the sediment column reducing the number of cysts at the sediment surface, available for the Mysida feeding activity. *M. balthica* was the most harmful species for the recruitment of *Bosmina*, whereas the nonindigenous *Marenzelleria* spp. did not affect ephippia. The action of limicole Polychaeta (*Marenzelleria* sp.) was found as not affecting the distribution and the survival rate of the Cladocera cysts (Fig. 5.15).

In many cases, however, cysts can be ingested but not digested and pass through the gut of deposit feeders without consequences for their viability. Both subitaneous

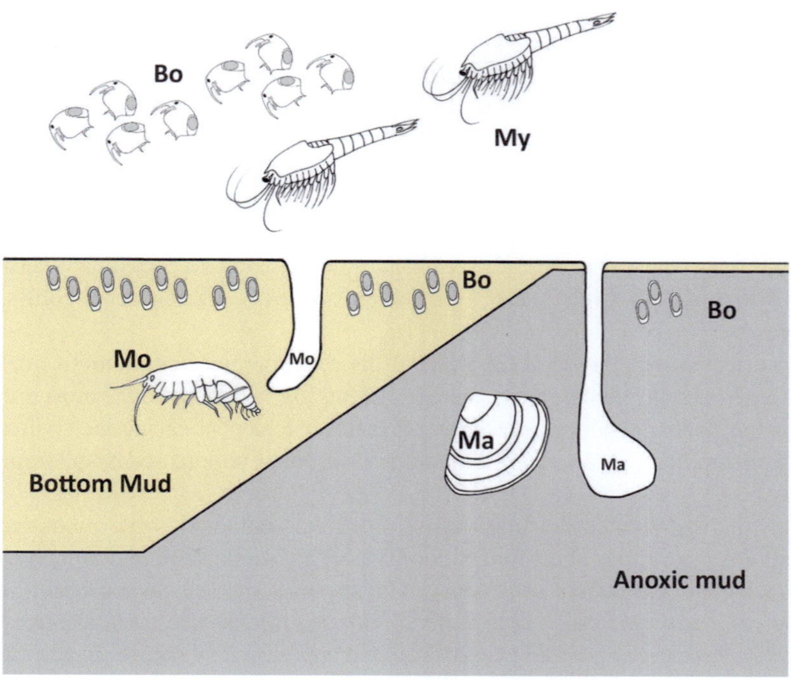

Fig. 5.15 Scheme of the benthic pelagic interaction for the survival and hatching of Cladocera (Bo, *Bosmina longispina maritima*) cysts (ephippia) in the Baltic Sea. Mo, *Monoporeia affinis* (Crustacea, Amphipoda), probably affects the germination of cysts in sediments. Ma, *Macoma baltica* (Mollusca Bivalvia) feeds on cysts at the sediment surface. My, *Mysis mixta* (Crustacea, Mysida) feeds on planktonic *Bosmina* and on its cysts when the planktonic form is absent (graphic composition, A. M. Miglietta)

and diapause eggs produced by the Calanoida *Labidocera aestiva* were capable of surviving passage through the guts of two benthic Polychaeta. Being preyed upon (with a consequent passage through the gut) could be a strategy for egg-carrying Copepoda to enhance their reproductive success at the population level because embryos contained in the egg sac of *Eurytemora herdmanni* are not digested by the fish *Menidia menidia*, and their hatch rate was enhanced.

Finally, the selective predation of egg-carrying females enhances the likelihood that non-mature females will survive and produce their own clutch of eggs.

Resting spores of Bacillariophyta have been found undamaged and even viable in faecal pellets of Copepoda Calanoida. Dinophyta cysts have repeatedly been found in faecal pellets from Polychaeta and Bivalvia. The gut passage does not substantially reduce the germination rates of Dinophyta cysts and may even enhance them, as demonstrated by the gut of a copepod functioning as a facilitator for the hatching of Dinophyta cysts when they arrive outside.

Limicole Polychaeta rework sediment, with substantial implications for the overlying water column, similar to the impact of soil re-workers on seed banks in terrestrial forests. The re-working includes also viable particles as Dinophyta cysts from the surface down to 7 cm of depth. This reworking leads to the rapid burial of newly arrived cysts, but at the same time, it allows the uplift to the surface of deeply buried cysts.

Whatever the destiny of cysts in the superficial sediment layer (whether or not they are digested by a cyst consumer), their subtraction (e.g., germination) is responsible for the characteristic vertical profile of their abundance in the sediment column. Indeed, abundance minima are typically found just at the most superficial layer. Some findings of abundance peaks at the sediment surface, on the contrary, can easily be attributed to a proxy period or episode of massive production from the plankton.

In any case, the sediment surface is what is most relevant to encystment/germination dynamics, showing strong fluctuations in cyst abundance according to the season.

Resurrection ecology is at the basis of the functioning of communities in confined environments. Although such environments are known to be the most productive on the planet, for a long time they were believed to depend on marine vivification for the restoration of biological communities seasonally suppressed by environmental stress.

The marine vivification hypothesis, however, contained some embarrassing speculations, assuming the colonization of a highly productive, although stressed, area (a lagoon) dependent on a less productive one (the sea), whose biota is not biologically adapted to such high variability of conditions, and periodical stress.

Today, Resurrection Ecology explains the reappearance of species in accordance with the seasons; that for many species, they do not depend on neighbouring environments (as in supply-side ecology) but mainly on the bottom sediments of a confined area where they spent the adverse season (Fig. 5.16).

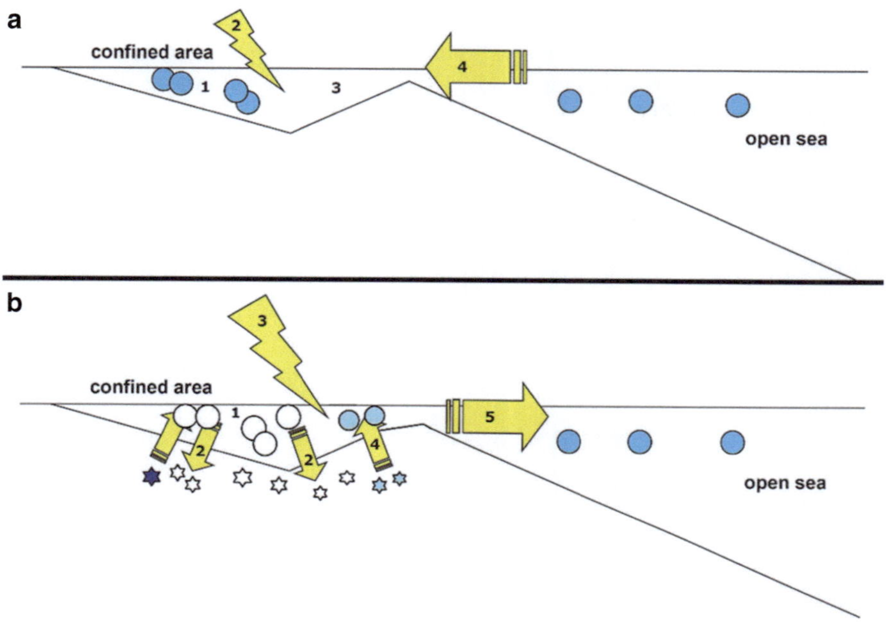

Fig. 5.16 (**a**) The marine vivification theory elaborated for coastal confined environments: for the restarting of confined plankton communities. (1) occurrence of plankton in suitable environmental conditions; (2) occurrence of an environmental situation that compromises the plankton survival in the confined environment; (3) disappearance of planktonic community from the confined environment; and (4) restoration of the confined plankton community is ensured by fuelling from the adjacent sea. (**b**) The resurrection ecology theory for the restarting of confined plankton communities. (1) Cysts and active stages co-exist in the same environment. (2) Swimming stages start the production of resting stages before the occurrence of harsh environmental conditions. (3) Harsh conditions occur after the disappearance of confined plankton community from the water column. (4) Restoration of the confined community by the sediment cyst bank. (5) Spill over and export of specimens and biomass in the adjacent open sea. (From Belmonte and Rubino 2019, modified by A. M. Miglietta)

Bathos = depth *Bacillus* = steak; *Phycus* = plant *Bi* = double; *Valva* (L) = shell, shield *Bios* = life *Calan* = from Irish caelan, slender *Dyno* = spinning; *Phycus* = plant *Epi* = over; *Pelagos* = open sea *Poly* = multiple; *Chaeta* = bristle, hair *Plankton* = vaguing; *Voros* = eater *Vivo* (L) = alive

Resurrection ecology in confined areas is a typical example of ecological resilience. The plankton of confined environments appear to be autochthonous, and the presence of similar species in neighbouring neritic areas is now believed to be the consequence of the export of individuals from the blooming populations of a confined area; the exact opposite of what was thought some time ago.

The export of cysts from production areas (in the euphotic epipelagos) involves other marine habitats as well.

In fact, cysts have been detected in deep-sea sediment or captured by deep sediment traps (in aphotic bathypelagos), and studies have even proposed a role for cysts via coastal-deep sea connections. Cysts produced by coastal and epipelagic species accumulate in deep-sea sediment off the mouth of the Foix canyon (Spain) at more than 1000 m in depth.

This suggests that submarine canyons play a role in the re-proposal of deep resting stages in the epipelagic and coastal areas during the periodical upwelling, which at times occurs.

As with terrestrial forest seed banks, in order to make a decisive contribution to our understanding of plankton functioning in coastal areas, the ecology of marine cyst banks must evaluate input (fecundity/fertility, import), storage (short and long term), and subtraction (death, predation, recruitment, export) for each of the species involved.

The relative abundance of a cyst type is affected by the composition of the predator community that feeds on cysts in sediments. If the whole community is considered, such grazing-dependent mortality is likely to affect the size and viability of a seeded plankton community. The existence of short resting periods (e.g., in Tintinnina) has been suggested as responsible for reduced numbers of cysts in sediment of species that are abundant in the plankton.

Cysts are able to protect organisms from unsuitable situations. This is not true only for each location exposed to seasonal variations, but it is also the basis to understand how species can pass through geographic and/or ecological barriers, thus enlarging their geographic distribution.

It is a well-known principle of the biogeography that smaller organisms tend to have wider or even cosmopolitan distributions, a higher efficiency of dispersal, a lower rate of allopatric speciation, and lower rates of local and global extinction than do larger organisms.

Organisms smaller than 150 μm are generally cosmopolitan because their low body mass facilitates aerial dispersal. The lack of biogeographical separation, however, can be attributed to the large number of individuals in each small species. In this frame, we have to add the presence of cysts (common among small organisms).

A high percentage of micro-species in a natural system are inactive (from 20% to 80%), and dispersal of cysts (either in the air or in the water) contributes to the apparent cosmopolitan distributions of *taxa*. Among freshwater Protozoa, Chromista, and small Metazoa at least, the capability to produce cysts facilitates dispersal by air, dust, and migrating animals.

However, modern methodological approaches such as molecular biology and genetics have shown that the number of ubiquitous species is smaller than usually thought, even among smaller species. This, in any case, cannot exclude the large dispersal potential that resting stages give to species.

Although endemics have been recorded among marine Protozoa and Chromista, it is presumable that while biogeographical limits generally affect large organisms, they do not apply to microorganisms.

The ability to not suffer adverse conditions automatically allows cysts to pass barriers that are both geographical and ecological for the active stages.

Amphi = two sided; *Podos* = feet   *Bi* = two; *Valva* (L) = shell, shield   *Clado* = branch; *Chiro* = hand, horn   *Cosmos* = the universe; *Polys* = community   *Ephippium* = epi, over; hippos, horse. The saddle.   *Mysis* = blinking

The composition of the cyst wall and the physiology of the encysted organisms, still poorly understood, should be closely connected to the species' geographical distribution, more than any mobile capability of the active stage.

The possibility of cyst transfers between separate water bodies by passive transport in the guts of fish swallowed by migrating water birds was already proposed by Darwin to explain the wide distribution of freshwater organisms.

Indeed, the digestive enzymes of birds were found to negatively affect neither the viability nor the hatching rate of cysts. This framework has yet to be verified for marine environments, where birds cannot play the same role as in brackish waters or flooded plains. However, as already mentioned, benthic detritivores as well as fish can be vectors of plankton species whose cysts are resistant to digestion and can survive in their gut for days before being expelled with faeces.

This scenario demonstrates that cysts can be transported long distances from their deposition site in the gut of vagile, or even swimming organisms. The shape of the resulting dispersal kernel depends on the mobility of the swallowing animal, its behaviour (migratory or sedentary), the time spent by cysts in its gut, and its own trophic destiny—that is, whether it becomes the prey of other species. Consequently, cyst distributions in space are the result of a complex set of intertwined processes.

Travelling inside a vector is not the only possibility for cysts. Sculptures on their surface (Fig. 5.17) have been interpreted as favouring their clinging to floating or moving objects. This form of dispersal may account for the geographic distribution of species, including their appearance at previously unreported sites.

If the extra-long diapause (ELD) is added to the framework of geographic distribution, it is not obvious that a new *taxon* in an area (a non-indigenous species, NIS) corresponds to a new arrival. In fact, the newly appearing species might derive from the germination of ELD cysts, thereby introducing a species from a different time (the past) but from the same site.

In any case, although other possibilities cannot be excluded, the majority of NISs in the world today are attributed to dispersion via human vectors, such as in the ballast waters of ships navigating between continents. The latter mechanism, already many years ago, was believed to have been responsible for about 4000 NISs (most of them planktonic) dispersed throughout the world (Carlton and Geller 1993).

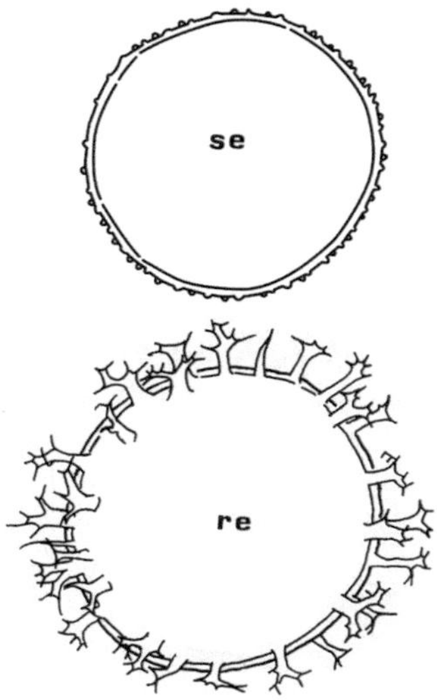

Fig. 5.17 Sculptures of *Pteriacartia josephinae* egg surface (subitaneous, on the top, and resting egg, on the bottom) (drawing, G. Belmonte)

As cysts, the majority of these stowaways survive the adversity of a long period in the ballast tanks, where they accumulate in the bottom sediments and rest for long periods. Although a true ecological community of active stages has been described as functioning in such tanks during the period of navigation, cysts represent the best way to travel. Ballast waters are the most extensively studied and verified of these vectors, partly due to the fear that toxic organisms may be transferred this way.

The dispersal of meso-zoo-plankton in accordance with the abovementioned mechanisms is still limited in comparison to the huge knowledge that is available for freshwater habitats.

Such a worldwide circulation might damage local biodiversity, in addition to distributing species that are dangerous for human health. Aquaculture farmers should avoid establishing their farms near the ballast water charge-discharge sites of oceanic ships. Only recently have attempts been made to conduct research on biocides that can be applied to resting stages contained in ballast waters or their residue. Considering that ballast waters are responsible even for transcontinental transfer, they represent for coastal marine biogeography what birds represent for continental wetlands biogeography.

When new species are recorded in ports, it is highly likely that they arrived with ships. The Calanoida *Paracartia grani,* well known from the North Atlantic, was first identified in the Mediterranean in the port of Malaga (Spain) in the late twentieth century. Subsequent reports of the species came only from ports in the whole Mediterranean Sea, and they were associated with the capacity to produce resting eggs, supporting the hypothesis that the species disperses via ships.

Navigation also seems to be the cause of dispersal for species living in apparently isolated habitats. Cladocera *Podon polyphemoides* and the Calanoida *Acartia*

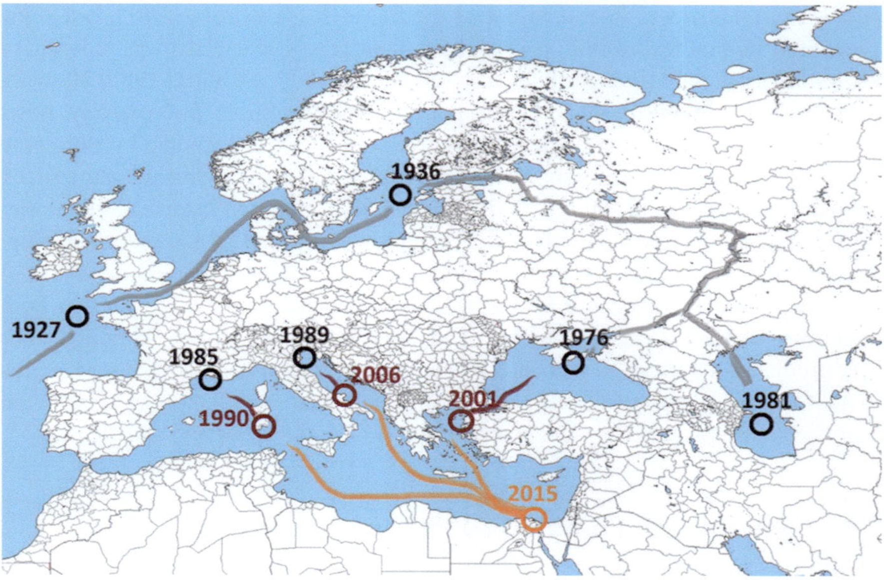

Fig. 5.18 The Sarmatian route of *Acartia tonsa* geographic expansion (grey tracks), and successive southwards invasions (red and orange tracks). The species was found in samples from the Atlantic Ocean (1849), but only in 1927 was it reported for the first time from European waters. Successively it extended its geographic distribution to the Baltic Sea. Only in 1985 was it reported for the first time in the Mediterranean. A study also reported the species from the Black Sea, but successively, the same samples were recognized as collected in 1976 (Belmonte 2021). Hence, the species did not arrive in the Black Sea from the Mediterranean but probably used the river road from the Baltic. Subsequently, the geographic range expanded southwards in the Mediterranean waters (original)

tonsa are well-known coastal marine species that have been reported from the Caspian Sea (a "lake") in 1957 and 1984, respectively. Their arrival in that isolated sea has been attributed to the existing navigable connection, via the Volga-Don canal, with the Azov-Black Sea system (Belmonte 2021).

Navigation via the Volga-Don-Neva has also been accepted as a possible way for the transfer of fauna from the Ponto-Caspian system (Black, Azov, and Caspian Seas) to the Baltic Sea along a "trans-Sarmatic route", now believed to be an important waterway of geographic dispersal (Fig. 5.18).

Although some species that have recently invaded the Baltic Sea from the Caspian (such as the Cladocera *Cercopagis pengoi*) are known to be highly tolerant of the fresh waters of rivers, they appear to have crossed the Sarmatian plain and invaded the Baltic Sea only recently. The construction of canals and the beginning of navigation inside them provide the best explanation for the recent northward dispersal of many Caspian species.

The Sarmatian route is not only a conduit for northward dispersal. In the case of the Calanoida *Acartia tonsa,* southward dispersal has been proposed, apparently bucking the current climate change trend, which is expected to drive species poleward. *A. tonsa* was first identified in the Baltic Sea (Brilinsky 1981), just a few years after its first record in European Atlantic waters. The species was reported in the Mediterranean basin more than 50 years later (Gaudy and Vinas 1985) allowing us to hypothesize an introduction through the Gibraltar Strait. Successively, however, it has been ascertained that samples from the Black Sea studied by Belmonte et al. (1994) were taken in 1976; hence, the species was already present in the Black Sea before its appearance in the Mediterranean Sea. Although its arrival in the Bay of Karadag (the Crimea and Black Sea) could be attributed to any incoming ship, the Sarmatian route, direct from the Baltic, is also a possibility (Belmonte 2021).

Hence, the river connection over the Sarmatian plain carries species in both directions, towards either the Baltic or the Black-Caspian basins. Although non-cyst-producing organisms (such as the Ctenophora *Mnemiopsis leydi*) also use this route, the resting eggs of *C. pengoi* and *A. tonsa* better enable them to survive the adverse conditions that they face along this 1000-km river journey.

5.4 Plankton in Confined Coastal Environments

Having the consciousness of the complexity of the lifestyle strategies for co-existing plankton organisms, when we approach the study of a coastal confined environment, we always have to consider the existence of induced and programmed dormancy (quiescence and diapause). In confined environmental conditions, planktonic species are smaller than those in adjacent marine environments because they adopt shorter life cycles to take advantage of shorter favourable periods and to better escape unpredictable variability in conditions.

Coastal confined environments, consequently, are characterized by two functional categories of species assemblage: one of active species that represent the realized Biodiversity of each site, and one that corresponds to a Potential Biodiversity, not expressed, hosting Temporary Inactive (dormant) protagonists. This last condition is the basis of the Resilience of plankton communities, in the context of Resurrection Ecology (Kerfoot and Weider 2004).

In each approach to the comparative evaluation of the confined coastal diversity and/or ecology, we have to consider that the sampling plan has to be accordingly tuned in terms of mesh size used, timing of collection, and space representativeness.

In coastal brackish waters the first evident datum is a plankton autochthonous in its composition, only marginally influenced by the neighbouring sea. The bulk of brackish water meso-zooplankton was composed by Copepoda Acartiidae and meroplankton. Many other forms are present although in small numbers, also representing contaminations coming from the open sea. In Mediterranean coastal lakes, *Acartia margalefi* (Copepoda) perfectly represents one of the rules of the confined plankton: species are smaller than in the open sea.

A = not; *Cartos* = combed *Auto* = by itself; *Chthon* = land *Meio* = intermediate; *Benthos* = staying on the bottom *Meros* = partial; *Plankton* = vaguing

A. margalefi, in fact, was in the past considered a "dwarf" form of *A. clausi*, until the discovery of its genetic originality. These very similar species (belonging to the same genus) are morphologically different only for body size (Belmonte and Cavallo 1998).

The small species (*A. margalefi*) is typical of confined environments where it realizes crowded populations during the period June–July and sometimes exports individuals to the open sea.

The large species (*A. clausi*) is typical of the open sea and not present in the very confined environments (if not as the result of temporary contamination) (Fig. 5.19).

The body size is also evidently linked to seasons and to environmental conditions. At the arrival of the summer season (and of connected hypoxic problems in the shallow water column of coastal brackish lakes), the body size reduction is replaced by a complete disappearance of the species.

Acartia margalefi co-exists with other Acartiidae species in confined environments, but avoids competition by alternating with them along the seasonal succession of the year.

This species, more than others, perfectly defines the adaptation to coastal confined environments: it is autochthonous, small-sized, producing seasonally large swarms of individuals, seasonally absent (producing resting stages), and involved in species succession with other similar species to avoid competition.

Biodiversity in confined environments is never contemporaneously expressed due to the necessity to avoid competition.

The environmental variability (inconstancy) of coastal confined brackish waters (corresponding to biological variability) is probably the weak point of the system, which easily hosts new arrivals even from distant sources (the so-called Non Indigenous Species, NIS).

In fact, the plankton of Lake Varano, southern Italy, resulted completely different in studies at 18 years of distance thank to the arrival of NISs (Belmonte et al. 2011). *Acartia tonsa* (Copepoda), *Aurelia* sp. (Cnidaria), and *Engraulis* eggs (Pisces) characterized the 2007–2008 plankton, but they were completely absent during 1989–1990.

In the framework of the variability of plankton composition with space and time, a good example comes from the Taranto Sea System (Southern Italy).

A study was planned in 2004–2006 (Belmonte et al. 2013) taking in consideration some of the "confinement rules": (1) Two mesh sizes were used for plankton nets (200 and 80 μm) to contrast the body size effect on the results of species collection along the confinement gradient (smallest species in the more confined portions). (2) A longer collection period was planned (a time lap of 2 years) to recognize a minimum of inter-annual variability. (3) Four stations were fixed along an evident confinement gradient to compare plankton from different confinement situations (Fig. 5.20).

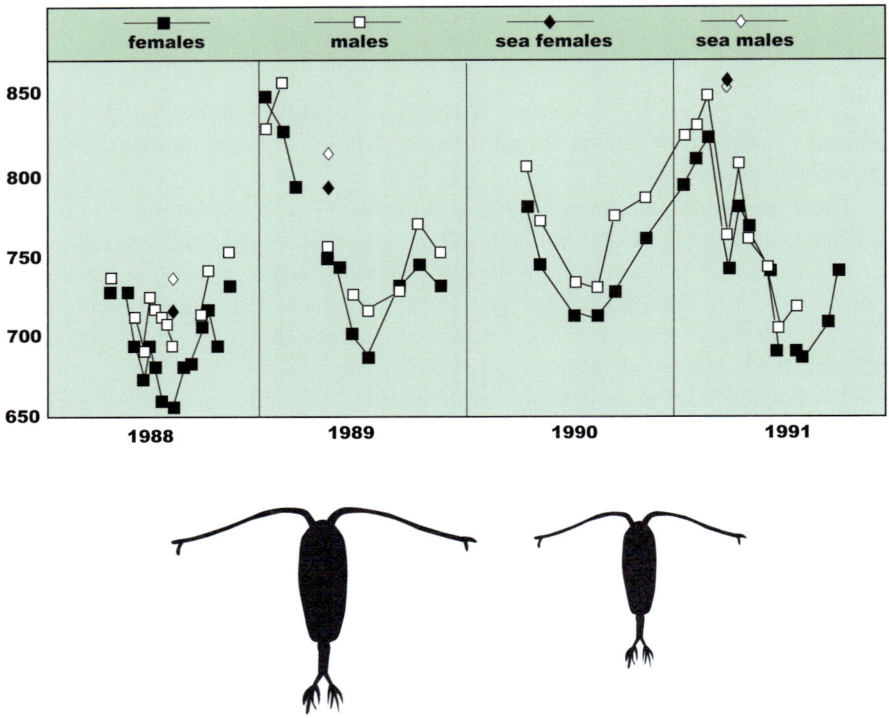

Fig. 5.19 Size variation in *A. margalefi* with time and space. Size is inversely correlated with temperature. (From Belmonte and Cavallo 1998, modified by A. M. Miglietta)

5.4.1 Community Composition Along a Confinement Gradient

The complex study of the Taranto zooplankton was, at the end, based on the analysis of a total of 576 samples related to 96 sets (dates) of abiotic characteristics (Temperature, Depth, Dissolved Oxygen, pH, Conductivity) of the water.

The analysis of abiotic features not only depicted the average differences between single stations but also showed evident grouping of month values, with the station in the open sea more stable or uniform than the station in the innermost part of the system.

As regarding the zooplankton, abundance fluctuations were synchronic in the four stations, but the amplitude of fluctuations was correlated with the confinement grade and reached its maximum intensity in the innermost station (Fig. 5.21).

A total of 346 *taxa* (240 holoplankton, 93 meroplankton, 13 from benthos) were identified in the whole period, but not "together", with each sampling date giving a total of only half (or even less) of this total. The station of the open sea (the Gulf)

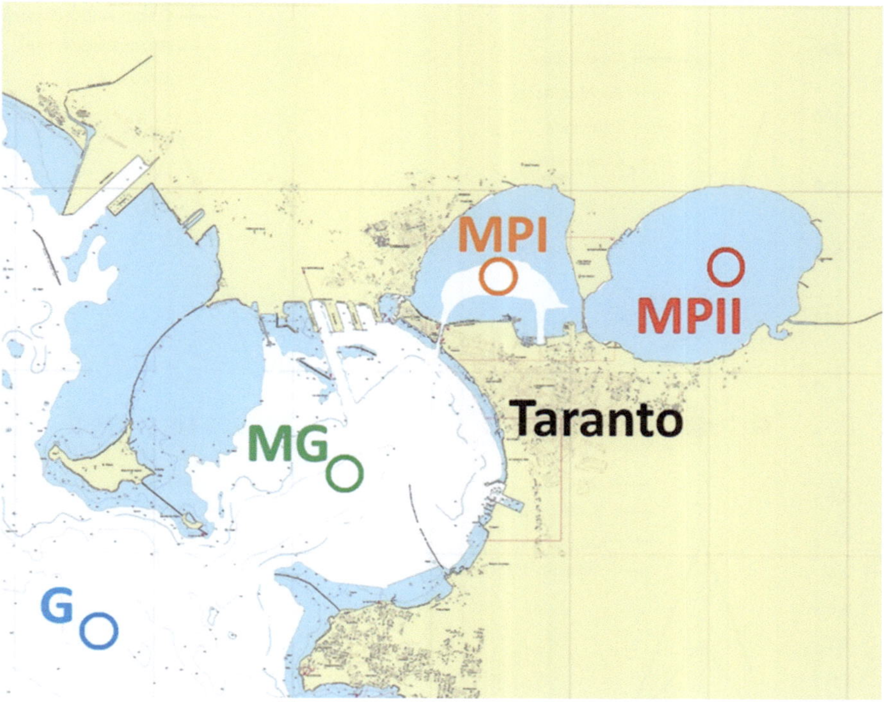

Fig. 5.20 Map of the Taranto sea system, with location of sampling stations considered in 2004–2006, in progressively confined situations, from the Gulf (G, open sea) to the innermost area of Mar Piccolo (MP II). Colours of the stations are also represented in correlated graphics (Figs. 5.21 and 5.23) (original)

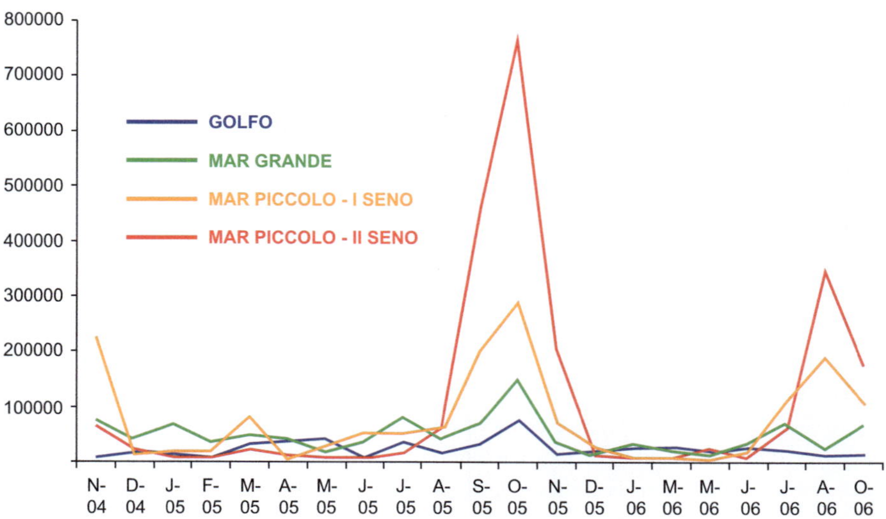

Fig. 5.21 Abundance fluctuations of total zooplankton in four stations of the Taranto Sea System. The abundance peaked synchronously at the four stations, but the intensity of peaks (abundance of individuals) was higher at the internal stations than at the external ones. (From Belmonte et al. 2013; modified by A.M. Miglietta)

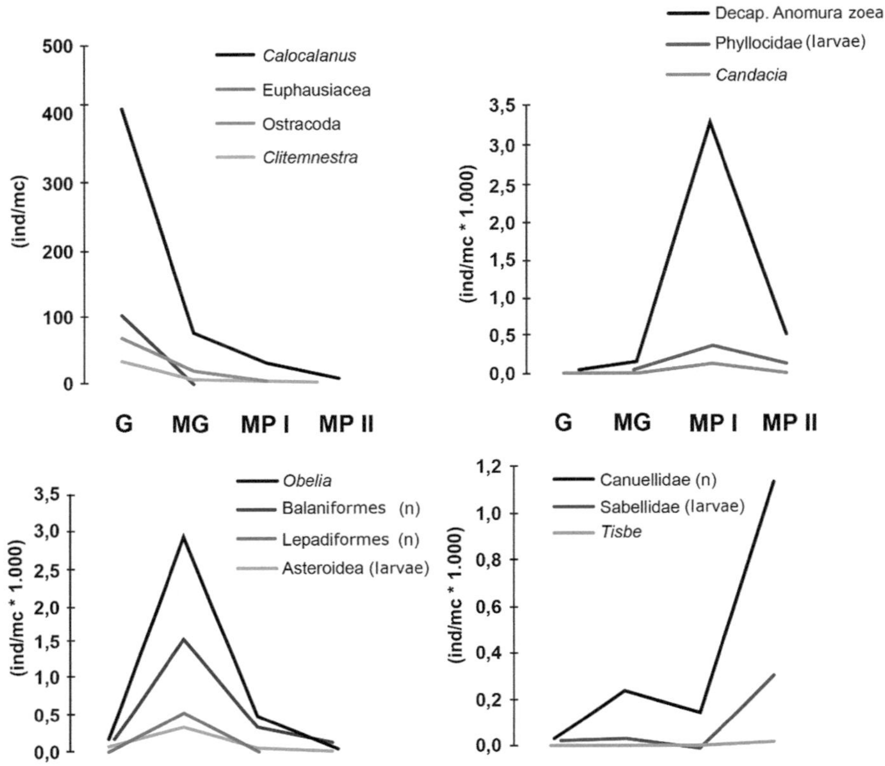

Fig. 5.22 Localized taxa were considered those peaking in one station of the sea system (75% of their presence was sampled in only one station). G, MG, MPI, MPII, stations considered. In each station, a series of taxa are shown as examples (graphic composition of A. M. Miglietta)

was characterized by the lowest abundance of individuals and the highest *taxa* richness.

The variability of the system was confirmed by the scarcity of species present everywhere and in every season. Perennial *taxa* (present along the whole studied period) were rare (67 = 20%), and even rarer (37 = 10%) were ubiquitous ones (found in all the studied stations) (Fig. 5.22). The very characteristic of the system was a high percentage of seasonal *taxa*, and of those linked to just one or two adjacent stations.

The percentage of *taxa* linked to one season grew in correlation with the confinement degree (Fig. 5.23).

The most interesting result, however, was in the growth of *taxa* richness according to the length of the period considered (and to the sampling effort carried out). If only one date is considered (this case derives from an average of 24 months), the open sea station (G, less confined) shows a *taxa* richness threefold that of the innermost station (MPII, the most confined). If the comparison is made on the total of the 24 months, the species number of the open sea G station doubles, and that of the

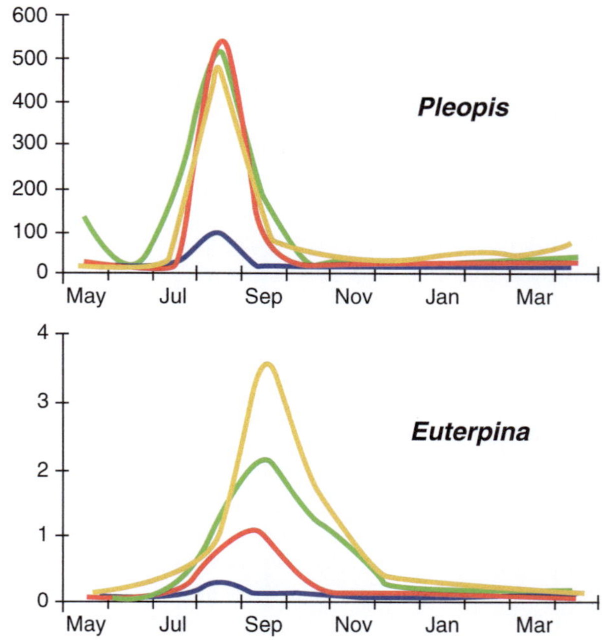

Fig. 5.23 Seasonal taxa were considered those peaking in a relatively brief period of the year (75% of their presence covers no more than 3 months). *Pleopis* (Cladocera) in August and *Euterpina* (Harpacticoida) in October are shown as examples (graphical composition of A.M. Miglietta)

Table 5.1 Average number of taxa for growing considered periods. Taranto Sea System, 2004–2006 (Belmonte et al. 2013)

Time interval considered	Number of sampling periods with same length	G	MG	MPI	MPII
1 month	21	162	122	80	53
6 months	4	248	193	140	105
1 year	2	288	224	173	144
2 years	1	318	268	198	173

innermost stations MPII triplicates, with the final result that the difference in *taxa* richness between the two extreme situations seems to diminish if long time windows are considered. This probably means that the smallest species of the MPII could show fast life cycles and short-term presences in the plankton, thus requiring a longer sampling effort (or a more frequent sampling rhythm) to intercept short-living species casually disregarded in single-year-based collections (Table 5.1).

What we know of the confined plankton, however, also authorizes us to consider that the marine seed bank of the bottom sediments could not represent its richness (injecting active stages in the water column) at each year cycle. Recent data show that the cysts interested in exchanges with the overlying water column account for only 10% of the cyst bank consistency. A large part of the stored diversity (in cyst form) remains unexpressed each year, and we need more years to perceive such an expression. The percentage of plankton unexpressed biodiversity (in the form of cysts in sediments) should be higher in the innermost part of the system (the most confined one), and, over many years, also the last unexpressed player will find its occasion of a presence in the plankton.

The 1990 data, opportunely weighted to be compared with those of 2004–2006, allowed us to also establish a variation in plankton composition of 13–15% in the more open stations and of 30% in the more confined ones. This changing world could be the result of long-term changes in the climate, and consequently, the system; thus, nobody can ensure us that the reason is in the presence of an unexpressed portion of biodiversity possibly larger than the expressed one. At least in one case, however, the weakness of a standard approach to the study of plankton has been demonstrated: in 1990 data, a species of Hydrozoa (*Eugymnanthea inquilina*) was not found among the 30 hydrozoan medusae reported in that study (Belmonte et al. 2001).

The species was reported in the study 15 years later (2004–2006) but the species was not indicated as a new arrival because a paper by Piraino et al. (1994) of 10 years earlier, dealt with *E. inquilina* from Taranto. The persistence of the plankton stage (the medusa) in the water column is about 15–20 days, and a sampling plan with collections executed monthly was probably not enough to collect such a short lasting plankton stage, although the species is present in the environment (in this case, as a polyp colony). This is an evident case where the investigation in benthos supplies information for the composition of plankton. In another example, the presence in the plankton of Actinotrochae larvae (typical of confined areas), communicates the existence of adults in the benthos somewhere of a phylum (Foronida) never reported by benthonologists in the Taranto sea bottoms.

Harsh conditions in the confined portions of coastal systems can evolve up to the disappearance of the water itself, as in saline ponds, obliging plankton to rest in a seasonally dry mud and/or arid situation. A small saline area along the coast of the Gulf of Taranto, represents a typical example of the problem. In 1998–2000 it was studied with a monthly collection of plankton when the water was present (Moscatello and Belmonte 2004). During a dry period, in addition, a sampling of sediment muds was executed to study the cyst bank and its consistency. In the 2 years of plankton collection, with visits of the site on 13 different dates, a total of 10 *taxa* were recognized. The study of the mud sediments collected on only one occasion allowed the recognition of 24 morphotypes of cysts. A study carried out on the plankton, some years later demonstrated that the plankton of the saline was completely different if compared with that of a few years before (Fig. 5.24).

The conclusion coming from these examples is that confined coastal areas have their own plankton (autochthonous, not derived from periodical vivifications from the open sea) and store them, during stressed periods, in sediment cyst banks. The plankton biodiversity of confined coastal sites is not contemporaneously expressed at each moment, but time is necessary to allow all the encysted *taxa* (waiting in the sediment banks) to express their active stages in the water column. Consequently, the study of benthic-plankton interactions (and of the resurrection ecology) highly affects our capability to understand the existence of unexpressed parts of biodiversity.

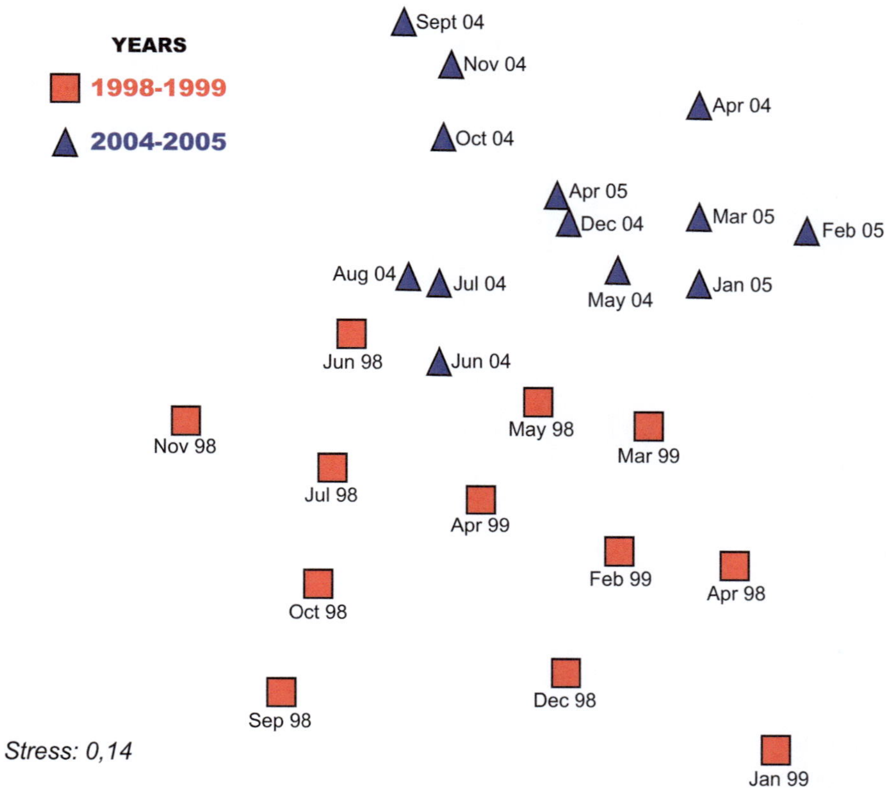

Fig. 5.24 Variability in abundance of plankton of *Vecchia Salina* (Gulf of Taranto coast, Italy); nMDS plot of 2 different years. (From Moscatello and Belmonte 2004 modified by A. M. Miglietta)

The study of the plankton dynamics in highly variable environments could also be expressed in space, other than in time. In fact, cysts are not homogeneously distributed in the horizontal extension of sediment cyst bank (as ascertained in the Mar Piccolo, Taranto) and different areas show different abundances and cyst compositions (Ferraro et al. 2017) (Fig. 5.25). Also, this suggestion calls for the adoption of a more intense sampling effort (in time and in space) to correctly evaluate the not-homogenously distributed biodiversity in confined environments.

A more recent study on the sediment cyst bank of Mar Piccolo, Taranto, allowed one to perceive the vast complexity of the scenario even if observed from the only sediment point of view. The sediment cyst bank of that site stores a quantity of resting stages (about 100 taxa), which is only superficially interested in exchanges with the water column (sinking and germination of cysts). In a variable, confined, environment, the strategy of species is to trust more on the sediment reservoir for the permanence in the habitat, than to stay in the plankton.

In a recent study, Rubino and Belmonte (2021) found that the cyst morphotypes in the sediment were more abundant than the plankton active stages potentially able

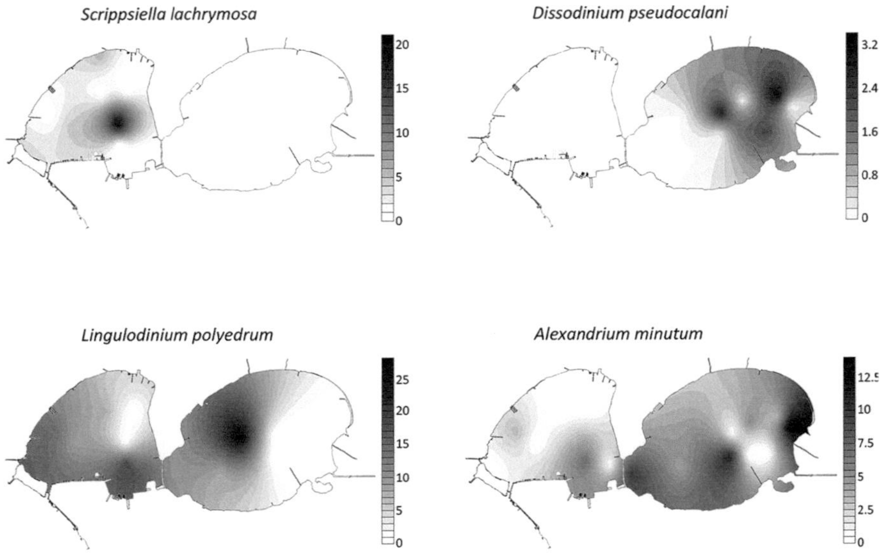

Fig. 5.25 Space differences in cyst bank abundance of four different Dinophyta, in the Mar Piccolo (first and second inlets) of Taranto. The scale of abundance (n. cysts g^{-1} of surface sediment) is indicated on the right of each distribution map. (From Ferraro et al. 2017; modified by A. M. Miglietta)

to produce them. If the presence of cysts in the sediment represents the production rate and/or depends also on the low germination rate in a habitat, this rate of active or encysted plankton could in some way represent the unpredictability of that environment and/or seasons, with variability (unpredictability) directly associated with the cyst bank richness.

5.4.2 Submarine Caves as Confined Environments

It is not easy to find research regarding the pelagic life inside submarine caves; notwithstanding, in this habitat a pelagos exists surrounded by substratum in any direction. Obviously, the pelagic world (and lifestyle) will be more suitable for organisms whose small size allows them to stay afloat, profiting from the water viscosity, without necessarily having contact with the benthos. In addition, the environment of a submarine cave is also hydrodynamically calm, if compared with the outside environment, and continuously dark in the innermost parts distant from the entrance.

The composition of the benthos is only exceptionally different from that which grows under open waters, and this simply suggests that cave animals derive from outside living colonizers (and possibly also deliver propagules in the outside world).

Such communication, for sessile species, can only be realized by propagules moving in the water column. Here, the term "propagule" will be preferred to the

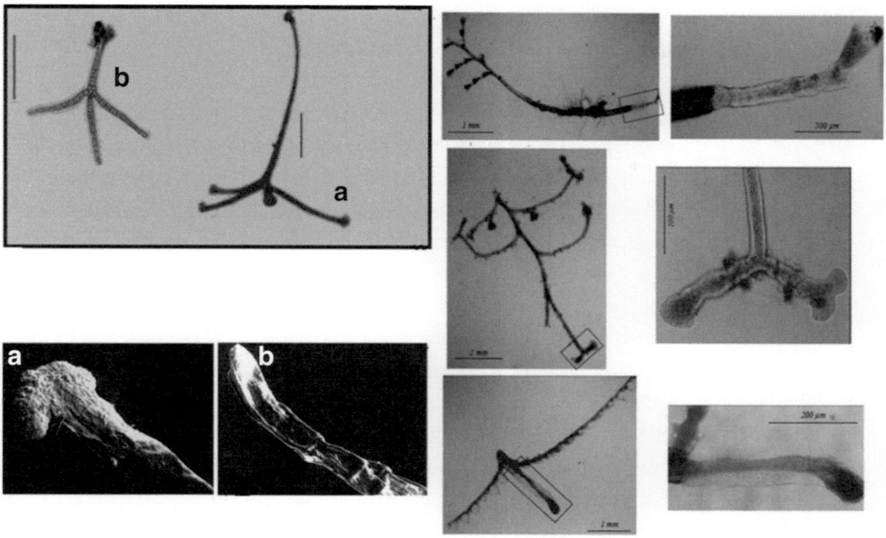

Fig. 5.26 Asexual specialized propagules. (a) *Halecium pusillum* (Hydrozoa); (b) *Sphacelaria* sp. (Phaeophyta). Up on the left, a photograph of the entire propagule. Down on the left, a detail (via SEM) of extremities of propagules a and b. On the right, examples of regeneration occurring on three Hydrozoa colony fragments. (From Pati and Belmonte 2018, modified by G. Belmonte)

explicit "larvae" because recent studies demonstrated that the space distribution of sessile species is not based only on the swimming larva. In fact, even in the absence of such a swimming stage, the geographic distribution and gene flow of sessile organisms are not compromised.

The Supply Side Ecology (Lewin 1986) studied lateral exchanges of larvae between adjacent sessile populations of discrete (not colonial) organisms. Colonial organisms, however, show a general trend in the reduction of the larva's importance and probably use other solutions to disperse their genome in the environment.

Sediment traps collocated at the bottom of coastal sites and periodically controlled, captured fragments of colonial species more than larvae searching for a settle. Such fragments, in addition, are not dead material ("detritus") but, due to the characteristics of the colonial organization, they contain viable specimens or parts, able to asexually regenerate a colony when settled on a suitable benthic substrate. Together with fragments, for some species, specialized propagules were also present, appositely adapted to float and to travel in the water currents (Pati and Belmonte 2018). Also, in this case, they are not larvae (produced by a sexual event of fertilization and preceded by embryonic development) and are called, for this reason, "asexual specialized propagules" (Fig. 5.26).

Box 5.2 Submarine Caves Genesis and Localization

Marine caves are not biologically cognate with the underground caves that we sometimes visit as tourists. These last derive from a chemical erosion (dissolution) of calcareous limestone operated by rainwater whose pH (around a value of 7.00) is generally aggressive on rocks (solid Calcium carbonate) and operates its dissolution at different levels of intensity (karstic phenomenon). Such different intensities of erosion produce a differential cavitation of the rocks, and local dissolutive processes are enhanced. Flooded cavities are water chambers, and the process of rock dissolution accelerates at such points. When the underground water reservoir is abundant (after a rainy season), the deepest cavities host water under pressure, and the erosive process is enhanced by this condition.

The marine water (pH around 8.00) is aggressive against the solid Calcium carbonate, only starting at very high pressures (recordable at about −3000 m).

Being the fresh water (from rain) less dense than the saline water (marine), the only rocks interested in karstic erosion due to the aggressive pH are those over the sea level.

In the geological history of the planet, however, the sea level was never fixed at a certain value for long periods, and the rocks could be exposed to karstic erosion accordingly with such a level oscillation. During the last glacial period of the planet (which lasted about 1 million years and ended just 20,000 years ago), the accumulation of ice at the two poles, subtracted water from the oceans and lowered their level by about 120 m from the sea level of today.

This exposed the rocks to rainwater and a lot of karstic erosion was possible during that period (Fig. 5.27).

Underground cavities represent a sort of water container system that springs outside over the sea level or just a few meters below. Wells and springs along a rocky coast represented the exit points of long underground pipelines of water. Such stone pipelines, when abandoned by water that flows at lower levels when karstic erosion has the possibility to continue its excavation activity, represent the caves and grottoes sometimes offered to visitors as nature monuments.

This process happened also during the last glacial period, and karstic erosion arrived to dissolve rocky banks down to 120 m below the current sea level. When the glacial period terminates, however, the ice caps at the poles melted and water returned to the oceans. The sea level rose to the current situation, and all the cavities produced during the glacial period, today stay below the sea level, invaded by marine water that does not continue the erosive process.

Such submarine caves are defined as hydrologically "fossil" because not active as freshwater pipelines, and they have become shelter sites for marine life.

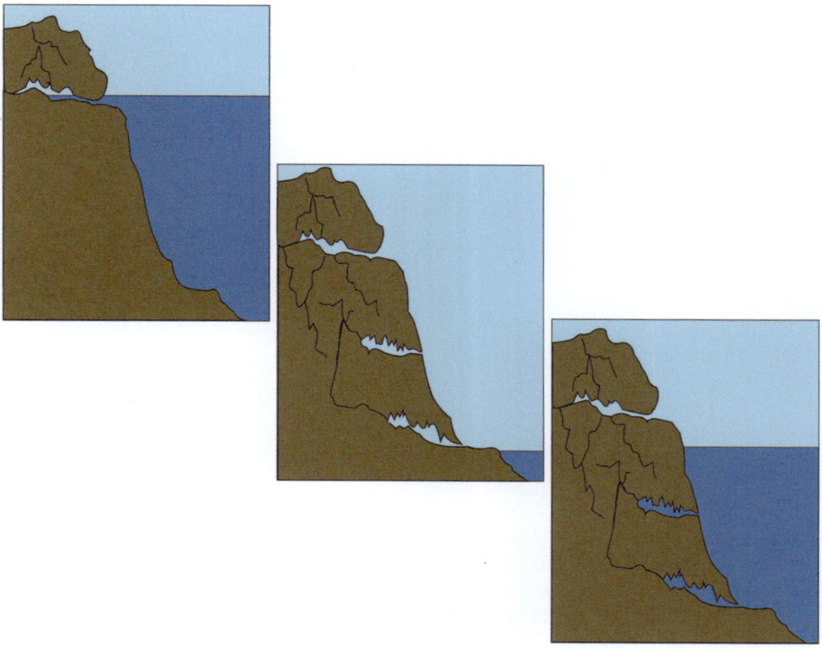

Fig. 5.27 Scheme of formation of submarine caves. At the left top, rocks interested by karstic activity are only those above sea level. In the centre, the diminution of the sea level (e.g., during the last glacial period) exposes more rocky layers to the karst erosion. On the bottom, right, the rise of the sea level determines the submersion of ancient karstic caves (original)

Finally, most of the geographic distribution of sessile organisms is due to their traveling capacity aboard rafts and floating materials, as sessile colonies, and not as floating fragments, asexual specialized propagules, or larvae.

This complex frame of traveling propagules has to be investigated to understand the preferential connection adopted by sessile benthos inside and outside the submarine caves.

In addition, the submarine cave habitat is suspected to be more affected by the substratum vicinity (at 360°) in terms of benthos-derived materials present in the water column.

The vicinity of the substratum in all directions affects the composition of the pelagic community in a submarine cave. Such an affection is responsible for particular characteristics of the pelagic

Community, which receives the name (not only in submarine caves) of hyperbenthos (or hypoplankton) just to underline their distinctive composition and space localization in relation to the true plankton.

Intense efforts to evaluate the plankton composition in submarine caves were carried out only rarely. In a cave in the centre of the Mediterranean Sea, the plan for sample collection was organized taking into account different parts of the studied

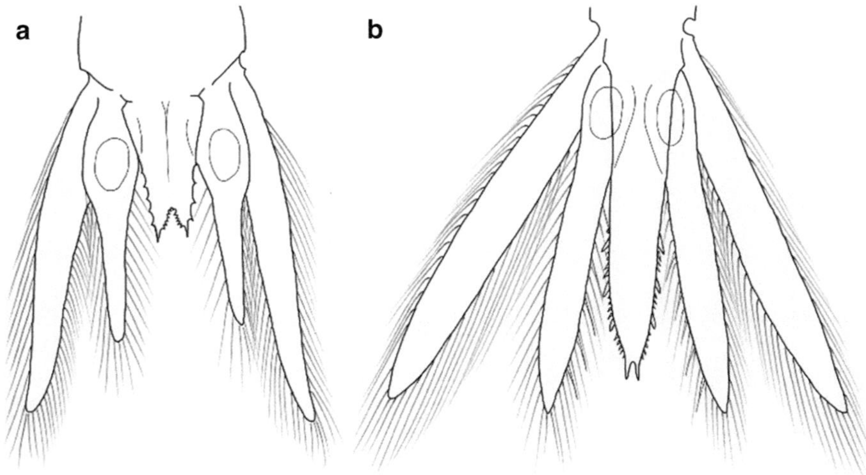

Fig. 5.28 The telson (diagnostic character) of two species of Mysida co-existing in the submarine cave "il Ciolo" (southern Italy). (**a**) *Hemimysis margalefi*; (**b**) *Siriella jaltensis*. The two uropodia close to the telson show the statocystS, which are typical of Mysida. (drawings not at the same scale; *H. margalefi* is half of *S. jaltensis* in size) (original)

system (thanks to the length of the submarine cave), different seasons, different daytimes (day and night), and using two mesh sizes of plankton net.

The 126 plankton samples collected in the study of Moscatello and Belmonte (2007) gave the following robust results for the plankton of a single submarine cave:

- High species richness (232 *taxa*).
- Numerical prevalence of Mysida populations (2 species).
- Taxonomical prevalence of Copepoda (Harpacticoida, in the cave).
- Prevalence of asexual propagules (specialized, and fragments) on larvae.
- Differences in the *taxa* distribution along the cave axis.
- Horizontal daily migration inside–outside the cave of some components.

Of the 232 taxa collected, only 50% can be attributed to the holoplankton. Mysida, the most evident component of the cave plankton, represented the hyperbenthos. Copepoda, responsible for the most diversity, were represented by 105 *taxa*. Contrarily to lagoon-confined environments, however, Harpacticoida (40 *taxa*), not properly holoplankton, were the characterizing copepods of the cave, where Cyclopoida (20 taxa) and Calanoida (37 taxa) were evidently less important than in the outside open habitat.

Mysida, numerically abundant and evident inside the cave, were present with the species *Hemimysis margalefi* and *Siriella jaltensis* whose co-existence is favoured by different body sizes and behaviours (Fig. 5.28). Both species are absent from the plankton outside the cave, during the day. *H. margalefi*, the smaller Mysida, is absent from all the outside samples (also during the night), thus allowing us to consider it a cave species. *S. jaltensis*, the largest Mysida, concentrates its population in the first part of the cave (close to the entrance) during the day, and goes outside during the night.

Horizontal migration (inside–outside the cave) is an important way of trophic transfer and communication between the cave and the outside environment. *S. jaltensis* moves exclusively between the first part of the cave (close to the entrance) and the outside environment. *H. margalefi*, on the contrary, never goes outside, but its population also executes horizontal movements between the innermost part of the cave (at about 100 m from the entrance) and that close to the entrance on a daily basis.

This horizontal migration is synchronized with that of *S. jaltensis*; thus, when *S. jaltensis* stays outside (during the night), *H. margalefi* occupies the position close to the entrance. When *S. jaltensis* stays inside the cave (during the day, close to the entrance), *H. margalefi* occupies the innermost part of the cave.

This synchronic behaviour corresponds to a transfer of biomass (and energy) from the productive outside part of the coastal system to the cave habitat, with an involvement of the innermost part (Fig. 5.29).

This important component (Mysida of hyperbenthos) of the submarine cave zooplankton has been recognized in many other submarine caves, and it seems a general rule of the ecological functioning of a system (the submarine cave) otherwise deprived of primary producers because of the absence of light.

In addition to the Mysida component (at 19% of the total number of specimens), the plankton of submarine caves was also characterized by non-holoplankton categories (meroplankton, fragments, and benthic derived material) for an additional 33% of the total specimens and/or items. In details, fragments (asexual propagules) were present elsewhere, but relatively more important in the internal part of the cave, where larvae resulted significantly reduced or absent, and in the evident opposite situation if compared with samples collected outside.

The innermost part of the cave was characterized by Harpacticoida (*Amonardia phyllopus*, *Tisbe* spp., *Paradactylopodia brevicornis*), Mysida (*Hemimysis margalefi*), and algal fragments.

At the entrance of the cave, the plankton was characterized by Harpacticoida of different species (*A. phyllopus*, *Tisbe* spp., *P. brevicornis, Harpacticus* sp.), Mysida (*Siriella jaltensis*), and algal fragments.

In the environment just outside the cave, other than algal fragments, a few Harpacticoida (*A. phyllopus*, and *Euterpina acutifrons*) and more Calanoida were found, with absence of Mysida.

Regarding the asexual propagules (specialized or simply fragments), studies carried out on the settlement process (Denitto et al. 2007) allowed us to understand what propagules can be considered "loose material" (detritus) and what propagules have a role in recruiting new benthic organisms.

Cycle = circular; *Ops* = eye *Holo* = complete, totally; *Plankton* = vaguing *Mysis* = blink *Harpacticon* = rapacious predator

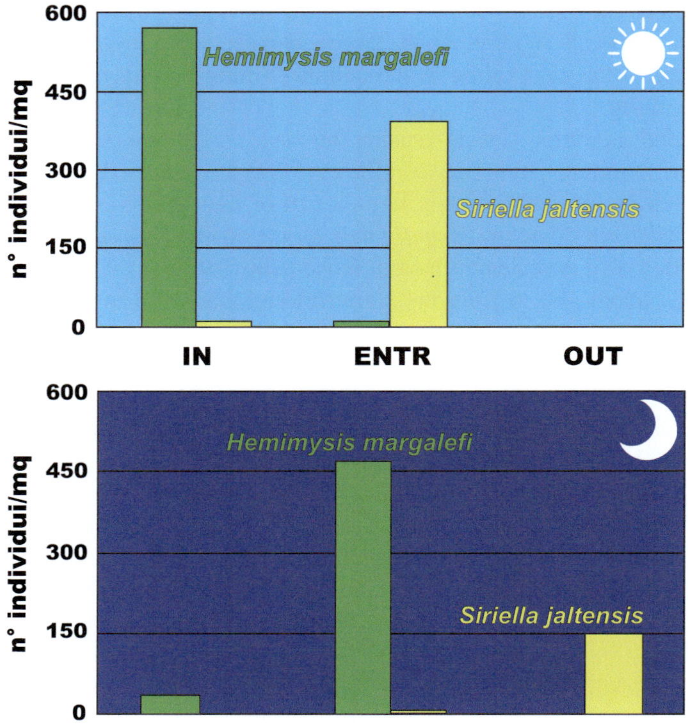

Fig. 5.29 Diagram showing the horizontal movement of Mysida species in the cave "il Ciolo" (southern Italy). IN, innermost part of the cave (at 60–90 m from the entrance); ENTR, part of the cave close to the entrance (at 10–40 m from the entrance); OUT, part of the outside marine environment (as a control). The Mysida species are *Hemimysis margalefi* and *Siriella jaltensis* (graphic composition of A. M. Miglietta)

Asexual propagules (unspecialized = fragments, and specialized) in the cave plankton were significantly more abundant than sexual ones (larvae), possibly having a heavy role in the recruitment of benthic populations.

Propagules *taxa* inside the cave were the same as those present outside, thus suggesting a possible exchange of material between the two parts, independently from that due to the daily migration of Mysida. How much the cave benthos recruitment depends on this propagule availability, and how much the cave benthos delivers propagules to the outside benthos, is still to be understood.

Submarine caves, however, are refuge areas for a large part of animal diversity, and even fish (nekton) use them as nurseries for their juveniles. The number of species (although without vegetal organisms) is significantly high (667 species in only 107 caves around the Salento Peninsula (Onorato and Belmonte 2018); 2369 species in 3051 caves of the Mediterranean Sea (Gerovassilieu and Bianchi 2021)) and suggests a role of biodiversity protection for these environments, with possibilities to host also mesopelagic species whose propagules find shelter in the caves of the euphotic layer, migrating upwards from dark depths during night or seasonally. The role of biodiversity protection, species refuge, and nursery for juveniles, are to be inserted in the frame of exchanges between caves and the outside world, added to the daily horizontal migration of the hyperbenthos.

References

Belmonte G (2021) Acartiidae Sars G.O., 1903. ICES identification leaflets for plankton. Leaflet n 194, 29 pp

Belmonte G, Cavallo A (1998) Body size and its variability in the copepod *Acartia margalefi* (Calanoida) from Lake Acquatina (SE Italy). Ital J Zool 64:377–382

Belmonte G, Rubino F (2019) Resting cysts from coastal marine plankton. Oceanogr Mar Biol Annu Rev 57:1–88

Belmonte G, Mazzocchi MG, Prusova IY, Shadrin NV (1994) *Acartia tonsa*: a species new for the Black Sea fauna. Hydrobiologia 292(293):9–15

Belmonte G, Castello P, Piccinni MR, Quarta S, Rubino F, Geraci S, Boero F (1995) Resting stages in marine sediments off the Italian coast. In: Elefteriou A et al (eds) Biology and ecology of shallow coastal waters. Olsen & Olsen Publication, Fredensborg, pp 53–58

Belmonte G, Miglietta AM, Rubino F, Boero F (1997) Morphological convergence in resting stages of planktonic organisms: a review. Hydrobiologia 355:159–165

Belmonte G, Fanelli G, Gravili C, Rubino F (2001) Composition, distribution and seasonality of zooplankton in Taranto seas (Ionian Sea, Italy). Biol Mar Mediterr 8:352–362

Belmonte G, Scirocco T, Denitto F (2011) Zooplankton composition in the lake Varano (Adriatic Sea coast, Italy). Ital J Zool 78:370–378

Belmonte G, Vaglio I, Rubino F, Alabiso G (2013) Zooplankton composition along the confinement gradient of the Taranto Sea System (Ionian Sea, south eastern Italy). J Mar Syst 128:222–238

Brilinsky JM (1981) Report on the presence of *Acartia tonsa* Dana (Copepoda) in the harbour of Dunkirk (France) and its geographical distribution in Europe. J Plankton Res 3(2):255–260

Carlton JT, Geller JB (1993) Ecological roulette: the global transport of nonindigenous marine organisms. Science 261:7

Denitto F, Terlizzi A, Belmonte G (2007) Settlement and primary succession in a shallow submarine cave: the spatial and temporal benthic assemblage distinctness. Mar Ecol PSZNI 28(Suppl 1):35–46

Ferraro L, Rubino F, Belmonte M, Da Prato S, Greco M, Frontalini F (2017) A multidisciplinary approach to study confined marine basins: the holobenthic and merobenthic assemblages in the Mar Piccolo of Taranto (Ionian Sea, Mediterranean). Mar Biodivers 47:887–911

Gaudy R, Vinas MD (1985) Premiere signalization en Mediterranee du copepode pelagique *Acartia tonsa*. Rapp Comm int Mer Medit 29:227–229

Gerovassilieu V, Bianchi CN (2021) Mediterranean marine caves: a synthesis of current knowledge. Oceanogr Mar Biol Annu Rev 59:1–88

Giangrande A, Geraci S, Belmonte G (1994) Life-cycle and life-history diversity in marine invertebrates and the implications in community dynamics. Oceanogr Mar Biol 32:305–333

Guelorget O, Perthuisot J (1992) Paralic ecosystems. Biological organization and functioning. Vie et Milieu/Life Environ 42(2):215–251. ffhal-03044531f

Hairston NG Jr, Van Brunt RA, Kearns CM, Engstrom DR (1995) Age and survivorship of diapausing eggs in a sediment egg bank. Ecology 76:1706–1717

Kerfoot WC, Weider LJ (2004) Experimental paleoecology (resurrection ecology): Chasig Van Valen's Red Queen hypothesis. Limnol Oceanogr 49:1300–1316

Lewin R (1986) Supply-side ecology. Science 234:25–27

Maps F, Runge JA, Leising A, Pershing AJ, Record NR, Plourde S, Pierson JJ (2012) Modelling the timing and duration of dormancy in populations of *Calanus finmarchicus* from the northwest Atlantic shelf. J Plankton Res 34:36–54

Moscatello S, Belmonte G (2004) Active and resting stages of Zooplankton and its seasonal evolution in a hypersaline temporary pond of the Mediterranean coast (the "Vecchia Salina", Torre Colimena, SE Italy). Sci Mar 68:491–500

Moscatello S, Belmonte G (2007) The plankton of shallow marine cave "Grotta di Ciolo" (Salento Peninsula S-E Italy). Mar Ecol PSZNI 28(Suppl 1):47–59

Nehring S (1997) Dinoflagellate resting cysts from recent German coastal sediments. Bot Mar 40:307–324

Onorato M, Belmonte G (2018) Submarine caves of Salento peninsula: faunal aspects. Thalassia Salentina 39:47–72

Pati AC, Belmonte G (2003) Disinfection effcacy on cysts viability of *Artemia franciscana* (Crustacea), *Hexarthra fennica* (Rotifera), and *Fabrea salina* (Ciliophora). Mar Biol 142:895–904

Pati AC, Belmonte G (2018) Asexually generated propagules from Subtidal Sessile Benthic Organisms. Pro Aqua Farm Mar Biol 1:180002

Pati AC, Belmonte G, Ceccherelli VU, Boero F (1999) The inactive temporary component: an unexplored fraction of the meiobenthos. Mar Biol 134:419–427

Piraino S, Totaro M, Geraci S, Boero F (1994) Ecology of the bivalve inhabiting hydroid *Eugymnanthea inquilina* in the coastal sounds of Taranto (Ionian Sea, SE Italy). Mar Biol 118:695–703

Rubino F, Belmonte G (2021) Habitat shift for plankton: the living side of benthic-pelagic coupling in the Mar Piccolo of Taranto (Southern Italy, Ionian Sea). Water 13:3619

Rubino F, Saracino OD, Moscatello S, Belmonte G (2009) An integrated water/sediment approach to study plankton (a case study in the Southern Adriatic Sea). J Mar Syst 78:536–546

Sazhina LI (1968) O zhimuyushchik yaitzak morskik Calanoida. Zool Zhurnal 47:1554–1556. (in Russian)

Stabili L, Miglietta AM, Belmonte G (1999) Lysozyme-like and trypsin-like activities in the cysts of *Artemia franciscana* Kellog, 1906. Is there a passive immunity in a resting stage? J Exp Mar Biol Ecol 237:291–303

Viitasalo S (2007) Effects of bioturbation by three macrozoobenthic species and predation by necto-benthic mysids on cladoceran benthic eggs. Mar Ecol Prog Ser 336:131–140

Wall D, Dale B (1968) Modern dinoflagellate cysts and evolution of the peridiniales. Micropaleontology 14(3):265–304

6 Pelagic Lifestyles in the Open Sea

6.1 Pelagos Vertical Partitioning

Pelagic waters can be vertically divided by horizontal borders, which can be simplified with the identification of a surface and a bottom and other intermediate limits that separate different water masses in **horizontal layers**. The **thermocline** is the most important of these separation borders. Lighter (less dense) water masses stay over the denser ones. The **density** of marine water is affected by salinity (surface waters also have a lower salinity because they receive freshwater from rivers and rain) and overall by temperature (cold water is generally denser than warm water). Wherever surface conditions produce a rise in water density, the formed denser water (more saline and colder) will tend to sink under the lighter water (less saline and warmer).

There exists a depth (variable) where surface waters remain separated and well distinguishable from the deep waters. This limit occurs between 80 and 200 m of depth. Such a layer is defined as **thermocline** (because the water temperature of the upper water-mass is evidently different from that of deep water-mass and the transition between the two parts is not progressive but abrupt). A **halocline** (level separating waters with different salinities) is somewhere recognizable as well, and an **oxycline** (change of oxygen concentration) too (Fig. 1.2).

These layers separate environments (air and water, water and sediment, or waters with different densities) with different physical properties and act, in the open ocean, as physical barriers for passively moving materials (Fig. 6.1).

Either organic or inorganic, dead or alive, suspended materials slow their motion or even brake when passing through such barriers, thus generally accumulating in their correspondence.

Hence, in the pelagos, we will find the accumulation of material (and plankton) in correspondence with such physical barriers separating water masses (Bakun 1996).

The accumulation of organic material practically defines areas of high food availability.

G. Belmonte, *Elements of Pelagos Biology*,
https://doi.org/10.1007/978-3-031-74280-4_6

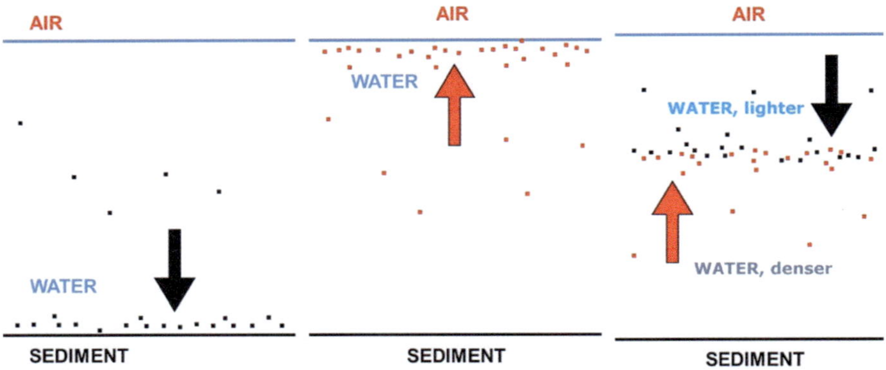

Fig. 6.1 Scheme of water borders, separating water masses in the vertical sense. Black elements (and arrows) are destined to sink down. Red elements (and arrows) are destined to buoy up. (Courtesy G. Belmonte)

Barriers realized by physical discontinuities exist not only to brake vertical flows of material but also to separate horizontally adjacent water masses, thus adding complexity to the geography of the open ocean and justifying the existence of biogeographic regions and provinces.

The spawning habitats of some pelagic species (we know them mainly because they are of economic importance) often do not coincide with the maximum concentration of food particles for larvae (as is common along water fronts). This suggests that such species search and find environmental situations different from those where a high presence of food and energy to grow is available, probably to avoid the predator attraction, which is stimulated by the accumulation of organic materials in such restricted areas.

In any case, any anchovy larva that experiences growth in an **oligotrophic** area and has access to only the average food concentration of those waters is doomed to starve. Only exceptionally fortunate larvae, which locate themselves in small patches of anomalously high food concentration, can survive and develop a new generation of adult anchovies.

Cline = jump, passage *Epi* = over, the taller part; *Pelagos* = open sea *Halo* = salt, saline *Oxy* = oxygen concentration *Thermo* = temperature

The anomalously high food (material) concentration areas are those where two different water masses are in contact, without mixing.

Such **water fronts** are collection points where the physical energy of the systems coincides with the enhancement of biological traits. For example, the sea floor, the most obvious of such border layers, represents an interface where the particles sinking down along the water column can accumulate on the sea floor, subtracting themselves from the overstanding water volume.

In this case, the physical force (gravity) drives the sinking of particles, and the physical obstacle (the sea floor) organizes their collection.

In the same way, although with a different efficacy, the buoyancy opposite to the gravity force is responsible for the accumulation of particles at the surface (the sea-air interface).

The thermocline is another interface where physical forces can accumulate particles (and food), concentrating biological activities and functioning. Water fronts have the same role, separating neighbouring water masses on a horizontal frame.

The water fronts are determined by the contrast between waters of different densities. In any case the denser water-mass glides beneath the less dense water-mass. However, before the reaching of a new equilibrium and the stop of the process, the different water masses will coexist side by side in some areas and will realize conditions for material accumulation at the interface. Due to the physical rule that denser waters layer glide beneath the less dense ones, the side of the denser water among two water masses is directed downward, tending to drive denser water to stay beneath the less dense water. During such a contact, part of the denser water modifies its density, but this is not enough to impede the down-gliding process. The side of the less dense water that touches the denser mass increases in some way its density and, together with the denser counterpart, starts to sink down. This subtraction of water from the contact surface of the two water masses attracts a new water supply from each respective side, and this determines the accumulation of material just at the interface.

Tiny, slowly swimming organisms of plankton accumulate in this way in the slowly sinking water-front interface. This undoubtedly represents a stimulating situation for the attraction of larger nekton species in search of food.

It must be said, however, that converging and sinking flow is probably too slow to be perceived by larger organisms. They, consequently, cannot perceive the direction of water masses into an interface; this notwithstanding, many pelagic organisms are suspected to have evolved a behaviour that has the only justification of favour the meeting (before or later) with such material concentration areas. They simply converge toward floating debris and materials. Fish may simply feel comfortable in the vicinity of drifting objects (and dolphins appear easily when a boat is in their area), but the most reliable interpretation is that the evolution of such a comfort lies in the fact that it is the only possibility that fishes have to direct themselves, following drifting objects, towards a convergent zone, where food should be abundant (Fig. 6.2).

The widespread attraction of fish to drifting objects has led to the realization of **fish aggregating devices** (**FADs**) in some coastal regions around the world (Kingsford 1999). FADs have been recognized as truly causing an enhancement of tuna growth in a Hawaiian settlement.

Oligos = a few elements; *Trophe* = food

Box 6.1 Fish-Aggregating Devices (FADs)

Fish-aggregating devices (FADs) are man-made buoys or rafts that attract and aggregate fish and help fishermen catch more fish. The FADs (among other indirect benefits):

1. Increase fishing efficiency.
2. Increase the catch per effort unit.
3. Reduce fishing time (and costs).
4. Reduce pressure on coastal ecosystems.

Fishermen use the localization of logs and debris accumulated at water fronts close to large river mouths because the concentration of large predator (tuna) schools has the maximum probability (Fig. 6.3).

The water fronts are sometimes very extended. In the case of sub-Antarctic circulation, or of the Gulf Stream, we have two other very evident cases of material and food accumulation. At the sub-Antarctic front, for example, the food abundance (living biomass) is so high that it fosters whales, the largest animals on the planet.

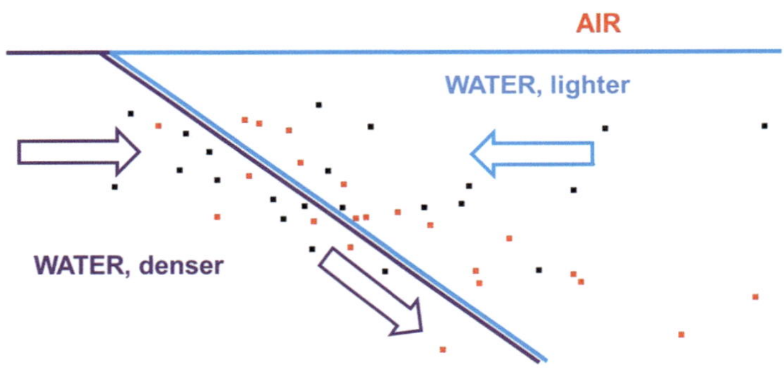

Fig. 6.2 Scheme of a front zone with accumulation of material that attracts large animals. (Courtesy G. Belmonte)

6.2 Techniques and Methods

In the open sea and ocean, the vertical dimension of the environment assumes more importance than in coastal and confined areas.

The collection methods, based on the use of bottles collecting water or plankton nets filtering water, must consider the importance of the vertical distribution of plankton, and they must be tuned accordingly. The vertical dimension of each geographic point (station) imposes the adoption of collection devices properly projected and equipped for a distinct or fractionated collection along water columns of hundreds or thousands of meters.

Fig. 6.3 Scheme of artificial fish aggregating devices, FADs. (1) Buoy linked to a floating "habitat mat" and anchored to the bottom; (2) FAD anchor on the sea floor; (3) Habitat mat floating and anchored to the buoy, the truly attractive device for fish (**fish calamite**) (graphic composition A. M. Miglietta)

One such device is the **multinet/multisample BIONESS** equipped with 9–12 plankton nets (generally having a mesh size of 200 μm) commanded remotely (from an oceanographic ship) and able to collect samples at precise intervals of depth from the bottom up to the surface. Generally, a BIONESS tool is also equipped with a multiparametric probe able to communicate to the surface the water characteristics and some important layers, such as the maximum chlorophyll depth or the thermocline, where the collection of plankton samples could be interesting.

Another sampling device common for open ocean scientific cruises (but not exclusive to the open sea) is the **neuston net** (in some versions called the "manta" net for its similarity with a ray form fish), which is projected to collect plankton from the just surface layer (not more than 50 cm below the sea surface).

Both the vertical distribution and horizontal surface distribution of plankton are generally not of interest in repetitive collections for the representativity of communities. In fact, both oblique hauls of BIONESS between two different depths and horizontal towing of manta nets are performed with the vessel navigating (at approximately 2–4 knots) and for a relatively long time (10–30 min). This allows the collection of plankton from large sea areas, and the consequent representativeness of each collection makes the replications superfluous.

BIONESS and manta nets are associated with open sea research because their use is linked to the availability of a ship or vessel due to their volume and size and the difficulty in managing them from small boats.

6.3 Passive and Active Movement of Pelagic Organisms

6.3.1 Neuston

Naumann (1917) was the first to recognize that at the surface of freshwater ponds, there exists a particular assemblage of organisms adapted to this particular and narrow layer.

Bacteria, Euglenidae, Chlamydomonadidae, Amebae, and other minute organisms inhabiting the surface film of small ponds and puddles received the collective name of **neuston** (Fig. 6.4).

Zernov (1934) noted that a similar distinctive assemblage of surface organisms was also present in marine waters. He established that the upper 5 cm of the sea contain great masses of Metazoa, Protozoa, and saprophytic Bacteria.

The bacterio-neuston, to tell the truth, is hundreds or thousands of times more abundant than the bacterio-plankton of the water column beneath.

The sea surface is a very extended habitat (more than 70% of the planet's surface), this notwithstanding, it could be classified, according to our perception of

Fig. 6.4 On the top left, neuston net designed to collect neuston from the surface 50 cm layer (Lab of Zoogeography and Fauna, University of Salento, Italy). On the bottom left, the Neuston net is washed on board to collect the sample concentrated in the terminal bottle. On the bottom right, one neuston sample was collected on a sieve to eliminate water. On the top, right, a multinet (BIONESS) for repeated collections of zooplankton at different depths (University of Messina, Italy) (Minerva Uno ship, Italian CNR). (Photographs G. Belmonte)

nature problems, as an **"extreme" environment**. In fact, in the first 10 cm of the sea surface, the solar light entering the water is already reduced to only 50% due to the absorption properties of the water. This means that the sea world is normally less illuminated than the very surface layer and, on the other hand, that the surface layer is subject to intense illumination (with a daily variation) if compared with the whole marine habitat. In addition. What has to be punctuated is that a large part of light radiation (wavelength around yellow–red) does not pass the first 5 cm. The same absence of penetration also characterizes ultraviolet radiation. Thus, the surface layer of oceans receives light wavelengths (comprising UV radiation), which are practically excluded in the whole marine environment (Fig. 6.5). Consequently, we believe that neuston life evolved differently from whole marine life to avoid problems with the particular environmental situation.

The sea surface is also affected by episodic pulses of freshwater (rain and river discharges) and by atmospheric storms and winds, other than temperature variations on a daily and yearly basis, which do not directly affect the deepest layers.

Such a scenario (temperature and salinity intense variations, UV radiation) is accomplished by human interference (with navigation) and pollution acting on surface waters first.

The neuston can be identified with one of the interface layers (between the sea and the atmosphere) where materials accumulate. In general, organic material is sparsely distributed in sea water (from 50 to 500 times more abundant than living material), where it comes from marine biological communities but also from land environments through river inflows and wind transport. Materials coming from land and derived from the decomposition of marine communities for a relatively long time are hosted at the sea surface before becoming heavier and/or reduced by microbial decomposition and sinking down to the sea floor.

Small insects and pollen grains from land are very common wind-derived neustons, even at many miles offshore, where they are able to remain floating for weeks before obeying the gravity force and sinking to the sea floor.

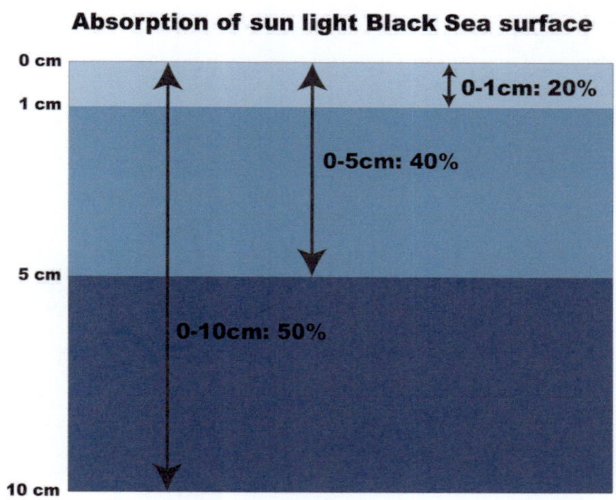

Fig. 6.5 Scheme of sun light extinction with depth in the Black Sea (first 10 cm). (Graphic composition A.M. Miglietta)

Other Arthropoda reach the water surface as dead material. Among these dead organisms are crustaceans of many species coming from the water below the surface, even plankton. Due to the presence of a complete cuticle covering for individuals, the *postmortem* degradation process acting on internal flesh starts to produce gas that remains trapped by the cuticle. This trapped gas determines the ballooning of the dead plankton specimens towards the water surface, where they stop. The organic material accumulating in this way at the sea surface is called **anti-rain** and is one of the main factors responsible for the accumulation of organic material in the neuston compartment (Fig. 6.6).

The anti-rain and the general accumulation power of any surface separating two physically different fluids (air and water, in the case of neuston) realize a concentration of organic matter at the ocean surface and attract, for this reason, many organisms in search of food. The existence and reliability of such a food source also dictated the evolution of morphology and behaviour in some aerial species (the bird with the scissor beak, for example; Fig. 6.7).

Anti = contrary

Rhincops nigra (Aves) feeds exclusively on this material-rich layer and has evolved a beak with the inferior part longer than the superior one, just to be efficacious in collecting neuston when the bird glides a few centimetres over the sea surface.

The availability of food also attracts swimming species from the water column beneath and their predators.

To tell the truth, some specialized neuston organisms (Cnidaria Siphonophora, *Velella*, and *Porpita*) are predators and live exclusively in this layer (even protruding out of the water, representing the **pleuston**).

Other neustonic species (Gastropoda *Janthina* and *Glaucus*) are specialized neustonic superpredators of *Velella* and *Porpita*, thus confirming the large

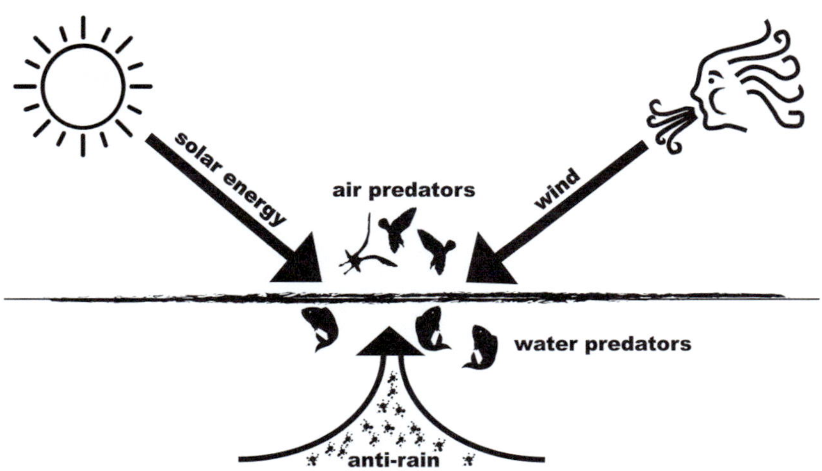

Fig. 6.6 A scheme of neuston multisource derivation. (Graphic composition A.M. Miglietta)

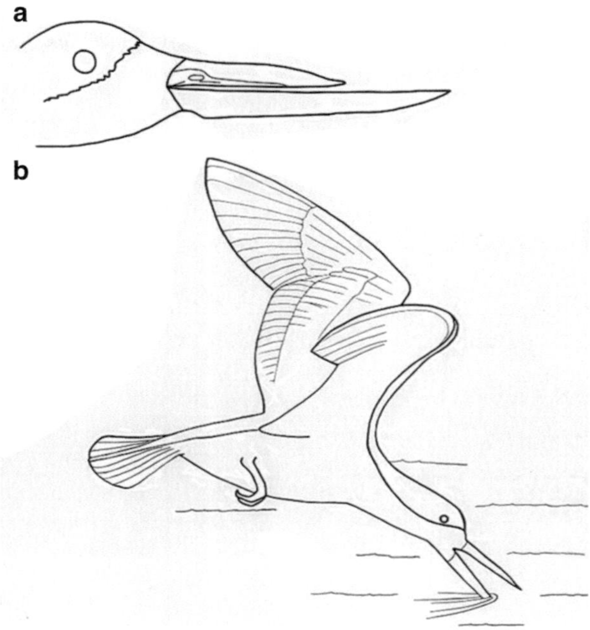

Fig. 6.7 *Rhincops nigra*, (**a**) bird explicitly adapted to collect neuston from the sea surface. a, detail of the asymmetric beak; (**b**) representation of the neuston collecting behaviour. (Courtesy G. Belmonte)

availability of energy (in terms of food) that is present in the neuston to sustain such a long trophic chain and the role of **superpredators** (trophic levels 3–5) for such strictly neustonic species (Fig. 6.8).

All these neustonic species are prey of gliding birds (specialized) or upcoming fish. The abundance of trophic resources, evidently, is probably not a limit, and this availability ensures the reproduction, growth, and recovery of neustonic species.

The predation pressure from below (the water column) has to be of the same order of magnitude as that exerted by the aerial organisms. In fact, an escape behaviour implying a dive in the sea is possible for many seston species prosecuted by their aerial predators, but many neuston species (e.g., Calanoida Pontellidae, or flying fish Exocetidae, or penguins and seals) evolved adaptations to jump out and to stay on floating ice, or in the air for a short time, to escape the aggression coming from marine predators.

Apart from living organisms (represented briefly by the previously mentioned cases) and inert particles (the detrital component of the anti-rain), a third element of the neuston probably better justifies the high nutritional content of the surface layer. This third component is represented by fish and invertebrate **eggs**, inert but energetically rich particles. Many eggs of fish species can be found in the neuston, but they were considered abortive and part of the "anti-rain". The suspect that some species evolved buoyant eggs in their life cycle, however, exists, and this could mean that this delicate stage of the life cycle probably finds convenience in locating in the neuston, a place where UV radiation, salinity, and pH variations probably perform in opposition to the lives of possible profiteers (e.g., Bacteria) and select only a few specialized predators to evolve distinct adaptations to profit from the available resources there.

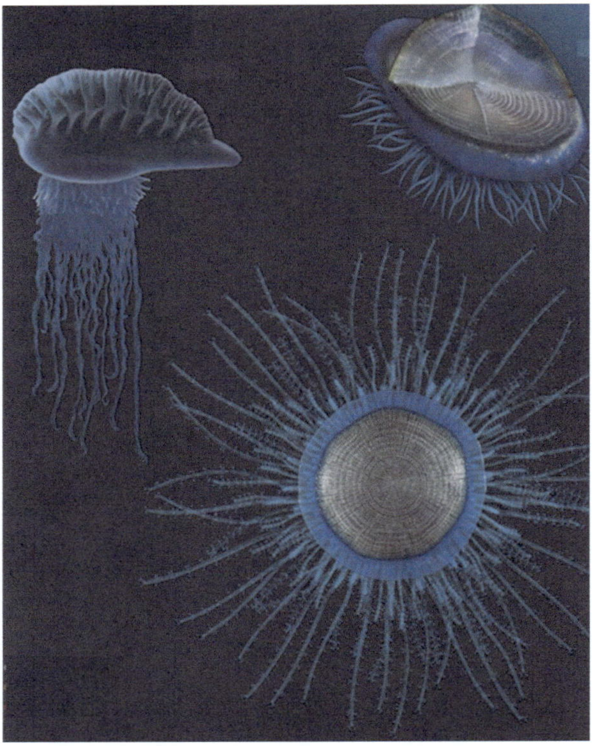

Fig. 6.8 Neustonic Hydrozoa (Cnidaria); on the top left, *Physalia*; on the top right, *Velella*; and on the bottom, *Porpita*. (Drawings A. Gennari)

6.3.2 Rafting and the Drifting World

The floating material at the sea surface, although temporary and, in any case, destined to sink to the sea floor with time, represents a key element of pelagic ecosystem functioning. Each floating fragment, debris, or litter can represent a raft, other than a food source, for other living organisms. Such a category of organisms does not show specialization or adaptations to pelagic-neustonic life, and they are simply castaways on **drifting rafts**. They are, from this point of view, plentiful benthic organisms, but the drift on board a raft allows them to passively disperse as plankto-pelagic items.

The presence of floating material at the sea surface is so obvious and common that entire groups of animals evolved their lives exclusively on them (e.g., the goose barnacles, Cirripedia Lepadomorpha, Fig. 6.9).

This situation, recently considered an exclusive product of human activities and of their incapacity to manage detritus and debris, is a natural aspect of a complex and interacting world. The strength of atmospheric episodes (such as hurricanes, typhoons, or, exceptionally, tsunamis) highly influences the final amount of land-derived material that can be transported and given to the sea.

Planktonic stages (larvae) of benthic species are particularly ready to profit from any available solid substrate where they settle and undergo metamorphosis, leading

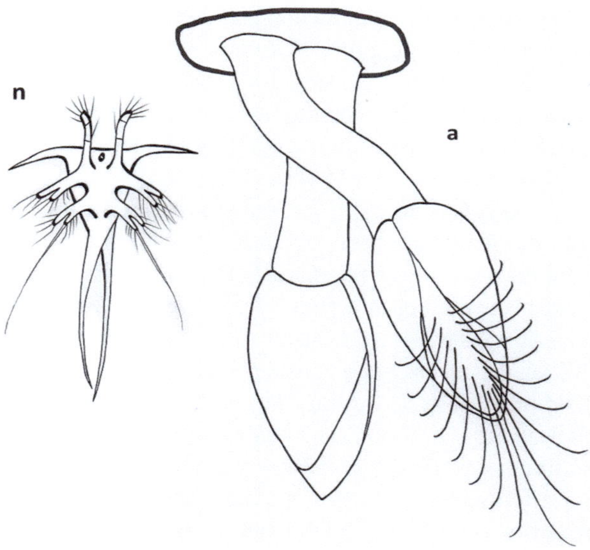

Fig. 6.9 Goose barnacles (Crustacea, Cirripedia, *Lepas*) as planktonic larvae (n, nauplius) and benthic adults (**a**) with cirri protruding outside the shell-like carapace to collect plankton food. (Courtesy G. Belmonte)

to the benthic sessile stage. If such a solid substrate is not fixed but floats, the settled organisms will represent a way for the species to travel inside oceanic currents, thus expanding its geographic distribution and/or recovering local demographic collapses without the involvement of larvae.

Planktonic stage durations should be related to the dispersal range and the geographic distribution of different species. However, many benthic species without a planktonic stage have been found to be successful inhabitants and/or colonizers of extended geographic areas than co-generics with a planktonic stage (Johannesson 1988; Boero and Bouillon 1993). Modular (colonial) organisms, although sessile, appear to not be dependent on the larva for their distribution over a wide geographic scale; in fact, the colonial condition appears to be associated with the loss of a long swimming larval stage. Jockiel (Jokiel 1990) proposed the rafting of small sessile colonies on floating pumice as the solution for the exchange of reef species propagules among volcanic islands in the Pacific Ocean. Bryan et al. (2012), in addition, described more than 80 species as interested in >5000 km of rafting in 7–8 months on volcanic pumice derived from an eruption in the Tonga archipelago. Their findings emphasize the rafting of small benthic colonies on pumice as important to the geographic distribution of marine benthic organisms.

(For a visual perception of an oceanic pumice raft, see https://www.facebook.com/rtenews/videos/a-giant-pumice-raft-has-been-spotted-in-the-pacific-ocean-near-vavau-tongapumice/782138495284/)

Larvae are not the only ones responsible for the species dispersal, and their large production in some species, when compared with their extremely low survival rate, represents an embarrassing loss of genetic material at each reproductive event. The high number, however, acquires importance if considered as a strategy to better intercept floating objects available at sea in a timely manner. Travelling rafters could, in this way, connect benthic populations at distant sites.

Seasonality in reproduction could even be a preadaptation of species if tuned to the periodical availability of floating materials at sea (e.g., when these derive from run-offs at each rainy season). Bryan et al. (2012) also proposed an adaptive response of benthos to earthquakes connected with unpredictable volcanic eruptions by coupling the spawning of coral species with pumice availability in the water.

Since the early sunrise of the planet, oceans receive materials from land regularly and as a consequence of intense weather events. Volcanoes throw pumice directly into the sea (if they are submarines) or indirectly if they emerge; winds transport land dust all around the planet surface (that is, at 70% ocean), glaciers scrape the land surface lying in a river or directly into the sea, what they collect, and rivers transport materials from the land to the oceans continuously and with high intensity, at least during each rain episode. Such material transport is commonly studied in the field of geology and is linked to problems of land erosion and sediment accumulation on the sea floor. Some studies of paleoecology, however, explicitly recognize rafting events in stratigraphic sequences. Materials of any size and type are destined for a long floating existence if they are buoyant. The final destiny of the buoyant material is to sink far from its origin site, often after long periods of pelagic existence.

Floating materials can be of many types, but the most evident and deserving attention are surely the vegetal materials, which also reach great volumes if based on the presence of wood logs. In contrast to volcanic pumice, the presence of **organic floating debris** is widespread in any part of the planet; a river mouth exists everywhere; and it also offers organic material as a food supply for rafters. Engineers of coast management report quantities (volumes and weights) of log subtracted from forests on land and transferred by rivers to the sea where they accumulate along the coasts close to the river mouth, enforcing the coastline in capturing and maintaining sediments, and rising problems to the management of the coastline by man.

Rivers, volcanoes, wind, and ice are responsible for the transport of materials to the ocean, but submarine vegetation could be added to the floating material that is able to transport benthos for long distances.

Thiel and Gutow (2005a) recognized **three main types of rafts** according to their permanence and availability in coastal areas: (a) frequent, (b) intermittent, and (c) episodic. All that review is focused on the ecological implications of such a phenomenon and its reflections on species genetics. What appears not to be duly considered, however, is the evolutionary importance it has to justify any intervention of management in the frame of nature conservation. The destiny of floating debris is to sink to the sea floor because sessile benthos settle on them, with time increasing the weight of the floating structure. In the case of organic rafts, microbes and rafters degrade organic matter over time, contributing to a decrease in its floatability.

The substrate composition inevitably affects the composition of the **rafting community**.

Mainly, suspension feeders are hosted on abiotic rafts (**pumice**), while grazing and boring species abound on rafting **macroalgae** and **wood**.

Mouths of large rivers have always been considered a danger for sailing (and for this reason, carefully avoided by ship routes) just for the possible presence of floating debris of large size, sometimes intertwined in small floating islands. Such large rafts have enormous importance in transferring marine benthos organisms (sessile and vagile) at level of their submerged parts.

Similar to **teleplanic** larvae, the rafting of sessile benthos could be considered responsible, although not the only reason, for the wider geographic distribution of species (Thiel and Hayes 2006).

The production of large quantities of larvae of some species (apparently a squandering of resources) could be explained by the precise advantage coming from the **increased possibility of intercepting floating vectors** and settling sedentary stages on them, with a sessile moving phase able to disperse the population propagules more efficaciously than larvae.

Entire and large communities, such as the ***Sargasso*** community, are based on floatability and rafts, thus testifying to the evolution over long periods of the rafting habitus.

Macro = large; *Tele* = distant; *Planos* = blowing *Sargassum* = (Portuguese) marine grape

If the production of large quantities of larvae could be considered useful for intercepting floating objects, the predominance of brooding and/or directly developing species on rafts is a fact.

If the spawning of larvae could have rafting dispersal as a by-product, we must understand what behaviour the directly developing species evolved to settle on floating objects in the absence of a pelagic swimming stage.

The presence of man, undoubtedly disturbing natural processes, has played a role only in the recent historical period (approximately 10,000 years).

The Neolithic agriculture revolution used warm temperate regions and forest habitats as the best places to realize human settlements and obtain products from fertile terrains. Forest woods, in addition, were the main resource for building materials and heat production. The abandonment of nomadism (generally considered to have occurred 10,000 years ago) corresponded to an abrupt increase in deforestation.

The elimination of temperate forests, however, was never completed until historically recent times when human demographic growth and the increased demand for heat energy by the industrial revolution surpassed the natural renewal rate of vegetal communities.

In the recent historical period, as a consequence, woods and logs started to diminish their presence in **river runoff** due to the reduction of the forested surface on land consequent to the massive use of land by human agriculture. In the Mediterranean area, a temperate climate region heavily impacted by human presence, ice does not transfer material from land to the sea, and pumice is produced by a small number of volcanoes and only episodically. In the Mediterranean area, consequently, the most important natural element that transfers materials from land to sea has been river runoff. However, in addition to forest reduction, transport from land to sea has also been reduced by the building of dams in very recent times (the last 100 years).

Whatever we can consider the Mediterranean situation, the energy that rivers produce coming down from mountains to sea has justified the realization of 537 large dams plus 8463 small barrages only in Italy in the last 100 years (Belmonte 2019). This capillary barrage of rivers (even in a not-extended country such as Italy), useful for energy production, water reservoirs, and land protection, inevitably produced a sharp diminution of materials transported by rivers into the sea.

Coastlines and dunes have been endangered for many years, and continuously managed to avoid their disappearance, mostly as a consequence of the lack of useful support for suspended particles (both large and small) from river runoff. Small suspended particles are simply diminished as a consequence of the building of dams and barriers on rivers' paths, but logs and trees have been abruptly reduced, and the drifting around of marine raft communities should be somehow negatively affected. (see photographs of dam collecting trees after the *tempesta vaia* in northern Italy: https://www.vigilfuoco.it/aspx/notizia.aspx?codnews=75831).

The modern situation is, consequently, a new level of ecological connectivity, with the disappearance or alteration of actors and actions that lasted over a million years, with enhanced effects in closed basins such as the Mediterranean Sea.

An additional alteration of the ancient, natural situation is not only the reduction of floating material but also the **variation in the composition of floating materials**. In short, a combination of increasing vector diversity and a changing climate sets the stage for a new era of rafting in the world's oceans. If wood logs have been drastically reduced in recent history, it is noteworthy that the realization of marine vessels by humans added floating surfaces to the system (at least during the last 3000 years in the Mediterranean area), with a drastic increase in the last 300 years. The Italian marine fleet, in the centre of the Mediterranean, currently offers floating surfaces to the sea, with approximately 100,000 vessels (of any size) continuously moving between ports.

The use of engines, in place of sails, to propel the navigation produced, in modern times, the possibility to sail in any direction, independently from surface currents and winds, thus sensibly altering ancient natural routes for possible rafters hosted on the hulls. The existence of **marine fouling on the hulls** of ships and boats is an evident proof of the possibility of benthos easily settling on floating objects of any type.

An experimental study by Carlton and Hodder (1995) on wooden vessels sailing along the North Pacific coast of America ascertained that slow-moving, fouled sailing vessels of relatively long residence times in ports may have significantly altered (homogenizing) the distribution of marine and estuarine organisms. These last, in detail, are subtracted by sailing vessels from the isolation due to the particular habitat (brackish water inlets) and biological traits (absence of planktonic stages).

There are no studies of the aboriginal distribution of species before the onset of coastal shipping (which is the ancient type of sailing). The current distribution of organisms along continental margins is consequently not "natural", as we generally admit, and the Mediterranean Sea is a typical locus of such a situation. The slow transport of organisms from one bay to another by vessels over the centuries would suggest the assumption that modern day distributions of organisms along a coastline are not the same as aboriginal distributions.

In a study on NIS introduced on Galapagos Island, the increase in recorded NIS is evident (from 6 to 53 species in the last 5 years) and attributed to the increase in vessel-based communication with other parts of the world (mainly South America).

Globally, shipping is the most important vector of introduction of marine nonindigenous species (NIS); both larger commercial vessels and smaller recreational boats act in the process of NIS introduction via ballast water and hull fouling.

In the Mediterranean Sea, **shipping and boating** occur within a complex network of port structures, thus facilitating the spread of organisms. Moreover, the abundance of artificial hard structures in ports provides suitable habitat for NISs, accelerating their local affirmation and spreading process.

The denied attention to the role of floating material as rafts for benthic ordinary dispersal impedes, today, a historical perception of how and if the quantity of floating material varies in the oceans.

A final element must be added to the already complex frame of the **variation of the floating material composition** in the recent historical period. If ships and boats were exclusively built with wood until two centuries ago, now the used materials exclude wood, but only in exceptional and rare cases. In addition, modern metallic or polymeric hulls are continuously cleaned and/or covered with anti-fouling paintings to impede the settlement of any kind of benthic community.

Curiously, and contrary to the past, boats and ships today play a role in transferring plankton living forms (possibly as resting stages) in the ballast waters, more so than benthos forms as fouling on the hull.

The final revolution in floating debris composition has been the discovery and synthesis of **plastic materials** (widespread in the last 70 years) that are not biologically degradable and persist in marine habitats for long periods before sinking and accumulating on the sea floor. This long-lasting floating material represents an important novelty (and enormous trouble) for the ocean of the future.

Natural elements also concur with the alteration of rafting material in the ocean in temperate areas; this is the case for ongoing climate change (indirectly sustained by human activities), which produces weather phenomena of high energy and consequent run-off power.

Although the pumice is present to represent inorganic rafts, continuing its role in the rafter distribution, it is not clear how the new inorganic rafts (plastics, overall) and weather situations are affecting the settlement and/or the survival of rafters or what differences there are with the traditional wood substratum (today diminished).

Management of coastal sites (specifically, marine protected areas) should monitor mechanisms of import and export of biodiversity. Barrage of rivers and insertion of new materials among floating debris could have an effect even heavier than, or synergizing with, ongoing climate change.

The adaptation of species to raft availability is disregarded in reviews by marine biologists on reproductive strategies and larval dispersal due to the absence of studies. The reduction in floating organic debris in the recent historical period, as a consequence of deforestation and dam building, is an unconsidered but probably heavily affecting element, endangering the equilibrium of benthic communities' distribution and their affirmation on large spatial scales. The injection of new materials (metals and plastics, in the last 200 and 70 years, respectively) derived by modern human activities into the ocean could not be enough to balance the loss of floating wood caused by deforestation and river damming because species (if not generalists) are probably not adapted to these new substrata.

Thiel and Gutow (2005b) listed a total of **1205 marine species** that have been reported as **rafters**. These, however, are related exclusively to the rafting materials of the last 100 years, and nothing can be compared with the evolutionary past, thus impeding us from understanding the role of humans in altering the rafting material composition.

Oceanography, geology, and coastal management developed, however, a large basis of knowledge to ensure a successful description of the particular aspect of species dispersal mediated by rafting material. In addition, the recent interest of marine scientists in floating plastics ensures the availability of methods and materials to better understand this suspected important role in the genetic connectivity and evolution of benthic communities.

6.3.3 Episodic Rafting Occasions

In 2011, an earthquake generated a **massive tsunami** in East Japan that originated a transoceanic biological rafting event. Coastal marine species were transferred **over 6 years** on objects deriving from the **run-off action** of the tsunami on Japanese land (Carlton et al. 2017). The material transferred from land to the sea by the run-off action, and the marine organisms that used such a material as rafts travelled thousands of km **across the Pacific Ocean** and were continuously stranded on the North West American shores in successive years. In fact, Japanese floating materials dragged by the tsunami in 2011 were found **on the Oregon and Washington state shores** from June 2012 to February 2017. In this episode, a total of **289 species** (belonging to 16 *phyla*) were recognized as "Japanese colonizers" of North America (see debris from the Japan tsunami in Oregon: https://www.bbc.com/news/magazine-19812373).

Although episodic, tsunamis of similar intensity were registered in the same area of Japan in 1933 and 1896, and we can be sure that they produced other episodes of organism export. The 2011 episode, however, can be considered different from the preceding episodes because only this last episode offered plastic material for the tsunami runoff.

The nonbiodegradable material (plastic) probably poses settlement problems to some species but offers an incredible advantage: being not consumable by rafters (at least a part of them), the plastic can decompose for hundreds of years. For this reason, this new material is able to go around for times longer than those of "traditional" wood pieces (reduced, with time, by microbial degradation and/or organism actions).

Plastic materials were already known at the beginning of the twentieth century, but objects completely invaded modern life starting in 1960 with the discovery of polymeric materials by Nobel Prize winners (for Chemistry) G. Natta and A. Ziegler. The success and global use of such synthetic polymers (in different forms) are currently the most important problems for conservation of the natural environment. Plastic materials cannot be degraded but can be reduced into smaller and smaller fragments, thus entering all living organisms with still-unknown effects. Today, plastic particles are a constant component associated with plankton samples.

Polyethylene, polypropylene, nylon, and paintings were the most represented categories of artificial material (plastics) found in the neuston from the Central Mediterranean Sea (Suaria et al. 2016). Sometimes such a plastic material is concentrated along surface current streams that distribute it even at great distances from the source point.

In a study on the stranded material over a beach of the south Adriatic (Central Mediterranean Sea), seeds of a freshwater species were recognized that entered the sea through a river mouth at hundreds of km from the site of collection (Fig. 6.10), and plastic debris (the main component of inorganic litter) were identified as coming from at least 15 different countries (not only "Mediterranean") (Mandic et al. 2022). These findings clearly showed that the plastic contamination of oceans today is not a problem of simple management because many and distant countries can be involved in plastic removal and/or recycling.

Plastic is not the only inorganic material representing possible rafts for the ocean environment. Aside from wood logs (organic material and "food" source for Bacteria and the connected consumers), in the story of the planet, there has always been also the presence of mineral, inorganic, rafts. Such rafting stones were (and are) produced by volcanic eruptions and have the name **pumice**.

Pumice results from the intense alveolation and quenching of silicic magma.

The buoyancy of pumice clasts depends on their size, shape, alveolation, permeability, and temperature when introduced into seawater. Centimetre-to meter-thick rafts of pumice clasts are observed when large volumes of buoyant pumice are rapidly released into or onto water. The routes followed by pumice rafts are dependent on the source of pumice clasts, initial raft dimensions, oceanic currents and wind direction, nearby coastlines, and the ability of pumice clasts to remain buoyant (Jutzeler et al. 2014, 2021).

Although known to coastal and maritime people since antiquity, sightings of pumice rafts remain uncommon and spectacular. Before satellite imagery, strandings of pumice clasts on coasts and reports of floating pumice rafts from ships were the only signs of eruptions from uninhabited islands and seamounts. The sources of pumice rafts previously recognized were shallow seamounts and domes, subaerial pyroclastic flows and fallout, shore erosion, fluvial transport, and volcanic-triggered landslides.

Fig. 6.10 Floating seeds (Tn) of a freshwater species (*Trapa natans*) from the Po River basin, found along the shores of the Salento Peninsula, at hundreds of km of distance. Black areas indicate the geographic distribution of *T. natans* in Italy. The star on the southeast peninsula of Italy indicates the site of collection of floating seeds on the beach. (Courtesy G. Belmonte)

A large amount of sea-rafted pumice reached the eastern coast of Australia in October 2002, approximately 1 year after a felsic, sub-marine explosive eruption at a previously unknown volcano of the Tonga archipelago (along the Tofua volcanic arc) (Fig. 6.11).

The eruption produced pumice rafts that first arrived and stranded in the Fiji archipelago on November 2002, approximately 1 month after the eruption: the only record of products from this submarine explosive eruption.

Pumice became stranded along at least 2000 km of the coastline of eastern Australia and was deposited on beaches. Typical amounts of pumice initially stranded on beaches were **500–4000** single **clasts per m^2**. Pumice was beached below maximum tidal/storm surge levels and was quickly **reworked back into the ocean**, such that the concentration of beached pumice rapidly dissipated at a few weeks from the initial stranding, and little record of this stranding event now exists.

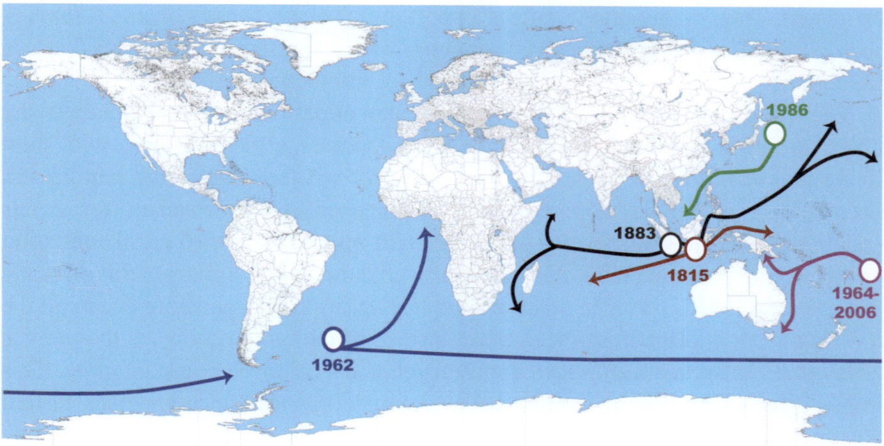

Fig. 6.11 Routes of the erupted pumice in the main volcanic episodes of the last 200 years. (Graphic composition A.M. Miglietta)

Most stranded pumice clasts ranged in **size from 2 to 5 cm** in diameter; the largest measured clasts were 20 cm in Fiji, and 10 cm in Australia. This suggests that **clasts reduce their volume with time thet they permain in the water**.

Pumice stranded on beaches in Fiji approximately 1 month after the eruption were free of fouling. Clasts stranded along Eastern Australia were mostly fouled by a variety of organisms.

The dominant organisms in order of abundance on pumice were algae, goose barnacles (Cirripedia), Polychaeta Serpulidae, Bryozoa, Cnidaria Madreporaria (*Pocillopora* sp., *Porites* sp.,

Cyphastrea sp.), Bivalvia (*Saccostrea* sp.), other pelagic Gastropoda (*Janthina* sp.). The size of some rafted coral colonies suggested growth of at least 12 months.

The abundance and variety of fouling *taxa*, coupled with the long dispersal trajectory (3500 km) and period of pumice floatation (1 year), confirm the importance of sea-rafted pumice as a **long-distance dispersal mechanism** for marine organisms.

Billions of rafting pumice clasts can be generated in a single small-volume eruption, and the geological implications for the transport of sessile *taxa* over large distances are significant.

An avenue for future research is to examine whether **speciation events and volcanism are somehow linked**; the periodic development of globalism for some *taxa* (e.g., Madreporaria, Gastropoda, Bryozoa) may correlate in time and/or space with voluminous volcanic events capable of producing silicic pumice-rich pyroclastic material.

The pumice as a vector for sedentary or sessile marine organisms is classified in the group of **geological rafts**.

Despite the small volume of the explosive eruption from the submarine Volcano in the Tonga arch, more than 10^{10} single pumice clasts (rafts) were produced, and for the 1984 Home Reef eruption, 10^{14} single clasts forming the pumice rafts were estimated. Thus, innumerable rafting opportunities occur for organisms following such pumice-producing explosive eruptions, with a transport mechanism that can function for years and travel many thousands of kilometres.

The space-time potential of pumice to remain afloat fundamentally changes the dispersal range or limitations for many marine species, especially those with short living planktonic larvae (generally, lecithotrophic). Although floating pumice has been shown to be an important long-distance rafting mechanism for corals (Madreporaria), the **coincidence of pumice raft appearance with coral spawning** events is also a critical factor. It is noteworthy that the eruption and generation of the pumice rafts in this instance preceded the late spring coral spawning events in the SW Pacific. It is clear from the size and nature of the biota rafting on the pumice clasts that transport times allowed colonization for up to one year or more. This implies that many organisms attached to pumice outside the immediate eastern Australian zone were subsequently transported along the eastern coast of Australia. The trajectory analysis indicates the New Caledonia–Vanuatu area as a probable source of that rafting biota (Fig. 6.12).

Furthermore, it has been noted that some Madreporaria colonies can attain the reproductive size while still attached to the pumice.

Of potential importance is that while biosecurity concentrates efforts with respect to shipping, the contents of ballast, and the use of antifouling paintings, pumice rafting may be a vector that circumvents these quarantine precautions. An example of this is the Serpulidae *Hydroides elegans* observed fouling the pumice clasts, which is designated as a marine pest for Australia.

Speciation events and volcanism may be linked such that the periodic development of globalism for some *taxa* (e.g., Madreporaria, Gastropoda, Bryozoa) may correlate in time and/or space with particular volcanic events. It is clear that silicic explosive volcanism is necessary to (1) produce geologic (pumice) raft material; (2)

Fig. 6.12 Pumice injection in the open ocean by a submarine volcano. (Graphic composition A.M. Miglietta)

have significant extent (e.g., continent-scale) and volume to have global significance and impact; and (3) be proximal to ocean basins for the pumice material to be transported.

Silicic explosive eruptions have produced some of the largest volume deposits ever known, and it is being increasingly recognized that there are continental-scale (1500 km long) and long-lived volcanic provinces capable of producing 106 km^3 of silicic pumice-rich pyroclastic material that can be injected into oceans. Such provinces may thus reduce endemism for some opportunistic species via large-scale pumice rafting processes.

Another significant recent pumice rafting and long-distance dispersal event occurred across the south west Pacific following the 2006 eruption of **Home Reef Volcano in Tonga**. In that episode, **more than 80 species** and a substantial biomass underwent a 5000 km journey in 7–8 months. Different microenvironmental conditions on the pumice pieces, caused by the relative stability of clasts at the sea surface, promoted diversification in biotic recruitment.

A survey of recent volcanic eruptions reveals that pumice rafts have occurred in all the oceans throughout the Holocene but are particularly high-frequency events in the Pacific Ocean.

As a consequence, pumice rafts are very important recruitment and dispersal devices based on the following:

1. Their frequentation of marine coastal environments (enhanced in importance if coinciding with spawning episodes).
2. The abundance of pumice that can survive long-distance transport.
3. Raft velocities, which at the surface (pushed by winds) are approximately twice the mean ocean current velocity.
4. The substantial biomass and biotic diversity hosted can be rafted thousands of kilometres.

Pumice rafting of organisms, while temporally random over short time frames, is a consistent and effective measure of transporting organisms over large distances and across deep ocean basins. Given the volume of pumice mobilized, pumice rafting is a massive transit process unrivalled by any other rafting substrata.

Volcanic eruptions appear to have an elevated frequency in the southwest Pacific, and historically, pumice rafts have occurred approximately once **every 10 years,** promoting an enhanced connectivity for shallow marine species in this region.

For the southwest Pacific, pumice rafting is not only a recent phenomenon (the modern Tonga–Kermadec Arc has been active since at least two million years). The success of these dispersal events in the southwest Pacific is further enhanced by the raft trajectory, along which many coral reef habitats exist, representing suitable and similar habitats involved into the connection of distant benthic populations.

The geographic distribution of species in tropical shallow sea is strongly controlled by ocean surface current patterns and enhanced by geologic events. Pumice rafts, as they enter tropical East Australian waters and then move both north and south following the East Australian Current, provide lines of internal communication for the Great Barrier Reef World Heritage Area. Recruitment largely occurs from oceanic reefs and the main reef-building organisms of Madreporaria, Bryozoa, and calcareous algae, all of which are abundant on pumice rafts. This offers natural restocking for reefs damaged by either natural or anthropogenic causes.

The 2012 eruption of the Havre caldera volcano (between Fiji and New Zealand) was the first to unambiguously demonstrate that also deep submarine (4700 m below sea level) silicic eruptions can create pumice rafts. An earthquake of 18 events over 12.5 h occurred at the 5 km wide Havre caldera volcano on July 17, 2012. The Moderate Resolution Imaging Spectroradiometer (MODIS) on 18 and 19 July showed an atmospheric plume and a thermal hot spot.

In the following days, the raft drifted northwest and was separated from its then-inactive source. In mid-August, the raft was up to 60 cm thick and composed of highly vesicular, rounded pumice clasts.

Until 6 August, MODIS images showed water discoloured by the presence of ash and possible chemical compounds dissolved from ash; pumice clasts collected in mid-August (3 weeks after the eruption) were already rounded. Comminution is most efficient on angular pyroclasts, implying that secondary ash from the pumice rafts was chiefly produced during the first 3 weeks of drifting, particularly the first week. Over the first 3 weeks, the pumice rafts spread over a total area of 110,000 km^2, and the estimated thickness of raft-generated ash beds expected to have accumulated on the sea floor averages 0.1–0.2 mm, up to 15 mm thick, depending on the abrasion rate over many hundreds of square kilometres.

On 9 August 2019, two vessels encountered a newly formed pumice area between Tonga and Fiji. This floating layer of pumice extended to 195 km^2 and consisted of a 15–30 cm-thick layer of freshly erupted pumice clasts. That pumice raft was 2 days old and produced by the eruption of a submarine volcano ca. 200 m below sea level, 50 km northwest of Vava'u in the Tonga archipelago.

The initial raft progressively fractioned into smaller portions, some of which interested Lakeba Island on September 2, 460 km west from its source. Hundreds of smaller rafts reached the Fiji archipelago on September 19, and most of them were stranded on the east shores of the Yasawa Islands for 2 months until being released back into the open ocean to continue their northwest-directed route through the South Equatorial Current.

After formation, pumice rafts progressively thin by dispersion on wider areas, pumice abrasion, and loss of pumice clasts through progressive waterlogging and overloading by biota and/or become stranded on coasts and islands. Pumice can float for months to many years and form an unusual flotsam that has been recognized since antiquity (Fouqué 1879).

Largely unrecognized, **hazards from pumice rafts** range from local nuisance to heavy effects on economies. Large pumice rafts, for example, blocked harbours for months and diverted maritime traffic. At a small scale, such areas of "dirty" ocean surfaces can produce serious problems for the engines of boats and ships entering and clogging the ways of water for engine cooling.

Pumice rafts also negatively interfere with the natural environment. Rafts block sunlight at the very first surface, impacting sea life at shallow depths, heating the air–sea layer, and impeding gas exchanges in the upper ocean.

What occurred (in terms of pumice injection into the sea) in historical eruptions is not reported by any chronicle. Thus, in the frame of a southwest Pacific phenomenon, the Mediterranean appears marginal, This notwithstanding, the record of powerful historical volcano eruptions (e.g., Santorini, in the Aegean Sea; the Vesuvius in South Italy) are available also for the Mediterranean Sea. The Mediterranean area is the scenario of crustal plate collision, exactly the same as the scenario of southwest Pacific tectonic arcs. The accumulation of pumice is still evident on what remains on Santorini Island (explosive eruption of approximately 1500 BC) and what produced the conservation of Pompeii, Stabia, Ercolano, and Oplonti (burrowed by 6–8 m of pumice and dust in the eruption event of 79 AD), preserving what is today the largest archaeologic site in the world. We can simply imagine that a similar quantity of pumice (either from Vesuvius or from Santorini), or a part of it, fell out directly on the sea, offering an immense possibility of rafting to many organisms and contributing to homogenizing the species diversity in the whole Mediterranean basin. What we can add, to conclude with the examination of the Mediterranean scenario, is the extreme rarity of such eruptions. After its explosive eruption in 79 AD, Vesuvius performed an even more intense explosion in 472 AD and thirty others to date. The last visible activity is dated to 1944; thus, the modern aspect of the mountain is the consequence of this relatively long period (80 years) of rest.

Environmental hazards from pumice rafts include modification of the food web by impeding photosynthetic activity with **shading** and by fostering and carrying invading exotic species as rafters.

Attrition during rafting may generate widespread **ash beds** with facies that share some similarities with water-settled aerially transported ash. Marine ash beds are extremely useful stratigraphic markers since they can be dated and correlated over thousands of square kilometres. In addition, the thickness and dispersal of ash in oceans are often correlated with eruption style and magnitude and used to infer distance and direction to the source, as in on-land studies.

The significance and origins of marine ash beds and pumice clasts on the sea floor must be considered with caution. Pumice rafts will produce ash by a secondary process of comminution, whereby adjacent pumice clasts within the raft gently but repeatedly collide with one another.

If not washed onto high ground by tsunamis or large storms, pumice is quickly abraded in the tidal zone and rarely preserved.

Long-drifting clasts have a survival time useful for the colonization process by marine life, particularly in tropical latitudes. If stranded on high ground, pumice clasts may be reworked by surficial water, resulting in further modification of their shape and depositional facies.

6.3.4 Constant Rafting Occasions

Other than being unpredictable, volcanic eruptions are also not frequent; in contrast, flooding episodes and short-duration storms could be seasonal, thus offering a basis for their predictability. Floating materials derived from land are not the only possibility for rafting organisms to stay at the sea surface. An entire world (extended as half of Europe) generated by the sea exists of floating algae, genus *Sargassum*, which offers rafting possibilities to typically benthic or benthic derived species.

Pelagic forms of the brown algae (**Phaeophyceae**) *Sargassum* occupy a large area in the centre of the North Atlantic Gyre (the "Sargasso Sea"). Ranging between approximately 20°–40° N and 30°–75° W, the Sargasso Sea is bordered by the Gulf Stream to the West, the North Equatorial Current to the South, and a Weak Circulation to the East. This floating ecosystem is located in the centre of an oligotrophic area in terms of plankton production, but it has higher concentrations of zooplankton and P nutrients than the surrounding waters. The Sargasso Sea is also interested in the subtropical convergence zone (STCZ), a frontal area characterized by high levels of productivity (Butler et al. 1983; Laffoley et al. 2011).

Sargassum floating assemblages on the surface realize massive rafts tens of meters in diameter, forming contiguous features possibly extending 200 km or more.

Two floating species of ***Sargassum***, *S. natans* and *S. fluitans*, dominate algal collections in the Sargasso Sea. Both species are equipped with buoyant, **gas-filled bladders** and are considered pelagic and reproduced by vegetative fragmentation. *Sargassum* macroalgal rafts provide important **feeding and spawning grounds** for pelagic fish and seabirds (Coston-Clements et al. 1991).

They represent a high-productivity ecosystem along open-ocean migratory routes for endangered humpback whales and sea turtles. *Sargassum* rafts have been designated as "essential fish habitat", a status that shares international management recommendations.

The first comprehensive studies of *Sargassum* abundance started in 1933.

More recent studies using satellite imagery, have revealed that the distribution and abundance of *Sargassum* in the Atlantic are highly variable over space and time. In 2011, for example, the ocean-scale buildup of *Sargassum* biomass has been 200-fold higher than the previous 8-year average (Fig. 6.13).

A diverse community of organisms is associated with floating *Sargassum*, including both sessile epibionts and mobile fauna, ranging from microbes to fish. A 1991 study reported **over 100 species of fishes and four sea turtles** associated with pelagic *Sargassum,* and a 2010 study reported a total of **over 145 invertebrate species**.

Several Porifera, Fungi, Bacteria, Diatomeae, and Protista have also been reported, and **ten fish and invertebrate *taxa* are currently considered endemic** to *Sargassum*.

Associated *taxa* in the Sargasso field in the middle of the Atlantic Ocean have been documented to be more abundant than in the adjacent Gulf Stream and possibly more diverse than in some other areas with *Sargassum*-type rafts, such as the East China Sea, New Zealand, coastal Ireland, and Iceland.

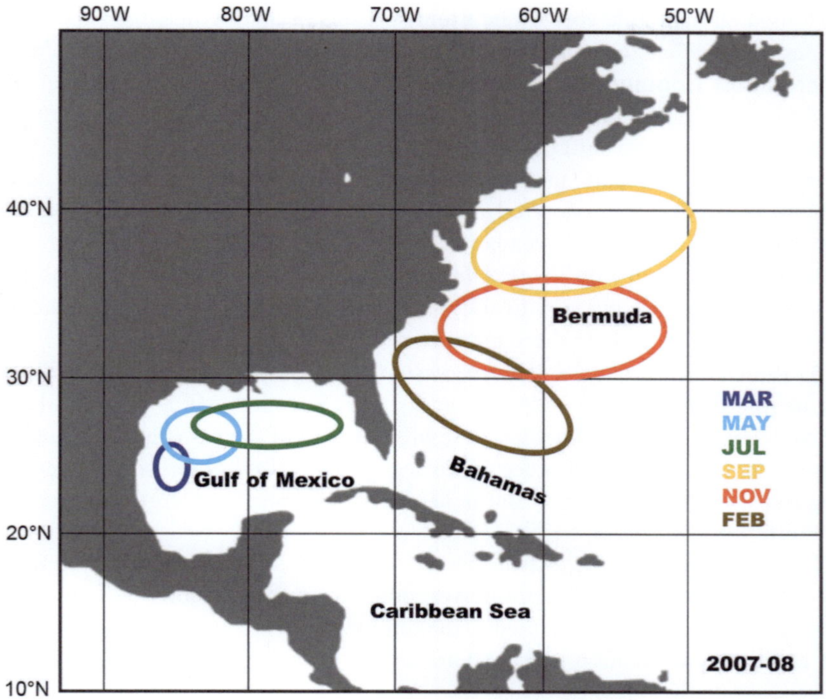

Fig. 6.13 Seasonal variation of the geographic positioning of the Sargasso sea in 2007–2008. (Graphic composition A.M. Miglietta)

Rising concentrations of atmospheric CO_2 are resulting in global warming and increased acidity of the ocean, including in the Sargasso Sea. Ocean acidification and the concomitant decrease in carbonate saturation of the water could be correlated with the reduced formation rates of the calcareous skeletons of some pelagic organisms, such as Mollusca and Foraminifera. Decreasing pH has also been hypothesized to be responsible for the decreased reproduction and increased mortality in Crustacea, including Copepoda, Eufausiacea, and Decapoda. In contrast, elevated concentrations of CO_2 can stimulate the growth of macroalgae, including Phaeophyceae.

Photosynthetic activity by phytoplankton and brown algae can buffer low-pH conditions. Taken together, these findings suggest that *Sargassum* fields might fare better than its faunal associates due to the impacts of climate change progress, influencing the community composition living within *Sargassum* rafts.

Given rising sea surface temperature (SST) and acidity, it has been hypothesized that the diverse biotic community associated with *Sargassum* has changed in macrofauna species composition, diversity, and evenness when compared to that of many years ago (Fine 1970). Globally, the magnitude and direction of biogeographic range shifts correspond with those of temperature changes, which are northward-directed in the Sargasso Sea.

Compared to mobile species, neustonic organisms are predicted to experience a greater range change in response to surface warming. As such, the **Mobile Macrofauna Community Structure** (MMCS) collected from 2011 and 2012 would be similar to those of historic samples collected from more southern sites, signalling a northward shift of rafting community associates.

In the last 30–40 years, in the Sargasso Sea, near Bermuda, a northward shift in a key isotherm (22.5 °C) occurred, coupled with a slight increase in wind speed and wave height during extreme weather events.

MMCS associated with *Sargassum* in a restricted geographic area (Sargasso Sea 25°–29° North) deviated significantly (data 2011–2012) from those communities documented historically in the same area.

At the basin-wide scale, the *Sargassum* MMCS appears to have exhibited a fair degree of variability over time and space. Recent MMCS were unlike any historical communities, whether compared to sites of the same latitude and geographic region or the broader historical dataset including sites across the Sargasso Sea, Gulf Stream, south of the subtropical convergence zone, and northern latitudes.

The conclusion of these facts is that the *Sargassum* MMCS is **a rich community but it has a variable composition over time**. From another point of view, we can state that species possibly adapted to the raft existence are an extremely large number, and each colonization is characterized by chance and driven by which species arrived first on the raft (Fig. 6.14).

Fig. 6.14 Fish beneath the *Sargassum*. The entire *Sargassum* community functions as an FAD for pelagic fish (painting Dawn Witherington)

In the late 1960s, *Sargassum* in the Gulf Stream and Sargasso Sea supported similar MMCSs, a pattern that was not observed in the next 10 years. Samples from these two historic collections shared only two of their respective top five most abundant species.

Amphi = in two directions; *Podos* = feet *Bryo* = musk; *Zoos* = animal, soul *Epi* = over; *Bios* = life *Lepas* = small shield *Spiro* = curved, spiral; *Orbis* = orbit *Phaeo* = brown; *Phycus* = plant, Alga *Sargassum* = neo-Latin from Portuguese "grape"

The *Sargassum* MMCS sampled near Florida exhibited seasonality in the 1970s, returning to a starting community composition each year, but in 2012, communities fell very far outside this circular pattern.

Regarding **sessile epibionts**, low coverage was recorded for calcareous **Bryozoa**, Polychaeta (***Spirorbus***), Crustacea (***Lepas***), crustose **coralline** algae, and noncalcareous Cnidaria (primarily **Hydrozoa**).

Bryozoa, crustose coralline algae, and Cnidaria each exhibited average coverages of approximately 10% or less. *Spirorbus* and *Lepas* demonstrated much lower coverage (>1%).

The feeding ecology of Sargasso MMCS has been generalized as being dominated by **suspension feeders**, then grazers, and browsers, followed by predators, and finally by detritus feeders. **Recently**, however, such a simplified pattern has been revisited.

Low coverage by sessile filter feeders has been found, together with **an especially high fraction of** the predator shrimp ***Latreutes fucorum*** (Crustacea), a loss of several soft-bodied predators of sessile fauna, and low numbers of grazers and detritivores.

Half of the MMCS present in historic Sargasso community but absent from recent collections are soft-bodied predators: Mollusca (**Nudibranchia**) and **Platyhelminthes** that prey on sessile fauna (epibiota).

While Platyhelminthes can be generalist predators and scavengers, many Nudibranchia have species-specific feeding morphologies and ecologies. The epibiont community observed recently might have been insufficient to support a large panel of predators. Concurrently, the disproportionally high abundance of *Latreutes fucorum*, which can prey on mobile macrofauna, suggests a top-down control of mid-level consumers, herbivores, and/or detritivores.

Amphipoda can exert indirect impacts on community composition and species richness through selective grazing activity and modification of their macroalgal habitat.

Sargassum that is visible by satellites appears to circulate in the open ocean and occasionally exhibits high interannual variation in estimated biomass (by satellite) and circulation patterns. Typically, this floating habitat appears to be generated from a **geographic source** in the **Gulf of Mexico** each spring, transferred through the

Florida Strait into the Gulf Stream during summer, and then circulated with prevailing currents to pass northeast of the Bahamas in February. However, the latitude at which suitable conditions for *Sargassum* can be found during each calendar month varies (Gower et al. 2013).

Given the very different circulation histories of the *Sargassum* habitats sampled in different years, we were not surprised to see variability in macrofauna communities over time and space. In 2010–11, the distribution, abundance, and/or circulation patterns of suitable conditions for *Sargassum* in the Atlantic were anomalous compared with those of the 2003–2010 period and exceeded the variability of those factors in living human memory.

In 2010, *Sargassum* biomass was very low, perhaps because of adverse effects caused by pollution coming from the "Deepwater Horizon" oil spill and cleanup efforts in the Gulf of Mexico.

In 2011, the major geographic source of *Sargassum* biomass was a region offshore of the Amazon delta rather than the Gulf of Mexico, and unprecedented masses washed ashore in the Caribbean Sea or drifted eastwards at low latitudes directed to Africa. The majority of that *Sargassum* biomass never appeared to reach the Sargasso Sea region, having never met with the Gulf Stream. In August 2011, high *Sargassum* biomass was situated in the Caribbean Sea, but very little was visible around the Bahamas archipelago. These **seasonal and yearly differences** in advection of *Sargassum* would contribute to variations in recruitment source populations and temperatures experienced by the inhabiting vagile macrofauna and sessile epibiota.

Offshore rafting communities can be simplified in comparison with coastal benthic counterparts. As such, conclusions drawn from one time or geographic area in the Sargasso Sea might not necessarily offer valid comparisons for another region or time. These characteristics might confound our ability to evaluate changes in *Sargassum* macrofauna communities over time. Therefore, large-scale oceanographic circulation could have significant implications for understanding *Sargassum* community variation in both long-and short-term cycles.

When interpreted together with the broader literature about climate change impacts on marine life, the results support the idea that **different species** and entire animal communities **do not show the same sensitivity** to pH and response **to ocean warming**. However, it is possible that increasingly acidic conditions of the Sargasso Sea have contributed to the observed reduction in coverage by sessile epibionts, especially Bryozoa such as ***Membranipora***, which is lightly calcified. Calcified animals might experience difficulty building exoskeletons and shells and attaching to *Sargassum* in lower pH sea water.

Sargassum with low epibiont coverage and its associates might experience increased UV exposure because of reduced overgrowth and reduced localized supply of nutrients from epibiont waste production. If low coverage also contributes to reducing sinking rates, because clumps do not sink from being overgrown and weighed down, then the rate of biomass and food supply of *Sargassum* to the deep sea would also be reduced.

Scavengers in the Sargasso Sea's deep waters (> 5000 m) consume fallen *Sargassum* and epibiota, thus being impacted by changes in surface *Sargassum* communities.

The diversity and species composition of macrofaunal communities associated with *Sargassum* might be **inherently unstable**. The diversity, evenness, and MMCS of the rafting communities in the Sargasso Sea showed **little consistency over time or space**. In some marine communities, stability has been **attributed to a lack of local stressors**, a trait that does not apply to the Sargasso Sea.

Local and basin-wide stressors include increasing wind disturbances, SST, acidification, and large shifts in phytoplankton communities and nutrient cycling, in addition to the widening of the geographic range of tropical species, and increasing ship traffic.

Fisheries have significantly altered the abundance of large animals (Pisces and Cetacea) in the Sargasso Sea. The same currents that transport *Sargassum* and larvae might also serve to concentrate pollutants known to occur in the region, including hydrocarbons, heavy metals, plastics, and oil spill cleanup chemicals.

A recent attempt produced a value of the *Sargassum* habitat in terms of ecological services.

Membrana = film, membrane; *Poro* = hole

Even if approximated and subjected to the already mentioned variations in time and space, the values of each service reach high scores, clearly showing the importance of a distant, unknown, and historically unused habitat (values expressed in thousand USD/year) (Pendleton et al. 2014).

1. Recreational fishery in Bermudas	600
2. Commercial fisheries	100,000
3. Atlantic whale watching touristic industry	500,000
4. Commercial eel harvest in North America and Europe	66,000
5. Turtle-based tourism in the Caribbean area	15,000

6.3.5 Oceanic Rafts and Island Biogeography

Floating material representing a raft for pelagic travels can also reach conspicuous sizes under favourable conditions and constitute "**floating islands**" when portions of them emerge visibly above sea level.

The term island, in this case, is used to indicate the rafting of terrestrial organisms in the ocean.

Floating islands do indeed exist in the fresh waters of six of the seven continents and sometimes in the oceans between them (Powers 1911; Houle 1998).

These islands are kept buoyant by the light, spongy tissues of certain aquatic plants, by gases released into the soil by decomposing vegetation, or by both of these reasons. Floating islands have attracted the attention of ancient authors, such as the Roman writer **Pliny** the Younger, who has left us an evocative description of the floating islands in **Lacus Vadimonis,** now a marshy pond known as Lago di Bassano, on the bank of the Tiber approximately **60 km north of Rome**.

He wrote: "*No boats are allowed on the lake, as its waters are sacred; but several floating islands swim about it, covered with reeds, rushes, and whatever other plants the fertile marshy ground nearby and the edge of the lake produce. Each island has a peculiar shape and size, but the edges of all of them are worn away by their frequent collisions with the shore and one another. ... Sometimes they cluster together and seem to form a small continent; sometimes they are dispersed by the shifting winds; ..... Sheep, seeking grass, proceed not only to the shores of the lake but also upon these islands, unaware that the ground is mobile, until, far from the shore, they are alarmed to find themselves surrounded by water, as though they have been suddenly placed there*". Floating islands commonly rise in newly flooded reservoirs. This happens if the flooded area has plenty of decomposing vegetation (producing oil material and/or gas).

Once the bottom of the reservoir is filled with certain types of peat, it becomes buoyant. In shallow parts of the reservoir, less than 2 m deep or so, this buoyancy can tear away sections of peaty soil up from the bottom of the reservoir, and they rise to the surface as floating islands.Floating islands also form during floods of the great tropical rivers when large masses of aquatic vegetation or chunks of their banks are torn away and carried downriver.

The **Congo** in Africa is one such river, and floating islands that came down the Congo were reported in the sea 240 km from the river's mouth.

Floating islands have also been reported in the **Sepik** River in Papua New Guinea following monsoon rains. The islands are called "***Lik Lik Aislans***" in pidgin English and can be up to 100 m across with living trees on them.

The Rio Paranà and Rio de la Plata in South America also generate floating islands; during the flooding period, the rivers are crowded with floating islands called ***camalotes***, which are matted masses of water hyacinth.

A famous episode at *Convento de San Francisco* in Santa Fe, Argentina, which is located on the Rio Paranà, involved the killing of two friars by a jaguar travelling on a *camalote*, and arrived at Convento on 18 April 1825.

In the flood of 1905, the Rio de la Plata at Buenos Aires was covered with ***camalotes*** as far as the eye could see, some half a mile long and 100 ft. wide, others just a few feet in diameter. As they came down the river, these islands hit moored ships and tore them from their anchor.

The passengers that are hosted on floating islands included species of tropical snakes, a deer, a puma, parrots, and monkeys. An Indian baby was found on a floating island that came ashore near Rosario, and although he was weak from hunger and exposure, the kid was brought back to health. Of course, floating islands that come down rivers end up at sea; many are quickly destroyed by the waves, but others survive for some time. The Congo and the Amazon rivers have the right water mass and seasonal strength to inject floating islands (hosting continental flora and fauna) in the middle of the ocean, towards an uncertain destiny for passengers that, in this situation, can be considered cast away. When in the ocean, however, the floating island could start to be invaded by marine propagules and to offer shelter (as FAD) to pelagic fish.

Box 6.2 Floating Islands at Sea

An article in the 8 November 1908, edition of *The Washington Post* reported that a US cruiser in the Caribbean north of Honduras encountered an island that they soon discovered was floating (this is certainly one of the largest floating islands ever seen at sea).

The paper reported: "*It proved to be a little island about three quarters of a mile around and a quarter wide. In shape, it was long and narrow, with a thick growth of vines and bushes reaching down to the water's very edge. Three tall coconut palms grew in the middle of it. No life of any kind was on the island, nor was there any water, though instead of being sandy or rocky as such islands are usually, the soil was rich, dark and very moist. After gathering the coconuts, the sailors returned to the cruiser, which, oddly enough, seemed much further off and considerably more to the southwest than when they left her. Then, it just dawned on them that they had been visiting one of the floating islands so often heard about but seldom seen in the South Atlantic*".A story published in several newspapers in June and July of 1902, gives a remarkable account of two floating islands spotted in the Caribbean Sea. The Norwegian ship *Donald*, steaming from Banes, Cuba, on its way to Philadelphia, encountered a floating island approximately 48 km from the island of San Salvador. As reported by the newspaper *Daily Chief (Perry, Iowa*), 15 *July 1902, p. 3* "*... High up in the trees was a small colony of mischievous monkeys, and as we got nearer they shied a number of coconuts at us. After a lot of trouble, we secured two of the attacking simians and at least a dozen cocoanuts. Then, we took to our boats, boarded the steamer, ordered full steam ahead, and soon the strange floating island was lost in haze. However, another surprise was in store for us the following day, when we passed within glass sight of another singular floating object just off the port bow. The lookout sung out 'Land ahead'. This amazed me, for I knew according to the chart land was not miles near. Nevertheless, curious from the previous day's experience, I determined to solve this further mystery of the sea, so I gave orders for the ship to steam close to what I now made out to be another floating island. Again, I had a boat lowered, and with the same crew, we landed on the island. We found it to be an exact duplicate of the day before, with an exception: instead of monkeys, we found a large survey of parrots of most brilliant plumage. Among them was one who was evidently the patriarch of the tribe, and I do not exaggerate when I say that the aged fellow could cuss in two languages. He was evidently a lost pet. We took him and a couple of his fellows aboard the steamer and soon left the floating island in the distance*".

These accounts are of particular **interest to biogeography**. They support the theory that floating islands have been important in the dispersal of terrestrial plant and animal species across the oceans and are important in explaining evolutionary links among geographically distant areas.

Natural floating islands are composed of vegetation growing on a buoyant mat of plant roots or other organic material. They typically occur when growths of cattails, bulrush, sedge, and reeds extend outwards from the shoreline of a wetland area. As the water gets deeper, the roots no longer reach the bottom, so the oxygen in their root mass determines their buoyancy, and the surrounding vegetation supports the retention of their top-side-up orientation. The area beneath these floating mats is exceptionally rich in aquatic lifeforms. Eventually, storm and/or flooding events tear whole sections free from the shore, and the islands thus migrate around following water currents and capturing surface winds.

In the Brazilian Amazon, floating islands form in lakes on the floodplains of white-water rivers. They are known as ***Matupá*** and are of variable size.

Matupás form initially from the agglomeration of aquatic vegetation, and over time, they can accumulate a substrate of organic matter sufficient to grow forest patches of several hectares in area, and up to 12 m in thickness.

Soil from *matupás* is especially fertile and is frequently gathered for use in vegetable agriculture. In some cases, crops are planted directly onto *matupás*, representing an incipient agricultural experiment. *Matupás* are also considered a strategic habitat for fishing, mainly for arapaima (***Arapaima gigas***).

The large importance of such a natural arch for terrestrial species distribution, is assumed as much as possible when floating islands enter the ocean, but the importance that such floating material could represent for marine ecosystems is completely ignored.

When at sea, submerged parts of floating islands start to be conquered by benthic organisms, and pelagic fish are attracted, for sure, by these floating and shadowing rafts (as FADs) in the hope of reaching, with them, frontal oceanic areas where to find food.

In contrast to pumice material (generally produced in the ocean and used as rafts by marine organisms), river floating islands bring terrestrial diversity in the ocean, thus offering an opportunity for long-distance travel to terrestrial species normally not migrating across the sea.

The difficulty of transoceanic migrations for terrestrial species depends on the characteristics of each particular *taxon*.

According to direct observations, a floating island was first seen in July 1892, 400 km east of Cape Code (39° N, 65° W), and seen again in September of the same year, 600 km east of Newfoundland (45° N, 42° W). Floating islands were also reported on the coast of Brazil and were interpreted as coming from the Niger and Congo Rivers in the west Africa coast.

Biogeographers often reach the conclusion that their study groups of land vertebrates arose on a continent and migrated to another one on floating islands, but none of these studies have attempted to demonstrate how the ancestors could have survived such a transoceanic journey. Simons (1976) examined whether the migration time of benthic organisms across the Atlantic could be of any help in explaining the passage of platyrrhine monkeys in South America (hypothesizing that these are derived from African ancestors).

The long-distance dispersal of marine organisms is driven by currents, but it is probably not affected by surface winds. It has been demonstrated that actual winds are a much stronger force than actual currents, and they might have been the most important factor in floating island velocity. Consequently, **floating islands could perform a transoceanic journey in a shorter time than the water masses of a surface current**. Overall, the average number of days of any transoceanic journey approximated 14 days and never exceeded 26 days even when palaeo-currents were not favourable. These values are reasonable only if they are lower than the starvation limit of the migrating terrestrial fauna in such conditions. What is difficult to evaluate is the duration a given species is able to survive without nutritional resources and particularly freshwater. These factors are highly variable and *taxon* dependent. The effects of water deprivation have been evaluated among marsupials, camels, rodents, sheep, goats, and many others. Among the numerous physiological effects that have been noted, water deprivation increases the activity of supraoptic neurons, the concentration of vasopressin, osmolality, Sodium, and plasma renin activity but decreases urinary flow, plasma volume, and, most importantly, food intake.

One general conclusion from these studies is that lizards (Reptilia) and Mammalia occupying dry habitats are better adapted to conserve their body water and to tolerate greater losses of body water than species occupying more humid habitats. In Mammalia, resistance to water unavailability seems to increase proportionally with body weight. For example, *Octodon degus*, a Chilean rodent with a mean body weight of 210 g, can survive up to 13.4 days without water, while the smaller 31 g sympatric *Oligoryzomys longicaudatus* can survive only 4 days in total water deprivation conditions.

A similar relationship might exist in Reptilia, which, in addition, are known to endure greater metabolic stresses than Mammalia.

In any case, the absence of freshwater and food experienced by rafters at the time of transoceanic travel should be sensibly shorter than the current time. In ancient times, Africa and South America were separated by a "narrower" Atlantic Ocean that started to divide the two continents approximately 100 million years ago.

Assuming that water and survival food might be available on a floating island for 2 days (or at least an entire day) and considering that a 210 g rodent can survive up to 13.4 days without water, one can conclude that the travel proposed is plausible for such a kind of Mammalia (approximately 200 g) and most land Reptilia.

With time, the distance among continents became a problem due to the continuous opening of the ocean floor, and its crossing time became superior to the survival time of rafting mammals, which simply terminated passing from one continent to another.

If this was the story of terrestrial mammals, however, we have to imagine that terrestrial invertebrates (more resistant to starvation, and probably able to profit of raft material) maintained their possibility of dispersal for longer geological periods, and marine invertebrates practically continued to use undisturbed such a possibility.

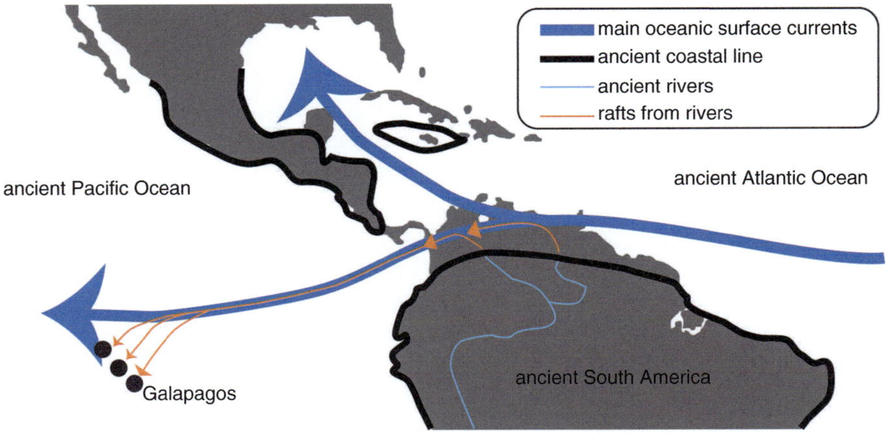

Fig. 6.15 Rafting islands in a possible paleogeographic scenario (more than 3 My ago). Before the uplift of the Panama isthmus (3–2 My ago), a subequatorial oceanic current flowed from the Atlantic to the Pacific Ocean. At the level of northern South America, large rivers (before changing their course) discharged possible floating islands hosting forest biodiversity. Some of the Galapagos islands had already emerged in those times and were thus invested in by these faunal arrivals. The bold line delimits the approximate coastline of that time. White areas correspond to modern geography. (Courtesy G. Belmonte)

The interpretation of palaeogeography greatly helps us understand the colonization of oceanic islands. In the paradigmatic case of **Galapagos**, terrestrial colonizers arrived from South America with no doubt. What is interesting is the elevated endemism level of the Galapagos fauna.

As Darwin started to note, in the case of **18 species of finches** (Passeriformes), they are well distinguishable from each other for their remarkable diversity in beak form and function. The closest known relative of the Galapagos finches is the South American *Tiaris obscurus*. Apart from the Cocos finch (from Cocos Island), the other species are found only on the 13 islands of the Galapagos archipelago. The **speciation process** is still in progress, and modern genetic analyses admit a **start of speciation at approximately two million years ago**. This datum agrees with the geological age of Galapagos, oceanic-volcanic islands having an age between three million years and 600,000 years. However, the most interesting palaeogeographical fact that sustains the hypothesis of colonization from the continent is the **uplifting of the Panama Isthmus** (completed approximately three–two million years ago), which created a barrier to the westwards-directed equatorial current coming from the Atlantic Ocean. This current could lead floating islands eastwards from the large river mouths of South America (in those times, the Amazon, probably in a unique water system with Orinoco, flowed on the north coast of South America).

This geographic scenario justifies the arrival of South American colonizers and the successive evolution of an endemic Galapagos biota when such an arrival terminated with the closure of the Panama Isthmus (Fig. 6.15).

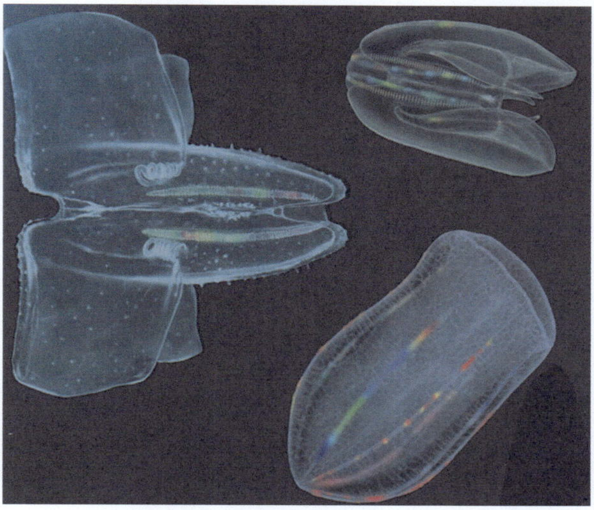

Fig. 6.16 Ctenophore comb jellies. On the left, *Leuchotea*; bottom right, *Beroe*; top right, *Mnemiopsis*. (vertical meridians with iridescence are comb rows). (Drawings A. Gennari)

6.4 Active Dispersal: Swimming Behaviour

Swimming evolved a number of times in unrelated lineages of marine organisms.

Supposed jellyfish fossils occur in the **Ediacaran** (more than 570 million years ago), but true free-swimming animals appear in the Early to Middle Cambrian. These are mostly related to Arthropoda and include ***Anomalocaris***, which swam by means of lateral lobes in a fashion reminiscent of today's cuttlefish. The most primitive swimming style, however, is that based on **cellular cilia**. Apart from unicellular organisms swimming by undulating a few long cilia (flagella or *undulipodia*) or **metachronically** moving cilia (in large numbers), the Metazoa **Rotifera** still use this primitive solution to swim. In general, all small organisms adopt such a solution, as in all larvae, and ciliary swimming is absent only where cilia are absent for phylogenetic constraints (e.g., among Crustacea). **Comb jellies (Ctenofora)** use giant cilia (grouped to form **combs**) to move. They move with the mouth onward, contrary to medusae. Ctenophores are usually equipped with eight **comb rows** for swimming (Fig. 6.16).

Each comb consists of thousands of **unusually long cilia**, up to 2 mm. Many combs are associated in a linear (meridian) disposal that starts from the apex of the animal, where anal pores are present, around a central apical organ with **statocysts** for perception of the body position in the water.

When trying to escape predators, Ctenophora can accelerate to six times their normal speed.

Some species reverse direction as part of their escape behaviour by reversing the power stroke of the comb plates on each meridian.

Anomalo = not normal; *Caris* = shrimp *Cilium* (L) = eyelash *Cteno* = comb; *Fero* = bearing *Ediacara* = site close to Adelaide, S Australia *Medusa* = one of the Gorgonians, with snakes at places of hair *Meta* = terminal; *Chrono* = time *Rota* = wheel; *Fero* = bearing *Stato* = condition; *Cyst* = spherical object

Box 6.3 Ultrastructure of Cilia and Flagella

In the transverse section of each cilium, a series of nine **microtubule doublets** is recognizable in an external ring around a central pair of microtubules. The doublets are formed by two tightly associated microtubules. Such an association is so tight that one of the two microtubules is not complete. The central pair of microtubules does not twist when the beat direction reverses (Fig. 6.17).

This provides the most important proof for the existence of a switching mechanism at the basis of the cilium governing the direction of ciliary movement. The direction of movement is determined by the flexion direction of the cilium. The flexion direction is due to a gliding movement of a microtubule on the other in each doublet.

The doublets of a side glide on each other to determine a shortage of that cilium side; contemporaneously, the doublets of the opposite side glide to allow an elongation. This determines the flexion of the cilium, i.e., a beat to move water, and the consequent movement of the cell.

Among radiata jellyfish (Cnidaria), the main form of swimming is the contraction of cup-shaped bodies. All **medusae** are free-swimmers using **jet propulsion.** These animals fill a muscular cavity and successively squirt out water to propel them in the opposite direction (Fig. 6.18).

Jet propulsion has two basic designs (the front or anterior part is intended where the movement is commonly oriented):

1. Water from the rear and expulsion from the rear (Cnidaria medusae),
2. Water from the front and expulsion from the rear (**Thaliacea Salpidae**).

Chordata Salpidae move with the mouth ahead, pumping water out of the back opening after the passage of the water through a sieve where food particles are collected.

Because of the expanse of the contracting cavity, the animal's velocity fluctuates, accelerating while expelling water and decelerating while introducing it.

All **Cephalopoda** can move by jet propulsion, a very energy-consuming way to travel compared to the tail-based propulsion used by fish. The relative efficiency of jet propulsion decreases as animal size increases.

The stop-start motion provided by the jets, however, continues to be useful for providing bursts of high speed, at least for capturing prey or avoiding predators.

This propulsion makes Cephalopoda the fastest marine invertebrates.

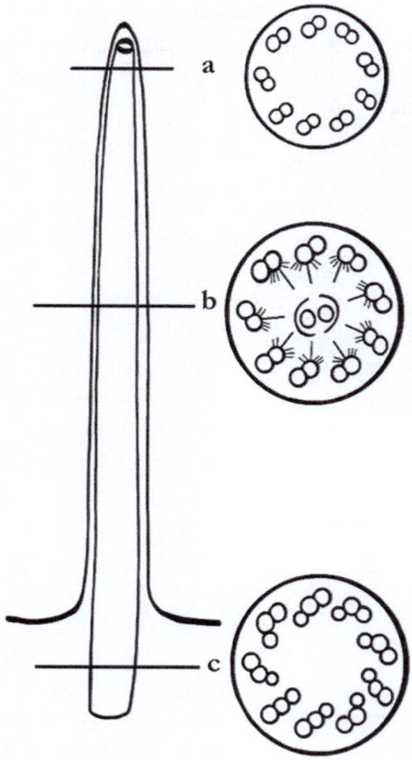

Fig. 6.17 Transverse section of a cilium. (**a**) distal section without central microtubules in the axial structure; (**b**) central microtubules separated, surrounded by nine pairs of microtubules, each armed with dynein extensions; (**c**) section at level of the basal body, nine triplets of microtubules are evident. (Courtesy G. Belmonte)

Fig. 6.18 Three scyphomedusae (Cnidaria Scyphozoa). Left, *Pelagia*; top right, *Chrysaora*; right bottom, *Drymonema*. Among the large animals of the ancient past, the jet propulsion performed with the cup (umbrella) was probably the best way to swim (drawing A. Gennari)

Oxygenated water is taken into the **mantle** cavity to exchange respiration gases with gills; the spent water is expelled through the **funnel** (a fold in the mantle) through a muscular contraction of the mantel (Fig. 6.19). The direction of the expelled water can be controlled somehow by differently orienting the funnel. Most Cephalopoda are neutrally buoyant, thanks to the presence of a **chambered shell**.

Teutidea swim more slowly than fish but use more power to generate their speed. The loss in efficiency is due to the amount of water the squid can accelerate out of its mantle cavity.

In most **Bilateria**, there are many methods of swimming. The arrow worms (**Chaetognatha**) undulate their finned bodies like fish. Crustacea and Polychaeta usually swim by **paddling** with special swimming legs (**pleopodia**, **peraeiopodia**, or antennae in Crustacea; **parapodia** in Annelida). In Crustacea, the entire abdomen, equipped with flattened appendages (the **uropodia**) beside the terminal telson, can be quickly flexed under the body, thus determining a powerful rowing pulse of the animal in the opposite direction.

Swimming Gastropoda (as the sea angels), flap fin-like structures derived from the foot.

An anatomical solution that is very original and allows large organisms to swim is the **tail**. A typical tail (a part of the body posterior to the anal opening) is present only in **Chordata** and **Chaetognatha**.

Tail-based swimming of fish is obtained by exerting force against the surrounding water by the alternate contraction of muscles on each side of the body to generate waves of flexion from anterior to posterior extremities.

Most fish generate thrust using lateral movements of their body in addition to the tail and caudal fin. However, there are also a large number of species that move mainly using their median and paired fins (Fig. 6.20).

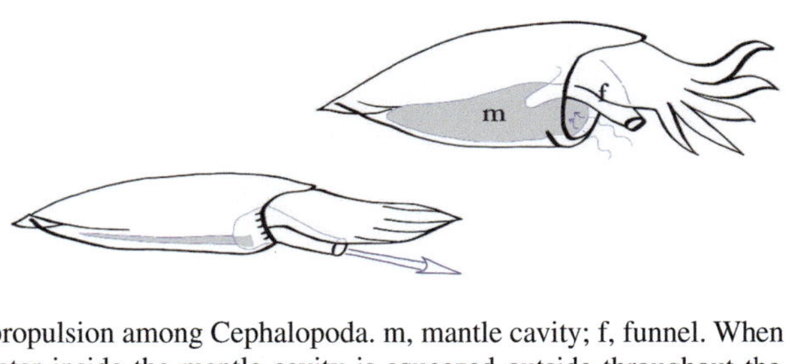

Fig. 6.19 Jet propulsion among Cephalopoda. m, mantle cavity; f, funnel. When mantle muscles contract, the water inside the mantle cavity is squeezed outside throughout the funnel, and the animal moves in the opposite direction. (Courtesy G. Belmonte)

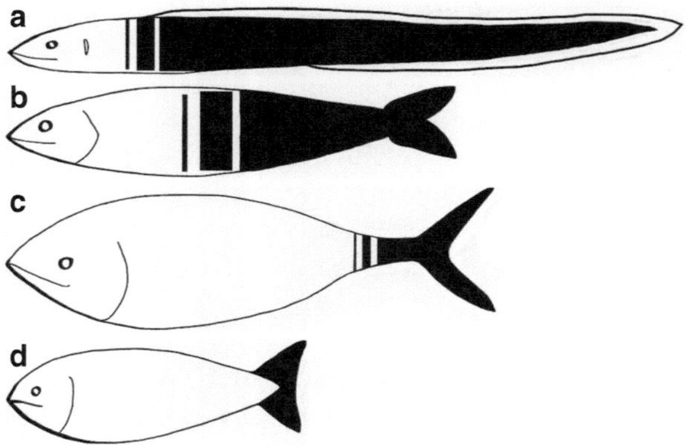

Fig. 6.20 Scheme of different wave propulsion in fish bodies for swimming. (**a**) anguilliform; (**b**) carangiform; (**c**) tunniform; (**d**) ostraciiform. (Courtesy G. Belmonte)

6.4.1 Anguilliform Locomotion

In long, slender fish, generally compressed (laterally flattened), a wave of muscular contraction with little increase in the amplitude is observable as it passes along the body.

6.4.2 Subcarangiform Locomotion

It distinguishes a more marked increase in wave amplitude along the body axes, with the vast majority of the work being done by the rear half of the fish.

6.4.3 Carangiform Locomotion

It is a stiffer and faster-moving action compared with the previous groups. The vast majority of movement is concentrated in the very rear of the body and tail.

6.4.4 Thunniform Locomotion

Characteristic of high-speed long-distance swimmers. Here, virtually all the lateral movement is limited in the caudal peduncle and the tail.

6.4.5 Ostraciiform Locomotion

no appreciable body wave when they employ caudal locomotion. Only the fin itself oscillates (often very rapidly), not exclusively the tail fin.

Cephalo = head; *Podos* = foot *Chaeto* = hook, needle; *Gnato* = mandible arch, jaws *Pleon* = abdomen *Peraeion* = torax *Para* = close to, similar *Uro* = tail

Ocean sunfish (***Mola mola***), in addition (note that they do not have a "tail"), and many small fish use fins other than the caudal one for swimming as well as for steering and dynamic lift. This action, in any case, produces a slower moving result.

Fish with electric organs, such as those in **Gymnotiformes**, swim by undulating their fins while keeping the body still, presumably so as not to disturb the electric field that they generate.

Bone and muscle tissues of fish are denser than water. To maintain their depth position, some fish increase buoyancy by means of a **gas bladder** or by storing oils or lipids. Fish without these features perform dynamic lifts orienting their pectoral fins during the caudal fin propulsion, in a manner similar to the use of wings by airplanes and birds.

Sharks are obliged to the **dynamic lift**; notice their well-developed pectoral fins. These fish must swim continuously to stay afloat, and they are incapable of swimming backwards or hovering.

A combination of **two pectoral fins** or **anal** and **dorsal** fins in couples can also be observed. In the **Rajiformes** swimming movement (dorso-ventrally flattened fish), thrust is produced by vertical undulations along large pectoral fins. In the **Amiiformes** movement, undulations are relative to the only long dorsal fin, while the body axis is held straight and stable. In the **Gymnotiformes** movement, undulations are relative to a long anal fin (upside down the amiiform body plan). In the **Balistiformes,** both anal and dorsal fins undulate.

Regarding the contemporary use of impaired fins (dorsal, anal) or paired fins (pectoral, pelvic), a **Tetradontiformes** movement is distinguished when dorsal and anal fins are flapped as a unit, either in phase or opposing one another. The ocean sunfish *Mola mola* is an example.

In **Labriformes** movement, the oscillatory movements of pectoral fins can be classified as drag-based or lift-based, in which propulsion is generated as a reaction to drag produced by dragging the fins through the water in a rowing motion.

A balance between **stability** and **manoeuvrability** is needed for swimming among fishes.

Swimming relies on more caudal body structures that can direct powerful thrust by only pushing water rearwards.

Body-caudal fin swimming is, therefore, stable and useful for large migration patterns to maximize efficiency over long distances and periods. Propulsive forces in **median-paired fin (MPF)** swimming are generated by multiple fins located on either side of the body that can be coordinated to execute elaborate turns. As a result, MPF swimming is well adapted for high manoeuvrability and is useful in elaborate escape patterns.

Fish do not rely exclusively on one single locomotion mode but are rather "generalists", choosing among many available behavioural techniques.

A particular way to swim is produced by fish species of the Family **Exocoetidae**, for whom, more correctly, we should use the verb "to fly".

These species glide directly over the surface of the water without ever flapping their "wings". Flying fish have evolved abnormally large pectoral fins that act as air foils and provide lift when the fish launches itself out of the water. Additional forward thrust and steering forces are created by dipping the **hypocaudal** lobe of the tail fin into the water and oscillating it very quickly.

Of the 64 extant species of flying fish, only two distinct body plans exist, each optimizing two different behaviours.

In the ***excocoetus*** or **monoplane body plan**, only the pectoral fins are enlarged to provide lift. Fish with this body plan tend to have a streamlined body, long, narrow wings, and higher wing loading, making these fish well adapted for higher flying speeds (Fig. 6.21).

Flying fish with a monoplane body plan launch from the water at high speeds and at a large angle of attack (sometimes up to 45°). In this way, monoplane fish are taking advantage of their adaptation for high flight speed, while fish with biplane design exploit their lift production ability during take-off.

Because flying fish are primarily aquatic animals, they are heavier than other habitual fliers, resulting in higher wing loading and lift-to-drag ratios for flying fish compared to a comparably sized bird.

In the ***Cypselurus*** model, a **biplane body plan** is present, involving not only the enlarged pectoral but also the pelvic fins. They also tend to have depressed bodies, which increase the total lift. As a result, they are well adapted for maximizing flight distance and duration.

Cypselurus species have broader "wings" than *Exocoetus*. In flying fish with the biplane design, the hypocaudal lobe remains in the water to generate thrust even after the trunk clears the water's surface and the wings are opened with a small angle of attack for lift generation.

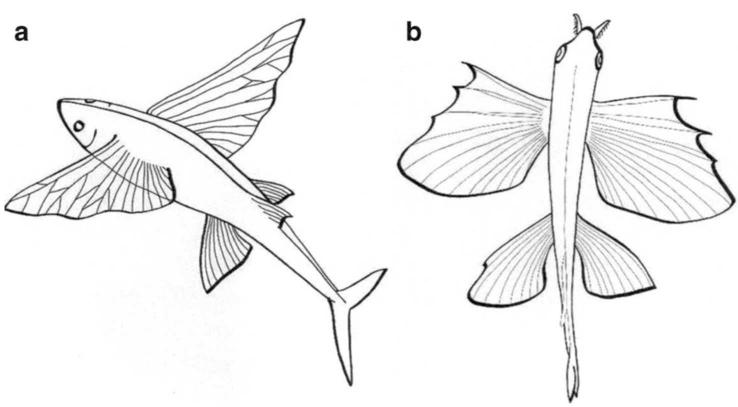

Fig. 6.21 Scheme of Exocetidae flying. (**a**) *Exocoetus* or monoplane body plan; (**b**) *Cypselurus* or biplane body plan. (Courtesy G. Belmonte)

Many non-fish vertebrates show peculiar swimming behaviours linked to their derivation by organisms evolved on lands. Many Tetrapoda re-evolved the ability to swim or have indeed returned to a completely aquatic lifestyle.

Primarily or exclusively aquatic animals have re-evolved from terrestrial Tetrapoda multiple times: examples include Reptilia (also extinct), Mammalia, and Aves.

Not properly aquatic animals (as humans) can also swim. Dogs swim recreationally. Umbra, a world record-holding dog, swam 3.7 nautical miles (6.4 km) in 73 min, placing her in the top 25% of long-distance human swimming competitors.

The fishing cat is a wild species of cat (*Prionailurus viverrinus*) that has evolved adaptations for an aquatic or semiaquatic lifestyle. Tigers and jaguars are the only large cats known to enter into water without problems. White bears (*Ursus maritimus*) are top predators familiar with marine water from which they obtain food. Birds can either be aquatic (as penguins, **Sfenisciformes**) or simply capture fish underwater (as many **Procellariiformes**). In both cases, they commonly swim flying (they use the wings which are the anterior legs), but in **Suliformes,** wings are maintained tight along the body, and the underwater propulsion to reach some fish food is realized by the diving velocity and movement of posterior palmate feet. Many Reptilia of the ancient oceans used the tail to propel water.

Exo = external; *Cetus* (L) = fish or *Koitos* = expulsion *Hypo* = downwards positioned; *Cauda* (L) = tail

They also reached large body sizes; thus, the tail was strong and long. In cases of the development of a true caudal fin, the ventral lobe was longer than the dorsal lobe, and it contained the termination of the vertebral column. When the tail does not develop a bilobed caudal fin, the whole tail functions as a propelling organ and undulates horizontally (this is the case of Crocodilia and **Hydrophidae**).

In species with a reduced tail, the swimming is due to the beating movement of the anterior legs as in a flight (e.g., in turtles) (Fig. 6.22). As some birds do, some marine Mammalia (**Pinnipedia Phocidae**) swim by using the anterior legs (i.e., they "fly") and use the posterior ones to manoeuvre changes of direction.

Sea Lions (**Pinnipedia Otaridae**), on the contrary, move horizontally the posterior part of the body using the finned posterior legs alternatively during oscillation and operated direction changes by using the anterior legs.

Cetacea evolved a caudal fin, but contrary to Pinnipedia, where the posterior part of the body is moved horizontally, they completely lose the posterior legs (**Mutica**) and vertically move the tail fin, which is oriented perpendicularly to that of fishes.

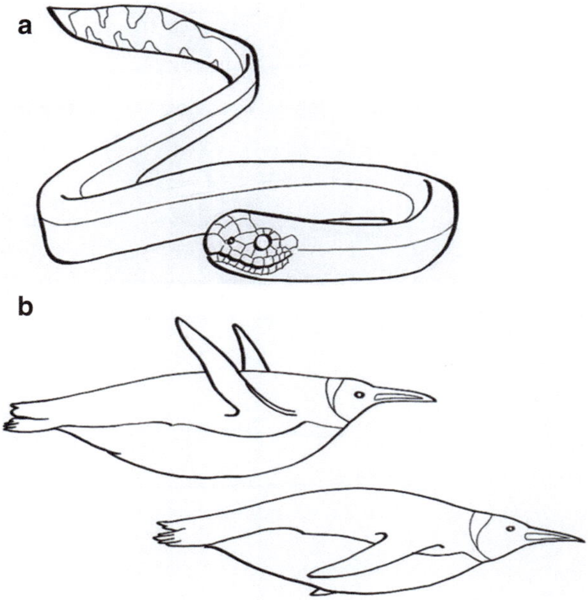

Fig. 6.22 Swimming style in Reptilia and Aves. (**a**) Hydrophidae; (**b**) Sfenisciformes. (Courtesy G. Belmonte)

6.4.6 Moving Between Different Habitats

Individual organisms moving in 3D ocean space determine and define the dynamics of the geographic distribution of their species.

Cypsela = botanical structure that allows the wind-driven dispersal of seeds; *Ura* = tail *Cetus* (L) = large fish *Mutica* (L) = wheat ear deprived of corn loin. Synonym of deprived of art *Phoca* = seal *Procella* (L) = violent storm *Sfeniscus* = wedge *Sula* = (Fin/Swed/Nor) marine bird

When long-distance movements are dictated by instincts, these movements are subjected to a genetic program that evolved to allow the perfect replication of migrating routes by successive generations.

Movements referable to a choice among alternatives, or representing intrinsic pulses without alternatives, belong to a behavioural phenotype typical of each species. A predominance of instinctive activities is typical of short-lived species, and more intense experience-based activity is obvious in long-lived species. In the last case, long-living organisms, as humans, have time to learn from their existence (this learning goes under the definition of **experience**) and add their learning to basic instinctual behaviour.

In the ocean, the small fish and crustaceans are generally more affected by the genetic component of the behaviour than the large fin fish and cetaceans.

A set of events that are generally classified in the frame of "instinct" stay beyond the preparation and execution of periodic (daily or seasonal) and long-distance

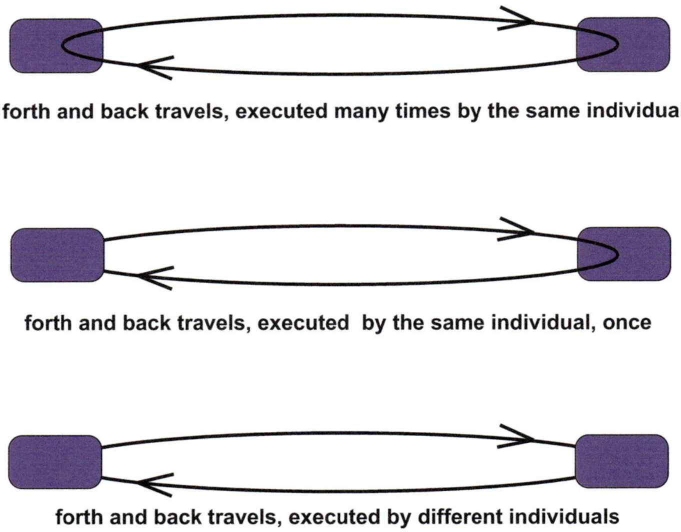

Fig. 6.23 Scheme of migration travels for each individual. (Courtesy G. Belmonte)

migrations of large species (Dingle and Drake 2007). The constraint succession of behaviours that can define a migration comprehends:

1. The perception of external stimuli.
2. Their translation in biochemical and physiological modifications.
3. The execution of the migration also elaborating stimuli coming from a changing environment.
4. Termination of travel only at arrival (see classification of **migratory behaviours** in Fig. 6.23).

Such a kind of **cyclic behaviour** has the same rhythm (and sometimes overlaps) as reproductive behaviour in sexual animals, a field where other categories of answers are present, as those consequent to emission and perception of messages (sounds, images, movements, odours) shared by individuals of the same group in the frame of "**intraspecific communication**".

"Day-by-day" instinctive behaviours are more common and diffuse; they are characterised by daily cycles of stimuli and answers and better represent the "**routine**" of existence.

They are behaviour cycles of sleep-vigilance or feeding, but migration can also be executed on a daily basis. Any movement of a living organism in the space can be considered in the field of behaviour, studied by ethology, and the particular biological rhythms staying beyond this group of behaviours are studied by **chronobiology**.

The presence of regulatory mechanisms grouped in the two categories of **oscillatory clocks** and **hourglasses** does not correspond to a simple regulatory mechanism

Box 6.4 Chronobiology

Chronobiology studies the so-called **biological clocks**: molecular and physiological settings that synchronize the complex assemblage of activities in each living organism.

Biological clocks can be subdivided into two large groups/types: **oscillatory clocks and hourglasses**.

An oscillatory clock is a type of pacemaker; it dictates the time rhythm for the execution of metabolic-physiologic-behavioural processes, which are inevitably cyclic. These (**endogenous**) cycles are generated and managed by oscillatory clocks and are independent of external (environmental) stimuli. The derived behavioural rhythm is termed **Infradian** (less than one cycle per day), if based on a cycle longer than 24 h, **circadian** if based on a cycle of approximately 24 h, and **ultradian** (more than one cycle in one day) if based on a cycle shorter than 24 h.

Biological clocks can be regulated (tuned) on the basis of time scales:

- Environmental (temperature-compensated).
- Biological (not temperature compensated).

An internal clock of the **hourglass type** measures time intervals; it indicates the start and the end of a process, but it does not dictate a rhythm or a cycle.

Such a kind of clock needs environmental signals to start. The most common processes driven by a biological hourglass in each individual are **embryo development** and the **germination** of plant seeds. Among cyclical phenomena, the switch on and the termination of single parts could be due to a biological hourglass, as in the case of **lethargy**, **migration**, or **menstruation**.

because mixed oscillating and hourglass internal clocks can develop, and multiple periodicity (endogenous and exogenous) can interfere in each biological process.

Among **ultradian cycles**, rhythmic phenomena are independent of the circadian cycle, but they probably interact with it in some way. Periods of 1.5–3 h are very common and are fundamentally based on the principle of avoiding spending energy continuously and adopting rest–recharge phases for each metabolic–physiologic performance.

Spontaneous activity perceivable in space (movement) is a parameter useful in chronobiology studies. The simple record of movements thus can show the existence of an endogenous rhythm and indicate its periodicity.

The registration of the movement activity and of its periodicity produces an **actogram,** which is the first element for the study of behavioural performance.

As each biological clock is characterized by rhythm persistence and is persistent in each species, techniques used for its description require analyses of more or less complex spectra (e.g., the **Fourier analysis**) (Fig. 6.24). The analyses of behaviour spectra tend to discover infinite opportune sinusoidal functions (= elementary movements) whose sum forms the final spectrum.

Chronos = time; *Bios* = life; *Logos* = discussion Circa = around, approximately; *Dian* = day *Ethos* = stay in the world; *Logos* = discussion *Infra* = numerically smaller than; *Dian* = day *Intra* = internal

In some underground environments, sunlight is absent, thus impeding organisms from tuning their activity on day–night alternations (the common pacemaker of the life rhythm on land).

The absence of light, however, is of fundamental importance in identifying the existence of **endogenous rhythms** (not connected with the day/night alternation) in the behaviour of living organisms.

Some species have been particularly studied from this point of view. Cognate *taxa*, with representatives classifiable as **troglophylous** (not obligatorily linked to live in caves) or **troglobiont** (obligatorily linked to stay all their lives in a cave) also provided comparative information about the presence and intensity of both types of rhythmic behaviour (exogenously dictated or endogenously tuned).

The adaptation of aquatic species to the dark environment is quite constant and shows the lack of pigmentation, the reduction or lack of eyes, other than long *sensilla,* as a remediation for the lack of visual sensitivity.

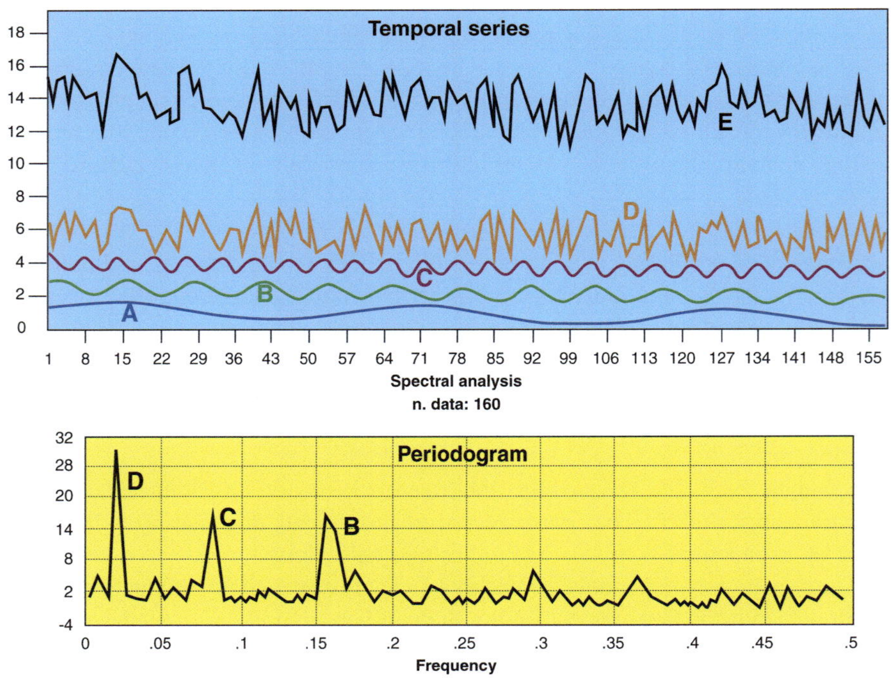

Fig. 6.24 A scheme of Fourier analysis. Top, single rhythms (**a–d**) composing the graphic (**e**) Bottom, time periods intervals typical of each rhythm. (Graphic composition A.M. Miglietta)

High resistance to starvation and a minimum number of egg-embryos for reproductive events are also common characteristics.

One such group of aquatic organisms is that of Mysida. *Spelaeomysis* is a genus typical of dark underground waters, and it has been used to set up scientific apparatuses and plan investigations on the presence and intensity of endogenous rhythms.

In the experimental use of underground aquatic species, they have to be preliminary acclimatized to the apparatus that will host them during experiments. Such an acclimatization lasts also some days, and it has the purpose of desynchronizing each individual before its employment.

Spelaeomysis bottazzii is a **stygobiont** (aquatic equivalent of subaerial **troglobiont**) and shows a pattern (**spectrum**) of activities (the **actogram**) very similar to that of non-aquatic animals (insects and mammals), and some generalizations can be listed:

- The circadian rhythm is not the only rhythm for the activity of animals.
- The ultradian components do not depend from the circadian ones.
- The ultradian components are not regulated by **photoperiod**, but they show higher amplitudes when the circadian is isolated from any photoperiod influence.
- They are temperature compensated.
- They stay either in **troglophylous** (insects and mammals) or in troglobionts (**stygobionts** in the case of aquatic organisms such as Mysida).

The behaviour of Mysida is particularly considered when these animals perform daily migrations inside-outside submarine caves. Mysida involved in such **horizontal migrations** are not stygobionts as the cited *Spelaeomysis*, and they do not show adaptation to the absolute dark.

They can be defined as **Stygophylous** and represent a form of daily migration of plankton, not vertical between different water layers but horizontal between the innermost part of a marine cave and the outside environment.

Ultra = numerically bigger; *Dian* = day

Many studies have clarified the migration behaviour of different Mysida and *Siriella jaltensis* in detail. Such a species, in addition, is absent from the cave habitat during the winter season; thus, it shows double migratory behaviour: daily horizontal during its stay in caves (from spring to autumn) and seasonal vertical when it leaves and reoccupies the cave at the beginning and at the end of the winter.

In a submarine cave of the central Mediterranean Sea, two species of Mysida perform daily horizontal migration: *Siriella jaltensis* and *Hemimysis margalefi*. The second species migrates between the innermost part of the cave (at 120 m from the entrance) and the entrance, but without going outside electing *H. magalefi* as the species more linked to caves than *S. jaltensis*. An actogram executed on *H. margalefi* in the laboratory allowed us to find its biological rhythm.

This behaviour synchronized with that of *S. jaltensis*, allows the two species to be responsible for a sort of recruitment of biomass produced outside, until the innermost part of the cave at more than 100 m from the entrance.

Some specimens collected in a submarine cave were acclimatized at constant values of 19.5 °C, 24 h light, and fed with yeast for 7 days before they were inserted in a small aquarium under an infrared video camera, organized to allow the registration of specimen positions every 30 min for a total of 5 days in complete darkness. The sequence of records was submitted to a Fourier analysis, which undoubtedly gave an activity rhythm of 23 h for *H. margalefi* (Fig. 6.25).

S. jaltensis did not show any rhythm, thus suggesting that it moves (inside and outside the cave) in response to the presence-absence of sunlight outside the cave, while *H. margalefi* (which does not perceive the light, inhabiting the innermost part of the cave) moves due to an endogenous clock with a cycle of approximately 1 day (circadian).

Acto (L) = executed, done; *Endo* = inside; *Genus* = generated *Spelaeo* = cavern; *Mysis* = blink *Styge* = underground water (from an ancient river of the hell); *Phylein* = attracted; *Biont* = living *Troglos* = cave; *Phylein* = attracted; *Biont* = living

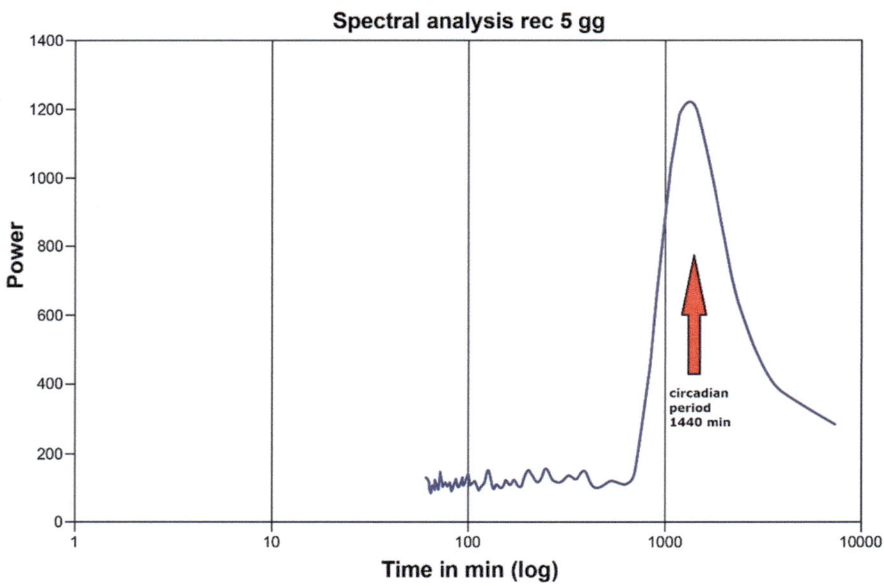

Fig. 6.25 Circadian biological rhythm of activity of *Hemimysis margalefi* from a submarine cave of the Mediterranean Sea. (Courtesy G. Belmonte)

6.4.7 Horizontal Migration

Having the freedom to move inside a 3D water world, fish and other nekton use this possibility to change their geographic position for optimization of resource acquisition and/or reproductive purposes. When such movements cover long distances or interest different geographic areas, or different water masses, we can speak of **migration**.

Migration is the field of persistent and active movements of animals, typically from one habitat to another, and is often caused by the spatial and temporal variations in resources and risks.

Through migration, individuals may increase their likelihood of encountering opportunities for feeding and growth, survival, and reproduction. As a result, migrants may enhance their fitness compared to nonemigrants, assuming that the benefits of migration outweigh the costs.

An accurate description of migration usually requires tracking the displacement of migrants across longitudes, latitudes, and altitudes or depths over time.

However, defining migration as a movement between habitats allow migratory trajectories to be simplified into one or two spatial dimensions along which implicated habitats exist.

In this regard, migratory trajectories of most terrestrial, aquatic, and avian species are usually described using two-dimensional geographic coordinate systems that disregard the vertical dimension. The vertical trajectory of migration becomes significant when migrants exhibit pronounced diving or flying behaviour during their geographic (horizontal) excursions.

For some animal migrations, however, the vertical dimension is the actual behaviourally (actively) controlled component and, consequently, the only descriptor of the migratory trajectory. These "**vertical migrations**" are widespread among zooplankton that swim up and down the water column.

Photo = light *Styge* = mythic underground water; *Bios* = life; *Phylein* = attraction

A round trip, executed by the same individual, many times during its existence, can be a daily migration (generally based on trophic necessities and performed between rest and feeding areas), as in the case of vertical movements of plankton, or an annual migration (generally based on reproductive reasons), as in the case of the best known habitat shifts (mountins-plaibns; tundra-savannah). In both daily and annual migrations, each involved individual is required to perform the migration many times in its life span. There are, however, some cases where individuals perform just one single round trip, notwithstanding a long-life duration (even of 25–30 years). Such migration movements are carried out by each single individual, and for this reason, a genetic basis is more recognizable that orders and shunts the individual in reaching the geographic target without errors and, in any cases, without memory of it.

A final example is the round trip completed by successive generations and not by the same individual. The latter case is that of shifts between the habitat of

Box 6.5 Migrations

Migration has been recognized as a behaviour genetically programmed only in recent times. The orienteering capacity of many species (the pigeons studied by Darwin, for example) and the seasonal appearance of many of them have been under the eyes of all naturalists since ancient times.

The first scientific attention to the problem of migration and orienteering started at the beginning of the twentieth century when the phenomenon was defined (Thompson 1936) as: "individual movements between different *habitats*". The definition contained the identification of a fundamental character: the existence of alternate directions (go and come back). Harden-Jones (1968) was the first to propose a genetically based action, speaking about the "obligation" of individuals to come back home. Baker (1978) tried to simplify an evidently complex field and proposed considering "migration" every movement between two space units.

The genetic program at the base of each migration movement and of its evolution was too evident, and only two years later, Dingle (1980) brought back the problem on a genetic/innate basis, defining migration as "a specialized movement, appropriately evolved to distribute synchronously individuals in the space". To complete his studies, Dingle (1996) arrived to establish a sort of table of stone for migratory movements.

The migration behaviour must be characterized by:

1. Persistence.
2. Order.
3. Unalterability.
4. Start and finish.
5. Redistribution of resources.

juveniles and larvae and that of adults, as in the case of Amphibia. They do not migrate between different geographic sites, but they do so between different lifestyles. The come back to aquatic life is performed by eggs and larvae of successive generation.

The already cited migration of Mysida between marine caves (during the day) and outside (during the night) is a typical example of **daily migration**.

Additionally, in subaerial environments, troglophylic species (bats) go outside the caves in the evening to feed on insects and go back inside the dark caves to rest during the day.

Typical **seasonal (annual) round trips** are the migration of tuna fish reproducing in the Mediterranean Sea (Fig. 6.26). They stay widely dispersed in the north east Atlantic Ocean and Ponto-Mediterranean basin, from Norway to the

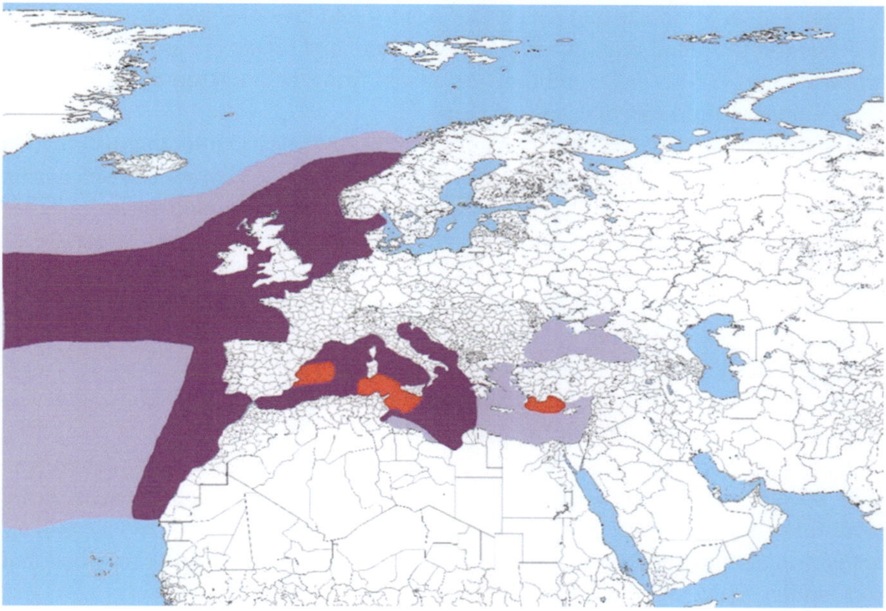

Fig. 6.26 Geographic distribution of bluefin tuna (*Thunnus thynnus*). Pink, area of presence; violet, feeding area; red, reproductive area. (Courtesy G. Belmonte)

Black Sea in July–November of each year. In December–April, they start to concentrate in only the Mediterranean basin. In May–June, all tuna stay concentrated in the western–central Mediterranean, where three reproductive areas were identified: one small area between Baleares and Catalonia, one large area between Sardinia, Sicily, and Tunisia, and one small area on the southern coast of Turkey.

The Mediterranean tuna (*Thunnus thynnus*) is the eastern population of a North Atlantic geographic distribution whose western individuals reproduce inside the Gulf of Mexico. The two populations segregate only for reproductive reasons; during the rest of the year, they stay completely mixed in the North Atlantic.

Due to the recognition of such a cyclic migrating behaviour, which leads to the concentration of large populations, and to the fact that this species executes migration in very surface waters, the migrating behaviour of tuna fish is known from humans who carried out tuna captures until ancient times (approximately 4000 years ago), as testified by paintings on ancient Greek pottery. Already approximately 1000 BC Phoenicians started to use mobile nets. In 700 AD, Byzantine fishermen used fixed nets, and in 1200 AD, the German emperor Frederic II established the first *tonnara* in Sicily.

Eggs and larvae are also in the surface waters, and juveniles are particularly predaceous (on the small prey of zooplankton). Eggs and larvae are small and **lecitotrophic** for at least 3 days, and they are absent from the plankton of the Alboran, Lion, Ligurian, Adriatic, and Aegean Seas.

Larvae take 10 days to reach a length of 1 cm. Juveniles smaller than 1000 g, on the contrary, have been collected in all the corners of the Mediterranean Sea during July–October in the last 200 years of monitoring. Postlarvae and early juveniles have large and teethed mouths other than large eyes and are predators of copepods. Adulthood is reached at 8–9 years in the Gulf of Mexico population, and at 3–4 years in the Mediterranean population. Individuals aged 15 years are common, and they arrive at a weight of 300 kg.

Large tuna feed on anchovies; they need to ingest 10% of their body weight each day to stay well. Adults generally stay far from the reproductive areas and down to 1000 m below sea level during July–April.

Due to the reduced gas bladder, tuna burn energy swimming continuously and have adaptations to such a lifestyle. The heart is relatively large, the muscles are plentiful of **myoglobin** (a protein with a high capacity for Oxygen), and a capillary net is available to enhance the exchange of warmth and gases between blood and surrounding tissues.

The vascular system and body mass in tuna are good enough to maintain their body temperature up to 10 °C higher than the external environment.

The swimming activity, fuelled by an efficacious availability of Oxygen by blood, is based on intermittent accelerations deriving from powerful undulations at the level of the caudal fin and the peduncle and is facilitated by the hydrodynamic body and the gregarious behaviour.

Many marine species other than tuna can perform long-distance migrations when individuals reproduce in areas very far from those where they grew.

Lecitos = egg/embryo energetic reserve; *Trophe* = food *Mios* = muscle; *Globine* = globular protein

In such species, individuals are obliged to migrate at least once in their life in a single roundtrip, which foresaw death at the end.

This condition suggests that the capacity for orientation is not learned by co-migrating "teachers", but it is based on an inherited behaviour.

Eels (genus *Anguilla*) reproduce at the bottom of the Sargasso Sea. Larvae of *Anguilla rostrata* go up rivers of North America, and larvae of *Anguilla anguilla* go up rivers of Atlantic Europe (reached even after 3 years of planktonic lifestyle as ***Leptocephalus***).

The two species could genetically separate from each other due to a different collocation of the reproduction area in the Sargasso Sea (*A. rostrata* in the western part, *A. anguilla* in the eastern part). The newly born larvae (*Leptocephalus*), as a consequence, are not able to find their home way, but are separately distributed by currents diverging from the Gulf Stream (Fig. 6.27).

The long-distance migration of eels is not only a problem of km; the same individual must pass a period as larva (*Leptocephalus*) in plankton before becoming an animal thriving on or close to the bottom. The adult stage is reached after many years of life in freshwater rivers.

At maturation, females evidently enlarge their body, passing a weight of 1500 g and reaching a length of 75 cm. In this condition, they descend the rivers, receive fertilization by males at the level of brackish water before entering the sea, and perform a long migration to the Sargasso Sea, where fertilized eggs are released and hatch leptocephali.

The reversed behaviour (life in the sea and reproduction in freshwaters) is that well known of the salmon (***Oncorhynchus gorbuscha***) (Fig. 6.28).

The possibility to return in the same birth place, at years and thousand kilometres of distance, is so interesting that many scientists proposed an intervention of the casus in the frame of round trip migrations. In fact, the return of eels in rivers of North America, or in those of Atlantic Europe, should not rely on the genetic memory of individuals but simply on their geographic positioning at birth in the Sargasso Sea. Individuals captured by the western side of the Gulf Stream in central north Atlantic are destined to north American Rivers, and those present in the main stream will arrive in Europe.

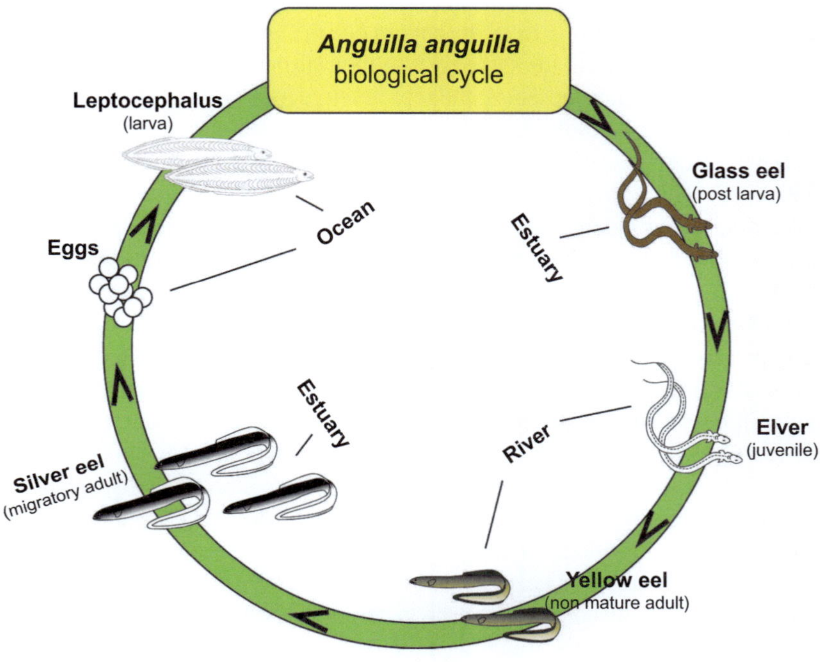

Fig. 6.27 Migration track of *Anguilla* (graphic composition A.M. Miglietta)

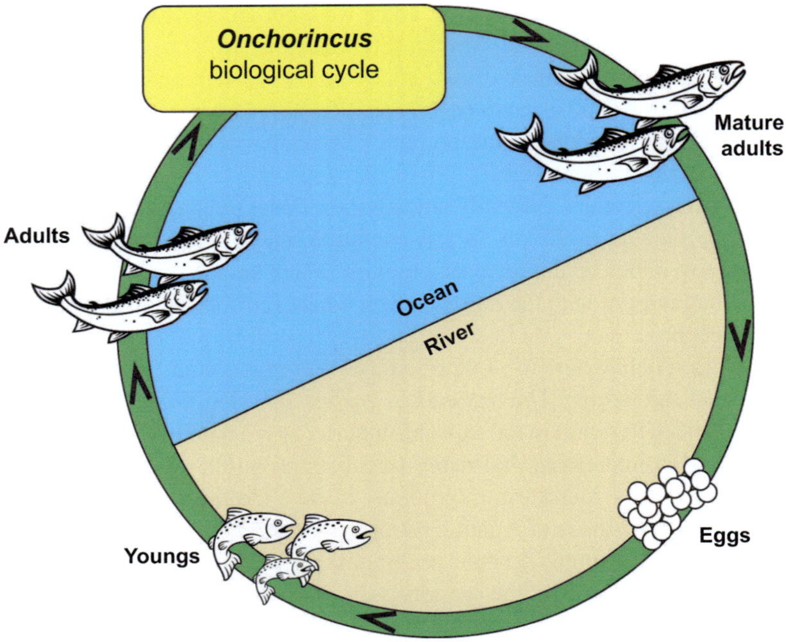

Fig. 6.28 Migration circuit of salmon (graphic composition A.M. Miglietta)

For the Pacific salmon, it could be the same: individuals have the possibility to meet a river mouth migrating from the central North Pacific in any direction towards the North, and the return in the same river of their birth is only a fascinating legend. Other cases have been proposed: in the reproductive migration of *Pleuronectes platessa* in the North Sea, individuals searching for a coastal site where to reproduce find, moving from the centre of the North Sea, a coastal situation in any direction at a relatively short distance (less than 1000 km).

However, recent studies have demonstrated that salmons actually return to the same river (after up to 15 years of marine life) where they were born. We must remember, in addition, that the eels directed in North America belong to a different species (*Anguilla rostrata*) from those directed in Europe (*A. anguilla*), thus excluding any possibility of remixing of populations at each year recruitment.

At least in sea–land migrations, a role of the geographic changes of the planet during millions of years of species (and their behaviour) evolution could be possible.

At least in eels, the search for a marine reproductive site (or the conquer of a freshwater habitat by a marine species that, after, retained the necessity to reproduce in marine waters) probably started in a geological period with a "small" Atlantic Ocean, when the distance between land rivers and ocean centre was minimal. Generation after generation, eels retained the instinctual execution of the round trip, progressively longer by some cm at each generation, until the current situation with reproductive area thousands of km from the river mouths.

Complex and long migrations, such as that of *Anguilla,* suggested to Myers (1949) the schematization of migratory behaviour in water habitats:

Oceanodromic, entirely carried out in marine waters (*Thunnus thynnus*).

Potamodromic, entirely carried out in fresh waters.

Diadromic, carried out between fresh and marine waters.

Anadromic, if egg deposition and hatch occur in fresh waters (*Oncorhynchus gorbuscha*)
Catadromic, if egg deposition and hatch occur in the sea (*Anguilla anguilla*)
Amphidromic if egg deposition and hatch are not the reason for travel

Some species of terrestrial-derived *taxa* live into the sea but return to the land to lay eggs (**thalassogeodromic** migration):

penguins, albatross and procellarids (Aves).
horse shoe crabs (Chelicerata).
turtles (Reptilia).

Their egg laying areas are generally close to the sea (because they are not adapted to conduct a life or simply to move on land, but the emperor penguin cares for its eggs in Antarctica, at sites distant hundreds of kilometres from the ocean.

Some species (of marine-derived *taxa*) live on land, but they return to the sea to lay eggs (**geothalassodromic** migration):

Crustacea Decapoda (*Birgus latro*).

The frame of animal migration is accomplished by the existence of spatial movements that are not true migrations. These movements of individuals are generally unidirectional (without a return, as in savannah grasshoppers) and involve many specimens but not all (some of the population components remain in the native site). The native site is never completely abandoned, and the newly invaded areas do not support invasive migrants for long times. The species geographic distribution, after a certain time, returns to the geographic area occupied before the "migration" but not thanks to a return of migrated individuals who die in new territories without descendants.

These population movements received the name "migration" before the definition was discussed and defined.

Thus, grasshopper migrations have a root in the Bible, and modern Lessepsian migration (another "improper" migration) is a term proposed by Por (1971) before the adoption of a precise definition for the phenomenon (Dingle 1996) (see Box 1.1, and Fig. 1.15).

6.4.8 Vertical Migrations

The concept of zooplankton vertical migration originated in the nineteenth century thanks to the observation of G. Cuvier (1817) on lake zooplankton. Field investigations conducted successively indicated that these periodic movements are a result of active migration, and the term "**vertical migration**" was established in the literature by the early twentieth century.

Due to the difficulty of tracking planktonic individuals across space and time, much of the current evidence for zooplankton vertical migrations comes from observing vertical changes in their population assemblages over time.

The predictions and interpretations of zooplankton vertical behaviour from population observations are more accurate at higher population sizes and obviously when vertical behaviours among individuals are largely synchronous. However, asynchronous vertical migrations also exist and likely signify the existence of among-individual variability in migratory behaviour.

Russell (1927) classified **zooplankton vertical migrations** based on the timing of the migration (periodicity), areas over which it takes place (e.g., geographical regions and spawning habitats), and biological attributes, such as ontogeny and sex.

His classification, on the basis of periodicity, proposes two types of vertical migrations: shorter-term **diel vertical migration** (**DVM**) with a periodicity of approximately 1 day, and long-term **seasonal vertical migration** (**SVM**) with a periodicity of up to 1 year (Bandara et al. 2021). In general, DVM represents the widespread tendency of zooplankton to occupy deeper waters during the day and near-surface waters during the night. Unlike DVM, SVM is largely confined to seasonal environments, such as high-latitude aquatic habitats and upwelling systems, and reflects the tendency of zooplankton to occupy different vertical habitats during different periods of the year. Since the vertical distribution of the migrants usually varies across developmental stages, **SVM** is also described as an **ontogenetic vertical migration**.

Despite the generality of the above classification, reverse diel and seasonal vertical migrations also exist.

In *Le Règne Animal*, Cuvier (1817) made what is probably the first written record of zooplankton DVM. In a clear shallow freshwater lake, in Europe, he noted that *Daphnia* (Cladocera) retreated to deeper waters during midday and ascended to near-surface waters in the evening. However, diel migrations of zooplankton in deeper lakes and the open ocean were not studied in detail at the time, probably due to the lack of appropriate sampling techniques to trace deep vertical movements of plankton.

This limit was addressed in the late nineteenth century, when T. Fuchs (1882) started with the separate sampling of different depth layers, and showed that the daytime disappearance of marine pelagic Crustacea from near-surface waters was due to their sinking to deeper layers.

Amphi = both sides; *Dromos* = circuit *Ana* = upwards; *Cata* = downwards; *Gea* = the mother land; *Lepto* = thin, small; *Cephalus* = head; *Oncho* = hook; *Rhynchus* = nose; *Gorbuscha* (Russian) = fish name *Ontos* = being, existence; *Genetic* = based on generation, race *Thalassa* = one of the Greek names of the sea

DVM is a widespread behaviour among many marine zooplankton *taxa* across all latitudes, and it is one of the most-studied patterns of the animal behaviour (Cushing 1951) (Fig. 6.29).

The most common DVM involves a night-time ascent to the epipelagos and a daytime descent to mesopelagos or deeper waters.

Active vertical migration (especially synchronized DVM) is considered less prominent at polar latitudes, where diel changes in irradiance are minimal during 24-h daylight and polar nights. Coordinated vertical migration is not perceivable in the Arctic during the period of midnight sun, probably due to a coarse subdivision of the vertical column studied.

Copepoda avoided the well-illuminated upper 40 m during the day and migrated into the 20–40 m layer at night where the highest Chl*a* concentration was located, and they probably found food.

During the day, light intensities reached minimal levels below 40 m, and migrating below this depth may thus offer Copepoda a refuge from visual predators. A "metabolic–demographic advantage hypothesis" states that migrating Copepoda may gain a metabolic advantage by leaving food-rich surface waters for deeper layers.

However, in polar areas, the water temperature actually increases with depth within the observed migratory layers. Thus, it is unlikely that any metabolic advantage exists for migrants, and the avoidance of visual predators is likely to be the main driver of DVM. In subpolar areas the upper 20 m consist largely of meltwater, with lower temperatures (approximately −1.5 °C) and salinities (approximately 32.3) than below the thermocline at 20–30 m (2–3 °C and 34–35 ppt, respectively).

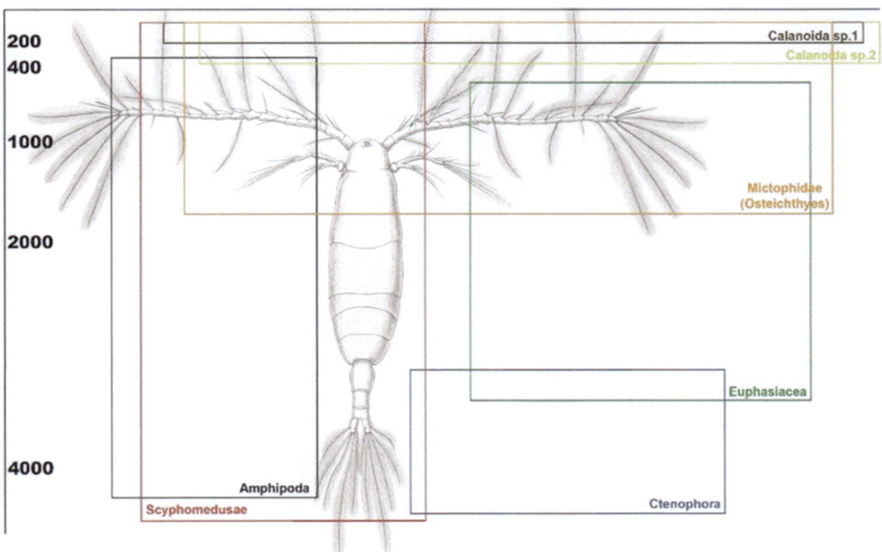

Fig. 6.29 Scheme of the daily vertical migration DVM of Crustacea and other plankton. A copepod calanoid is represented in the theatre out of scale, just in representation of one of the main actors of DVM (graphical composition A. M. Miglietta)

The main reason for avoidance of the extreme superficial epipelagos is likely a combination of little food and increased vulnerability to predation by visual predators, such as Amphipoda *Themisto* spp. and polar cod (*Boreogadus saida*).

Older *Calanus* individuals remain deeper in the water column both during day and night. Abundances of copepodid CV and adult (CVI) individuals in the upper 200 m were very low for both *C. finmarchicus* and *C. glacialis*. Deep sample collection showed that these individuals were concentrated deeper than 200 m, indicating that a larger part of the population had already descended to such overwintering depths at the end of August.

The DVM in *Pseudocalanus* spp. has a magnitude of 20–40 m. *Pseudocalanus* and the CI/CII of *Calanus* are similar in size (0.5–1.2 mm), indicating that they may avoid the same type of predator specialized to prey on this size class. Predators that tend to hunt prey of this size include Amphipoda and Euphausiacea (Macdonald 1927).

A comparatively rare variation in the standard DVM is the **twilight DVM**, where ascent occurs during dusk and descent occurs around midnight (midnight sinking).

Several proximate causes for such a behaviour have been proposed. A passive response that results from the general slowing of swimming speed that occurs in complete darkness cannot be excluded. Alternatively, satiation could be the cue for sinking, possibly operating through the slowing of swimming speed and/or the increase in body density of satiated specimens. The fact that sinking occurs at midnight in complete darkness makes it unlikely that visual predation has a role, but **tactile predation** may play a role since motion has been shown to increase the risk of predation in a number of studies. One of the biggest problems in determining the exact cause of midnight sinking is the fact that in many instances, it is difficult to detect any sinking at all.

The acoustic Doppler current profiler (ADCP) is a recent technology that is able to obtain extra information on vertical migration behaviour. A study used a 300 kHz ADCP over a continuous period in the summer of 1999 in the Clyde Sea to investigate the vertical migration behaviour of zooplankton, in particular the copepod *Calanus finmarchicus* and the krill (Euphausiacea) *Meganyctiphanes norvegica* and *Thysanoessa raschii*.

C. helgolandicus sometimes occurs in the North Sea together with *C. finmarchicus* and exhibits a similar lifestyle. *Thysanoessa raschii* and *Meganyctiphanes norvegica* (Euphausiacea) are two of the most abundant krill species in the north Atlantic Ocean. Within this geographic range, *T. raschii* is principally a neritic species, and *M. norvegica* inhabits both on-and off-shelf regions (especially shelf breaks). Both *M. norvegica* and *T. raschii* are strong vertical migrators, and both exhibit midnight sinking behaviour.

M. norvegica is omnivorous and is a predator of *C. finmarchicus*, especially at earlier copepodid ages (Tarling et al. 2002). *T. raschii* is more herbivorous by comparison.

A study demonstrated that after an initial movement to the surface at approximately 17:00 h, a regular exchange of *Calanus finmarchicus* between the surface and deeper layers produced no variation in the biomass distribution. Only when krill arrive do all *C. finmarchicus* respond by moving down together thus subtracting themselves from the surface layer. In a second scenario, *C. finmarchicus* remains in the surface layers, either continuing to feed or pausing between feeding bouts until the arrival of krill.

In both cases, Euphausiacea (krill) seem to be the reason for the mass midnight sinking of Calanoida.

Satiation-driven descent was advocated by other studies. However, none of them considered the presence of potential predators and their respective times of arrival. Evidence that **krill actively prey on copepods** such as *Calanus* is clear. Generally, however, the consumption levels of *Calanus* were lower than those of other nonmigrating copepod species, reflecting the ability of *Calanus* to avoid predation through the midnight sinking. Although it is likely that *Calanus* descends satiated (with food in their stomachs), it is clear that descent was an adaptation to avoid predation rather than a post feeding moment.

C. finmarchicus is particularly widespread throughout most of the north Atlantic Ocean and is capable of migrating over several hundred meters and attaining vertical velocities approaching 0.01 m sec^{-1}. Observations of midnight sinking behaviour in this species are common. The species is primarily herbivorous but also relatively flexible in adapting its feeding style to environmental conditions. It shows 1–3 generations per year, according to location, during which the majority of age instars occupy vertical layers above 50 m. However, approaching winter, a large percentage of the population remains at a pre-adult condition (CV) and enters a state of diapause. These individuals may stop performing diel vertical migration and remain in the deeper layers until the following spring.

For herbivorous zooplankton, Daily Vertical Migration is typically a strategy that allows feeding in the food-rich upper pelagos during the darker hours when light-dependent (visual) predators are absent or scarce, and the phytoplankton has completed its reproductive cycle under sunlight (Fig. 6.30).

In some cases, zooplankton avoid the photosynthetically active layer because some phytoplankton (Diatomeae) release, during photosynthesis, metabolic by-products in the water that can be toxic to zooplankton.

Finally, light intensity also seems to play a role because the single frontal eye of Calanoida is probably adapted to low light intensities (those in the deep waters during day), thus commanding the vertical movement of copepods in search of a constantly low light intensity.

In contrast to the above classic patterns, reverse DVM (daytime ascent and night-time descent) likely enables zooplankton to evade light-independent (tactile) invertebrate predators that, on their side, perform classic DVM to escape their own visual predators (Fig. 6.31).

Combining his own observations with those of other Nordic colleagues, A. Weismann (1876) summarized the ability of lacustrine Daphnidae to evacuate the pelagos seasonally and occupy the bottom sediments as resting eggs. Although this phenomenon was unknown in the marine realm at that time, somebody noted the tendency of some marine Crustacea to disappear from the upper pelagos in the summer–autumn months and reappear in the following spring. To investigate this further, depth-stratified net sampling in the Mediterranean Sea carried out at the end of the

Fig. 6.30 Daily vertical migration of zooplankton (graphic composition A.M. Miglietta)

nineteenth century showed that the seasonal disappearance of Cnidaria medusae and Crustacea from the upper pelagos was due to their migration to depths below 1000 m.

By the early twentieth century, scientists started to use the term "**seasonal vertical migration**" (SVM) to refer to such a kind of migrating behaviour. Although not as widespread as DVM, SVM is a behavioural strategy common among many planktonic inhabitants of seasonal environments.

Some aquatic environments are characterized by predictable degradation of habitat quality in the (upper) pelagos during part of the year. In polar environments and coastal upwelling systems, this occurs through seasonal depletion of primary production due to light or nutrient limitation.

As the conditions in the near-surface waters deteriorate, zooplankton descend (passively sinking or actively swimming down) to deeper waters or to the sediment and may enter a state of inactivity and suppressed development (or produce resting stages) termed dormancy (Belmonte and Rubino 2019). Dormancy is sometimes commanded to terminate at the predictable onset of favourable environmental conditions, and zooplankton ascend again to the upper pelagos. This periodic seasonal movement in and out of the upper pelagos characterizes the SVM.

Based on the different diapause strategies of zooplankton, SVM can be classified into (1) low-amplitude (depth) SVM of resting stages that undergo diapause in or on the bottom sediment and (2) high-amplitude SVM of late-juvenile and adult stages that "**overwinter**" or "**oversummer**" in the deeper pelagos (Fig. 6.31).

In polar settings where the productive season is typically short, herbivorous Calanoida of the genera *Calanoides*, *Eucalanus*, *Calanus*, and *Neocalanus* are well

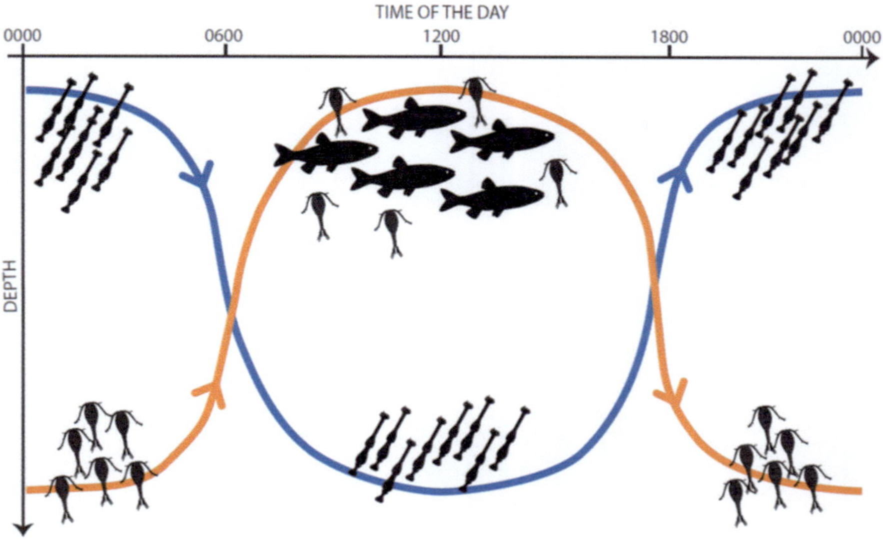

Fig. 6.31 Daily vertical migration of Copepoda, Chaetognatha, and fish. (From Bandara et al. 2021, modified by A. M. Miglietta)

known to perform SVM. Juveniles of these Calanoida often thrive in warmer, food-rich, near-surface waters and grow and develop rapidly during the productive period (spring and summer). Seasonal changes in irradiance, nutrient concentration, and grazing pressure tend to limit pelagic primary production during late astronomical summer and following autumn. Consequently, the growth and development of species with relatively larger body sizes (e.g., *Calanus glacialis* and *C. hyperboreus*) become time-constrained, and the life cycle is not completed within a single productive season. Although species with smaller body sizes (e.g., *C. finmarchicus* and *Eucalanus bungii*) may complete several generations within a single productive season, overwintering becomes inevitable for the generation thriving during the shortage of food supply towards autumn. Older developmental instars of these Copepoda (e.g., copepodite ages CIII, CIV, and CV) migrate to deeper waters with accumulated lipid reserves for overwintering. These overwintering specimens remain in deep waters with limited physical and physiological activity and wake-up to ascend back to the near-surface waters to complete their life cycle at the onset of the following productive season.

SVM in the Arctic is a widespread trait in zooplankton, and seasonality in food availability drives seasonal vertical strategies of species. The interspecies relationships are stronger between consumers (herbivores, carnivores, and successive predators) than inside the producer community.

Some polar herbivorous Copepoda start their seasonal descent sometimes well before the termination of pelagic primary production. The actual cues that drive the descent to overwintering depths in the prediction of the termination of pelagic primary production are suspected to be the perception of photoperiod together with cues related to developmental stage and their reserve levels.

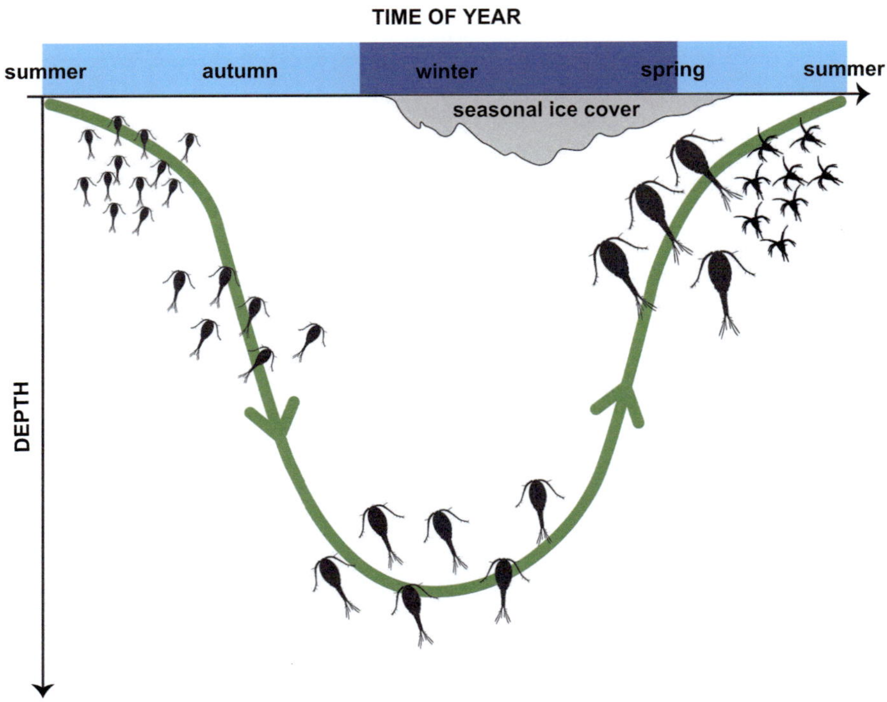

Fig. 6.32 Scheme of seasonal vertical migration (SVM) in polar seas. (From Bandara et al. 2021; modified by A.M. Miglietta)

SVM reversed patterns characterized by a summertime seasonal descent are underlined by **over summering** strategies. Over summering has been documented for the temperate copepod *Calanus sinicus* in the Chinese Yellow Sea and the high-latitude copepod *C. finmarchicus* in the north American Gulf of Maine.

SVMs are also observed among carnivorous zooplankton, such as Cnidaria medusae, Euphausiacea, and Chaetognatha, and generally correspond to the vertical trajectories of their herbivorous preys along the water column. SVM patterns of herbivorous zooplankton are often followed by pelagic predators both invertebrates and vertebrates.

Energy supplied by lipids accumulated in the upper pelagos by overwintering *Calanus* flows along trophic chains and may be consumed by apex predators, such as marine Mammalia and Aves, up to humans.

Vertical net hauls in the Firth of Clyde captured large numbers of younger northern krill *Meganyctiphanes norvegica* in shallow waters in spring and summer. In the late winter, most of the krill population descended below 100 m, and their gut largely contained *Calanus finmarchicus*.

Additionally, also the seasonal vertical distribution of the Chaetognatha *Pseudosagitta gazellae* is due to the vertical distribution of their copepod preys (David 1956). Similarly, statistically significant associations existed between seasonal vertical distributions of prey and predator zooplankton *taxa* across three adjacent trophic levels (Fig. 6.32).

Seasonally migrating zooplankton play a prominent role in the deep-ocean Carbon sequestration and distribution process. The lipid reserves accumulated by zooplankton in spring and summer are carried down into the deep ocean during their autumn–winter descent phase. At these depths, stored lipids are typically metabolized at a rate accounting for ca. 25% of the basal metabolic rate. Despite the reduced metabolism, a 6–8 month-long overwintering period can exhaust a substantial fraction (44–93%) of the stored lipids. Finally, only a part of the overwintering population ascends to the near-surface waters in the following spring, while the rest are either consumed by deep-dwelling predators or die due to starvation, disease and other sources of mortality.

The lipid-based Carbon flux driven by the SVM (the **lipid pump**) is thus a largely unidirectional process that transfers **Carbon** directly from epipelagos to the deep ocean with little loss on the way (Visser et al. 2017). The contribution of the lipid pump in the North Atlantic to the total particulate organic Carbon flux has been estimated as 2.0–6.0 g C m^{-2} $year^{-1}$ for *C. finmarchicus* and 3.5–6.0 g C m^{-2} $year^{-1}$ for *C. hyperboreus*.

References

Baker RR (1978) The evolutionary ecology of animal migration. Hodder And Stoughton, London

Bakun A (1996) Patterns in the ocean. Ocean processes and marine population dynamics. California Sea Grant College system, and Centro de Investigaciones Biologicas del Noroeste. La Paz, BCS, Mexico, p 323

Bandara K, Varpe Ø, Wijewardene L, Tverberg V, Eiane K (2021) Two hundred years of zooplankton vertical migration research. Biol Rev 96:1547–1589

Belmonte G (2019) Floating objects of non marine origin, rafting of marine organisms, and the interfering role of man in Mediterranean Sea. Int J Oceanograp Aquac 3(2):000168

Belmonte G, Rubino F (2019) Resting cysts from coastal marine plankton. Oceanogr Mar Biol Annu Rev 57:1

Boero F, Bouillon J (1993) Zoogeography and life cycle patterns of Mediterranean hydromedusae (Cnidaria). Biol J Linnean Soc 48(3):239–266

Bryan SE, Cook AG, Evans JP, Hebden K, Hurrey L et al (2012) Rapid, long distance dispersal by pumice rafting. PLoS One 7(7):40583

Butler JN, Morris BF, Cadwallader J, Stoner AW (1983) Studies of *Sargassum* and the *Sargassum* community, vol 22. Bermuda Biological Station Special Publications, pp 1–307

Carlton JT, Hodder J (1995) Biogeography and dispersal of coastal marine organisms: experimental studies on a replica of a 16th-century sailing vessel. Mar Biol 121:721–730

Carlton JT, Chapman JW, Geller JB, Miller JA, Carlton DA, Treneman NC, Steves BT, Ruiz GM (2017) Tsunami-driven rafting: transoceanic species dispersal and implications for marine biogeography. Science 357(6358):1402–1406

Coston-Clements L, Settle LR, Hoss DE, Cross FA (1991) Utilization of the *Sargassum* habitat by marine invertebrates and vertebrates—a review. NOAA technical Memorandum NMFSSEFSC-296, p 30

Cushing D (1951) The vertical migration of planktonic crustacea. Biol Rev 26:158–192

Cuvier G (1817) Le Règne Animal. Chez Deterville, Paris

David PM (1956) The distribution of *Sagitta gazellae* Ritter-Zahony. Discov Rep 27:237–277

Laffoley DDA, Roe HSJ, Angel MV, Ardron J, Bates NR, Boyd IL, Brooke S, Buck KN, Carlson CA, Causey B, Conte MH, Christiansen S, Cleary J, Donnelly J, Earle SA et al (2011) The

protection and management of the Sargasso Sea: the golden floating rainforest of the Atlantic Ocean. Summary Science and Supporting Evidence Case, Sargasso Sea Alliance, p 44
Dingle H (1980) Ecology and evolution of migration. In: Gauthreaux SA Jr (ed) Animal migration, orientation, and navigation. Academic Press, New York, pp 1–101
Dingle H (1996) Migration: the biology of life on the move. Oxford University Press, New York
Dingle H, Drake VA (2007) What is migration. Bioscience 57(2):113–121
Fine ML (1970) Faunal variations on pelagic *Sargassum*. Mar Biol 7:112–122
Fouqué F (1879) Santorin et ses eruption. Masson, Paris
Fuchs TH (1882) Über die Pelagische Flora und Fauna. Verhandlungen der KaiserlichKöniglichen Geologischen Reichsanstalt 7:49–70
Gower J, Young E, King S (2013) Satellite images suggest a new *Sargassum* source region in 2011. Remote Sens Lett 4:764–773. https://doi.org/10.1080/2150704X.2013.796433
Harden-Jones FR (1968) Fish migration. Edward Arnold, London, p 325
Houle A (1998) Floating islands: a mode of long-distance dispersal for small and medium-sized terrestrial vertebrates. Divers Distrib 4(5/6):201–216
Johannesson K (1988) The paradox of Rockall: why is a brooding gastropod (Littorina saxatilis) more widespread than one having a planktonic larval dispersal stage (L. littorea)? Mar Biol 99(4):507–513
Jokiel PL (1990) Long-distance dispersal by rafting: reemergence of an old hypothesis. Endeavour 14(2):66–73
Jutzeler M, Marsh R, Carey RJ, White JDL, Talling PJ, Karlstrom L (2014) On the fate of pumice rafts formed during the 2012 Havre submarine eruption. Nat Commun 5(1):3660. https://doi.org/10.1038/ncomms4660
Jutzeler M, Marsh R, van Sebille E, Mittal T, Carey RJ, Fauria KE, Manga M, McPhie J (2021) Ongoing dispersal of the 7 august 2019 pumice raft from the Tonga arc in the southwestern Pacific Ocean. Geophys Res Lett 47:e1701121. https://doi.org/10.1029/2019GL086768
Kingsford MJ (1999) Fish attraction devices (FADs) and experimental designs. Sci Mar 63(3–4):181–190
Macdonald R (1927) Food and habits of *Meganyctiphanes norvegica*. J Mar Biol Assoc U K 14:753–784
Mandic M, Gvozdenović S, De Vito D, Alfonso G, Daja S, Ago B, Cela E, Ivanovic A, Zoto A, Malovrazic N, Beli E, Ingrosso G, De Leo F, Pestoric B, Lule A, Vata F, De Rinaldis A, Carpentieri C, Bode A, Nazaj S, Durmishi C, Paladini G, Peraš I, Raičević M, Fraissinet S, Boero F, Piraino S (2022) Setting thresholds is not enough: beach litter as indicator of poor environmental status in the southern Adriatic Sea. Mar Pollut Bull 177:113551
Myers GS (1949) Usage of anadromous, catadromous and allied terms for migratory fishes. Copeia 1949:89–97
Naumann E (1917) Undersokningar ofver fytoplankton och under den pelagiska regionen forsig gaende gyttjeoch dybildningar iron visa sydoch mellansvenska ubergsvatten. Kungl Srenska Vetensk Akad Handl 56(6):1–163. (in Swedish)
Pendleton L, Krowicki F, Strosser P, Hallett-Murdoch J (2014) Assessing the economic contribution of marine and coastal ecosystem Services in the Sargasso Sea. NI R 14-05. Duke University, Durham, NC
Por FD (1971) One hundred years of Suez canal-a century of Lessepsian migration: retrospect and viewpoints. Syst Zool 20:138–159
Powers S (1911) Floating Islands. Pop Sci Monthly 79:303–307
Russel FS (1927) The vertical distribution of plankton in the sea. Biol Rev 2:213–262
Simons EL (1976) The fossil record of primate phylogeny. In: Goodman M, Tashian RE, Tashian JH (eds) Molelcular anthropology. Plenum Press, New York, pp 35–62
Suaria G, Avio CG, Mineo A, Lattin GL, Belmonte G, Moore CJ, Regoli F, Aliani S (2016) The Mediterranean plastic soup: synthetic polymers in Mediterranean surface waters. Sci Rep 6:37551
Tarling G, Jarvis T, Emsley S, Matthews J (2002) Midnight sinking behaviour in *Calanus finmarchicus*: a response to satiation or krill predation? Mar Ecol Prog Ser 240:183–194

Thiel M, Gutow L (2005a) The ecology of rafting in the marine environment. I. The floating substrata. Oceanogr Mar Biol Annu Rev 42:181–264
Thiel M, Gutow L (2005b) The ecology of rafting in the marine environment. II. The rafting organisms and community. Oceanogr Mar Biol Annu Rev 43:279–418
Thiel M, Hayes PA (2006) The ecology of rafting in the marine environment. III. Biogeographical and evolutionary consequences. Oceanogr Mar Biol Annu Rev 44:323–429
Thompson AL (1936) Recent progress in bird migration—a review of the literature, 1926-35. Ibis 13(6):472–530
Visser AW, Grønning J, Jonasdottir SH (2017) *Calanus hyperboreus* and the lipid pump. Limnol Oceanogr 62:1155–1165
Weismann A (1876) Zur Naturgeschsargassumichte der Daphnoiden. In: Siebold CTV, Kölliker AV, Ehlers E (eds) Wissenschaftliche Zoologie. W. Engelmann, Leipzig, pp 51–112
Zernov SA (1934) General Hydrobiology (in Russian). Biomedgiz, Moskow, Leningrad, p 504

Pelagic Lifestyles in the Dark

7

7.1 Hyperbenthos

Hyperbenthos is a term given to the community of small animals living in the water layer close (generally, at less than 1 m) to the sea floor (Mees and Jones 1997). The species assemblage of hyperbenthos is characteristically different from either benthic or pelagic communities, and it includes typical species (autochthonous), other than plankton-derived species at their lower habitat extension (mostly on a seasonal migration basis), or benthos-derived species mostly entering the water column daily (Arkett 1984; Armonies 1989).

In all the investigated marine habitats, there was a general tendency for an increase in biomass concentration at the **benthos-water column interface**. Collecting such organisms is not easy due to their mobility, and conventional sampling devices (for plankton or benthos) are not adapted to give reliable descriptions of the hyperbenthic community. Conventional sampling, as a consequence, is substituted by many alternative approaches (Omori 1969). For soft bottoms, sledges are used, while in hard bottom benthos, traps are preferred.

In a pioneering study, Bieri and Tokioka (1968) revealed that the first 20 cm above the sea floor contained several times more organisms (Amphipoda, Mysida, Decapoda) than the above-standing 20 cm. In the case of Cumacea, the presence was described as exclusive to the first 20 cm layer. In a successive study, it was established that the bottom proximity at a site 350 m below sea level affected not only the concentration of individuals (increasing their numbers) but also the species richness in comparison with the layer just 50 cm above.

Due to their elevated content of Mysida (Clutter 1969; Carleton and Hamner 1989), considered the principal consumers of plankton cysts, the hyperbenthos could affect the plankton composition and represent an important food resource for demersal fish.

Hyperbenthic communities are known from the deep sea, where such a fauna is usually termed "benthopelagic plankton" or "demersal plankton". The zone in

G. Belmonte, *Elements of Pelagos Biology*,
https://doi.org/10.1007/978-3-031-74280-4_7

which hyperbenthos occurs is referred to as the "benthopelagic" or "benthic boundary layer" (Childress et al. 1989).

In the deep ocean, the sea floor can be viewed as a collector halting the sinking of particles coming from the above water column.

Most pioneer collections from the deep sea hyperbenthos produced descriptions of new species.

Copepoda (mostly Harpacticoida) are 80% of the specimens collected in the hyperbenthos, and other abundant organisms are **Mysida**, **Isopoda**, **Amphipoda Gammaridae**, **Ostracoda**, and **Chaetognatha**. Large representatives of many taxa were rare, but this could be a consequence of their ability to escape from the sampling device. In contrast to shallower areas, no differences exist between day and night. Abundances range from 2 to 230 ind. 100 m^{-3}.

Three depth intervals, each characterized by a specific hyperbenthic fauna, were identified at the Arcachon Plateau (Atlantic France) (Comet et al. 1983). The hyperbenthos of the upper part of the plateau (a muddy-sand substratum at depths of 350–500 m) was characterized by high concentrations of **Isopoda Asellota**. At intermediate depths (mud floors at 600–1100 m), **Amphipoda** dominated, sometimes accompanied by **Cumacea**. In Canyon proper (>1100 m), Isopoda again predominated, accompanied by significant numbers of **Pycnogonida** and **Tanaidacea**.

Abundances in the lower layer grow with bottom depth (2700 ind. 100 m^{-3} at 430 m; 4500 at 710 m; 600 at 1030 m; 700 at 3060 m) and are always higher than in overlaying water layers.

The Catalan Sea slope (western Mediterranean) was sampled between 400 and 1860 m deep using a small plankton net attached to a beam trawl and a Macer-GIROQ sledge.

On the upper slope, Amphipoda dominated the hyperbenthos, whereas Cumacea became progressively more abundant with increasing depth and dominated the lower slope.

Coral reefs appear to harbour unique and abundant hyperbenthos not found in the surrounding waters. Elements of this reef assemblage include resident, swarming, or demersal zooplankters.

The most common taxa are Copepoda, Amphipoda, Cumacea, Mysida, Ostracoda, Tanaidacea, Decapoda larvae, Polychaeta, Chaetognatha, Larvacea, fish larvae, and Mollusca veligers.

In a subtidal sand flat in the Gulf of California, Copepoda, Amphipoda, and Isopoda have been reported as predominant. Caridea, Mysida, **Cephalochordata**, and **Decapoda** larvae were collected only occasionally. All these animals remain in proximity to the sea floor during the day, and there is some discussion as to the extent of their upward migration into the water column.

Amphi = both sides; *Pous Podos* = feet *Cephale* = head; *Chorda* = flexible axis, vibrating wire *Chaeta* = needle, hard bristle; *Gnathos* = mouth, jaws *Cope* = oar; *Pous Podos* = feet *Cuma* = antique city of southern Italy, site of the mythic "sybil" *Deca* = ten (10); *Pous Podos* = feet *Pycno* = dense; *Gonos* = angle, knee *Hyper* = over the top *Ostracon* = body shield

Smaller animals have been reported to remain within 30–50 cm over the bottom, whereas larger individuals migrated throughout the entire water column.

The distribution, abundance, and behaviour of these animals vary with season, bottom type, lunar period, and daylight, and these stimuli are differently perceived by different *taxa*.

A study carried out in a shallow bay in North Carolina (Cahoon and Tronzo 1992) showed that hyperbenthic zooplankton are taxonomically distinct from holozooplankton and that high abundances typically occur in the hyperbenthos.

Abundances of 160–8000 ind. m^{-2} per night, have been reported as emerging from the bottom sediment.

The seasonal and diel variations in the community structure of Mysida in the Davies Reef Lagoon (Great Barrier Reef) over 1 year showed day abundances higher than night abundances (Jacoby and Greenwood 1988).

Common features of the hyperbenthos include the following:

(a) Most species are typical.
(b) Most species are present in higher abundances than in either the overlaying water or in the sediment.

More recently, there has been recognition of a distinction between the truly hyperbenthic species and a variety of "visiting" or "immigrant" animals that can also be classified as endobenthic, epibenthic, or planktonic.

A higher diversity in the hyperbenthos, compared with the macrobenthos of the same area, has been demonstrated on many occasions, and many new species and high abundances of species have been recorded that were rarely collected by ordinary plankton tow nets. The hyperbenthic catches, in addition, comprised different species compared with macrobenthic samples from the same area.

The recognition of the hyperbenthos as a specific *habitat* has increased in recent years, such that many species (of a variety of *taxa*) are referred to explicitly as hyperbenthic. In addition to Mysida, probably the most typical hyperbenthic *taxon*, several new species, genera, and families of Calanoida Copepoda, a variety of new benthoplanktonic Chaetognatha, and an entirely novel hyperbenthic Ostracoda fauna are currently recognized.

Many *taxa* that were previously thought to be exclusively endobenthic, epibenthic, or pelagic are now found to have hyperbenthic representatives. Bentho-pelagic medusae are more numerous and diverse than previously believed, with abundances of up to 80 ind. m^{-2} on or near the bottom (Larson et al. 1992). Hydromedusae concentrate in the hyperbenthos during the day. Swimming Holothuroidea were also recorded in high concentrations in the hyperbenthos of abyssal depths. Decapoda larvae are more common on the bottom during the day than during the night, and some larval Polychaeta swim preferentially close to the bottom.

Since the body size of marine invertebrate larvae imposes limitations on locomotion, differences in ways of propulsion between life-history stages occur, and these affect habitat choice and the temporary presence of some species in the hyperbenthos. For example, Antarctic **Euphausiacea** change their community distribution from the pelagic zone to the upside-down hyperbenthos (under coastal fast ice) as a strategy to survive the winter (see Seasonal Vertical Migrations, Chap. 6) (Kawaguchi et al. 1986).

Among temporary hyperbenthos, an important presence is that of late-stage larval fish. **Fish larvae** are especially abundant in the near-bottom layers. The distribution of larval fish is controlled by both active behaviour and passive transport mechanisms resulting from a combination of biotic and abiotic factors. Biotic factors include seasonal abundances of adults in reproductive status, environmental preferences, availability of suitable food, presence/absence of potential predators, and larval behaviour. Abiotic factors include the oceanographic, hydrological, and climatological parameters of the area (e.g., temperature, salinity, stratification, turbidity, riverine contributions, weather, and water-flow patterns, rates, directions, and anomalies). The relative importance of passive and behaviourally mediated transport is still under discussion (Palmer 1988).

7.1.1 Small-Scale Zonation

Using nets mounted on sledges at different distances above the sea floor, several authors have investigated the vertical distributional patterns of hyperbenthos.

Swimming activity and preferred distances above the bottom appear to differ between sexes and between developmental stages of Mysida (Kaartvedt 1985). Both plankton and vagile benthic species differ in their vertical zonation in the near-bottom area. Generally, the highest number of species is found in the lowermost layers sampled (Fig. 7.1, Tables 7.1 and 7.2).

Diel activity close to the sea floor has been documented for several *taxa*. Generally, deep-living hyperbenthic Cumacea (Fig. 7.2) and Amphipoda are associated with sediment during the day, and swimming activity, with a variety of migration patterns, increases within the lower metres of the water column at night.

Direct observations of hyperbenthic Mysida (Fig. 7.2) show that different species display varying degrees of affinity for the substratum.

Many hyperbenthic species show **diel changes** in their vertical distribution. Mainly, these are nocturnal or tidally phased (especially in estuaries) vertical migrations. Many of the hyperbenthic Mysida found near the bottom during the day underwent nocturnal migrations in the above water, but even so, a significant

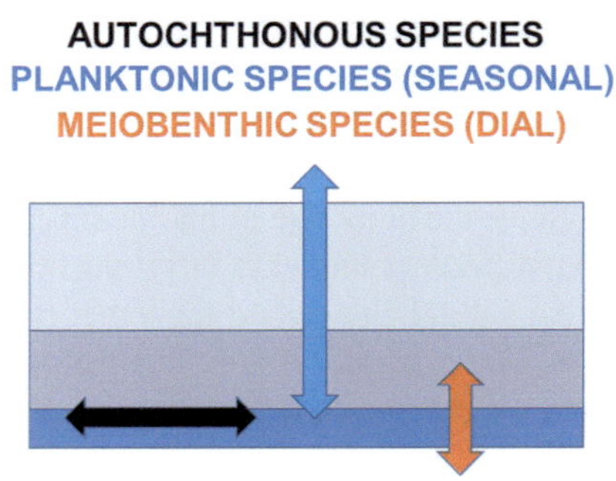

Fig. 7.1 Scheme of hyperbenthos organization in the 50 cm above the bottom (courtesy G. Belmonte)

Table 7.1 Abundances (number of specimens) of some taxa in layers at different distances from the bottom (from Bieri and Tokioka 1968)

	Number of specimens m^{-3}				
Water layer	Amphipoda	Mysida	Copepoda	Cumacea	Decapoda larvae
20–40 cm	0.50	0.05	1.50	0.00	0.01
00–20 cm	800.00	3.30	7.60	1.90	4.30

Table 7.2 Total abundance (number of specimens) and species richness (num. species) in layers at different distances from the bottom (from Huberdeau and Brunel 1982)

Water layer	Number of specimens m^{-3}	Number of species
106–142 cm	0.01–1.69	21
28–64 cm	6.80–10.65	87

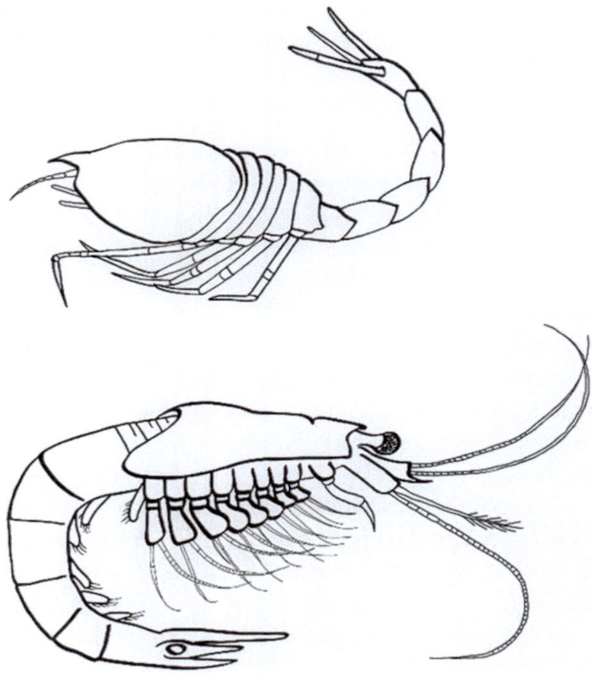

Fig. 7.2 Hyperbenthos components; up, Cumacea; bottom, Mysida (courtesy G. Belmonte)

proportion of the total population of each species remained in the hyperbenthos during darkness. The hypotheses to explain upward migrations refer to the availability of food in non-hyperthermic layers, and the return to deeper layers close to the sediment could be justified by a lower metabolism request, avoidance of visual predation, management of horizontal dispersion, and breeding migrations (Wittmann 1977).

Meiofaunal recolonization after sediment disturbance in coastal areas indicates that partial recovery occurs within a few tidal cycles (Sherman and Coull 1980). Recruitment for this rapid recovery may occur through the hyperbenthic temporary presence of meiobenthic populations. By analogy to the meiobenthos sensu stricto, the term meiohyperbenthos (animal <0.5 mm) has been proposed as opposed to macrohyperbenthos (animals larger than 0.5 mm or 1 mm). Periodic hyperbenthic behaviour has been reported for Copepoda Harpacticoida (mainly diatom eaters), encompassing both passive suspension and active emergence from sediment for rapid dispersion and recolonization. Considerable numbers of meiobenthic Platyhelminthes Turbellaria regularly enter the overlying water column during the night and, after distribution through water currents, can colonize new areas in a short while.

For most hyperbenthic *taxa*, basic information on relationships between length, weight and chemical composition is generally not available in sufficient detail for the study of **energy fluxes** through food webs. Secondary production estimates are scattered and, with few exceptions, are not detailed for any one species. Detailed production estimates are available for the following Baltic Mysida populations:

Mysis mixta (0.6 g C m^{-2} $year^{-1}$),
Schistomysis ornata (23 mg ash free dry weight m^{-3} $year^{-1}$),
Schistomysis kervillei (130 mg AFDW m^{-3} $year^{-1}$),
Schistomysis parkeri (52.6 mg AFDW m^{-2} $year^{-1}$),
Neomysis integer (300 mg AFDW m^{-3} $year^{-1}$).

The hyperbenthos, as a consequence of the above reported data, plays an important role in the coupling of benthic and pelagic food webs.

The highest hyperbenthic abundances have been reported from estuarine areas, eutrophic shallow coastal areas, and medium-to deep areas of continental slopes. In very turbid areas, where the hyperbenthos tends to peak, the clogging of feeding mechanisms and the risk of burial generally negatively affect the persistence of sessile guilds. Alternatively, below a certain threshold of food availability, mobility may become necessary to acquire sufficient food. Mobility may be a further advantage to anticipate stochastic events in highly unpredictable environments such as estuaries, for example, to flee adverse Oxygen conditions or high concentrations of pollutants.

Most Mysida are **omnivores**, feeding on detritus, algae, and zooplankton (Mauchline 1988). Since feeding is probably selective for different zooplankton species and size groups, Mysida have a potential role in driving the structure and composition of zooplankton. Most Mysida utilize organic detritus to a considerable extent and can be responsible for the remineralization of a substantial proportion of the detritus.

The study of the hyperbenthic fauna is encouraged by many authors in its role as a constituent of the diet for fish, and some of the earliest studies were specifically conducted for that purpose (Alldredge and King 1985).

Hyperbenthic microhabitat feeding guilds were recognized in the **Gobiidae** assemblage in Lough Hyne, Ireland, and Mysida feeding guilds were distinguished in several demersal fish assemblages. Meiobenthic *taxa*, notably Copepoda **Harpacticoida**, contribute significantly to the diets of many species of post-larval and juvenile fish. Mysida progressively replace Copepoda as a food source for demersal fish during their growth. Such an **ontogenetic trophic shift** has been observed in several species, including *Gadus morhua*.

Hyperbenthon often dominates the diet of commercially important fish and sustains high numbers of non-commercial demersal fish (e.g., Gobiidae), which are important prey for the larger size classes of the fish compartment.

In recent years, numerous invertebrates have been reported as consumers of hyperbenthos. Birds and marine mammals are also possible components of such a consumers' assemblage. For example, the analysis of the faecal matter of grey whales has revealed a preponderance of hyperbenthic Mysida cuticle fragments.

Cnidaria may play an important role in the hyperbenthos composition. For example, the annual average standing stock of the metazoan plankton in the Kiel Bight (German Baltic Sea) included 1% hyperbenthic Crustacea (Mysida and Cumacea) and 51% Cnidaria.

Juveniles of the squid *Sepietta oweniana* forage with the hyperbenthos of Norwegian waters, preferentially catching free-swimming Mysida. The diet of *Sepia officinalis* changes during development, and the smallest individuals feed on hyperbenthic Mysida. Part of the diet of deep-sea Octopoda also consists, to a large extent, of hyperbenthic organisms. Predatory Crustacea on hyperbenthos, often in commercially exploited stocks, include **Caridea** shrimp and rock lobsters.

Special types of equipment are needed to collect the hyperbenthic organisms, and species that are now regarded as rare may prove to be very abundant when proper collecting apparatus is used.

Large fish and epibenthic Crustacea, though they feed on the hyperbenthos and are thus an integral part of that community, can be erroneously considered accidental catches because they are not sampled efficiently by sledges, and this fact will seriously bias the biomass estimates.

Although the intensity of research on hyperbenthos has increased in recent years, more descriptive baseline studies are needed. Special attention should be given to areas where commercial fisheries are high and to nursery areas for commercially important species.

Although studies of the fluxes from the hyperbenthos to higher trophic levels are available from several areas, there is still a general need for studies on the fluxes, including lower trophic levels of the hyperbenthos, other than production and consumption estimates for the dominant species.

7.2 Bioluminescence and Communication in the Dark

The sunlight is absorbed by sea water, and only an average of 0.02% of its outside intensity is still perceivable at a depth of 100 m (Fig. 7.3). Thus, only the epipelagos are interested in the presence of sunlight with a cyclically variable intensity due to the different positioning of the sun in the sky (and alternation with night).

The **mesopelagos** (200–1000 m of depth), is also defined as the "twilight zone", and any remaining visible (from the animal point of view) sunlight can be considered absent below such a zone. Sunlight represents a limit for plant existence (and chlorophyll photosynthesis), but not for animals. Without enough sunlight, we have to exclude plants and photosynthetically active organisms from the deeper part of the ocean, which represents an available habitat only for **heterotrophic** organisms. This "limiting condition" is not an exception for life on the planet because it affects approximately 80% of the oceans (their average depth is 3800 m), and it can be considered typical and not exceptional for pelagic ocean lifestyles.

Among fish, the adaptation to deep sea conditions requires the reduction, up to the elimination, of the **gas bladder**, due to the intense hydrostatic pressure they have to manage, but eyes do not reduce or disappear, notwithstanding the darkness. As underground and cave fauna typically adapt with blindness to a situation of complete darkness, we can simply derive that the persistence of eyes in deep-sea animals suggests that light is somehow present in the deep sea.

The light able to form images and information useful for deep-sea organisms at more than 1000 m below sea level is obviously not derived from the sun. The enormous extension of the dark sea is interested in light emission of biochemical origin, a phenomenon termed **bioluminescence**. Most likely, such a large availability of habitat also justifies the large evolutionary success of bioluminescence (Lau and Oakley 2021), which has been estimated to have at least 94 independent origins across the tree of life.

During the night, the sea is also dark at the surface, and it is common to attend to the spectacular "milky sea effect" produced by bioluminescent Bacteria and/or Dinophyta covering up to a thousand square miles of sea surface. The production of light during the day is not visible, but it is also not useful; thus, organisms evolved **biorhythms** (internal clocks) to produce light messages only in the dark moments, when the luminous messages can be perceived.

In bioluminescence generation, **luciferin** (a pigment) reacts with oxygen to generate light, and **luciferase** (an enzyme) acts as a catalyst for the reaction, sometimes mediated by cofactors such as Ca ions or ATP (Fig. 7.4).

Box 7.1 Light produced by living organisms

Such a biochemical phenomenon produces light using energy derived from a chemical reaction. Other phenomena are involved in the production of light signals that can be visible to other organisms. **Phosphorescence** and **fluorescence** induce some particular substrata to glow if they are exposed to radiation (light) and re-emit it at a longer wavelength than the received one. **Iridescence** is the phenomenon of certain surfaces that appear to gradually change colour as the angle of view or the angle of illumination changes due to the different reflection angles they offer to the incident light. Both of these cases involve the reuse of the incident light. Bioluminescence (or chemiluminescence), on the other hand, produces autochthonous light.

Such a phenomenon is now intended to be widespread in dark pelagos organisms (Haddock et al. 2010), justifying the persistence of organs sensible to light messages in the deep (on the contrary, the underground fauna, adapted to the absence of light, is typically interested in the reduction and absence of eyes).

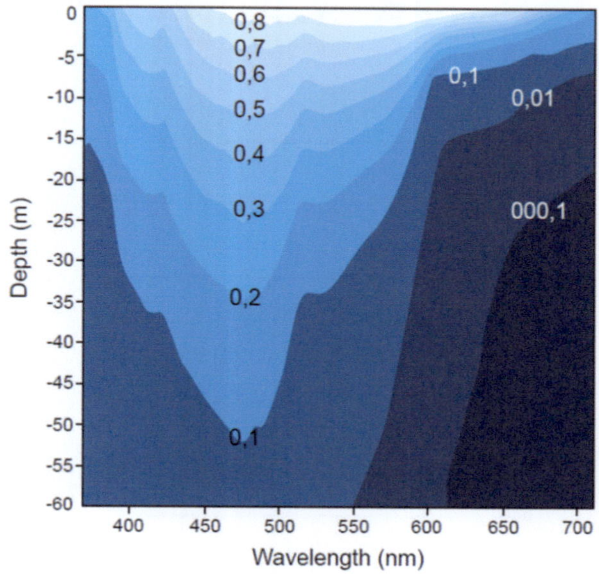

Fig. 7.3 The differential absorbance of sunlight wavelengths through the sea water thickness (graphical composition A. M. Miglietta)

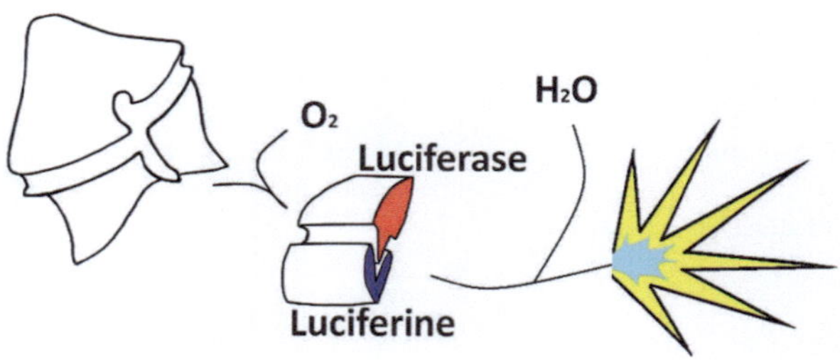

Fig. 7.4 a simplification of the bioluminescence reaction to produce light in marine organisms (graphical composition A. M. Miglietta)

The chemical reaction can occur either inside or outside the cell. In any case, the reagent apparatus is built by the cell organism of interest. Among Bacteria, bioluminescence is the expression of DNA portions (genes) controlled by an operon (**lux operon**).

Bioluminescence is a complex trait that has evolved dozens of times through either novel mechanisms or conserved toolkits in different, not only marine, taxa. Luminous organisms that share luciferin substrates like **coelenterazine** or **vargulin** to produce light, evolutionarily converge in biochemical pathways, producing bioluminescence.

Most marine light emissions are in the blue-green light spectrum, the wavelengths that pass farthest through seawater. However, some loose-jawed fish emit red and infrared light, and the polychaete *Tomopteris* emits yellow light. Due to the absorption capacity of seawater, we can hypothesize that blue-green bioluminescence emits messages destined to travel longer distances than red bioluminescence (Fig. 7.5).

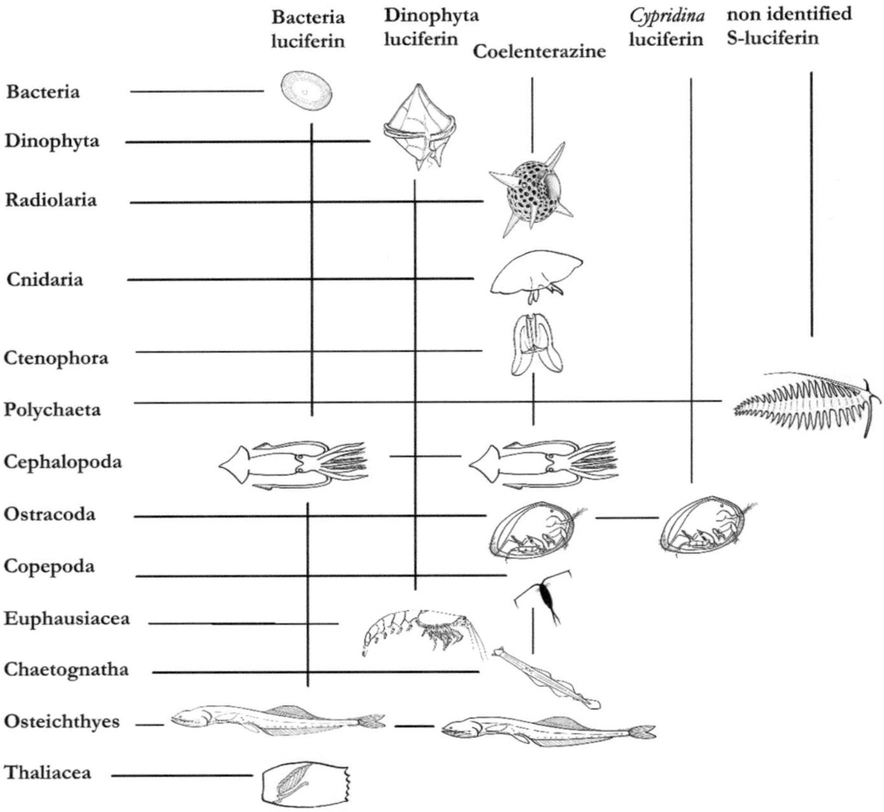

Fig. 7.5 The most common luciferins in marine animals (graphical composition A. M. Miglietta)

Aliivibrio fischeri is a gram-negative bacterium found everywhere in the marine environment, predominantly in symbiosis with various marine animals. Free-living *A. fischeri* are found in very low quantities (almost undetectable) under a heterotrophic lifestyle, relying on organic matter or molecules present in the water. These bacteria, rare in the water environment, are found in higher concentrations in symbiosis with certain deep sea life forms within special light organs or in the enteral (gut) microbiota of marine animals. Symbiotic relationships in fish and squid appear to have evolved separately.

The most prolific of these relationships is with the Hawaiian bobtail squid (***Euprymna scolopes***) (Fig. 7.6) (McFall-Ngai and Montgomery 1990).

Free-living *A. fischeri* in the ocean are captured and accumulated by the light organs of juvenile squid and fish. *E. scolopes* produces mucous as a reaction to the presence of peptidoglycans in the bacterial walls. The mucous produced around the light organ captures bacteria. The squid, in the meantime, produces a peroxidase (a microbicide enzyme), which uses a hydrogen peroxide as a substrate. *A. fischeri*, on its side, is equipped with a catalase that subtracts the peroxide before it matches the peroxidases, thus being able to remain in the new habitat. *A. fischeri* is continuously

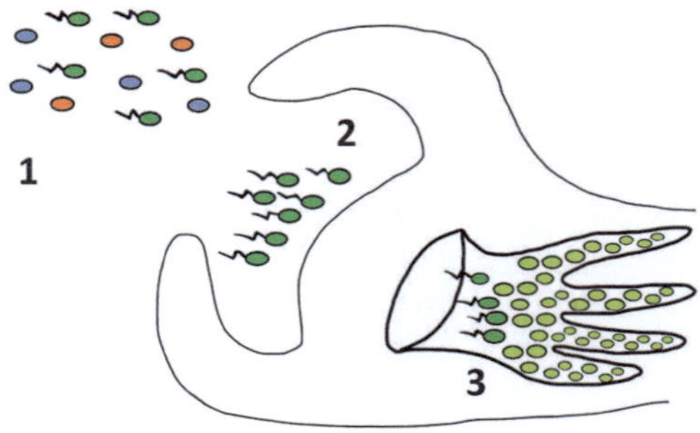

Fig. 7.6 Scheme of recruitment of Bacteria in the light organ of Hawaiian squid (*Euprymna scolopes*). (1) Bacteria community in the water; (2) Recruitment/entrance of only *Aliivibrio fischeri* in the light organ of *E. scolopes*; (3) *A. fischeri* lose the flagellum and stop inside the organ position where they will produce light (from (Royet et al. 2011) modified by A. M. Miglietta)

captured by *E. scolopes* mucus; here, however, it is the only microbe able to survive the molecular attack of the squid. This exclusivity allows *A. fischeri* to be the only survivor that is able to increase its concentration. When in the mucous (denser than the water), *A. fischeri* cells equip themselves with flagella to move in a dense medium and migrate to the light organ.

Lux Lucis (L) = light *Fero* = bring, bear

Due to the appearance of flagella, *A. fischeri* cells pass the ciliated ducts of the light organ, and successively, they swim towards a large theca, where they insert themselves among the epithelial cells. *A. fischeri* feeds on the host amino acids and sugar, and they completely colonize the available space 10–12 h after infection.

The **light organ** of certain squids contains reflective plates that intensify and direct the light produced due to the presence of **reflectins** (proteins). They regulate light to keep squid from casting a shadow on moonlit nights, for example. Sepiolidae squids expel 90% of the symbiotic bacteria in their light organs each morning in a process known as "**venting**". Venting is hypothesized to provide a free-living inoculum source for newly hatched squids. The bioluminescence of *A. fischeri* is caused by transcription of the lux operon, induced by population-dependent quorum sensing. The lux operon is a 9-kilobase fragment of the *A. fischeri* genome that controls bioluminescence through catalysis by the enzyme luciferase. The luminescence appears to be more active during the night.

A set of genes is responsible for the building up of the bacterial luciferin–luciferase system.

In *A. fischeri*, **five genes are involved** in the so-called **luxCDAB(F)E complex**:
lux A and lux B code for the components of luciferase,
lux CDE codes for a fatty acid reductase complex that makes the fatty acids available for the luciferase mechanism,

lux C codes for the enzyme acyl-reductase,
lux D codes for acyl-transferase,
lux E makes the proteins needed for the enzyme (acyl-protein synthetase).

Two genes (*LuxR* and *LuxI*) are involved in regulating OPERON. Several factors appear to induce and inhibit the transcription of this gene set (light emission).

Bacterial luciferin is a reduced riboflavin phosphate (FMNH2) that is oxidized in association with a long-chain aldehyde, Oxygen, and a luciferase. Luciferase produces blue/green light through the oxidation of reduced flavin mononucleotide and a long-chain aldehyde by O_2.

$$FMNH_2 + O_2 + R - CHO \rightarrow FMN + R - COOH + H_2O + Light$$

The reduced flavin mononucleotide (FMNH) is provided by the gene LuxG.

To generate the aldehyde needed in the reaction above, three additional enzymes are involved. The fatty acids needed for the reaction are pulled out of the fatty acid biosynthesis pathway by the enzyme acyl-transferase. Acyl-transferase reacts with acyl-ACP to release R-COOH, a free fatty acid. R-COOH is reduced by a two-enzyme system to an aldehyde.

$$R - COOH + ATP + NADPH \rightarrow R - CHO + AMP + PP + NADP+$$

Bioluminescence in Bacteria is **regulated by autoinduction**. An autoinducer is a transcriptional promoter of the enzymes necessary for bioluminescence reactions. Before the glow can be lit, a certain concentration of an autoinducer must be present. Therefore, for bioluminescence to occur, high concentrations of *A. fischeri* should be present in the host organism.

Krill (Euphausiacea) are bioluminescent animals equipped with **photophores**. The light is generated by an enzyme-catalysed reaction, wherein a molecule (luciferin) is activated by an enzyme (luciferase).

The luciferin of many krill species is a molecule (**tetrapyrrole**) **similar** but not identical **to the Dinophyta luciferin**.

The krill probably do not produce this substance by themselves, but they acquire it as part of their diet, which contains Dinophyta.

Photophores are organs devoted to the production of light. They are generally collocated on the ventral side of the individual, at the basis of the legs (Fig. 7.7).

The function of photophores in krill includes **communication for mating**, social interaction, or **orientation** and a form of counterillumination **camouflage** against the overhead ambient light.

Bathypelagic fish are dark or sometimes red and are equipped with a small number of photophores, generally used to entice prey or to attract a mate.

Some bathypelagic fish (flash-like fish) have a reflecting structure behind the retina in the eyes, which enables them to perceive the eyeshine in other fish when these are illuminated by light produced in photophores.

Among **Dinophyta,** bioluminescence is so common that the *taxon* also received the name Pyrrophyta (burning plants).

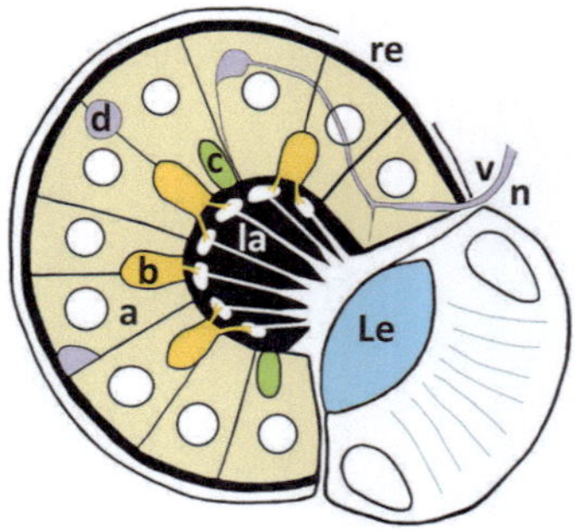

Fig. 7.7 Longitudinal section of a photophore of Euphausiacea. It is possible to recognize a **lantern** (la) made by **photocyte processes** (from b cells) and **refractive** rods. Light produced in the lantern is reflected by a **reflector** (re) and obliged to pass throughout a **lens** (le) before going outside. Apart from the b cells, large cells (a), small cells (c) and d-cells are present in the photophore. On each side of the lens, photophore vessels (V) and nerves (N) enter the organ. Capillaries branch off from the arteries and pass throughout both D-and C-cells before reaching the La. Nerves follow the capillaries and end at a sphincter-like structure at the base of the C-cells (graphical composition A. M. Miglietta)

The agitation of seawater containing dinoflagellates will stimulate light flash emissions. In fact, when the dinoflagellate perceives the presence of an unwelcome species (the "**burglar**"), such as a copepod, it emits a light flash (the "**alarm**"), which lasts 0.1–0.5 s. The flash is an attracting stimulus for an additional player (the "**police**"), which generally closes in looking for food.

Chloro = green; *Phyllos* = leaf   *Photos* = light; *Cytos* = cell   *Photo* = light; *Phore* = bear, bring

The police species is generally a predator of the burglar species (Ghabru et al. 2023). When the burglar species (a copepod) perceives the flash, it generally gives a jump because staying will make it vulnerable to the arrival of the police species (generally a predator of copepods).

Dinophyta luciferin is thought to be **derived from** chlorophyll and has a very similar structure. In *Gonyaulax*, at pH 8, the molecule is "protected" from luciferase by a "luciferin-binding protein", but when the pH is lowered to 6, luciferin reacts, and light is produced.

The production of light is due to the association of luciferin with a protein, luciferase, a catalyst. In the process of being oxidized, luciferin briefly exists in an excited state, after which it decays to its ground-level state, releasing energy in the form of photons (light) (Fig. 7.8).

Very curiously, **chlorophyll-like luciferin** works in a reaction where Oxygen and chemical energy are consumed to produce light, while in photosynthesis, light is the energy source from which molecular links are created, and Oxygen is made free.

Dinophyta are the most common sources of bioluminescence in the surface waters of the ocean. During dark nights, their presence is responsible for light

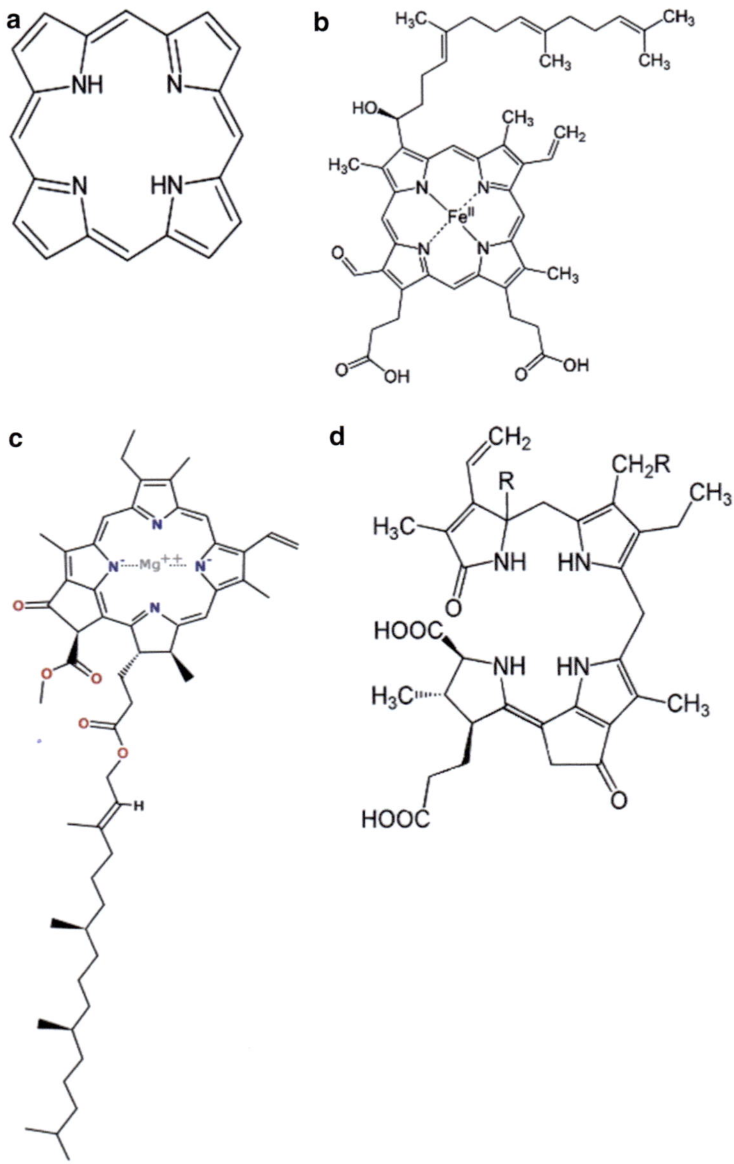

Fig. 7.8 Scheme of the tetrapyrrole ring of luciferin in Dinophyta (d) compared with similar molecules for the transport of Oxygen (c) and production of energy in photosynthesis (b). In a, the basic scheme of a tetrapyrrole ring (graphic composition A. M. Miglietta)

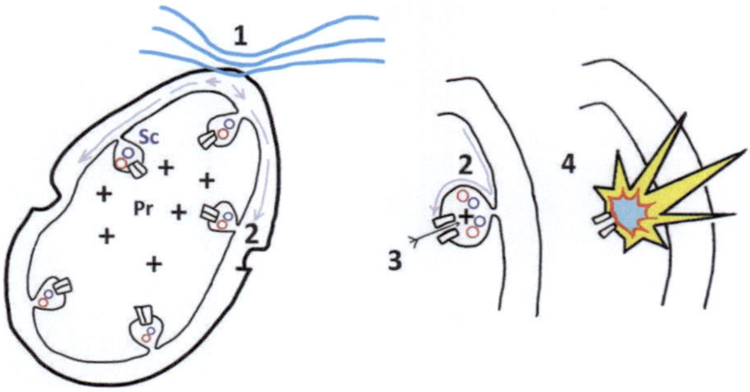

Fig. 7.9 Scheme of protonic switch-on of bioluminescence in Dinophyta. 1, wave pressure; 2, induced potential on internal membrane; 3, opening of Scintillon channels and entrance of protons in contact with luciferin–luciferase (circles of different colours inside each scintillon); 4, light emission from each scintillon. Pr, protons; Sc, scintillon (graphical composition A. M. Miglietta)

displays generated by breaking waves, swimming fish, or boats that excite cells to emit light. The wavelength of light emitted is approximately in the blue-green region of the visible spectrum (λ, approx. 450–520 nm). In most Dinophyta, bioluminescence is under **internal clock** control. At the end of the day, the luminescent molecules are packaged in vesicles (**scintillons**) that migrate into the cytoplasm.

An action potential is generated at the level of the internal membrane of scintillons. This action potential propagates throughout the cell, allowing protons to pass from the vacuole to the cytoplasm. This has the consequence of decreasing the pH of the cytoplasm and activating the chemo-luminescence in the scintillons (Fig. 7.9).

Many jellyfish are bioluminescent, especially comb jellies (Ctenophora), a taxon where more than 90% of planktonic species are known to produce light. Arguably, the most famous of all bioluminescent invertebrates is the Cnidarian ***Aequorea victoria***, which is the first species from which **GFP (green fluorescent protein)** was isolated, a discovery that went on to win the Nobel Prize (Fig. 7.10).

Aequorea victoria produces flashes of blue light through the quick release of Ca^{2+} ions, which interact with molecules of the photoprotein **aequorin**. The blue light is in turn transduced to green by GFP. Both aequorin and GFP became important in biological research (Shimomura 2005). In 1961, Shimomura and Johnson isolated aequorin and its small molecule cofactor coelenterazine from large numbers of *Aequorea* medusae at Friday Harbor Laboratories. In 1967, Ridgeway and Ashley microinjected **aequorin** into single muscle fibres of barnacles (Cirripedia) and observed transient Ca ion-dependent signals during muscle contraction. For his research into GFP, Osamu Shimomura was awarded the 2008 Nobel Prize for chemistry.

Bioluminescence is a form of intraspecific communication between animals, but it can also be used for defensive and offensive purposes (Table 7.3).

In **Ostracoda** (Fig. 7.11), a typical luciferin is present, different from those of Bacteria, Dinophyta, and Cnidaria, in addition to a kind of coelenterazine, and its use allowed a detailed study of **courtship behaviour** among these animals in the dark deep ocean (Morin 2019).

Fig. 7.10 *Aequorea victoria* (drawing A. Gennari)

Table 7.3 Different scopes of bioluminescence in marine animals

Category	Induced reaction	Taxa
Defence	Startle	Dinophyta, Cephalopoda
	Counterillumination	Many
	Misdirection (smoke screen)	Many
	Distraction (body pieces)	Cephalopoda, Polychaeta, Siphonophora
	Burglar alarm	Dinophyta, Cnidaria
	Attraction on sacrificial tag	Holothuroidea, Cnidaria, Polychaeta
	Warning advice	Cnidaria
Offense	Lure prey	Osteichthyes, Siphonophora
	Lure with external lights	Cetacea, Chondrichthyes
	Confuse movement in prey	Osteichthyes, Cephalopoda
	Illumination of the prey	Osteichthyes
	Reproductive mate attraction	Ostracoda, Osteichthyes, Polychaeta

Vargula hilgendorfii produces blue light. The luciferase enzyme consists of a 555-Aa-long peptide with a molecular mass of 61,627 u, while the vargulin (luciferin) has only a mass of 405.5 u. corresponding to **three Aa**: tryptophan, arginine, and isoleucine. The wavelength of the luminescence (448–452 nm) depends on the pH and salinity of the water, with the maximum being at 452 nm in seawater with a pH greater than 8 and a salinity of 35 psu. The reaction produces CO_2, oxyluciferin, and blue light. As an intermediate step, a 1,2-dioxetane ring is formed.

Approximately half of the known species Ostracoda Cypridinidae are luminescent, probably to deter predators, but field observations suggested that the

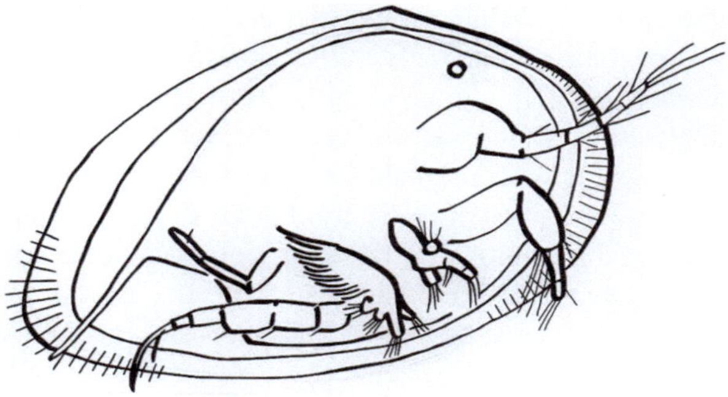

Fig. 7.11 Scheme of Ostracoda morphology. The right valve of the carapax has been deleted to show the internal specimen, with reduced segmentation and relative appendages (courtesy G. Belmonte)

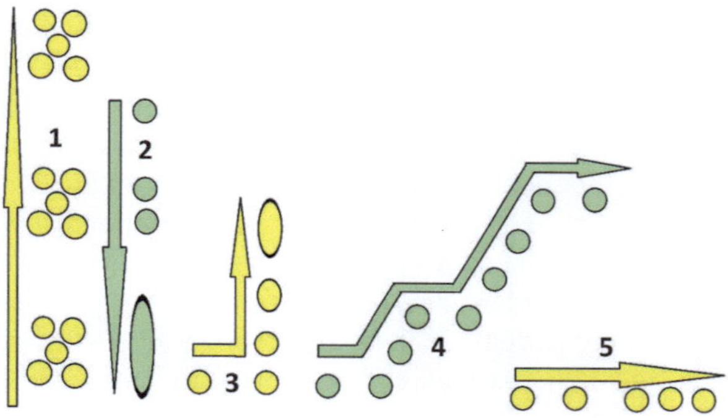

Fig. 7.12 Luminescent courtship displays (vertical upwards, vertical downwards, and lateral displays) of five Cypridinidae species at night over a shallow back reef near South Water Cay, Belize. Colours distinguish different displays. Circles and ellipses indicate the light spots and their duration (graphical composition A. M. Miglietta)

bioluminescence in Ostracoda might be more complex than a mere spewing light upon irritation; in fact, some of the luminescent species use their light in courtship displays. These courtship displays occur in a restricted geographic area (the Caribbean) and nowhere else. The initial discovery of these courtship displays occurred in 1980 (Fig. 7.12).

Only males produce displays to attract and guide sexually receptive females, which can emit light when disturbed but not during courtship displays. When sufficient ambient darkness is achieved, males enter the water column and repeatedly emit light pulses while swimming thus originating luminous rows, seeking to attract sexually receptive females, which target the light trail without luminescing, and swim to the male, whereupon they copulate with internal fertilization. The female then returns to the benthos and broods her fertilized eggs for the next few weeks, while the male continues his nightly quest, using his ritualized displays in attempts to attract and inseminate other receptive females.

Luminescent Cypridinidae occur in all oceans and at virtually all depths (surface to >4000 m), but only the Caribbean area has one monophyletic clade (more than 87 species), which has been found to also perform extraordinarily complex sexual courtship displays involving the ejection of pulses of light into the sea in near darkness. Depending on the species, males produce their displays above specific microhabitats within reefs and seagrass beds. Displays occur throughout the year as ambient light conditions drop below a critical near-dark threshold each evening, after sunset or moonset, whichever occurs later. The lone exception in the clade is a luminescent but nonsignaling species ***Vargula tsujii***, which occurs along the Pacific coast of North America from northern Baja California to central California. This peculiar distribution may represent a remnant of the signalling clade that was cut off during the uplift of the Isthmus of Panama 3.1–2.8 million years ago with subsequent loss of signalling by the Pacific species' ancestors. No signalling species have been found along the Pacific coast of the Isthmus of Panama or anywhere else in the world.

The displays themselves provide us with a kind of "fingerprint" that allows us to easily identify distinct species in the field. Within the signalling species, sexual selection for species-specific displays, along with restricted gene flow engendered by the limited dispersal capabilities of these Ostracoda, has almost certainly led to the evolution of a highly diversified species flock in the Caribbean.

Most luminescent Cypridinidae are small and mostly benthic, with adult males ranging from 1.2–2.3 mm and females ranging from approximately 1.6–3.0 mm. They have two large compound eyes, each with 16 *ommatidia*; the eyes of males are approximately 1.25 times larger than those of females. Except during their brief pelagic courtship display periods, individuals probably live within reefs, seagrass beds, or sediments near or below their courtship arenas. During their courtship periods, they enter the water column with their signalling grounds just above their reef or seagrass bed. Little is known about the feeding habits of species, but at least a few are scavengers, coming to traps baited with fish remains.

Signalling Ostracoda appear to mate and reproduce throughout the year and show no seasonality. Fertilization is internal, with sperm being transferred by a spermatophore (as usual for Crustacea). The male eighth limb functions as a copulatory organ, whereas in females, the eighth limb is reduced and includes a pair of knobs, probably offered for the grasping by males, and slits. The details of the transfer of sperm and accessory materials are largely unknown. The spermatophore appears to act as a mating plug and probably impedes other male copulations.

All offspring in one brood originate from the same father, and furthermore, females can store sperm and produce multiple broods from one single insemination. Females brood eggs into the dorsal camera (a kind of ephippium similar to that of Cladocera), where approximately 12 embryos are brooded. The mean sex ratio of each brood is mostly 1.0.

Gestation lasts for a total of approximately 3.5 weeks. The embryos develop into the mother's body for approximately 1 week before being positioned into the dorsal camera, where they spend an additional 2.5 weeks. At this point, the first instar offspring are released from such a marsupium as a crawl-away juvenile into the environment (Ostracoda have not the larva nauplius in the life cycle). New eggs develop in the female soon after the juvenile release. There is no pelagic larval stage, so dispersal is very restricted.

The luminescent reaction involves luciferin (**cypridinidine LH_2** = a tripeptide of tryptophan, isoleucine, and arginine, together forming a fused **imidazopyrazine** ring) + cypridinid luciferase (**cypridinid L-ase**), which is a protein, + O_2, in seawater. LH_2 and L-ase, along with mucus, are ejected separately through specialized nozzles from separate large secretory cells situated in the upper lip and **mix only when in sea water**. As a result, light is an **extracellular product** (and also external to the body, to be precise, in the present case), although molecules are released by cells in the body of the ostracod. The duration and intensity of luminescence vary depending on the relative proportions of LH_2 and L-ase, L-ase kinetics, and perhaps the relative amount of mucus. The light is blue, centred around λ_{max} of 462 nm, but such a value varies among species.

Predatory attacks, although rare, cause massive releases of luminescence from Ostracoda of either sex, all instars, and even late-stage embryos in the mother's dorsal camera. The light produced during predation events releases approximately fifty times more luminescence (as the total number of photons) than a courtship display. Fish (predators) usually immediately regurgitate ostracod prey when they release luminescence. The regurgitation behaviour suggests a startle response, unpalatability, and/or aposematism in response to light and ostracods.

Ostracoda appear to be unpalatable to the fish in all cases.

Courtship displays are a form of precopulatory mating selection. Sexually receptive females respond to species-specific male displays by swimming to intercept a signalling male in the water column but without luminescing themselves. The courtship arenas in the water column are equivalent to **leks** of terrestrial courtship behaviours.

When small plankton nets encounter an active display, usually only adult males and, at most, one receptive female are captured. This happens because mated males remain sexually active and continue to signal, but mated females do not continue to be sexually receptive and become benthic in short time, staring to brood eggs and embryos. As a result, because males remain signalling and mated females leave the courtship arena (lek), the operational sex ratio within the lek is highly skewed toward males even though the actual sex ratio in the population is near 1:1.

This sex ratio bias leads to strong **intrasexual competition**. Males produce various mating tactics, and even individual males will switch tactics during a single display event.

Massive numbers of displays occur in areas above a reef or seagrass bed due to hundreds of signalling males in each courtship arena at night. The highest abundances of luminous displays are observed in the absence of the moon in the sky. Displays take place just at the end of twilight during the 2 weeks following the full moon or later and later each night as the moon sets during the 2 weeks after the new moon. Initiation of displays is precise, within minutes, and appears to be driven by an **isolume** (particular ambient light intensity of the sky) as dusk or moonset occurs. Display initiation appears to be **consistently isolated** within species but varies slightly among species.

Displays consist of a series (**trains**) of light pulses, which together form coded luminescent spots, usually with decreasing pulse intensity and duration and decreasing interpulse intervals and distances from beginning to end. Light pulse trains are regularly repeated by males during the display period. Males swim rapidly (usually 8–11 cm s^{-1}) in a tight helicoidal pathway.

Each display is usually composed of two phases: **an initial** alerting or **species recognition phase** and a **final trill or tracking phase**. The initial (alerting/species recognition) phase usually consists of a few pulses that are brighter, of longer duration, and relatively widely spaced, while the secondary (trill/tracking) phase consists of many pulses that are dimmer, of shorter duration, and more regularly and closely spaced. The relative time allotted to the two phases often differs among species, with the initial phase dominating in some species, the trill phase dominating in others, and all situations in between being observed.

Displays vary among species. Train pulses can be directed vertical, oblique, or horizontal. Train lengths vary among species from only a few cm to many m. Train durations vary from short (< 5 sec) to long (nearly 1 min). Pulse numbers vary from one or two to hundreds per display. Displays can occur almost anywhere in the water column, from near the substratum to high in the water column, even up to the surface in shallow water. In any case, such pelagic behaviour is manifested by normally benthic species.

Shoals of pelagic luminescent *Cypridina* are known to be common in the northwestern Indian Ocean, including the Arabian Sea, the Malabar Coast of India, and the Maldives. The first published reports made the following observations between the Malabar Coast and the Maldives (ca. 8°47′N, 75°20′E) in 1754: "… [the sea] was covered over with small stars; every wave which broke about us dispersed a most vivid light, in complexion like that of a silver tissue electrified in the dark" (Morin 2019).

The most famous of all luminescent Ostracoda is *Vargula* (formerly *Cypridina*) *hilgendorfii* from shallow waters of Japan. It is locally known as ***umi-hotaru***, which translates to "sea-firefly".

The species was well known because specimens could be readily caught in substantial numbers in shallow water on fish carcasses placed on the sea floor.

Bioluminescence is an excellent, non-invasive indicator of chemical reaction occurrence. Harvey began his research work on bioluminescence of Coleoptera Lampiridae (fireflies) and Bacteria, but that changed abruptly when he travelled to Japan on his honeymoon in 1916 and met *umi-hotaru* (*C. hilgendorfii*) at the Misaki Marine Laboratory near Tokyo. This ostracod produces bluish luminescence when disturbed, and it was known in Japan that if they were dried under the sun, their luminescence would return simply by adding water. If the dried Ostracoda were kept free of moisture with a desiccant, the luminescence would be preserved for indefinite periods (decades!).

The first article by Harvey (1917) on the chemistry of light production by *umi-hotaru* came from that first visit to Japan. He returned to Japan and, critically, managed to set in place a procedure to receive regular shipments of dried *umi-hotaru* from Japan to Princeton. Harvey, in parallel with his ostracod research, also continued to study Bacteria, Lampiridae, Pennatulacea, Ctenophora, Hydromedusae, Pisces, Decapoda, and Polychaeta, and he wrote a compendious book *Bioluminescence* (Harvey 1952), which reviewed the known literature on bioluminescent species.

Harvey showed that the bioluminescent reaction was highly specific to Cypridinidae and that neither the luciferase nor the luciferin would cross-react with molecules of other luminescent taxa. The final structure of cypridinid luciferin was reported by Shimomura and associates, and the purification, characterization, cDNA cloning, and sequencing of cypridinid luciferase were accomplished between the early 1950s and 1989 by a number of researchers.

Shimomura et al. (1957), working in the laboratory of Yoshimasa Hirata at Nagoya University, Japan, produced the first crystals of pure cypridinid luciferin. They showed that **cypridinid luciferin** is a tripeptide with an imidazopyrazine ring). Shimomura never worked directly with Harvey, but he joined the laboratory of Frank Johnson for 3 years at Princeton in 1960–1963, where his focus shifted to the aequorin/GFP system in Cnidaria and ultimately earned him the Nobel Prize in Physiology or Medicine in 2008.

Cypridinid luciferase is a monooxygenase formed by a single acidic polypeptide chain of 555 Aa, with a molecular weight of approximately 62 kDa.

The cypridinid luminescent reaction involves only the substrate, the enzyme, and O_2 in a three-step process whereby (1) the luciferin, bound to the luciferase, is oxygenated to form a cyclic peroxide (dioxetanone), which (2) decomposes to oxyluciferin in an excited state (indicated by the asterisk in the reaction below) and releases CO_2, and (3) the excited electron in the oxyluciferin then emits a photon of blue light when it falls back to its ground state:

$$\text{Cypridinid luciferase} + \text{cypridinid luciferin} + O_2$$

$$\text{Cypridinid luciferase} - \text{cypridinid luciferin} - \text{dioxetanone intermediate}$$

$$\text{Cypridinid luciferase} - \text{cypridinid oxyluciferin}^{*} + CO_2$$

$$\text{Cypridinid luciferase} + \text{cypridinid oxyluciferin} + h\upsilon\left(\lambda_{max} \sim 462\,\text{nm}\right)$$

There are two other noteworthy light-emitting relationships involving **Ostracoda**. The first system involves marine fishes. While the cypridinid luciferin–luciferase system is unique to **Cypridinidae** Ostracoda, at least three groups of fishes (species of *Porichthys* of **Batrachoid**idae, *Apogon* of **Apogonidae**, and *Parapriacanthus* of **Pempheridae**) utilize cypridinid luciferin to emit their own light, but in conjunction with different oxygenases. These fishes most likely obtain the substrate from their ostracod prey by ingestion (they evolved the insensibility to ostracod luminescence and did not regurgitate their prey). The details of the respective enzymes in comparison to those of Cyprinidae are unknown and should be explored (Fig. 7.13).

A second system involves another group of Myodocopidae Ostracoda, the pelagic **Halocypridae**, which are also bioluminescent, but the reaction is **intracellular**, is of a different colour (green), and light emission occurs by a very different biochemical mechanism involving different enzymes and substrates. This system

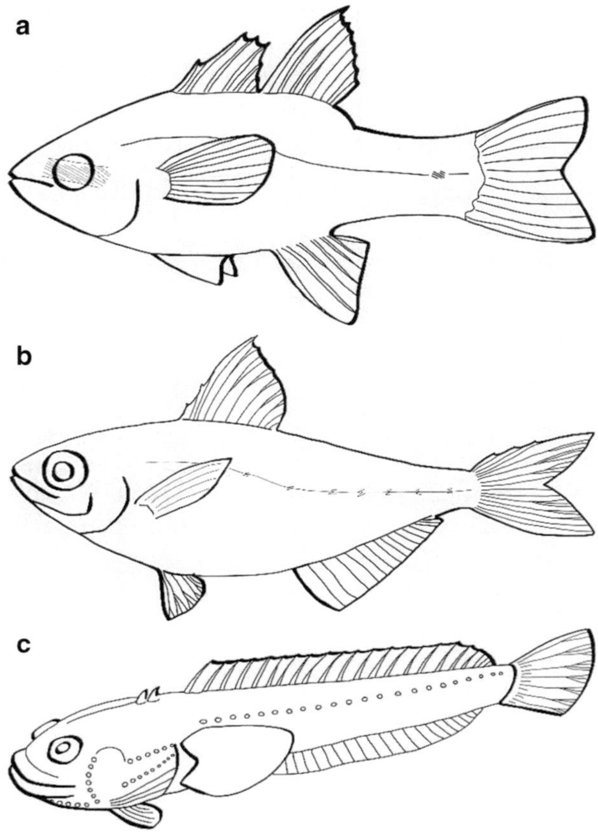

Fig. 7.13 (**a**) *Apogon*; (**b**) *Porichthys*; (**c**) *Parapriacanthus* (courtesy G. Belmonte)

represents an independent evolution of light emission in the same taxon (Ostracoda). Halocypridae use the widespread luciferin molecule **coelenterazine**, which, similar to cypridinid luciferin, also has a fused imidazopyrazine ring core but is composed of three distinctly different amino acids (two tyrosines and one phenylalanine) and clearly has a different evolutionary origin from cypridinid luciferin.

The entire family Cypridinidae now contains over **250** described species distributed in 30 genera. Approximately half the species are known to be luminescent, and the Caribbean species contain approximately half of the known luminescent species (most of which are known but yet undescribed).

Many species in the signalling clade are very difficult to distinguish from one another on a morphological basis but are relatively easy to separate based on their courtship behaviour and associated light emission.

Prior to 1980, only four species of luminescent Ostracoda, all assigned to the genus *Vargula,* were known from the Caribbean. These are currently designated *Photeros parasitica* and *P. harveyi* from Jamaica, *Kornickeria bullae* from moderate depths off the U.S. Virgin Islands, and *Vargula magna* from deep water (160–202 m) on the continental shelves of North Carolina and West Florida.

References

Alldredge AL, King JM (1985) The distance demersal zooplankton migrate above the benthos: implications for predation. Mar Biol 84:253–260

Arkett SA (1984) Diel vertical migration and feeding behavior of a demersal hydromedusan *(Polyorchis penicillatus)*. Can J Fish Aquat Sci 41:1837–1843

Armonies W (1989) Meiofauna emergence from intertidal sediments measured in the field: significant contribution to nocturnal planktonic biomass in shallow waters. Helgolander Meeresuntersuchungen 43:29–43

Bieri R, Tokioka T (1968) Dragonet II: an opening–closing quantitative trawl for the study of microvertical distribution of zooplankton and the meio-epibenthos, vol 15. Publications of Seto Marine Biological Laboratory, pp 373–390

Cahoon LB, Tronzo CR (1992) Quantitative estimates of demersal zooplankton abundance in Onslow Bay, North Carolina, USA. Mar Ecol Prog Ser 87:197–200

Carleton JH, Hamner WM (1989) Resident mysids: community structure, abundance and small-scale distributions in a coral reef lagoon. Mar Biol 102:461–472

Childress J, Gluck DL, Camey RS, Gowing MM (1989) Benthopelagic biomass distribution and oxygen consumption in a deep-sea benthic boundary layer dominated by gelatinous organisms. Limnol Oceanogr 34:913–930

Clutter RI (1969) The microdistribution and social behavior of some pelagic mysid shrimps. J Exp Mar Biol Ecol 3:125–155

Comet M, Bouchet J-M, Lissalde J-P, Sorbe J-C (1983) Donnees qualitatives sur le benthos et le suprabenthos d'un transect du plateau continental Sud-Gasconge. Cah Biol Mar 24:69–84

Ghabru A, Rana N, Verma G (2023) Bioluminesence: "The Living Light". In: Taria S (ed) Recent advances in molecular biology and Plant physiology, vol 4. AkiNik Publications, New Delhi, pp 204–227

Haddock SHD, Moline MA, Case JF (2010) Bioluminescence in the sea. Annu Rev Mar Sci 2:443–493

Harvey EN (1917) Studies on bioluminescence. IV. The chemistry of light production in a Japanese ostracod crustacean, *Cypridina hilgendorfii*. Muller Am J Physiol 42:318–341

Harvey EN (1952) Bioluminescence. Academic Press, New York, p 649

Huberdeau L, Brunel P (1982) Efficacite et selectivite faunistique comparee de quatre appareils de prelevement endo-, epi-et supra-benthiques sur deux types de fonds. Mar Biol 69:331–343

Jacoby CA, Greenwood JG (1988) Spatial, temporal, and behavioral patterns in emergence of zooplankton in the lagoon of heron reef, great barrier reef, Australia. Mar Biol 97:309–328

Kaartvedt S (1985) Diel changes in small-scale vertical distribution of hyperbenthic mysids. Sarsia 70:287–295

Kawaguchi K, Matsuda O, Ishikawa S, Naito Y (1986) A light trap to collect krill and other micronektonic and planktonic animals under the Antarctic coastal fast ice. Polar Biol 6:37–42

Larson RJ, Matsumoto GI, Madin LP, Lewis LM (1992) Deep-sea benthic and benthopelagic medusae: recent observations from submersibles and a remotely operated vehicle. Bull Mar Sci 51:277–286

Lau ES, Oakley TH (2021) Multilevel convergence of complex traits and the evolution of bioluminescence. Biol Rev 96:673–691

Mauchline J (1988) The biology of mysids and euphausiids. Adv Mar Biol 18:1–680

McFall-Ngai MJ, Montgomery MK (1990) The anatomy and morphology of the adult bacterial light organ of *Euprymna scolopes* berry (Cephalopoda:Sepiolidae). Biol Bull 179:332–339

Mees J, Jones MB (1997) The Hyperbenthos. Oceanogr Mar Biol Annu Rev 35:221–255

Morin JG (2019) Luminaries of the reef: The history of luminescent ostracods and their courtship displays in the Caribbean. J Crustac Biol 39(3):227–243

Omori M (1969) A bottom-net to collect zooplankton living close to the sea-floor. J Oceanograp Soc Japan 25:291–294

Palmer MA (1988) Dispersal of marine meiofauna : a review and conceptual model explaining passive transport and active emergence with implications for recruitment. Mar Ecol Prog Ser 48:81–91

Royet J, Gupta D, Dziarsky R (2011) Peptidoglycan recognition proteins: modulators of the microbiome and inflammation. Nat Rev Immunol 11:837

Sherman KM, Coull BC (1980) The response of meiofauna to sediment disturbance. J Exp Mar Biol Ecol 46:59–71

Shimomura O (2005) The discovery of aequorin and green fluorescent protein. J Microsc 217:3–15

Shimomura O, Goto T, Hirata Y (1957) Crystalline Cypridina luciferin. Bull Chem Soc Jpn 30:929–933

Wittmann KJ (1977) Modification of association and swarming in north Adriatic Mysidacea in relation to habitat and interacting species. In: Keegan BK et al (eds) Biology of benthic organisms. Pergamon Press, Oxford, pp 605–612

8 Diving of Lunged Animals

8.1 Apnoea in Reptilia and Aves

Among the animals that need to breathe air, although living in the water, the most adapted can be considered Reptilia (snakes and turtles) (Seymour 1982). Marine snakes are typical of the Indo-West Pacific area, plus only one species of the East Pacific. They have a posterior part of the body horizontally flattened (**compressed**) and swim undulating the whole body (Fig. 6.23). They are known to be particularly poisonous, but they are also interesting for their exceptional ability to avoid breathing air for long periods of time. This possibility is due to the development of an under skin capillary net in the area of the head, which can exchange respiratory gases between blood and water.

Additionally, some turtles have been found to be able to use the skin capillaries (of the mouth, in this case) to exchange respiratory gases directly with water. Both of these solutions bypass the problems derived from a lung apparatus that can exchange respiratory gases only with air and support extralong dives of the specimens.

This is not the case for birds (Aves), which remain dependent on lung-based respiration. Many species thrive their lives at sea and reach the shore only for reproduction. Some of these species are flying birds that dive down to some metres of depth to catch their food (fish).

However, an entire family, **Spheniscidae** (penguins) (Fig. 8.1), lost the capacity to fly and evolved to be more adapted to aquatic life. The necessity to catch fish, however, also obliges penguins to manage their lives by switching between air breathing and apnoea.

Penguins have a vision that is believed to be better underwater than in the air. Their wings are used no longer for flying but to help them swim and dive into the water. For this purpose, penguins' wing bones are fused straight rather than angled like those of flying birds; hence, they are rigid and powerful to move away water and not air. In contrast to other (flying) birds, penguins have solid bones, which are an important ballast in sinking down, and do not suffer any crash or fracture

G. Belmonte, *Elements of Pelagos Biology*,
https://doi.org/10.1007/978-3-031-74280-4_8

Fig. 8.1 Penguins; extinct (**a**–**c**) and still living (**d**–**f**) species. (Courtesy G. Belmonte)

determined by pressure when they stay at more than 100 m of depth (as in many cases it has been demonstrated) (Ponganis 2015).

The Oxygen reserve for long-lasting dives is in the high contents of myoglobin in the muscular apparatus (just where the Oxygen is needed by aerobic production of energy).

In any case, a life adapted to oceans and completely disconnected from any terrestrial link (as in the case of **Mammalia Cetacea**), notwithstanding the problematic necessity of breathing air, requires additional evolutionary solutions. Fundamentally, lunged animals have to resolve a series of problem categories: (a) where to find Oxygen for muscles when obliged to stop breathing; (b) how to manage high external pressures generated by depths; and (c) how to correct optical aberrations deriving from underwater vision.

Among subpolar birds, a third additional problem could be the water temperature (approximately 0–2 °C throughout the year). Aves, in fact, are warm-blooded organisms (with an internal temperature of 37–38 °C) and have to manage temperature homeostasis either when in rest (avoiding the cooling of the body) or when in activity (cooling the muscular machine producing energy and heat). To keep the body warm, they use an apparatus of very efficient feathers, short and packed together tightly, to exclude water from the skin and to create a smooth surface to lower drag when at sea. The feathers are coated with oil from a gland near the tail to increase the “waterproof” factor. This waterproof layer all around the body avoids cooling of the body and simultaneously allows better buoyancy.

Penguins, however, can raise their feathers to allow water to contact the skin and subtract heat when in excess (e.g., during a swimming activity to prosecute prey). This ability allows penguins to perform dives without energy expenditure to overcome the buoyancy and the correct dispersion of excessive heat produced during muscular activity (Kooyman et al. 1992).

Numerous tiny blood capillaries close to the skin surface on their wings also allow extra heat to escape by simply opening their wings. Penguins also release heat through their feet, which are not covered by feathers. According to the necessity of the organism, the blood vessels of these uncovered parts can be replenished with blood (in search of low temperatures to be distributed to the remaining part of the body) or, alternatively, can be restricted to diminish the blood presence and its cooling.

8.2 Apnoea in Marine Mammalia

As marine Reptilia and Aves, Mammalia also evolved adaptations for a marine lifestyle with the constraint of the use of atmospheric Oxygen for biochemical respiration of energetic molecules (Davis 2019). Due to this constraint, and more or less as in the case of reptiles and birds, marine mammals have to resolve two large categories of problems: (a) those deriving from the impossibility of breathing air when diving, and (b) those deriving from the alteration of physical properties of the environment increasing with depth (e.g., darkness and pressure).

Even some specimens of not-well-adapted species (such as cats, bears, and humans) can offer some examples. However, these examples are generally considered exceptions, and in the case of humans, they are considered useful for physiological and/or psychological studies. Champions in this field pass −110 m in apnoea and are able to breathe hold for up to 18 minutes.

Adaptations to diving of air-breathing marine mammals can be grouped into two categories (Williams and Ponganis 2021):

1. Adaptations to pressure.
2. Adaptations to breathe-hold.

Pressure problems deal with mechanical effects and the increased solubility of gas. One mechanical effect of increasing external pressure (directly proportional to the depth) is the collapse of air-filled spaces in the body, such as the lungs. For example, a 10 L lung at the surface (atmosphere pressure, 1 atm) has a volume of 5 L at 10 m below sea level (2 atm) and of <1 L at 100 m depth (11 atm and a volume of 1/11 of its original size). This notwithstanding, marine mammals (seals, dolphins, and whales) did not adopt an evolutionary reduction of their lungs and, on the contrary, are structured to perform highly efficient air-to-blood exchanges of respiratory gases. Air breathed in a single complete breathing act accounts for less than 30% of lung volumes in terrestrial mammals and more than 80% in dolphins (Fig. 8.2).

In marine mammals, **lung alveoli** collapse first during an increase in pressure. This collapse at the level of the district where Oxygen and CO_2 are exchanged between air and blood impedes the increase in gas solubility in the blood (determined by the pressure increase). Alveoli may also aid in lung reinflation due to specialized molecules (surfactants) when "normal" pressure is reestablished.

Another air-filled cavity in mammals is at the level of the middle ear, separated from the exterior environment by a tiny membrane (**eardrum**), which is particularly endangered in conditions under increasing pressure. Additionally, the middle ear cavity needs to be eliminated or filled during a dive, and this happens when the cavernous sinuses at the level of the middle ear walls engorge blood. The engorgement produces a diminution and finally the filling in of the air space posterior to the eardrum. This eliminates the risk of membrane laceration (Fig. 8.3).

The pressure can also directly affect nervous elements (cells, organs, and cell membranes), although in a not immediately perceivable mode or rarely perceived, because it is an issue (high-pressure nervous syndrome, **HPNS**) at very high depths (500–1000 m) and pressures.

The human nervous system makes free tremors at −150 m and convulsions at −500 m. Marine mammals and seabirds, however, do not experience HPNS, even if they are able to reach such depths. Elephant seals, sperm whales, and beaked whales routinely dive to depths of 500–1000 m (Tyack et al. 2006). There are no data on whether these animals do experimental HPNS, how they tolerate it, or how they avoid it. The range of dive depth for such species is that of normal behaviour. They do not dive to those depths to explore their dive limits (as humans do), but simply they dive regularly to 500–1000 m (and even more), probably to routinely find food. Humans do not perceive the abundance of fish at those depths (the fishery is not technologically equipped to reach those depths), but the evolution of the physiology

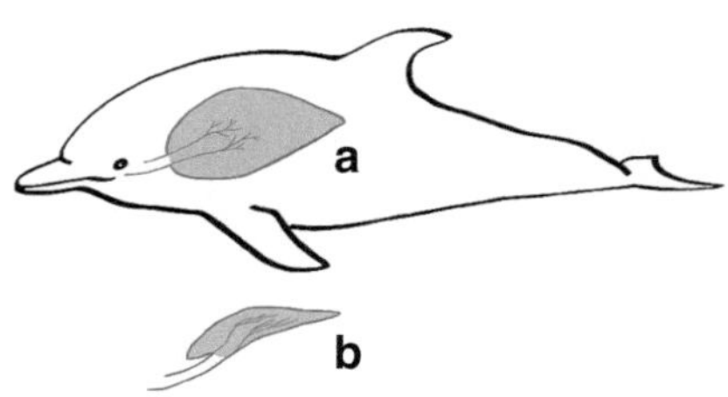

Fig. 8.2 Lung volume reduction with depth. (**a**) Volume at surface; (**b**) volume at depth of 50 m. (Courtesy G. Belmonte)

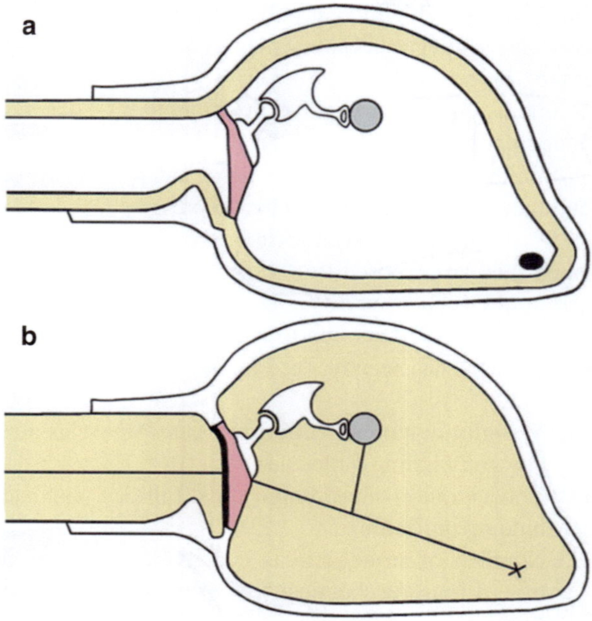

Fig. 8.3 Replenishment of the space for protection of the eardrum under high external pressure. Eardrum is coloured in pink. Beige is the colour of the skin of the external and median ear. (**a**) Subaerial condition; (**b**) underwater condition, with inflation by blood engorgement (hyperemia), of the skin to protect the eardrum. (Graphical composition A. M. Miglietta)

of large marine carnivores to perform such a kind of dive indirectly suggests that the bathypelagos have to be rich in prey.

Problems connected with breath hold require solutions at the level of metabolism, blood flow, and oxygen storage capacity.

At high partial pressures, increased concentrations of N_2 cause states of euphoria and delusion (**Nitrogen narcosis**) in humans. O_2 is toxic (produces blackout and death) at partial pressures greater than 1 atm (= −10 m).

When a scuba diver, breathing air from a bottle, returns to the sea surface, under decreasing pressure, the compressed air in the tissues (due to the high pressure) flows out of the tissues into the blood. If the diver does not give the gas time to slowly come out (decompress), the gas will form bubbles.

Marine mammals, unlike human divers, do not have bottles or other gas reservoirs and hence exclusively breath-hold dive. Some of them allow their lungs to collapse during the dive. As the lung collapses, air passes from the lung volume into the higher respiratory pathways (large bronchioles and trachea). At this level, the passage of gas in tissues cannot occur because the walls of the airways are impermeable to such a passage.

The decompression sickness **DCS**, however, has been reported in human pearl apnoea divers undergoing prolonged repetitive breath-hold dives.

Penguins, sea lions, and fur seals repetitively dive (in apnoea) over periods of many hours or days.

It is unclear how they avoid decompression sickness.

Among the repetitively diving animals, only *Tursiops truncatus* hosts relatively high levels of N_2 in muscle tissues, but it is unclear whether the levels of muscle N_2

are sufficiently high to cause bubble formation in the blood, whether there are any mechanisms to keep N_2 in the tissue, or whether there is any tolerance to N_2 bubbles if they form.

Interestingly, histological observations of bone necrosis in sperm whales are consistent with symptoms of DCS.

Similar observations of mosasaurs and plesiosaurs (extinct Reptilia) also suggest that they could have suffered decompression sickness.

Diving mammals are probably more susceptible to DCS (the so-called *bends*) when they are exposed to military sonars. The sonar stimulates bubble formation and/or elicits aberrant diving behaviour that results in acute and severe DCS.

In summary, adaptation to deep dives requires a powerful capacity for Oxygen storage. This has been reached by increasing:

(a) **Myoglobin concentration** in tissues (muscles are particularly rich in such an Oxygen-binding molecule).
(b) Concentration of **red blood cells** (fulfilled with haemoglobin, another Oxygen-binding molecule).
(c) Numbers of **mitochondria** in cells (where to use Oxygen as a receptor of electrons during the chemical process of energy storage in complex molecules), and.
(d) **Energy conservation** strategies.

Apart from the above-listed "increases" in physiological powers (in comparison with the physiology of terrestrial cognate *taxa*), a generally adopted strategy is the **reduction of capillary nets** that could be obstructed by denser blood (because they contain more red blood cells than terrestrial species).

Weddell seals save energy and Oxygen primarily by gliding rather than swimming during dives.

Partial pressures of Oxygen in the blood <30 mmHg (which cause blackouts to humans) do not cause problems for seals because they have much higher levels of haemoglobin in their blood.

Nearly all of the blood in Weddel seals is supplied to the brain, spinal cord, eyes, and adrenal glands. The blood supply to the muscles is limited or halted during long dives.

A seal's core body temperature is approximately 38 °C, and this temperature must be maintained even in polar waters (close to 0 °C) or on marine ice (largely lower than 0 °C). Seals can skip the capillary bed entirely (a potential problem for very dense blood), but they have the necessity to cool their body to avoid extra warmth. For this purpose, they can dilate special blood vessels that are near the surface of the skin, allowing the warm blood to reach the surface for cooling. The return of cooled blood to the internal body allows the extraction of heat produced during muscular contraction and/or metabolism (Table 8.1).

It is well known (Kooyman et al. 1976) that seals reduce the beat heart rate from 60–70 bpm to 15 bpm during a long dive (**Bradycardia**). If a seal runs out of O_2, it then converts glucose to lactic acid (switching glycolysis from the **aerobic** pathway to the **anaerobic** pathway or from respiration to fermentation). Weddell's and other

Table 8.1 Erythrocyte abundance in total blood volume (haematocrit) and blood volume in total body mass for different terrestrial and marine species

Haematocrit	Humans	41–50% (males)
		36–44% (females)
	Diving mammals	50–68%
Blood volume	Humans	7% of body mass
	Diving mammals	10–20% of body mass

seals even have **extra-big spleens** to store **erythrocytes** and release them during an Oxygen demand during a dive.

Bradycardia is in turn associated with an overall reduction in metabolism as the work of many organs is slowed. Keeping the blood pressure constant while the heart pump action is reduced requires a new adaptation at the level of the aorta, which, in these animals, is highly elastic.

The highly elastic aorta of seals stores part of the kinetic energy present in the blood at each heart pumping act and releases it over the intervening period, thus keeping the blood pressure constant at peripheric districts.

Seals do not take an enormous breath, but they do **hyperventilate** before a dive.

They store the Oxygen coming from hyperventilation in the Haemoglobin of their blood and in the Myoglobin of their muscles, and they expel the air from the lungs before the dive starts.

Seals have more blood than land animals of a similar size and more Haemoglobin to carry Oxygen also associated with a greater abundance of erythrocytes in the blood.

The muscles of diving vertebrates have greater concentrations of myoglobin (10–30 times greater than in their terrestrial relatives).

The highest values belong to longer, deeper-diving mammals such as elephant seals and sperm whales.

A diver lifestyle does not require only adaptations to pressure increases and Oxygen uptake. Apart from the anatomical modifications to allow a general aquatic lifestyle (fins, hydrodynamic bodies), some solutions evolved to help deep dives in the dark (Davis and Williams 2012).

Large eyes in seals are an underwater adaptation. **A flattened cornea** is associated with a spherical crystalline lens to correct vision alterations occurring when a mammal eye (with a hemispheric cornea and a flattened, lenticular crystalline lens) is immersed in sea water (Fig. 8.4).

In dolphins, the marine species have a generally flattened eye, which produces a flattening of the cornea and of the crystalline lens, but in freshwater species (such as the dolphin beluga), the lens is lenticular (flattened), and the cornea is flattened, meaning that the cornea corrects problems of light routes in the water (both fresh and marine), while the shape of the lens is used to correct the additional disturbance provoked by salt content and increased density of marine water.

Pupils of seals' eyes can open wide to let in more light while diving. Finally, to increase the vision efficacy in the dark of deep pelagos, seal retina receptors are only rods that work well in low light conditions (no cones to detect colours).

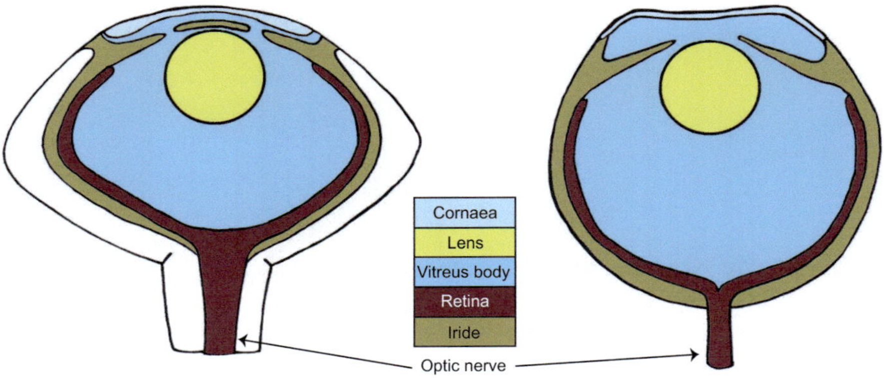

Fig. 8.4 Eye sections of marine mammals. Left, dolphin; right, fur seal. It should be noted that the lens is spherical, and the cornea has a flattened curvature, contrary to Mammalia living in subaerial conditions. (Graphical composition A. M. Miglietta)

Most likely, the same necessity to perceive objects (as food) in the dark pushed the evolution of the **melon organ** in dolphins. The melon is an ovoid-shaped fatty organ found in the forehead of all toothed cetaceans (Odontoceti) and is believed to be used in **echolocation**. The dolphin is able to generate sound in the form of clicks within the nasal sacs situated behind the melon. When the sound (produced in nasal sacks and amplified by the melon) goes outside and strikes an object, some of the soundwave energy is reflected back and perceived by the same emitting subject (Fig. 8.5). The system has evolved to echolocate fish prey. In fact, most fish of the epi and mesopelagos are equipped with hydrostatic gas bladders, and just such a cave organ (filled with gas) is responsible for the echolocation answer to the echo sound sent by dolphins all around.

Whatever the performance of diving mammals, they generally dive within their normal limits and do not approach extreme performances if not for emergencies.

They do not go deeper than where the prey are (generally, less than 200 m) but can exaggerate their diving if requested by feeding necessity.

Generally, deep dives require more time than shallow dives. This notwithstanding, however, the longest dives are usually relatively shallow, as the animal swims as little as possible, often passively sinking (to preserve energy). This probably reduces metabolism to the greatest extent possible to obtain the longest dive possible. (Table 8.2).

Whales of the family Balaenopteridae feed by engulfing entire schools of (small) prey. The whale then forces the water through its baleen plates, retaining thousands of small organisms (generally, krill) in its mouth. Such a feeding habit requires increased effort to swim with an open mouth through the water. Given their feeding behaviour, blue whales and their kin have a very costly method of feeding that consumes Oxygen rapidly. This condition imposes shallow and short dives compared with other Cetacea. In the end, the largest cetacean is not a better diver of its correlatives. On the other side of feeding behaviour, the sperm whale (*Physeter macrocephalus*) is a carnivore (the largest predator on the planet) and needs to perform

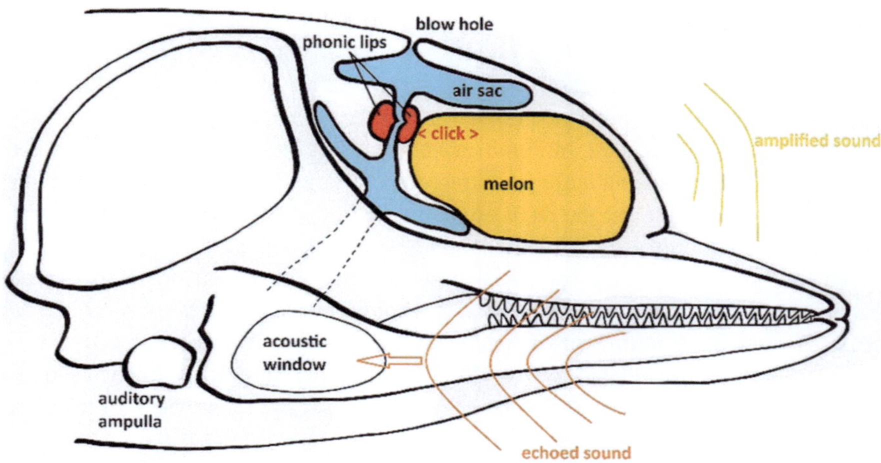

Fig. 8.5 Scheme of echo sound emission and perception by carnivore Cetacea. The sound is emitted by an apparatus equipped with phonic lips and melon and is perceived by a fat organ (acoustic window) in the inferior jaw of the animal, close to the acoustic apparatus (auditory ampulla). (Graphical composition A. M. Miglietta)

Table 8.2 Maximum depth reached by lunged species (only >300 m are considered)

300 m	Spotted seal (*Phoca largha*)
304 m	Leopard seal (*Hydruga leptonyx*)
324 m	Bajkal seal (*Pusa sibirica*)
325 m	King penguin (*Aptenodytes patagonicus*)
436 m	Grey seal (*Halichoerus grypus*)
452 m	Steller Sea lion (*Eumetopias jubatus*)
480 m	Bearded seal (*Erignathus barbatus*)
481 m	Harbour seal (*Phoca vitulina*)
500 m	Ringed seal (*Pusa hispida*)
	Walrus (*Odobenus rosmarus*)
536 m	California Sea lion (*Zalophus californianus*)
565 m	Hawaiian monk seal (*Neomonachus schauinslandi*)
	Emperor penguin (*Aptenodytes forsteri*)
568 m	Harp seal (*Pagophilus groenlandicus*)
597 m	New Zealand Sea lion (*Phocarctos hookeri*)
600 m	Ribbon seal (*Histriophoca fasciata*)
713 m	Crabeater seal (*Lobodon carcinophaga*)
792 m	Ross seal (*Ommatophoca rossii*)
904 m	Weddel seal (*Leptonychotes weddellii*)
1000 m	Pilot whale (*Globicephalus* sp.)
1016 m	Hooded seal (*Crystophora cristata*)
1280 m	Leatherback Sea turtle (*Dermochelis coriacea*)
1453 m	Northern bottlenose whale (*Hyperoodon ampullatus*)
1599 m	Blainville's beaked whale (*Mesoplodon densirostris*)
1733 m	Northern elephant seal (*Mirounga angustirostris*)
1777 m	Baird's beaked whale (*Berardius bairdii*)
2388 m	Southern elephant seal (*Mirounga leonina*)
2992 m	Cuvier's beaked whale (*Ziphius cavirostris*)
3000 m	Sperm whale (*Physeter macrocephalus*)

dives of even 2000 m. Krill is not its food, and fish of those depths have no gas bladder, which can betray their presence to an echolocation apparatus. Sperm whales, thus, do not need echolocation or large swarms of shrimp. They, however, maintain a good visual perception (even in the dark), evidently because they individuate, in the absolute dark, bioluminescent prey. It is thought that the most important prey source for such a large predator could be large and abundant, and the best candidates for this role are the large or giant squids, fast swimmers, school forming organisms, and bioluminescent inhabitants of dark, deep oceans.

The same reasons agree for a diet based on bioluminescent organisms of the bathypelagos for Cuvier's beaked whales (*Ziphius cavirostris*), because this cetacean performs dives until approximately 3000 m of depth (Schorr et al. 2014).

Dive depth is not related to dive duration or to the body mass of the diver. In analysing the diving of many species (Williams et al. 2011), we have to take into account the following:

1. Some animals exist that are simply poor divers. A possible example is the bearded seal, *Erignathus barbatus*, which weighs 350 kg but dives an average of 17 m in shallow Arctic waters and remains submerged for only 120 s.
2. Species equipped with larger spleens or larger blood volumes for their size tend to dive more deeply and for longer.
3. It is possible that the diving ability of certain species is simply unknown.

Some species may work well within their physiological limits because food is relatively easy to obtain or there is little competition for food.

The amount of time a species spends recovering and/or preparing for a dive depends not only on the time spent under water but also on the depth of the dive. This rule implies spending more time at the surface after deep or long dives to restore metabolic equilibrium and avoid stress generated by repetitive dives.

The metabolic cost of a dive also depends on air volume within the body and lipid presence among tissues because these two elements determine buoyancy and thus the energy costs of the different phases of a dive.

For positively buoyant animals, the descent phase of a dive is the most expensive.

However, the deeper these animals descend, the less buoyant they become owing to the compression of the internal air or gas.

Some animals adjust the Oxygen storage in their lungs depending on the duration of the dive.

They can affect their buoyancy and hence the cost incurred by it throughout the dive.

A deep dive is more expensive than a shallow one *for a given dive duration*. Consequently, divers tend to dive to a shallower possible depth on the basis of the trade-off between the rate of prey capture and the increased metabolic expense of deeper dives.

This agrees with the model of optimal foraging depth: "diving animals choose to forage at a depth that is always shallower than the depth at which prey abundance is highest".

Even if that species is physiologically capable of reaching the depth of maximum prey density, its diving behaviour might limit dive depth further to optimize foraging efficiency.

References

Davis RW (2019) Marine mammals: adaptations for an aquatic life. Springer Nature, Berlin

Davis RW, Williams TM (2012) The marine mammal dive response is exercise modulated to maximize aerobic dive duration. J Comp Physiol A Neuroethol Sens Neural Behav Physiol 198:583–591

Kooyman G, Gentry R, Urquhart D (1976) Northern fur seal diving behavior: a new approach to its study. Science 193:411–412

Kooyman GL, Ponganis PJ, Castellini MA, Ponganis EP, Ponganis KV, Thorson PH, Eckert SA, Le Maho Y (1992) Heart rates and swim speeds of emperor penguins diving under sea ice. J Exp Biol 165:161–180

Ponganis PJ (2015) Diving physiology of marine mammals and seabirds. Cambridge University Press, Cambridge

Schorr GS, Falcone EA, Moretti DJ, Andrews RD (2014) First long-term behavioral records from Cuvier's beaked whales (*Ziphius cavirostris*) reveal record-breaking dives. PLoS One 9:e92633

Seymour RS (1982) Physiological adaptations to aquatic life. In: Gans C, Pouch FH (eds) Biology of the Reptilia, physiological ecology, vol 13. Academic Press, Cambridge, p 51

Tyack PL, Johnson M, Soto NA, Sturlese A, Madsen PT (2006) Extreme diving of beaked whales. J Exp Biol 209:4238–4253

Williams CL, Ponganis PJ (2021) Diving physiology of marine mammals and birds: the development of biologging techniques. Philos Trans R Soc B 376:20200211

Williams CL, Meir JU, Ponganis PJ (2011) What triggers the aerobic dive limit? Patterns of muscle oxygen depletion during dives of emperor penguins. J Exp Biol 214:1802–1812

Human Interactions with Pelagos

9

9.1 Disturbance on a Small Scale

Echo sound is produced by military and/or research devices for exploratory purposes. Sound, under the name of noise, however, is produced by any kind of engine and human activity at sea.

The sound emitted by human activities in submarine waters is considered ocean **noise** and can interfere with, or impede the ability of some marine animals to hear natural sounds (Fig. 9.1).

Over the past couple of centuries, humans have progressively operated to alter the natural soundtrack of the ocean with shipping, industrial fishing, coastal building, oil drilling, seismic surveys, warfare, sea-bed mining, and sonar-based navigation (Simmonds and MacLennan 2005; Ramirez-Llodra et al. 2011; Jalkanen et al. 2022).

Although ocean noise pollution cannot be considered widespread in the ocean (mainly due to the large extent of the ocean domain), it can be just as harmful to the ocean environment as other kinds of pollution. In contrast to plastic pollution or fertilizer runoff, remedies for ocean noise pollution appear easy to find, and the derived damage is not persistent and can be reversed.

According to recent data, shipping alone has contributed to an estimated 32-fold increase in low-frequency noise along major shipping routes in the past 50 years, driving marine animals away from vital breeding and feeding grounds. The low-frequency noise generated by human activities can travel long distances in the water environment (this is the main reason for their danger) (Fig. 3.9).

The ocean noise sources are differently distributed among continents. Dynamite fishing is particularly common in Southeast Asia and coastal Africa, while in the North Sea it is still common (lasting for more than 70 years) the controlled detonation of bombs remained unfired from the Second World War. Underwater drilling to

This chapter is written by Genuario Belmonte and Giacomo Milisenda.

G. Belmonte, *Elements of Pelagos Biology*,
https://doi.org/10.1007/978-3-031-74280-4_9

Fig. 9.1 Scheme of the main sound sources generated by human activities (orange waves), or natural (yellow waves) that disturb ocean life (graphical composition A. M. Miglietta)

find and extract oil and gas from sea bottoms (and the seismic surveys used to localize new fields) is a common and continuous presence in the seas of Northern Europe (overall).

Even environmentally sustainable activities, such as offshore wind farms, increase water noise either during construction or during operation, although the noise derived from turbines is extinguished by a few hundred meters.

The immediate response to such noises is that pelagic animals abandon the noisy sites. The colligation wind farm conservation areas is, consequently, not so obvious in marine dominion, and this has to be considered in the planning of both energy farms and biodiversity conservation areas.

The prompt restart of natural communities at noise cessation, however, is a hopeful and concrete reality. During the COVID-19 pandemic of 2020, the stop of the free circulation and activities of the humanities globally interested 60% of the Earth's population for at least 1 month. The ocean noise (calculated) was reduced by 20%, and water life was noted to be visibly favoured by this stop of human activities (Belmonte et al. 2023a).

Vibrations produced by wind farms can represent a disturbance for water life, although in restricted areas (called "fields"). This disturbance is in opposition to the role of biodiversity facilitation that any artificial island can represent (see chapter on Fish Accumulation Devices, FAD) in the open ocean (Relini et al. 1994). The cost-benefit evaluation must consequently be well determined, possibly with an assessment of ecosystem services derived from the interested marine areas in the absence and/or presence of plants.

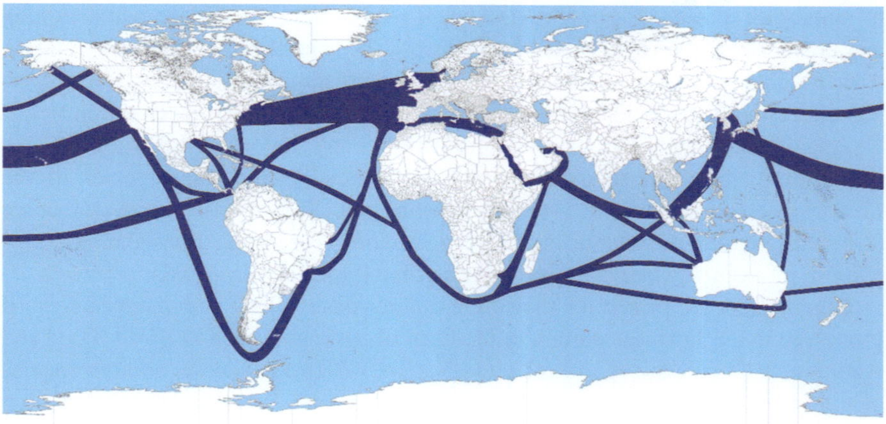

Fig. 9.2 Main navigation routes in the world's oceans. The thickness of the tapes corresponds to the intensity of the traffic. (Courtesy G. Belmonte)

Ship navigation has a heavy role in the biological alteration of marine communities by distributing nonindigenous species (NIS) among ports around the world (see chapters on migration and marine biogeography). In addition (as already indicated above), it is responsible for the physical disturbance due to the production of noise along the routes (Fig. 9.2).

Aside from such a physical alteration of the marine environment, we must also consider the breaking of the neuston integrity at the sea surface along the navigation routes. Navigation routes take into consideration the shortest lines passing through the oceans and seas and separating starting and arrival points (ports). Thus, the navigation routes are relatively narrow tapes of surface water interested in the intense routine transit of ships.

Most of the ocean surface is not interested in such standard trade routes, but the fishery navy is not linked to such waterways and continuously explores new directions, places, and times to successfully collect fish.

9.2 Disturbance at a Large Scale. Open Sea Fishery (Contribution of Giacomo Milisenda)

The unique and diverse marine environment of the Mediterranean Sea supports various faunal species (fish and mammals, overall) that play a crucial role in the marine food web (Bianchi and Morri 2000; Coll et al. 2010). The sustainable management of Mediterranean pelagic resources is of paramount importance due to their ecological, economic, and social value (GFCM (general fisheries Commission for the Mediterranean) 2017). Overfishing represents an unbalanced management of fish stocks and, together with habitat degradation and climate change, significantly threatens the health and stability of Mediterranean pelagic ecosystems (Tsikliras et al. 2015). Regarding only the pelagos dominion, the Mediterranean Sea is home to numerous species that play a significant role in marine ecosystem functioning and are vital for the region's fishing industry. As fisheries are an aspect of large-scale interaction between humans and the sea, overfishing is one of the most dangerous direct disturbances that humans can perform on natural marine communities.

Pelagic fish inhabit the water column without any important link with the benthos. Some of the most abundant and commercially important species of the Mediterranean Sea include Atlantic bluefin tuna (*Thunnus thynnus*), albacore (*Thunnus alalunga*), Atlantic mackerel (*Scomber scombrus*), European anchovy (*Engraulis encrasicolus*), and European sardine (*Sardina pilchardus*).

Bluefin tuna are prized for their high-quality flesh, which is consumed fresh if processed in the Japanese culture in the form of *sushi* and *sashimi* (FAO (Food and Agriculture Organization of the United Nations) 2018). Albacore, another species of tuna (*Thunnus alalunga*), is smaller in size and is mainly caught for canning purposes (ISSF (International Seafood Sustainability Foundation) 2021). Atlantic mackerel (*Scomber scombrus*) is smaller than tuna and is primarily caught for human consumption, fishmeal, and fish oil production (ICES (International Council for the Exploration of the Sea) 2019). European anchovies (*Engraulis encrasicolus*) and sardines (*Sardina pilchardus*) are small, planktivorous fish that form large schools and represent a vital component of the Mediterranean marine food web. They are the main target of small-scale fisheries and are typically processed into canned or preserved products.

Pelagic fishing in the Mediterranean is a major source of income, employment, and food security for millions of people in the region (GFCM (general fisheries Commission for the Mediterranean) 2017). The fishing sector provides direct employment to over 250,000 fishermen and indirectly supports many other industries, including processing, distribution, and marketing. The total annual catch of pelagic fish in the Mediterranean is estimated to be approximately **one million tons**, with *Engraulis* and *Sardina* constituting approximately 60% of the total catch (FAO (Food and Agriculture Organization of the United Nations) 2018). In addition to their direct economic value, pelagic fish also play a crucial role in supporting other marine industries, such as tourism and recreational fishing.

In the Mediterranean area, pelagic fisheries provide a vital source of income and cultural identity. Many traditional fishing practices have been passed down through generations and are integral to the social fabric of these communities. The sustainable management of pelagic fish stocks in the Mediterranean is crucial for maintaining the ecological balance of marine ecosystems and ensuring the long-term viability of the fishing industry and tradition (GFCM (general fisheries Commission for the Mediterranean) 2017).

9.2.1 Deep-Sea Pelagic Resources

Deep-sea pelagic resources for fisheries, represented by marine organisms living below the euphotic zone, are increasingly being targeted for exploitation due to the depletion of resources in coastal and epipelagic areas. The exploitation of deep-sea pelagic resources is characterized by unique challenges, including the development of specialized extraction techniques that are not traditional and, hence, still not perfect (Danovaro et al. 2014; Clark et al. 2016). In addition, the environmental and socio-economic impacts and the implementation of strategies to ensure sustainable use of these resources are still under assessment.

The extraction of deep-sea pelagic fish resources involves the use of advanced fishing techniques and technologies, such as deep-water trawling, longlining, and remotely operated vehicles (ROVs). **Deep-water trawling** involves dragging a large net along the seabed to catch demersal and hyperbenthic species. **Longlining** is another method used to target deep-sea species, consisting of a long fishing line with baited hooks attached at regular intervals, which can be deployed at various depths to catch specific species. **ROVs** are underwater robots that can be equipped with cameras and manipulators for sampling and collecting deep-sea organisms. These extraction techniques have enabled the exploitation of previously inaccessible deep-sea pelagic resources but have also raised concerns about their environmental impacts.

The exploitation of deep-sea pelagic fish resources has significant environmental and socio-economic impacts because deep-sea organisms are characterized by slow growth rates, low reproduction rates, and long lifespans, making them particularly vulnerable to overexploitation and slow recovery after disturbances. Deep-water trawling, in particular, has been linked to the destruction of deep-sea bottoms, such as cold-water coral reefs and sponge grounds, which are essential for the survival of many deep-sea species, other than for a still unexplored and unknown hyperbenthos (Bensch et al. 2008). These environmental impacts can have cascading effects on marine food webs, potentially leading to ecosystem collapse and the loss of biodiversity.

The socio-economic impacts of deep-sea pelagic resource exploitation include the potential for job creation and economic growth in coastal communities, as well as increased revenue from the export of high-value deep-sea fish and other resources. However, the sustainability of these benefits is uncertain, given the vulnerability of deep-sea ecosystems to overexploitation and the potential depletion of deep-sea resources.

To ensure the sustainable exploitation of deep-sea pelagic fish resources, several strategies can be employed, such as the adoption of precautionary management measures, or their implementation, the establishment of deep/pelagic **marine protected areas (MPAs)**, and the promotion of research and technological innovation. Precautionary management measures include the setting of conservative **catch limits**, the use of **selective fishing gear**, and the **spatial and temporal stops** to protect vulnerable species and habitats.

Research and technology can contribute to the development of more sustainable extraction techniques, as well as the improvement of scientific knowledge on deep-sea ecosystems and their response to exploitation pressures. Additionally, promoting international collaboration and the sharing of scientific data can facilitate the development of effective management strategies and conservation measures for deep-sea pelagic resources (Norse et al. 2012).

For pelagic fishing, the main techniques used in the Mediterranean include purse seining, midwater trawling, and longlining (Tsikliras et al. 2015). **Midwater**

trawling uses a conical net with a large mouth, towed behind a vessel and far from the bottom at varying depths (150–400 m) to capture pelagic species, such as hake (*Merluccius merluccius*), horse mackerel (*Trachurus trachurus*), and blue whiting (*Micromesistius poutassou*) (D'Onghia et al. 2003). This method can be selective when used correctly, minimizing the capture of nontarget species (Fig. 9.3).

Longlining consists of deploying a main fishing line with numerous baited hooks attached at regular intervals. The line can be set at various depths to target specific pelagic species, such as swordfish and tuna. This method is considered

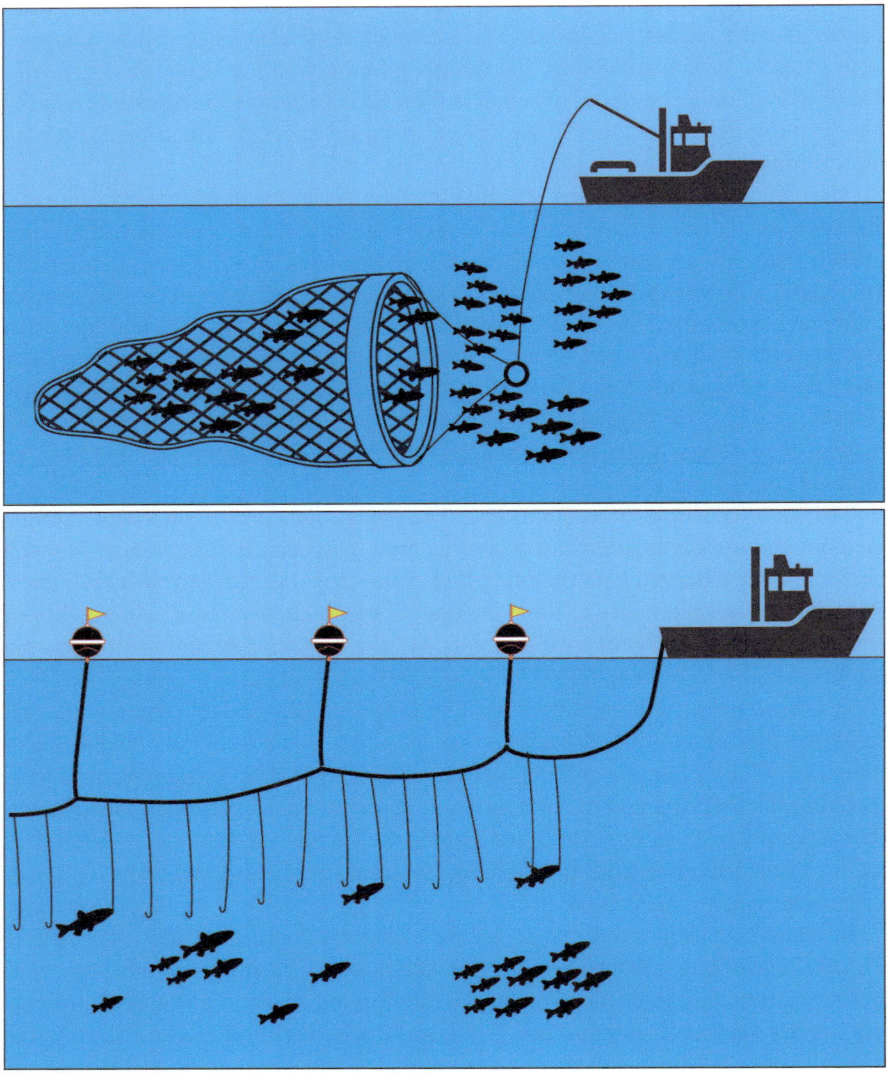

Fig. 9.3 Midwater trawling (top), and longlining (bottom) fishery (graphical composition A. M. Miglietta)

more selective than other fishing techniques, reducing the risk of bycatch, but only if performed at depths of more than 100 m to maximize the avoidance of dolphins and turtles (Fig. 9.3).

The regulation and management of pelagic fish resources in the Mediterranean are primarily overseen by the General Fisheries Commission for the Mediterranean (GFCM), a regional fisheries management organization under the Food and Agriculture Organization of the United Nations (FAO). The GFCM is responsible for developing and implementing management measures, such as catch quotas, minimum landing sizes, and closed areas, to ensure the sustainable use of pelagic resources.

National governments also play a role in the management of pelagic resources, implementing regulations, and monitoring compliance within their exclusive economic zones (EEZs). Cooperation among Mediterranean countries is essential for the effective management of shared pelagic resources, as many species are migratory and cross national boundaries.

To promote sustainable fishing in the Mediterranean, several strategies can be employed, including the adoption of ecosystem-based management approaches, the implementation of effective monitoring and enforcement mechanisms, and the promotion of research and innovation. Ecosystem-based management approaches recognize the interconnectedness of marine ecosystems and aim to balance the needs of both human societies and the environment. This includes managing fisheries in a way that considers the entire ecosystem rather than focusing on individual species.

Effective monitoring and enforcement mechanisms are essential for ensuring compliance with fisheries regulations and management measures. This may include the use of satellite monitoring systems, on-board observers, and regular inspections of vessels at sea and in ports.

Research and innovation can contribute to the development of more sustainable fishing techniques and gear, as well as improving scientific knowledge of pelagic ecosystems and their response to fishing pressure. This can inform the development of better management strategies and conservation measures. Additionally, promoting stakeholder involvement, such as engaging with fishers, local communities, and nongovernmental organizations, can help ensure that management measures are both effective and socially acceptable.

Education and awareness-raising initiatives can also play a significant role in promoting sustainable fishing practices among fishers, consumers, and policymakers. This may include the promotion of eco-labelling schemes, which certify sustainable seafood products, and the implementation of educational campaigns to inform consumers about the importance of choosing sustainably sourced seafood.

The study and management of pelagic resources, particularly those inhabiting the deep sea, require effective sampling methods to obtain accurate information on their abundance, distribution, and ecology.

Traditional sampling techniques for pelagic resources (Gartner et al. 2008) can be effective for capturing a variety of pelagic organisms, but they often suffer from issues related to selectivity, depth limitations, and potential damage to the environment or organisms being sampled (Cousins et al. 2019). These methods, however, are limited to the upper layers of the water column, mainly the epipelagos. For the very deep part of the water column at present, we cannot speak about "fishery" but have to use more appropriately the term "sampling" because we are still in the epoch of exploration of what is there.

Modern **sampling** techniques aim to address these limitations and improve the efficiency and accuracy of pelagic resource assessments. Examples of these techniques include the use of remotely operated vehicles (ROVs), autonomous underwater vehicles (AUVs), and environmental DNA (eDNA) analysis.

ROVs and AUVs allow for non-invasive, high-resolution sampling of deep-sea pelagic resources. These underwater vehicles can be equipped with cameras, sensors, and sampling devices, allowing for the collection of biological, chemical, and physical data without disturbing the organisms or their environment.

eDNA sampling involves the collection and analysis of genetic material shed by organisms into the environment. This non-invasive method allows for the detection and identification of pelagic species in a given area, providing valuable information on their presence and abundance without the need for their direct capture.

Advanced technologies, such as acoustic echolocation and biologging, have greatly improved our ability to monitor and study pelagic organisms, particularly in the deep sea.

Acoustic methods use sound waves to detect, locate, and quantify pelagic fish assemblages, offering a disturbing but not mortal approach to studying their distribution, abundance, and behaviour. These methods are particularly useful for studying deep-sea organisms that are difficult to collect using traditional techniques.

Biologging involves the attachment of electronic tags to individual organisms, allowing for the collection of detailed information on their movements, behaviour, and physiology. This technology has greatly advanced our understanding of the ecology and life history of many pelagic species, including deep-diving marine mammals and fish.

The study and management of deep-sea pelagic resources present several challenges, including the remote and harsh environment, the limited availability of funding and resources, and the often-elusive nature of these organisms. However, these challenges also represent opportunities for innovation, collaboration, and the development of new methods and technologies.

The growing recognition of the importance of deep-sea pelagos and their vulnerability to human activities, such as fishing and climate change, highlights the need for increased research, monitoring, and conservation efforts.

Future perspectives for the research, management, and sustainability of pelagic resources are promising, as advancements in technology and scientific understanding continue to develop rapidly. The increasing use of interdisciplinary approaches, innovative sampling techniques, and advanced monitoring technologies will help to overcome the challenges associated with studying deep-sea pelagic ecosystems and provide valuable insights into their ecology, biodiversity, and conservation needs.

Furthermore, increased international collaboration and investment in research and conservation efforts will be essential for ensuring the long-term sustainability of pelagic resources in a closed sea as the Mediterranean. This collaboration may include the sharing of data, resources, and expertise, as well as the establishment of regional and international conservation initiatives, such as marine protected areas and fisheries management plans (GFCM (General Fisheries Commission for the Mediterranean) 2019).

In conclusion, continued research, innovation, and collaboration are critical to the effective management and conservation of pelagic resources in the Mediterranean. By embracing these opportunities and addressing the challenges that lie ahead, we can work towards a more sustainable and informed future for these vital ecosystems and their resources.

The deep-sea environment, characterized by low temperatures, high pressure, and complete darkness, hosts a diverse array of organisms adapted to live in such conditions. The water column below 1000 m (the bathypelagos) is the habitat of a variety of fish, invertebrates, and other organisms that have never been used for human nutrition on a large scale.

Many of these species exhibit adaptations that enable them to survive in the deep sea, such as bioluminescence for communication and predation, large eyes or enhanced sensory structures to detect scarce prey, and energy-efficient modes of locomotion.

Deep-sea fish, such as lanternfish (Myctophidae), grenadiers (Macrouridae), and deep-sea anglerfish (Ceratiidae), are characterized by slow growth, low reproduction rates, and extended lifespans (Gage and Tyler 1991). Invertebrates, such as deep-sea shrimp and squid, also play an essential role in the bathypelagic food web, serving as prey for larger predators such as fish and marine mammals.

Deep-sea organisms represent a largely untapped resource for human nutrition. Some species, such as the orange roughy (*Hoplostethus atlanticus*) and the Patagonian toothfish (*Dissostichus eleginoides*), have already been targeted by commercial fisheries due to their high market value (Fig. 9.4) (Collins et al. 2010). These fish are rich in protein, omega-3 fatty acids, and other essential nutrients, making them a valuable addition to human diets.

Recently, *D. eleginoides* (the Patagonian tootfish) has become the target of an international fishery, and ships coming from Argentina, South Africa, and Norway collected approximately 100,000 tons of such a fish in only 1 year (1997). This exaggerated fishery has been so intense that the species is now at extinction risk. *D. eleginoides* is captured with longline devices, thus causing the by-capture (and death) of more than 140,000 albatross and other sea birds attempting to subtract the fish from the hooks.

In 2014–2016, the organization Sea Shepherd allowed the stoppage of six such ships, known for their illegal fisheries. In addition, a movie (*Ocean Warriors: Chasing the Thunder*) was realized during the chase. Illegal fishermen paid more than 46 million euros in penalties.

Similar to fish, deep-sea invertebrates such as shrimp and squid also provide a sustainable alternative to overexploited shallow-water species. However, the exploitation of deep-sea resources for human nutrition should be approached with caution, given the potential ecological impacts and the limited knowledge about the biology and population dynamics of deep-sea organisms.

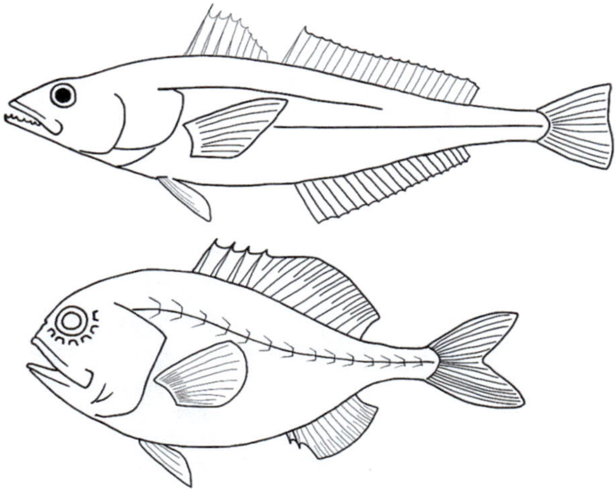

Fig. 9.4 Top, *Dissostichus eleginoides*; bottom, *Hoplostethus atlanticus*. (Courtesy G. Belmonte)

The sustainable exploitation of deep water-column organisms presents several challenges, including the difficulty of accessing these remote environments, the vulnerability of deep-sea species to overfishing, and the potential impacts on deep-sea ecosystems. Furthermore, the limited knowledge of the biology, distribution, and abundance of deep-sea organisms still hampers the development of effective fisheries management strategies.

Despite these challenges, opportunities exist for the sustainable use of deep water-column resources. Advances in deep-sea exploration and sampling technologies, such as remotely operated vehicles (ROVs) and autonomous underwater vehicles (AUVs), could improve our understanding of deep-sea ecosystems and inform management decisions.

Additionally, collaboration between researchers, industry, and policymakers could promote the development of sustainable deep-sea fisheries, incorporating precautionary approaches and adaptive management strategies. Establishing marine protected areas and implementing catch limits based on scientific data can help prevent overexploitation and ensure the long-term sustainability of deep water-column resources.

Bathypelagic organisms are still not an important food source for humans but are a vital food source for many large marine predators. Deep-sea lanternfish (Myctophidae) are known for their bioluminescent features and high abundance in the bathypelagic zone. They serve as a critical food source for various large predators, including tuna, billfish, and several species of marine mammals and seabirds (Choy et al. 2017).

Similarly, bathypelagic squid and crustaceans are consumed by deep-diving marine mammals, such as sperm whales (*Physeter macrocephalus*), beaked whales (Ziphiidae), and elephant seals (*Mirounga* spp.). These predators rely on bathypelagic prey to meet their energy and nutritional demands, as the deep-sea environment provides a constant, reliable, and abundant source of food.

Bathypelagic organisms occupy a crucial position in marine food webs, linking primary production at the ocean's surface with the deep-sea ecosystem (the largest of the planet). They transfer energy and nutrients from the upper layers of the ocean to the deep-sea environment through processes such as diel vertical migration and consumption by large deep-sea predators.

Diel vertical migration is a common behaviour among bathypelagic species such as lanternfish and krill. This movement allows them to feed on surface-dwelling plankton during the night and retreat to depths during the day, where they become prey for deep-diving predators.

Understanding the importance of bathypelagic organisms in the diet of large predators is essential for the conservation and management of marine ecosystems. Overexploitation of these prey species could lead to significant declines in predator populations and disrupt the balance of marine food webs.

As such, effective management strategies should consider the ecological role of bathypelagic organisms in marine ecosystems and incorporate this information into decisions about resource exploitation, habitat protection, and the establishment of marine protected areas. Moreover, monitoring the abundance and distribution of bathypelagic organisms can provide valuable insights into the health and resilience of marine ecosystems, helping to inform conservation and management efforts.

Bathypelagic organisms are crucial components of marine ecosystems, serving as a vital food source for large predators and contributing to the transfer of energy and nutrients between different ocean depths. Recognizing their ecological significance and incorporating this knowledge into conservation and management strategies is essential for the long-term sustainability and health of marine ecosystems.

9.3 Conservation Policies and Actions

Pelagic ocean is the largest environment on the planet and will probably be affected only partially by the ongoing changes in climate and human pressure that interest more and more other parts (the emerging ones) of the planet.

Apart from the enormous extension and volumetric dimension, the changing climate will have less importance for oceans than for land because of the thermal capacity of water, which is higher than that of the atmosphere. This notwithstanding, however, the changing climate and the growing human pressure will exert their conditioning action on the surface layer and on the life adapted to stay offshore.

In the past, oil spills and general oil product-based contamination of the sea surface alerted scientists and produced socio-political movements for better management of the environment and of the oceans. Today, the alarm does not derive from oil but from synthetic materials, known under the common name of plastics.

Additionally, from the protection point of view, what men can protect has to stay close to their life sites, and open seas have traditionally been disregarded from this point of view (Agardy 1997). Remote-observing satellites are now available that continuously check the sea surface. However, in this case, the check is only superficial, and nothing is known about the deep pelagos.

To confirm this trend in protection policies, in the Mediterranean, a long list of Marine Protected Areas has been instituted (31, of different levels of realization, only in Italy), but all along the coasts, and respecting the EU rule that the offshore border has to coincide with the 50 m isobath (García-Ruiz 2013). In one case, however, a Pelagos Sanctuary has been established with a management agreement between Monaco, France, and Italy, with the aim of protecting large mammals (Cetacea) in a large extension of the open sea in the northern Mediterranean, with Corse Island in its centre.

During the past decades, the concept of marine protected areas (MPAs) has gained considerable favour worldwide as a tool for the conservation of endangered marine species and their habitats, including upper-trophic predators such as marine mammals. Initially, spatially explicit designations for marine mammal conservation were introduced to protect Pinnipedia sites on land. Eventually, MPAs were established to protect important aquatic habitats, including calving grounds and foraging areas. The charismatic value of marine mammals in the public arena, combined with the widely diffused perception of their threatened status, eventually led to the establishment of large-scale sanctuaries encompassing the entire territorial waters and exclusive economic zones (EEZs) of individual nations, such as in Ireland (1991) and in Mexico (2002).

International cetacean sanctuaries encompassing entire ocean basins, such as the 1979 Indian Ocean Sanctuary and the 1994 Southern Ocean Sanctuary, were established exclusively to protect large whale populations from commercial harvest.

Increasingly, MPAs designed for the conservation of marine mammals and their habitats have developed into more complex regimes involving the implementation of ecosystem-level management schemes capable of addressing the broader range of threats affecting these populations. A variety of species throughout the globe, including Odontoceti and Mysticeti Cetacea, Pinnipedia, and Sirenidae, have already benefited from such protective measures.

The recent establishment of an international sanctuary that includes a high sea portion for Cetacea in the western Mediterranean provides a valuable illustration of how the current global effort to promote MPA networks in regional seas can be advanced. In 1991, a grassroots effort originating in Italy stimulated a proposal for the establishment of a large MPA in the northwestern Mediterranean Sea to protect Cetacea. This idea eventually led to a tripartite agreement on the creation of the "International Sanctuary for the Protection of Mediterranean Marine Mammals", also known as the "Pelagos Sanctuary" (Notarbartolo di Sciara et al. 2008) (Fig. 9.5).

The impetus for the sanctuary proposal was threefold: (1) recently acquired knowledge of the presence of important populations of Cetacea in the area; (2) awareness of the existence of serious threats to these populations; and (3) a lack of legal instruments to protect the Mediterranean high seas beyond the 12 nautical mile (22.3 km) buffer provided by the national territorial seas, where most of the habitats of these cetacean populations lie.

As early as 1899, Albert I of Monaco, a pioneer of oceanography, commented that he was more likely to sight cetaceans from the windows of his palace than during dedicated cruises to the Arctic Seas. The notion of cetacean abundance in that

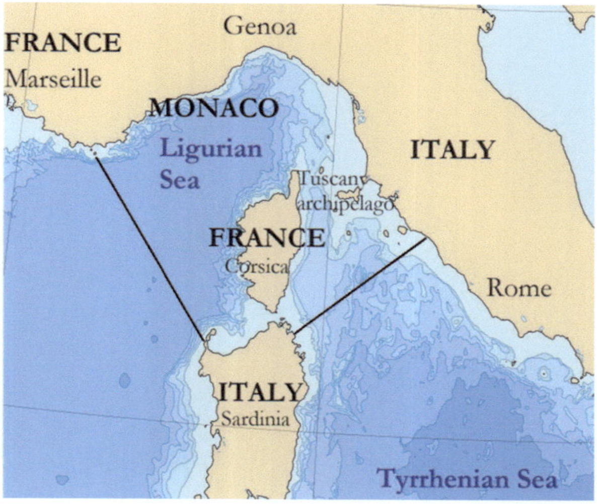

Fig. 9.5 Map of the Pelagos Sanctuary in the north-west Mediterranean Sea. (Courtesy G. Belmonte)

portion of the Mediterranean, however, was lost in the following century, when ecological knowledge asserted a generalized oligotrophy of Mediterranean offshore waters, with no appreciable upwelling and low primary productivity, thus unable to support significant resident populations of whales. The first research expeditions exclusively devoted to cetacean population ecology in the Mediterranean during the late 1980s revealed the presence of important habitats and populations within a wide marine area between Corsica and the continental coasts of France and Italy.

The productivity of the western Mediterranean is notoriously spatially heterogeneous and influenced by the effects of currents, the input of river runoff, and mixing by localized winds steered by topographic features. While the Mediterranean has long been considered an oligotrophic sea with low nutrients and ocean productivity, satellite-derived Sea-viewing Wide Field-of-View Sensor (SeaWiFS) imagery has revealed the existence of several regions of elevated chlorophyll *a* concentration (Chl*a*), a metric of the abundance of phytoplankton standing stocks in the euphotic zone of the water column. In addition to vast oligotrophic regions (Chl*a*: < 0.1 mg m^{-3}) in the Eastern Mediterranean basin, mesotrophic waters (Chl*a*, 0.1–0.3 mg m^{-3}) and eutrophic waters (Chl*a*, 0.3–1 mg m^{-3}) are found in the Aegean Sea, the Adriatic Sea, and the north western Mediterranean off the coasts of Spain and France (Fig. 9.6).

Hypertrophic coastal waters (Chl*a* > 1 mg m^{-3}) occur along the margin of the basin, in correspondence with main river discharge and/or urban settlements, but a large hypertrophic area can be recognized as corresponding to the whole Azov Sea (Russia). Areas with a high concentration of Chl*a* (0.3 < Chl*a* < 1.0 mg m^{-3}) can be individuated in the Ebro Delta (Spain), Gulf of Lions (France), North Adriatic (Italian side), Gulf of Gabes (Tunisia), Nile delta (Egypt), and a large portion of the North West Black Sea (Ukraine, Russia).

A dominant cyclonic (anticlockwise) surface current flowing northwards along Corsica and Tuscany and reaching the coast of Liguria and mainland France in a

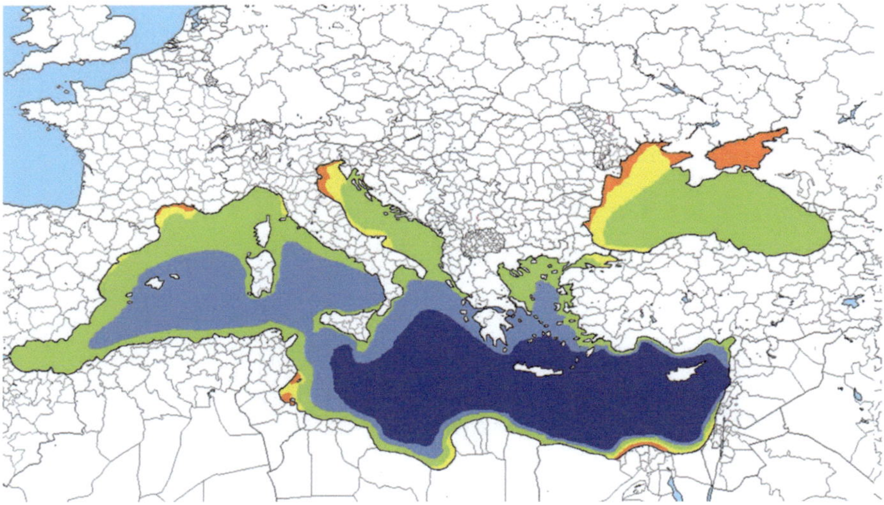

Fig. 9.6 Distribution of Chl*a* concentration in the surface waters of the Mediterranean Sea. Form maximum concentrations of orange areas (Chl*a* > 1 mg m^{-3}) to minimum concentrations of blue areas (Chl*a*: < 0.1 mg m^{-3}). (Courtesy G. Belmonte)

westerly direction creates a permanent water front separating coastal from offshore waters, associated with enhanced productivity and retention, generating intense biological activity. This phenomenon is seasonally reinforced by vertical water mixing and localized coastal upwelling, supplied with nutrients and other organic substances coming from rivers (mostly the Rhone) into the euphotic layer, where they fuel phytoplankton. Consequently, high levels of primary production, with Chl*a* concentrations >10 mg m^{-3} in spring, support a rich biomass of highly diversified zooplankton, including krill (*Meganyctiphanes norvegica*) (such a high concentration of Chl*a* interests a thick layer of the epipelagos and is not easily perceived by satellite sensors exploring just the surface). Zooplankton concentrations, in turn, attract various upper-trophic level predators, including krill-eating, fish-eating, and squid-eating Cetacea.

The western offshore portion of the Sanctuary consists of a uniform abyssal plain at depths of 2500–2700 m. The Pelagos Sanctuary contains deep-water and shelf-slope habitats suitable for the breeding and foraging needs of all the Cetacea species found in the western Mediterranean Sea, with *Balaenoptera physalus* and *Stenella coeruleoalba* as the most represented species in the sanctuary.

Approximately 3500 fin whales (*Balaenoptera physalus*) thrive in the West Mediterranean, and a considerable proportion of these concentrate in the Pelagos Sanctuary in the summer to feed on krill. Striped dolphins (*Stenella coeruleoalba*) are the most abundant Cetacea throughout Mediterranean offshore waters, with an estimated summer population size of 20,000–30,000 specimens within the Sanctuary. Six other Cetacea species are regular components of the Sanctuary fauna: sperm whales (*Physeter macrocephalus*), pilot whales (*Globicephala melas*), and

Risso's dolphins (*Grampus griseus*), which are frequent in both offshore and slope waters. Cuvier's beaked whales, *Ziphius cavirostris* prefer specific slope areas overlying submarine canyons; short-beaked common dolphins, *Delphinus delphis*, are found both in coastal and offshore waters, particularly in the southern part of the Sanctuary; and predominantly coastal bottlenose dolphins, *Tursiops truncatus*, frequent mostly the shelf areas surrounding Corsica, North Sardinia, the Tuscan Archipelago and continental France (Forcada et al. 1995; Panigada et al. 2005).

This faunal diversity coexists with very high levels of human pressure. Most of the coastal areas of the Sanctuary, particularly on the mainland, are heavily populated with large and medium-sized coastal cities, other than ports of major commercial and military importance and industrial areas. Furthermore, the entire Sanctuary coastal zone contains major marine tourism destinations, which include whale-watching tours, ferry traffic, and coastal runoff, especially during the summer months. As a consequence, diverse human activities can act positively (for protection purposes) and negatively (with disturbance) on cetacean populations of the sanctuary.

In recent years, the Pelagos Sanctuary has developed awareness concerning the extent and intensity of human activities, notably fisheries, as well as the navigation and transportation of hazardous substances, highlighting the serious threats to marine biodiversity. Pelagic driftnet fishing for swordfish, *Xiphias gladius*, had a particular success in Italy, where in the late 1980s over 700 vessels were licensed to use driftnets, each of them with nets averaging 20 km in length. Most of this fleet converged in the summer on the rich Ligurian Sea fishing grounds (Fierstine et al. 1997). This fishery caused high levels of accidental mortality in at least six Cetacea species. Such circumstances were given very high visibility by the national and international media, probably thanks to the existence of the sanctuary. Strong pressure was thus exerted by public opinion on the Italian government to restrain driftnet fishing efforts in the area. The Italian government was eventually spurred into action by rulings in 1989 and 1990 by the higher Italian courts and established a driftnet-free zone in a portion of the Liguria Sea in 1990, prohibiting Italian vessels, except a small number of local boats, from fishing with this device (Garibaldi 2015). Although the zone failed to include most of the cetacean habitat to be protected, it was legally significant because it discouraged vessels from arriving from distant locations to fish there. It was also the first introduction of a prohibition in the Mediterranean area with the aim of protecting Cetacea.

After years of institutional partiality towards short-term fishing interests, the tide eventually turned in favour of more farsighted environmental concerns and led to decisions by the European Commission first to limit the maximum net length and later to ban driftnets altogether from European vessels (1 January 2002).

Apart from fisheries, human pressure on the Pelagos Sanctuary is represented by an increase in maritime traffic, including an expanding commercial whale-watching industry, which clearly carries the risk of potential disturbance for Cetacea and increases the risk of collisions for larger species.

Finally, maritime transportation of hazardous substances to and from the coastal industrial hotspots of northern Italy and southern France posed significant threats to the marine environment, as demonstrated by a major spill off the port of Genoa in 1994, in which 144,000 tons of oil burned for days on the sea.

ACCOBAMS (Agreement on the Conservation of Cetaceans of the Black Sea, Mediterranean Sea, and Contiguous Atlantic area), which attempts to integrate spatial conservation measures with species-oriented measures in the Mediterranean and Black Seas, came into force in 2001.

In 2003, France established an ecological zone beyond the 12 nautical miles limit of the territorial sea (Law n. 346 of 15 April 2003), and in 2006, Italy adopted legislation (Law n. 61 of 8 February 2006), allowing the establishment of ecological zones with, among others, the purpose of protecting marine mammals and biodiversity.

In 1990, the Tethys Research Institute proposed to the "European Association Rotary for the Environment" a project for the establishment of an MPA encompassing the most important habitat for cetaceans in the region. The rationale behind the proposal entitled "Project Pelagos" challenged the mainstream legal notion of the time that establishing a protected area in the high seas was impossible. The proposal envisaged the creation of a Biosphere Reserve in the open sea off the Liguria–Corsica–Provencal coasts and the establishment of an international authority responsible for the sustainable management and conservation of the basin's natural resources. Prince Rainier III of Monaco granted the support of the Principality to the proposal and recommended the creation of a sanctuary for Cetacea in the Liguria–Corsica–Provencal basin through a trilateral agreement among France, Italy, and Monaco managed by an international body hosted by Monaco.

The Agreement, followed by a declaration, was signed in Rome on 25 November 1999, successively ratified by Monaco (2000), France (2001), and Italy (2002), and entered into force on 21 February 2002. The protected area was enlarged from the original proposal to encompass the waters between Corsica and Tuscany and included waters having the legal status of maritime internal waters, territorial seas, and high seas (today also ecological zones).

In brief, the adopted measures prohibit any deliberate taking (defined in broad terms as "hunting, catching, killing, or harassing of marine mammals, as well as the attempting of such actions") or disturbance of Cetacea. The parties also agreed to exchange their views with the objective of regulating and, if appropriate, prohibiting high-speed offshore races and regulating whale-watching activities for tourism purposes.

Despite the wish of the negotiators that the process be expeditious, the agreement entered into force almost 3 years after its signature. At an extraordinary meeting of the parties in 2005, it was agreed that Italy would host the headquarters (in Genoa), that the executive secretary would be a French national, and that the chairing of the scientific and technical committee would be the responsibility of the Principality of Monaco.

The Pelagos Sanctuary is a key area for Mediterranean Cetacea because it contains essential habitats for the diverse complement of species regularly sighted in the Western Mediterranean and supports large resident populations of several genetically distinct stocks. In addition to protecting localized summer-time cetacean aggregations, the effective management of the Pelagos Sanctuary will have broader implications for the conservation of the Corsican-Liguria-Provencal basin and the entire Mediterranean basin.

Protecting Ligurian Sea Cetacea will enhance regional conservation efforts by helping to maintain the integrity of a productive and high-value ecosystem of ecological and fisheries importance and will provide a stimulus and insight into ways of protecting far-ranging species within the entire Mediterranean basin. However, the effective management of the sanctuary will require the articulation of clear and quantifiable objectives and the sign of a monitoring system to assess whether these objectives are being met.

As a first step, three complementary sets of goals, which span a broad array of ecological factors, are proposed: (1) protecting threatened species and their habitats within the sanctuary; (2) protecting the ecosystems and food webs supporting these species and habitats; and (3) protecting highly migratory species in the wider Mediterranean basin.

As a flagship for future marine conservation approaches beyond territorial waters, the Pelagos Sanctuary sets precedents for how marine conservation actions can be implemented in high-use and heavily industrialized areas such as the Mediterranean Sea, how high-seas MPAs can be achieved through a collective agreement among like-minded nations, and how they can be supported by a regional framework. Nevertheless, the sanctuary implementation represents merely the beginning of a long iterative process whereby the sanctuary's objectives will be clearly articulated through a consensus process, with subsequent management continuously monitored and assessed to gauge its failures and successes. The progress towards effective management has started with the compilation of broad goals that are articulated in a management plan. These goals initially targeted the protection of marine mammal populations and their habitats but have been expanded to include broader ecosystem-level considerations relating to cetaceans' prey resources and the physical–biological processes responsible for enhanced productivity in the Ligurian Sea.

Goal 1: **Protection** of Pelagos **Cetacea** and their **habitats**. In particular, maintaining a favourable conservation status for the area's marine mammals, along the lines of the EU "Habitats Directive" (given that they are all listed in the Directive's Annex IV), also considering their wide-ranging habits, is not easy and requires well-designed marine zoning strategies capable of incorporating the natural history (e.g., ontogenetic habitat requirements, seasonal migration routes) of these species, as well as the inherent variability in their oceanic environment (e.g., changes in the extent and location of frontal systems and eddies). Even when guided by well-conceived management plans, the risk exists that MPAs remain little more than "paper parks" unless broader ecosystem-level and socio-economic considerations are taken into account at the time of implementation.

Goal 2: **Protection** of Cetacea **food webs** and **ecosystems**. A broad-based ecosystem-level management plan for the Pelagos Sanctuary will have to include an understanding of the ecology of the local Cetacea, as well as their life-support systems. Some would refer to this as classic habitat protection, but Harwood's (Hardwood Ecosystem Experiment 2017) characterization of protecting functional ecological units is preferred. In particular, when considering fin whales, a broader ecosystem-level approach would entail protecting the mechanisms that allow large euphausiid biomass to subsist. Thus, the Pelagos Sanctuary should be viewed as both a whale and a krill sanctuary, which should be capable of accommodating the inherent dynamics of marine food webs and oceanographic processes supporting the high localized productivity and prey aggregations of importance to highly mobile marine predators. This will involve addressing the need to monitor the abundance, distribution, and seasonality of the krill *Meganyctiphanes norvegica* throughout the frontal system.

Goal 3: **Protection** of highly **migratory Cetacea** in the Mediterranean Basin. The Pelagos Sanctuary management plan should consider the threats and protections that exist both within and outside its jurisdiction in tight connection with ACCOBAMS. In addition to addressing Cetacea movements and habitats within Sanctuary waters, a monitoring plan will require information on the threats and impacts these species face outside of MPA waters. In particular, a system for recording anthropogenic impacts and mortality (e.g., ship strikes, entanglements in fishing gear, strandings, mortality from military sonar, and fisheries bycatch) will be critical to evaluating the effectiveness of sanctuaries establishments. Together, the information on impacts and habitat use patterns will help Pelagos and ACCOBAMS managers interpret year-to-year changes in cetacean distribution and abundance in the face of oceanographic changes and consider such knowledge when proposing the development of larger networks of MPAs for cetaceans in the Mediterranean Sea.

In addition to the critical information on the threats and distributions of Cetacea needed for protecting these species and their critical habitats, a broader ecosystem-level understanding of the physical (ocean currents, frontal systems) and biological (productivity, prey distributions) processes supporting Cetacea aggregations in the Liguria Sea is needed to devise a longer-term management plan for the Pelagos Sanctuary.

Because lower-trophic-level planktonic organisms are more closely tied to physical processes responsible for ocean productivity and aggregation than upper-trophic-level nekton (e.g., squid and fish), these trophic linkages should be easier to outline and model for plankton feeding than for fish/squid-feeding Cetacea. Focusing on the fin whale, the most numerous zooplankton-feeding cetacean in the Liguria Sea, an assessment was made as to what extent the productivity of their summer-time foraging grounds varies temporally, from month to month and from year to year.

While the study area considered by existing data is restricted to the waters north of Corsica, these replicated systematic surveys provide a long-term perspective of fin whale (*Balaenoptera physalus*) distributions in the Liguria frontal system. Other larger-scale summer surveys have sighted fin whales within a broader area of the Liguria Sea, extending from southern France to the south west of Corsica. The area of fin whale concentrations covered 23,000 km^2 (26% of the entire sanctuary) and spanned depths from 32 to 2823 m. Overall, 80.9%, 18.9%, and 0.2% of this area consisted of abyssal plain (>2000 m), slope (2000–200 m), and shelf (<200 m) waters, respectively, reinforcing the suggestion that fin whales in the area occur off the continental shelf and slope. Having defined the "climatological" fin whale summer-time foraging grounds, the Chl*a* concentration within this region during four summer months (June–September) was calculated using time series of monthly (SeaWiFS) imagery (8 years: 1998–2005), with a spatial resolution of 9 km. A time series of 32 images (4 months and 8 years) was compiled, and a two-way analysis of variance (ANOVA) was performed to determine whether there was significant monthly and interannual variability in Chl*a* concentrations in this region. While no significant variability across months was detected, year-to-year changes were documented. However, the extent to which localized primary and secondary productivity and fin whale distributions respond to this variability is unknown. In particular, the degree to which the observed changes in primary producer abundance may eventually influence the upper-trophic-level predators in this system is not understood. Nevertheless, these remote sensing analyses highlight year-to-year changes in ocean productivity and water mass distributions. Thus, a comprehensive management plan should understand how these physical-biological interactions influence ecosystem structure and species distributions from year to year. For example, two critical issues pertain to the ability of the Pelagos Sanctuary to support resident Cetacea populations and to the connectivity between the Liguria Sea and other Cetacea foraging grounds elsewhere in the Western Mediterranean. These factors will influence the ability of resource managers to interpret trends in population abundance and demographic rates of local populations and the basin-wide movement patterns of these species. This understanding will be critical to assessing whether the Pelagos Sanctuary is achieving its conservation goals at regional (Liguria Sea) and basin-wide (Mediterranean) scales.

Perhaps the most serious threat to Ligurian Cetacea is the development by fishermen in France of modified *thonaille* or *courantille volante* fishing gear by adding floating anchors to gill nets as a way to overcome the driftnet ban. Despite this slight modification, this fishing gear is still considered by the European Commission as a driftnet and, as such, is prohibited. In spite of bans imposed by regional regulatory bodies such as the European Commission, the General Fisheries Commission for the Mediterranean, and the International Commission for the Conservation of Atlantic Tuna, driftnets are still used in the Sanctuary and cause significant levels of Cetacea mortality.

The second challenge is determining the extent to which the use of the Pelagos zone as a test zone for acoustic experiments, mainly by the military and scientists using the NATO Undersea Research Centre (NURC) in La Spezia, constitutes a threat to Cetacea and the ecosystem. The Pelagos parties raised their concern about these activities at their first meeting and started fruitful discussions with NURC. This collaboration is ongoing within the scope of the ACCOBAMS work on the impact of submarine noise on Cetacea. The third obstacle entails streamlining the management of what is inherently a complex multinational entity with overlapping jurisdictions. France, Italy, and Monaco all utilize the international tools quoted in the agreement preamble to protect Cetacea in the Pelagos region. A clear hierarchy between these initiatives must be established to avoid duplication and dispersion of conservation actions.

The fourth challenge to the effective management of the Pelagos Sanctuary entails the need for a framework to guide and assess the effectiveness of broader ecosystem-level objectives, which seek to extend conservation measures from protected species and their habitats to marine food webs and ecosystem-wide processes. It is essential that effective ecosystem-level management has a clear rationale and a firm knowledge base.

Effective ecosystem-level management of the Pelagos Sanctuary will require a multidisciplinary understanding of the environment (Liguria Sea) and its Cetacea populations. In particular, it is critical to gain an understanding of the physical processes that sustain high localized ocean productivity in the Liguria Sea and the trophic links supporting the food webs exploited by cetacean aggregations in this area. For instance, an estimated 1000 fin whales aggregate within the Pelagos Sanctuary in summer, each capable of consuming up to several hundred kg of Euphausiacea krill daily. Therefore, an understanding of the processes responsible for the formation of critical physical (e.g., the Ligurian front) and biological (e.g., Euphausiacea concentrations) features is essential to understanding how the Liguria Sea can support these upper-trophic marine predator changes interannually and over the long term. In particular, these dynamics will influence the criteria and the success of any zonation proposal because the location and extent of these habitat features vary in time and space. Furthermore, an improved understanding of the ecological significance of the Pelagos Sanctuary for marine mammals will require a detailed knowledge of the trophic interactions supporting these higher-level predators, including the spatial scales of their foraging, their local trophic requirements, and the dynamics of their prey. Habitat-use analyses will focus on identifying critically important foraging areas and migration routes, which will require more stringent protective measures (e.g., fishery closures, rerouting of shipping lanes to avoid vessel strikes).

Another critical aspect of MPA management entails the continued monitoring of the ecological and anthropogenic conditions within and outside the MPA, including the status of the protected resources, the patterns of human use in time and space, and the status and trends in existing and anticipated threats. Adaptive management will be particularly important in the Pelagos Sanctuary, given the dynamic nature of the frontal region and the highly migratory habits of Cetacea. The three broad ecosystem-level management goals listed above provide a tangible target for the management of this MPA, which will help steer its implementation plan and the design of a monitoring plan to assess its effectiveness. In addition to the systematic ecological and socio-economic research programme envisioned by the management plan, effective stewardship will require developing measurable and tangible objectives. These may include measures of biotic integrity (e.g., standing stocks, productivity) and environmental variability (e.g., oceanography, disturbance regimes), along with appropriate indicators of physical and biological variability at short (interannual) and long (climate change) temporal scales. By linking the broader management goals with a field monitoring programme, these quantitative and measurable objectives will help identify those habitats, processes, and threats "critical" to achieving the management objectives. At the same time, the work that is necessary if these goals are to be achieved will catalyse research, outreach, and education within the Sanctuary and in the broader Mediterranean Sea.

On the basis of the management plan, there are three general recommendations for developing a monitoring programme and associated quantitative metrics of success. First, we proceed with developing clear, specific, and quantifiable objectives for management. Along with the objectives, a monitoring plan for measuring management success against each objective should be articulated. The lack of clear metrics of success is a dangerous pitfall of MPA implementation because the inability to gauge the success of these management actions can result in disillusionment, loss of credibility, and community and industry backlash against established and future parks and marine zoning initiatives. Second, we developed a broad biogeographic and ecosystem-level analysis of Cetacea physiology and ecology, including their distributions, abundance, habitats, and community structure. This approach should quantify changing conditions within and outside sanctuary waters and consider genetic, stock, species, community, and ecosystem levels. Potential metrics could include changes in ocean productivity over time, studies of the density and distribution of Cetacea, research on the population and group structure (e.g., age classes) and reproductive rates of Cetacea, and investigations of the abundance and composition of their prey. Third, a programme was established to map the spatial and temporal distributions of the threats to Cetacea and the oceanographic processes supporting ocean productivity and prey availability within and outside of the Sanctuary. Potential metrics could include changes in bycatch rates and ship strikes, surveys of floating marine debris and derelict fishing gear, studies of pollutants in the food web and in cetacean tissues, studies of anthropogenic noise levels in the area, and monitoring of human activities (whale-watching, oil tanker and cargo vessel traffic, fishery distributions) within and outside the Sanctuary.

A wealth of information on oceanographic, ecological, and socio-economic aspects will be needed to integrate the Pelagos Sanctuary management plan with other aspects of Cetacea and ecosystem protections outside of its waters.

The development of a coordinated sanctuary management plan and a quantitative monitoring programme will be critical to ensuring the effective implementation of the Pelagos Sanctuary and the long-term conservation of Liguria Sea Cetacea. This dual approach will provide the information foundation needed for the long-term management of the sanctuary, as well as clear guidelines to assess the sanctuary's effectiveness.

9.4 Pelagos for Tourism: Public Aquaria and Whale Watching

An aspect of indoor tourism has historically developed the proposal of exotic or unreachable natural habitats in closed buildings where to host a large public (Belmonte and Miglietta 2022). Pelagic organisms (medusae, fin fish, and marine Mammalia) are commonly hosted in closed structures where people can observe

Fig. 9.7 Left, an internal tank of the aquarium of Sevastopol (Russia), with sturgeons. Right, the building of the Stazione Zoologica of Naples (the "Aquarium", as known by local people). (photographs G. Belmonte)

them alive. The proposal flourished in the second half of the nineteenth century when Naples, Monaco, and Sevastopol opened (in Italy, the Principality of Monaco, and Russia, respectively) aquaria of living marine organisms for the public (Fig. 9.7).

Those structures, to tell the truth, were realized to sustain the research on marine organisms carried out by scientists in the so-called Marine Biology Stations. Thus, their tourist attraction arrived as a by-product. The size of the ancient aquaria, however, was never enough to mimic the ocean environment, and large species were not commonly found in small tanks.

Regarding the scenario, fishes and mobile (swimming) fauna always had a preferential position in the water tanks, due to the life that the movement communicates, more than the static staying of sessile living organisms.

The necessity to host large specimens, or large schools of fish, and the positive reaction of the public encouraged the realization of large aquaria (sometimes termed "*oceanaria*" for their size), where large specimens can be hosted in vast water tanks and many members of the public can be received.

Thus, in a time span of 120 years, in Italy, the marine aquarium proposal passed from the pioneering *Stazione Zoologica* (opened in 1873 and still maintaining the original tanks) able to receive today 120,000 visitors per year to that of Genoa (opened, 1992), which repeatedly surpassed 1,000,000 visitors per year.

The enlargement of tanks and structures, with the consequent complication of management due to the needed technological solutions, is suggested by the high demand for and appreciation of large pelagic animals by the public. Sharks, dolphins, and sea turtles are attractive and stimulate economic investments.

Modern *oceanaria* are present on all continents and host, in many cases, hundreds of marine species and millions of visitors per year, thus representing important economic reference points for local societies other than the occasion of studies on hosted species (Fig. 9.8).

The enlargement of aquaria in modern times was probably driven by tourism-economy requests, but their complexity spontaneously requested a series of

Fig. 9.8 At the top, Lisboa Oceanarium (Portugal) (photograph A.M.Miglietta); on the bottom, San Sebastian Aquarium (Spain) (photograph G. Belmonte)

attention (for the management, for the public, and for the animals) that created information and made the aquaria actual research centres.

More recently, the tourist proposal has also involved the open sea. Many operators propose open sea fishing (fishery tourism, developed in Italy together with *itti*-tourism, in a legal form from 1999) to the public to receive an integration of resources for the use of their equipment, sometimes also higher than the resources obtained by the fishery itself.

A proposal not integrative of other activities is that of "whale-watching", where the public is accompanied in the open sea by tour operators to observe Cetacea in their natural environment. When tour operators are scientists or researchers who collect data on Cetacea behaviour, the action enters the frame of **research-tourism**, an activity that sustains the costs of data collection in the natural environment (the open sea in this case) (Belmonte et al. 2023b).

Aquaria, fishery tourism, and research tourism create an important supply to the human economy (tourism is the largest industry of the planet) but must be maintained in the frame of eco-sustainability to avoid becoming an additional aspect of the "human impact" on natural environments. What this tourism realizes in addition (and probably the most important aspect) is the education of visitors with respect to nature and species. These elements, in fact, represent a valuable resource to be preserved (in the specific case of the tourism economy), and finally, the actions are conducted by the best prepared operators on the market (biologists, fishermen, and researchers) that represent the best guarantee for the eco-sustainability and the resource safety of the proposals (Fig. 9.9).

9.5 Pelagos for Arts Inspiration

The pelagic domain has always inspired literature, with *The Odyssey* by Homer probably being the first best seller among those describing adventures and interactions between man and pelagic monsters. Pelagic life has always represented a frontier for human courage, sometimes uncrossable, as in the case of the fight of Captain Acab with the white whale "Moby Dick" in the romance of Melville (1851). The consciousness of an impossible struggle is also that of the old fisherman and the marlin in a romance, winner of the Nobel Prize (Hemingway 1952). The literary case of a man integrated with the deep pelagos is only that of Captain Nemo (Verne 1870), whose oceanic lifestyle is punctuated with normal interaction with the rest of humanity.

Singers and music players derived inspiration by literature and stories of pelagic life, but in the case of the rock star Frank Zappa (1940–1993), a music brane was not derived from a literary personage but directly from a small medusa dedicated to him (*Phialella zappai*) (Fig. 9.10).

Painters have always reproduced and represented nature in their artistic products, but pelagic and/or deep sea scenes are not common, probably because such a world has never been deeply explored by artists and because the deep pelagos proposal should not find understanding in the large public.

The beauties of planktonic organisms, however, captured many artists-scientists who illustrated their work realizing scientific representations of great artistic value in the past and in the present.

Fig. 9.9 Activity of whale watching in the Gulf of Taranto (central Mediterranean) managed by the "Jonian Dolphin Association", a team of researchers studying Cetacea population dynamics and behaviour. (With permission from Jonian Dolphin Association)

Ernst Haeckel, one of them, realized atlases of unicellular organisms that gave him prestige in addition to his already scientific reputation (Fig. 9.11 top, left); in one of his last books (Haeckel 1904), he proposed a collection of scientific representations of marine organisms and hosted drawings of other scientists, such as copepods, from the wonderful atlas of Giesbrecht (1892) (Fig. 9.11 bottom). Modern technologies finally assisted and supported the work of Alberto Gennari, who realized didactic drawings for the Marine Biology Museum of the University of Salento from 2005 to 2017 (Fig. 9.11 top, right).

Fig. 9.10 The 2004 calendar of Marine Biology Museum of University of Salento (Italy) dedicated to Frank Zappa and his rock music (drawing of A. Gennari). The hydrozoan species *Phialella zappai* was dedicated to Frank Zappa in 1987 by the marine biologist F. Boero). The silhouette of the guitar is proposed as the bottom for the polyp benthic stage of the medusa. The spots (representing the nematocysts) aligned long one of the tentacles of the floating medusa are the notes sequence of the prelude of a song that the artist wrote and dedicated to the marine biologist who described the species

Art dedicated to illustrating nature and its functioning (scientific illustration) became a powerful method of communication in all science fields because it allowed complex ideas, details, and theories to be presented in an interesting and informative way. Among complex things, the functioning of a living being and/or of a community of organisms are the best subjects to be represented by drawings. Pelagic life is the last one to be focussed on, but many attempts exist.

Valent artisans created wonderful representations of pelagic micro creatures.

A famous example was the glass pelagic animals of Leopold and Rudolf Blaschka (father and son) collections (Fig. 2.25b) distributed in all the main Natural History Museums of the world. The Blaschka realized different objects for different Museums: although they were able to produce wonderful and complex invertebrates (jelly plankton, overall), they did not limit their production to the same set of species but explored continuously the biodiversity to find always new species to represent, becoming collection keepers of nature forms, more than glass technicians.

A recent extension of the necessity to represent plankton forms (mostly microscopic, transparent, and/or delicate for preservative solutions) has been the intervention of resinous materials to be transformed by hands (Fig. 9.12) or by a 3D printer.

The evolution of instruments to visually explore the micro-world in the last 50 years allowed an incredible enrichment of our general perception of living organisms suspended in a water world.

Fig. 9.11 Drawing and painting reproduction of marine life. Top left, Siphonophora of E. Haeckel (1904); top right, a scene of fishery menaced by a "jelly sea" in the interpretation of A. Gennari (for the 2014 Calendar for the Marine Biology Station, University of Salento); bottom, Copepoda of W. Giesbrecht (1892) (see also Fig. 3.24)

Fig. 9.12 3D model of a copepod hanging from the roof of a room of the oceanarium in Stralsund (Germany). (Photograph G. Belmonte)

References

Agardy TS (1997) Marine protected areas and ocean conservation. R.E. Landes, Austin, TX

Belmonte G, Miglietta AM (2022) La Natura risorsa Turistica. Grifo Publ, Lecce, p 288

Belmonte G, Denti G, Rubino F (2023a) Planktonic Copepoda along the Confinement Gradient of the Taranto Sea System (Southern Italy) after Lockdown of Human Activities Due to the COVID-19 Pandemic. Water 15:2449

Belmonte G, Balestrieri R, Santacesarea FC, Fanizza C (2023b) Proposals of "open air" tourism, associated with scientific investigation in the wild, ready as psychological health care in the period of COVID 2019 pandemic. In: Trono A, Castronuovo V, Kosmas P (eds) Managing natural and cultural heritage for a durable tourism. Springer Nature, Switzerland, pp 337–250

Bensch A, Gianni M, Gréboval D, Sanders JS, Hjort A (2008) Worldwide review of bottom fisheries in the high seas. FAO Fish Aquac Tech Paper 522:1–145

Bianchi CN, Morri C (2000) Marine biodiversity of the Mediterranean Sea: situation, problems and prospects for future research. Mar Pollut Bull 40(5):367–376

Choy CA, Haddock SH, Robison BH (2017) Deep pelagic food web structure as revealed by in situ feeding observations. Proc R Soc B Biol Sci 284(1868):20172116

Clark MR, Althaus F, Schlacher TA, Williams A, Bowden DA, Rowden AA (2016) The impacts of deep-sea fisheries on benthic communities: a review. ICES J Mar Sci 73(suppl_1):i51–i69

Coll M, Piroddi C, Steenbeek J, Kaschner K, Lasram FBR, Aguzzi J, Ballesteros E, Bianchi CN, Corbera J, Dailianis T, Danovaro R, Estrada M, Froglia C, Galil BS, Gasol JM, Gertwagen R, Gil J, Guilhaumon F, Kesner-Reyes K, Kitsos M-S, Koukouras A, Lampadariou N, Laxamana E, Lopez-fe De La Cuadra CM, Lotze HK, Martin D, Mouillot D, Oro D, Raicevich S, Rius-Barile J, Saiz-Salinas JI, San Vicente C, Somot S, Templado J, Turon X, Vafidis D, Villanueva R, Voultsiadou E (2010) The biodiversity of the Mediterranean Sea: estimates, patterns, and threats. PLoS One 5(8):e11842

Collins MA, Brickle P, Brown J, Belchier M (2010) The Patagonian Toothfish: biology, ecology and fishery. Adv Mar Biol 58:227–300

Cousins NJ, Lindeque PK, Fileman ES (2019) A comparison of traditional and novel zooplankton sampling methods for environmental DNA metabarcoding. Environ DNA 1(4):342–350

D'Onghia G, Carlucci R, Maiorano P, Panza M (2003) Discards from deep-water bottom trawling in the eastern-Central Mediterranean Sea and effects of mesh size changes. J Northwest Atl Fish Sci 31:245–261

Danovaro R, Snelgrove PV, Tyler P (2014) Challenging the paradigms of deep-sea ecology. Trends Ecol Evol 29(8):465–475

FAO (Food and Agriculture Organization of the United Nations) (2018) The state of Mediterranean and Black Sea fisheries 2018. General Fisheries Commission for the Mediterranean, Rome

Fierstine HL, Walters V, Reyier EA (1997) The biology and fishery of swordfish, *Xiphias gladius*, in the eastern North Pacific Ocean. NOAA Technical Report NMFS 142:1–32

Forcada J, Notarbartolo di Sciara G, Fabbri F (1995) Abundance of fin whales and striped dolphins summering in the Corso–Ligurian Basin. Mammalia 59:127–140

Gage JD, Tyler PA (1991) Deep-sea biology: a natural history of organisms at the deep-sea floor. Cambridge University Press, Cambridge

García-Ruiz C (2013) In: Esteban A, Templado J (eds) Mediterranean marine protected areas: tools for conservation and sustainable management. IUCN

Garibaldi F (2015) Effects of the introduction of the mesopelagic longline on catches and size structure of swordfish in the Ligurian Sea (western Mediterranean). Collect Vol Sci Pap ICCAT 71(5):2006–2014

Gartner JV, Crabtree RE, Sulak KJ (2008) A multiple gear approach to deep-sea sampling: biomass and diversity of the fishes of the eastern Gulf of Mexico. Deep-Sea Res II Top Stud Oceanogr 55(1–2):188–201

GFCM (General fisheries Commission for the Mediterranean) (2017) State of Mediterranean and Black Sea Fisheries. FAO, Geneva

GFCM (General Fisheries Commission for the Mediterranean) (2019) Report of the twenty-first session of the Scientific Advisory Committee on Fisheries (SAC). GFCM: SAC21/2019/2

Giesbrecht W (1892) Systematik und Faunistik der pelagischen Copepoden des Golfes von Neapel und der angrenzenden Meeres-abschnitte. Friedlander & Sohn, Berlin

Haeckel H (1904) Kunstformen der Natur. Prestel Publ, New York, London

Hardwood Ecosystem Experiment (2017) Research activities. www.heeforeststudy.org

Hemingway E (1952) The old man and the sea. Charles Scribners' Sons, New York

ICES (International Council for the Exploration of the Sea) (2019) Report of the working group on widely distributed stocks (WGWIDE). ICES Sci Rep 1(56):1–410

ISSF (International Seafood Sustainability Foundation) (2021) Tuna stock status update, 2021: status of the world fisheries for tuna. ISSF Technical Report 2021-09. International Seafood Sustainability Foundation, Washington, D.C

Jalkanen P-J, Johansson L, Andersson M, Majamaki E, Sigray P (2022) Underwater noise emissions from ships during 2014–2020. Environ Pollut 311:119766

Melville H (1851) Moby dick or the whale. Harper & Brothers Publ, New York

Norse EA, Brooke S, Cheung WW, Clark MR, Ekeland I, Froese R, Gjerde KM, Haedrick RL, Heppell SS, Morato T, Morgan LE, Pauly D, Sumalia R, Watson R (2012) Sustainability of deep-sea fisheries. Mar Policy 36(2):307–320

Notarbartolo di Sciara G, Agardy TS, Hyrenbach D, Scovazzi T, Van Klaveren P (2008) The Pelagos sanctuary for Mediterranean marine mammals. Aquatic Conserv: Marine Freshwater Ecosystems 18:367–391

Panigada S, Notarbartolo di Sciara G, Zanardelli PM, Airoldi S, Borsani JF, Jahoda M (2005) Fin whales summering in the Ligurian Sea: distribution, encounter rate, mean group size and relation to physiographic variables. J Cetacean Res Manag 7:137–145

Ramirez-Llodra E, Tyler PA, Baker MC, Bergstad OA, Clark MR, Escobar E, Levin LA, Menot L, Rowden AA, Smith CR, Van Dover CL (2011) Man and the last great wilderness: human impact on the deep sea. PLoS One 6(8):e22588

Relini M, Orsi-Relini L, Relini G (1994) An offshore buoy as a FAD in the Mediterranean. Bull Mar Sci 55(2–3):1099–1105

Simmonds J, MacLennan D (2005) Fisheries acoustics: theory and practice. John Wiley & Sons, New York

Tsikliras AC, Dinouli A, Tsiros VZ, Tsalkou E (2015) The Mediterranean and Black Sea fisheries at risk from overexploitation. PLoS One 10(3):e0121188

Verne J (1870) Vingt mille lieues sous les mers. Biblioteque d'Education et de recreation. Hetzler et C Publ, Paris

Elements of Pelagos Biology

With a Focus on the Mediterranean Sea

List of Geographic Sites

Adriatic Sea
Aegean Sea
Agulhas
Alboran Sea
Aleutines
Amazon River
Amu-daria
Angola
Antarctic Ocean
Apennines
Aqaba
Argentina
Aral Sea
Ascension island
Atlit Yam, Israel
Atlantic Ocean
Australia
Azov Sea
Bab el Mandab
Baikal Lake
Balearic Sea
Baltic Sea
Bassano Lake
Benguela
Bering Sea
Bermuda archipelago
Biwa Lake
Black Sea
Boothia gulf

G. Belmonte, *Elements of Pelagos Biology*,
https://doi.org/10.1007/978-3-031-74280-4

Bosphorus, Istanbul
Bosphorus, Cimmerian
California
Canaries archipelago
Cape Andreas Kastros, Cyprus
Cape Cod
Caribbean Sea
Caspian Sea
Catalan Sea
China
Congo River
Corsica
Crimea
Denmark
Ercolano
Fanafjorden, Norway
Fayoum, Wadi Al-Hitan
Fiji islands
Finsterthaler
Florida Keys
Foix canyon
Formentera, Balearic
Frankhthi cave, Greece
Galapagos
Gibraltar strait
Great Barrier Reef
Guiana
Guinea
Gulf Stream
Hatteras cape
Hatay-Eufrates valley
Havre caldera
Hawaii
Humboldt
Indian Ocean
Indonesia archipelago
Iskenderun, Turkey
Japan
Kashmir
Khiwa
Krakatau
Kuroshio
Lakeba Island
Las Cuevas de Nerja, Spain
Latnjajaure

Elements of Pelagos Biology

With a Focus on the Mediterranean Sea

List of Cited Taxa

Abylopsis
Acartia clausi
Acartia margalefi
Acartia tonsa
Aequorea victoria
Aglaura
Agrostichthys
Aliivibrio fischeri
Ambulocetus natans
Amonardia phyllopus
Anguilla anguilla
Anguilla rostrata
Anomalocaris
Aphanius fasciatus
Apogon
Aptenodytes forsteri
Aptenodytes patagonicus
Apulemia
Arapaima gigas
Archaescrictius ruggieroi
Argonauta argo
Artemia franciscana
Asellota Isopoda
Aspidosiphon elegans
Asterionellopsis
Aurelia sp.
Auxis rochei
Bacteriastrum
Balaenoptera physalus

G. Belmonte, *Elements of Pelagos Biology*,
https://doi.org/10.1007/978-3-031-74280-4

Balistes capriscus
Basilosaurus
Berardius bairdii
Beroe ovata
Birgus latro
Bolinus brandaris
Boreogadus saida
Bosmina longispina maritima
Calanoides
Calanus finmarchicus
Calanus glacialis
Calanus helgolandicus
Calanus hyperboreus
Calanus sinicus
Calocalanus pavo
Callianira
Calocalanus pavoninus
Caretta caretta
Caulerpa racemosa var. *cylindracea*
Caulerpa taxifolia
Cephalochordata
Centropages typicus
Ceratiidae
Cercopagis pengoi
Cetorhinus
Chaetoceros
Calothrix rhizosoleniae
Chelonia mydas
Chrysaora
Chrysochromulina
Citharistes
Cladocora caespitosa
Clausocalanus arcuicornis
Clausocalanus furcatus
Clausocalanus lividus
Clausocalanus paululus
Clausocalanus pergens
Climacodium frauenfeldianum
Clupea harengus
Clupeonella
Corallium rubrum
Corophium volutator
Corycaeus
Crassostrea gigas
Crocosphaera watsonii

Crystophora cristata
Ctenocalanus vanus
Cumacea
Cyclothone signata
Cyphastrea
Cypridina
Cypselurus
Cystoseira compressa
Dactylosolen fragilissimus
Decapoda
Delphinus delphis
Dermochelys coriacea
Dictyocha
Dissostichus eleginoides
Distephanus
Emiliania huxleyi
Enewton harveyi
Engraulis encrasicolus
Engraulis japonicus
Erignathus barbatus
Eucalanus
Eucalanus bungii
Euchaeta acuta
Eudoxoides
Eumetopias jubatus
Euphausia
Euphausiacea
Euprymna scolopes
Eurytemora herdmanni
Euterpina acutifrons
Exocetus
Farranula rostrata
Fenneropenaeus merguiensis
Fialella zappai
Fritillaria
Gadhus morua
Gammarideae Amphipoda
Glaucus
Globicephala melas
Globigerina glutinata
Golfingia margaritacea
Goniomenus
Gonyaulax
Grampus griseus
Gymnodinium

Gyrodinium
Halichoerus grypus
Haloptilus longicornis,
Halosphaera viridis
Harpacticoida
Harpacticus sp.
Hemiaulus
Hemimysis margalefi
Heterocapsa
Hexaplex trunculus
Hippopotamus
Hippospongia
Histioneis
Histriophoca fasciata
Homarus americanus
Hoplostethus atlanticus
Hydra
Hydractinia
Hydroides elegans
Hydruga leptonyx
Hyemoschus aquaticus
Hyperoodon ampullatus
Illex argentinus
Imantonia
Indohyus mayor
Janthina
Kornickeria bullae
Kornickeria louisi
Labidocera aestiva
Laboea strobili
Labrus merula
Lagocephalus sceleratus
Lampyridae Coleoptera
Latreutes fucorum
Leedsichthys problematicus
Lensia
Lepas anatiphera
Leptocylindrus danicus
Leptocephalus
Leptonychotes weddellii
Lessardia
Liriope
Liza ramada
Lobodon carcinophaga
Lophelia pertusa

Lucicutia flavicornis
Macoma balthica
Macrouridae
Makaira nigricans
Marenzelleria sp.
Mecynocera clausi
Megalodon megalodon
Meganyctiphanes norvegica
Membranipora
Menidia menidia
Mesoplodon densirostris
Messapicetus longirostris
Microsetella rosea
Millepora
Minidiscus
Mirounga angustirostris
Mirounga leonina
Mnemiopsis leidyi
Mobula birostris
Mola mola
Monoporeia affinis
Muggiaea
Myctophidae
Mysida
Mysis mixta
Neocalanus
Neoceratium
Neomonachus shauinslandi
Neomysis integer
Obelia
Octodon degus
Odobenus rosmarus
Oithona plumifera
Oithona setigera
Oithona similis
Ommatophoca rossii
Oncaea media
Oncaea mediterranea
Oncorhynchus gorbuscha
Ondatra
Ornithocercus
Oligoryzomys longicaudatus
Ostracoda
Oxytoxum
Pagophilus groenlandicus

Pakicetus inachus
Paracalanus parvus
Paracartia grani
Paracartia latisetosa,
Paradactylopodia brevicornis
Parapriacanthus
Phaeocystis
Phaeoptyx pigmenteria
Phascolion convestitum
Phoca largha
Phoca vitulina
Phocarctos hookeri
Photeros annecohenae
Photeros harveyi
Photeros parasitica
Physalia physalis
Physeter macrocephalus
Pinna nobilis
Plagioselmis prolonga
Pleuromamma gracilis
Pleuromamma abdominalis
Pleuronectes platessa
Plotosus lineatus
Pocillopora
Podon polyphemoides
Polyprion americanus
Pollachius virens
Porites
Porichthys
Porpita porpita
Posidonia oceanica
Proboscia
Prochlorococcus
Prorocentrum
Protoperidinium
Psetta maxima
Pseudocalanus
Pseudo-nitzschia delicatissima
Pseudosagitta gazellae
Pteriacartia josephinae
Puffinus mauretanicus
Pusa hispida
Pusa sibirica
Pycnogonida
Reclutia

Regalecus
Rhincodon typus
Rhincops nigra
Rhizosolenia
Rhopalonema
Rhopilema nomadica
Richelia intracellularis
Ruditapes philippinarum
Saccostrea
Sagitta
Sardina pilchardus
Sargassum
Sargassum fluitans
Sargassum natans
Sarpa salpa
Schistomysis kervillei
Schistomysis ornata
Schistomysis parkeri
Scomber scombrus
Scophthalmus rhombus
Scrippsiella trochoidea
Sepia officinalis
Sepietta oweniana
Sfeniscidae
Siriella jaltensis
Skeletonema
Spelaeomysis bottazzii
Spirorbus
Sparus aurata
Spongia
Stenella coeruleoalba
Subeucalanus monachus
Synechococcus
Tanaidacea
Temora stylifera
Tetrapturus belone
Thalassionema frauenfeldii
Thalassionema nitzschioides
Thalassiosira
Themisto
Thysanoessa raschii
Thunnus alalunga
Thunnus thynnus
Tisbe sp.
Tomopteris

Trichodesmium spp.
Turbellaria
Tursiops truncatus
Vargula hilgendorfii
Vargula magna
Vargula tsujii
Velella velella
Xiphias gladius
Zalophus californianus
Zygophyseter varolai
Zyphius cavirostris

Plates[1]

[1] The 33 photographic plates were contributed by Prof. Otto Larink.

G. Belmonte, *Elements of Pelagos Biology*,
https://doi.org/10.1007/978-3-031-74280-4

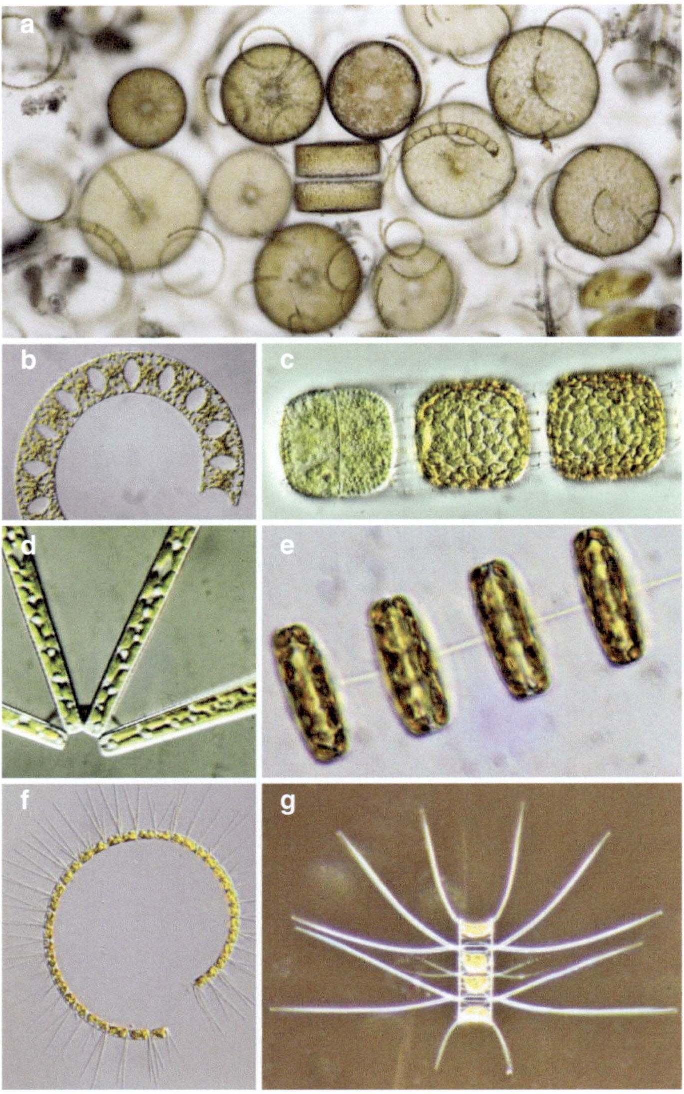

Plate 1 (**a**) *Coscinodiscus* sp. and different diatoms. Width of photo ca. 2 mm. Sylt, Germany 2015. (**b**) *Eucampia zodiacus* (Bacillariophyceae, Hemiaulales, Hemiaulaceae), width 38 µm. Sylt, Germany 2015. (**c**) *Stephanopyxis turris* (Bacillariophyceae, Melosirales, Stephanopyxidaceae), width 45 µm. Sylt, Germany 2015. (**d**) *Thalassionema frauenfeldi* (Bacillariophyceae, Thalassionematales, Talassionemataceae), length of cells 60 µm. Sylt, Germany 2014. (**e**) *Thalassiosira* sp. (Bacillariophyceae, Thalassiosirales, Thalassiosiraceae), diameter 38 µm. New Brunswick, Canada 2012. (**f**) *Chaetoceros curvisetus* (Bacillariophyceae, Chaetocerotanae, Chaetocerotaceae), cell diameter 18 µm. Sylt, Germany 2015. (**g**) *Chaetoceros decipiens* (Bacillariophyceae, Chaetocerotanae, Chaetocerotaceae), length of bristles up to 100 µm. New Brunswick, Canada 2012

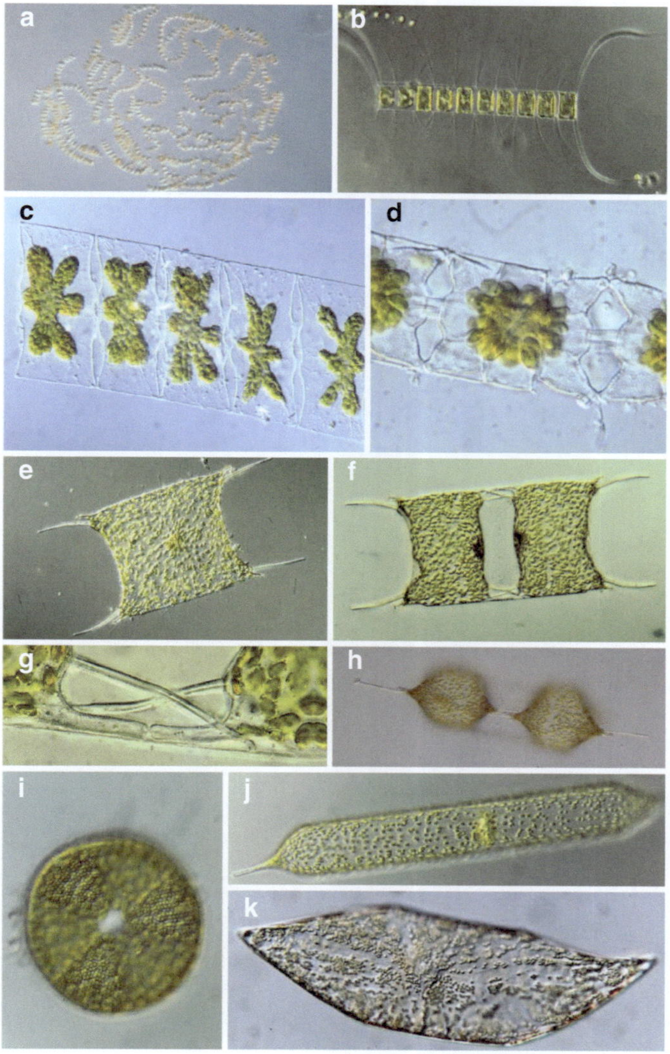

Plate 2 (**a**) *Chaetoceros socialis* (Bacillariophyceae, Chaetocerotanae, Chaetocerotaceae), diameter of colony 600 µm. Helgoland, Germany 2015. (**b**) *Chaetoceros decipiens* (Bacillariophyceae, Chaetocerotanae, Chaetocerotaceae), diameter of cells 20 µm. Sylt, Germany 2016. (**c**) *Mediopyxis helysia* (Bacillariophyceae, Mediophyceae), width of cell 135 µm. Sylt, Germany 2013. (**d**) *Odontella aurita* (Bacillariophyceae, Triceratiales, Triceratiaceae), diameter 33 µm. Sylt, Germany 2013. (**e**) *Odontella sinensis* (Bacillariophyceae, Triceratiales, Triceratiaceae), width of cell µm, Sylt, Germany 2014. (**f**) *Odontella sinensis* (Bacillariophyceae, Triceratiales, Triceratiaceae), cells after division, width of cells 195 µm. Sylt, Germany 2014. (**g**) *Odontella sinensis* (Bacillariophyceae, Triceratiales, Triceratiaceae), length of short processus 28 µm. Sylt, Germany 2014. (**h**) *Odontella sinensis* (Bacillariophyceae, Triceratiales, Triceratiaceae), length of hort processus 20 µm. Sylt, Germany 2014. (**i**) *Odontella sinensis* (Bacillariophyceae, Triceratiales, Triceratiaceae), lateral view, length 180 µm. Sylt, Germany 2015. (**j**) *Actinoptychus senarius* (Bacillariophyceae, Coscinodiscales, Heliopeltaceae), diameter 100 µm. Sylt, Germany 2015. (**k**) *Proboscia indica* (Bacillariophyceae, Rhizosoleniales, Rhizosoleniaceae), diameter 40 µm. Sylt, Germany 2014

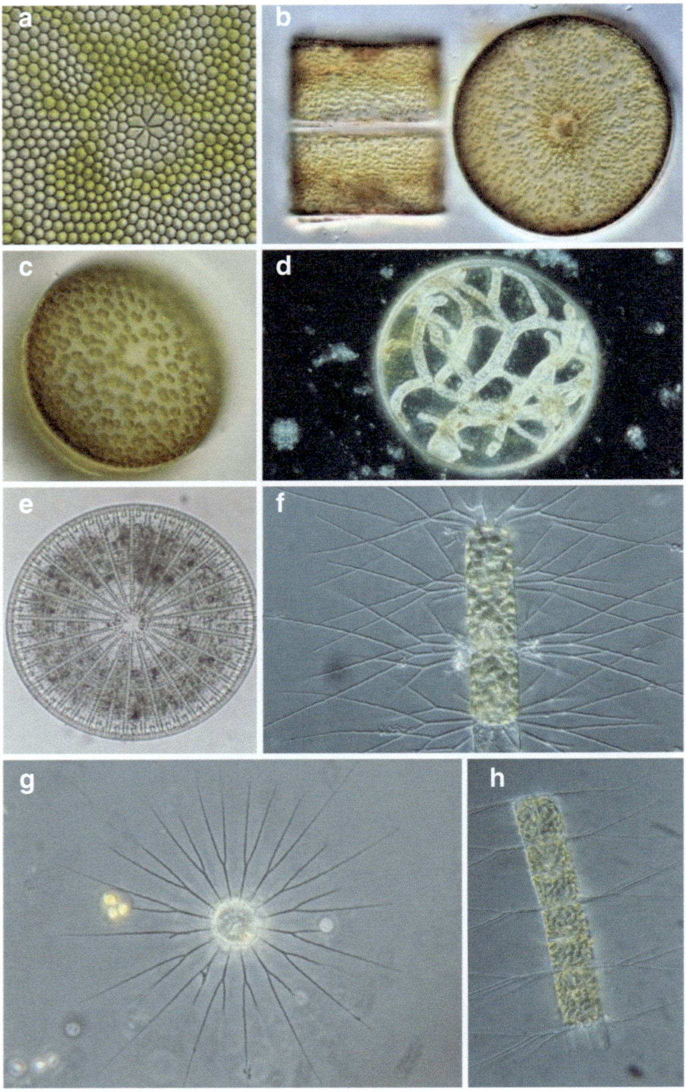

Plate 3 (**a**) *Coscinodiscus radiatus* (Bacillariophyceae, Coscinodiscales, Coscinodiscaceae), width of photo ca. 80 μm. Sylt, Germany 2014. (**b**) *Coscinodiscus radiatus* (Bacillariophyceae, Coscinodiscales, Coscinodiscaceae), diameter 370 and 240 μm. Helgoland, Germany 2015. (**c**) *Coscinodiscus granii* (Bacillariophyceae, Coscinodiscales, Coscinodiscaceae), diameter 195 μm. Sylt, Germany 2014. (**d**) *Coscinodiscus wailesi* (Bacillariophyceae, Coscinodiscales, Coscinodiscaceae) with *Lagenisma coscinodisci* (Oomycota, Lagenismatales, Lagenismataceae), diameter 330 μm. Helgoland, Germany 2014. (**e**) *Arachnodiscus* sp. (Bacillariophyceae, Arachnodiscales, Arachnodiscaceae), diameter 180 μm. Cape Town, South Africa 2011. (**f**) *Bacteriastrum hyalinum* (Bacillariophyceae, Chaetocerotanae inc. sedis, Chaetocerotaceae), diameter 15 μm. Helgoland, Germany 2021. (**g**) *Bacteriastrum hyalinum* (Bacillariophyceae, Chaetocerotanae inc. sedis, Chaetocerotaceae), diameter 15 μm. Helgoland, Germany 2021. (**h**) *Bacteriastrum hyalinum* (Bacillariophyceae, Chaetocerotanae inc. sedis, Chaetocerotaceae), diameter 15 μm. Helgoland, Germany 2021

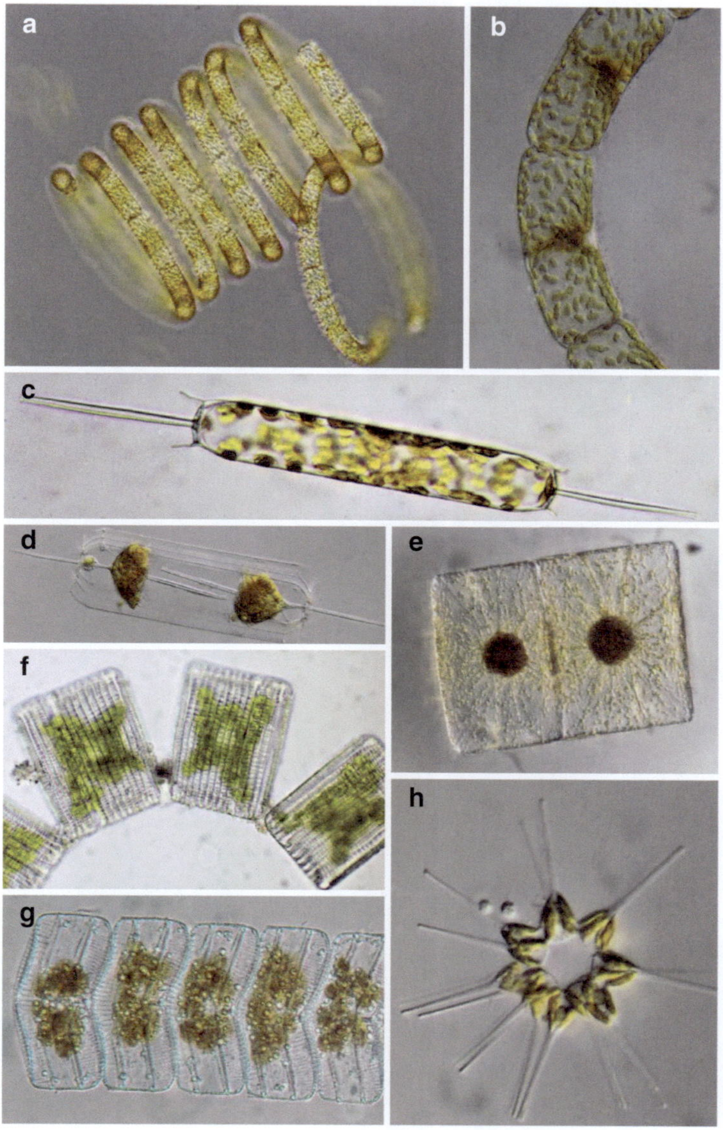

Plate 4 (**a**) *Guinardia striata* (Bacillariophyceae, Rhizosoleniales, Rhizosoleniaceae), diameter of spiral 420 µm. Sylt, Germany 2015. (**b**) *Guinardia striata* (Bacillariophyceae, Rhizosoleniales, Rhizosoleniaceae), cell diameter 40 µm. Sylt, Germany 2015. (**c**) *Ditylum brightwellii* (Bacillariophyceae, Lithodesmiales, Lithodesmiaceae), diameter 15 µm. New Brunswick, Canada 2015. (**d**) *Ditylum brightwellii* (Bacillariophyceae, Lithodesmiales, Lithodesmiaceae), length 110 µm. Sylt, Germany 2013. (**e**) *Striatella* sp. (Bacillariophyceae, Striatellales, Striatellaceae), width of cells 160 µm. St. Cyr, France 2018. (**f**) *Rhabdonema* sp. (Bacillariophyceae, Rhabdonematales, Rhabdonemataceae), length of cells 90 µm. Cape Town, South Africa 2011. (**g**) *Achnanthes* sp. (Bacillariophyceae, Achnanthales, Achnanthaceae), width of cells 70 µm. Helgoland, Germany 2015. (**h**) *Asterionella glacialis* (Bacillariophyceae, Fragilariales, Fragilariaceae), length of cells 38 µm. New Brunswick, Canada 2015

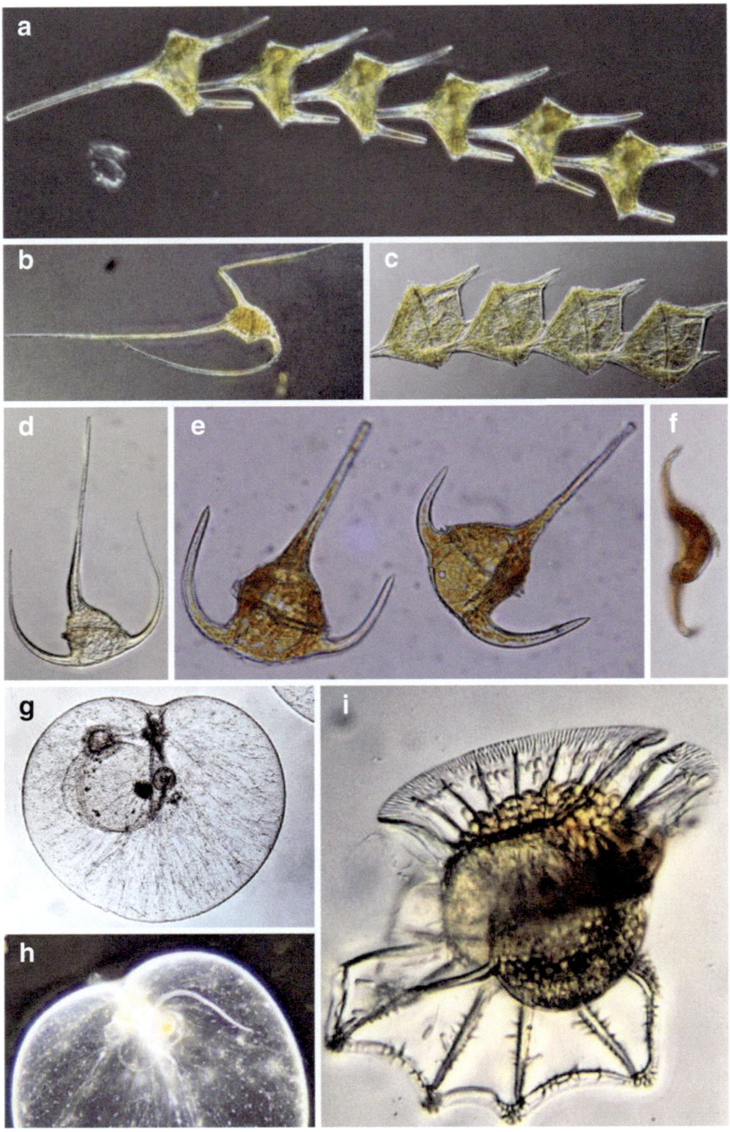

Plate 5 (**a**) *Tripos candelabrum* (Dinoflagellata, Gonyaulacales, Ceratiaceae), diameter 95 µm. Marbella, Spain 2017. (**b**) *Tripos hexacanthus* (Dinoflagellata, Gonyaulacales, Ceratiaceae), length 150 µm. Lanzarote, Spain 2019. (**c**) *Tripos pentagonus* (Dinoflagellata, Gonyaulacales, Ceratiaceae), diameter 90 mm. Marbella, Spain 2017. (**d**) *Tripos contortus* (Dinoflagellata, Gonyaulacales, Ceratiaceae), length 150 µm. Lanzarote, Spain 2019. (**e**) *Tripos* sp. (Dinoflagellata, Gonyaulacales, Ceratiaceae), width 160 µm. New Brunswick, Canada 2015. (**f**) *Tripos* sp. (Dinoflagellata, Gonyaulacales, Ceratiaceae), apical view, width 160 µm. New Brunswick, Canada 2015. (**g**) *Noctiluca miliaris* (Dinoflagellata, Noctilucales, Noctilucaceae), diameter 700 µm. Helgoland, Germany 2018. (**h**) *Noctiluca miliaris* (Dinoflagellata, Noctilucales, Noctilucaceae), with apical tentacle, diameter 900 µm. Helgoland, Germany 2012. (**i**) *Ornithocercus steinii* (Dinoflagellata, Dinophysiales, Dinophysaceae), diameter 180 µm. Lanzarote, Spain 2022

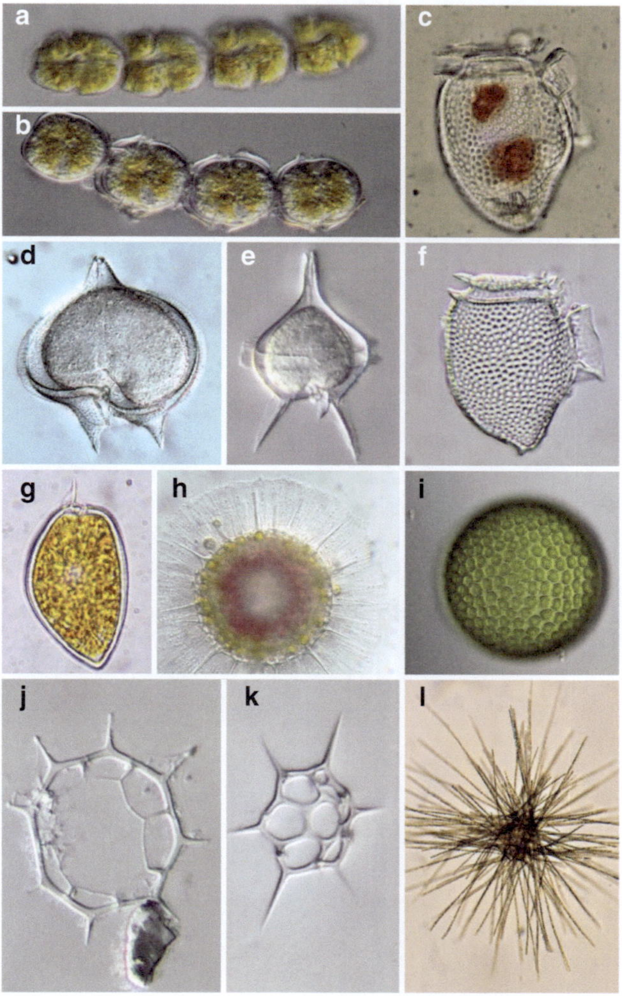

Plate 6 (**a**) A*lexandrium* sp. (Dinoflagellata, Gonyaulacales, Ostreopsidaceae), diameter 50 µm. Marbella, Spain 2017. (**b**) *Gymnodinium catenatum* (Dinoflagellata, Gymnodiniales, Gymnodiniaceae), length 180 µm. Marbella, Spain 2017. (**c**) *Dinophysis norwegica* (Dinoflagellata, Dinophysiales, Dinophysaceae), height 65 µm. New Brunswick, Canada 2012. (**d**) *Protoperidinium depressum* (Dinoflagellata. Peridiniales, Protoperidiniaceae), diameter 110 µm. Helgoland, Germany 2015. (**e**) *Protoperidinium diabolum* (Dinoflagellata. Peridiniales, Protoperidiniaceae), height 35 µm. Marbella, Spain 2017. (**f**) *Dinophysis norwegica* (Dinoflagellata, Dinophysiales, Dinophysaceae), height 55 µm. New Brunswick, Canada 2015. (**g**) *Prorocentrum* sp. (Dinoflagellata, Prorocentrales, Prorocentraceae), length 45 µm. New Brunswick, Canada 2015. (**h**) *Rhizosphaera* sp. (Radiozoa, Polycystinea, Entactinaria, Rhizosphaeridae), diameter 270 µm. Palamos, Spain 2009. (**i**) *Halosphera viridis* (Chlorophyta, Pyramimonadales, Pyramimonadaphyceae), diameter 350 µm. Lanzarote, Spain 2022. (**j**) *Octactis octonaria* (Ochrophyta, Dictyochales, Dictyochaceae), diameter 40 µm. Marbella, Spain 2017. (**k**) *Dictyocha speculum* (Ochrophyta, Dictyochales, Dictyochaceae), diameter 60 µm. Sylt, Germany 2015. (**l**) *Trichodesmium* sp.(Cyanobacteria, Cyanophyceae, Oscillatoriales, Phormidiaceae), diameter 2 mm. Lanzarote, Spain 2019

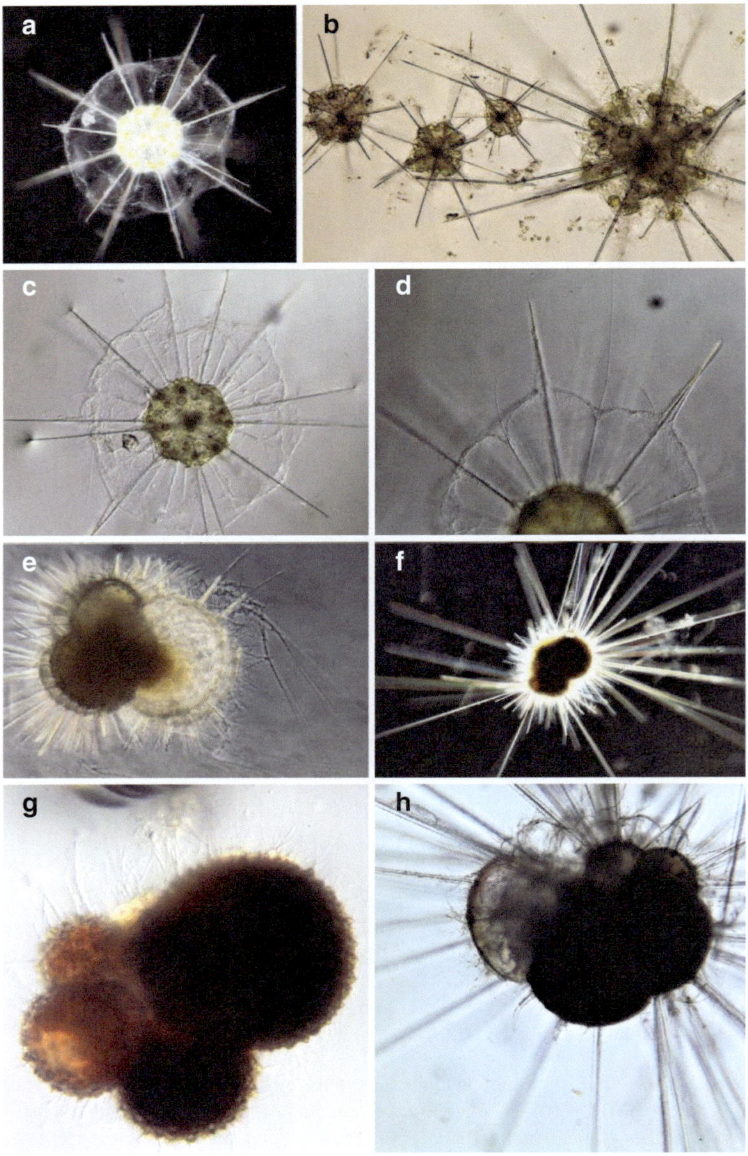

Plate 7 (**a**) *Acanthometra* sp. (Radiozoa, Arthracanthida, Acanthometridae), diameter 220 µm. Marbella, Spain 2017. (**b**) *Acanthometra* sp. (Radiozoa, Arthracanthida, Acanthometridae), diameter 50–220 µm. Marbella, Spain 2017. (**c**) *Acanthometra* sp. (Radiozoa, Arthracanthida, Acanthometridae), diameter 220 µm. Marbella, Spain 2017. (**d**) *Acanthometra* sp. (Radiozoa, Arthracanthida, Acanthometridae), diameter 220 µm. Marbella, Spain 2017. (**e**) *Globoturborotalita rubescens* (Foraminifera, Rotaliida, Globigerinidae), diameter 180 µm. Lanzarote, Spain 2022. (**f**) *Globigerinella siphonifera* (Foraminifera, Rotaliida, Globigerinidae), diameter of shell 350 µm. Lanzarote, Spain 2022. (**g**) *Trilobatus sacculifer* (Foraminifera, Rotaliida, Globigerinidae), length 300 µm. Lanzarote, Spain 2019. (**h**) *Hastigerina pelagica* (Foraminifera, Rotaliida, Hastigerinidae), diameter of shell 450 µm. Lanzarote, Spain 2022

Plate 8 (**a**) Zoothamnium sp. (Ciliophora, Sessilida, Zoothamniidae), expanded colony, length of colony 1.5 mm. Palamos, Spain 2009. (**b**) *Zoothamnium* sp. (Ciliophora, Sessilida, Zoothamniidae), contracted colony, diameter 220 µm. Palamos, Spain 2009. (**c**) *Favella ehrenbergi* (Ciliophora, Oligotrichea, Choreotrichida, Ptychocyclididae), length 90 µm. Sylt, Germany 2016. (**d**) *Codonella* sp. (Ciliophora, Oligotrichea, Choreotrichida, Codonellidae), length 110 µm. Marbella, Spain 2017. (**e**) *Dictyocysta elegans* (Ciliophora, Oligotrichea, Choreotrichida, Dictyocystidae), diameter 65 µm. Marbella, Spain 2017. (**f**) *Tintinnidium* sp. (Ciliophora,Oligotrichea, Choreotrichida, Tintinnidiidae), length 100 µm. Sylt, Germany 2013. (**g**) *Xystonella* sp. (Ciliophora, Oligotrichea, Choreotrichida, Xystonellidae), length 280 µm. Lanzarote, Spain 2022. (**h**) *Tintinnopsis* sp. (Ciliophora,Oligotrichea, Choreotrichida, Codonellidae), length 110 µm. Marbella, Spain 2017

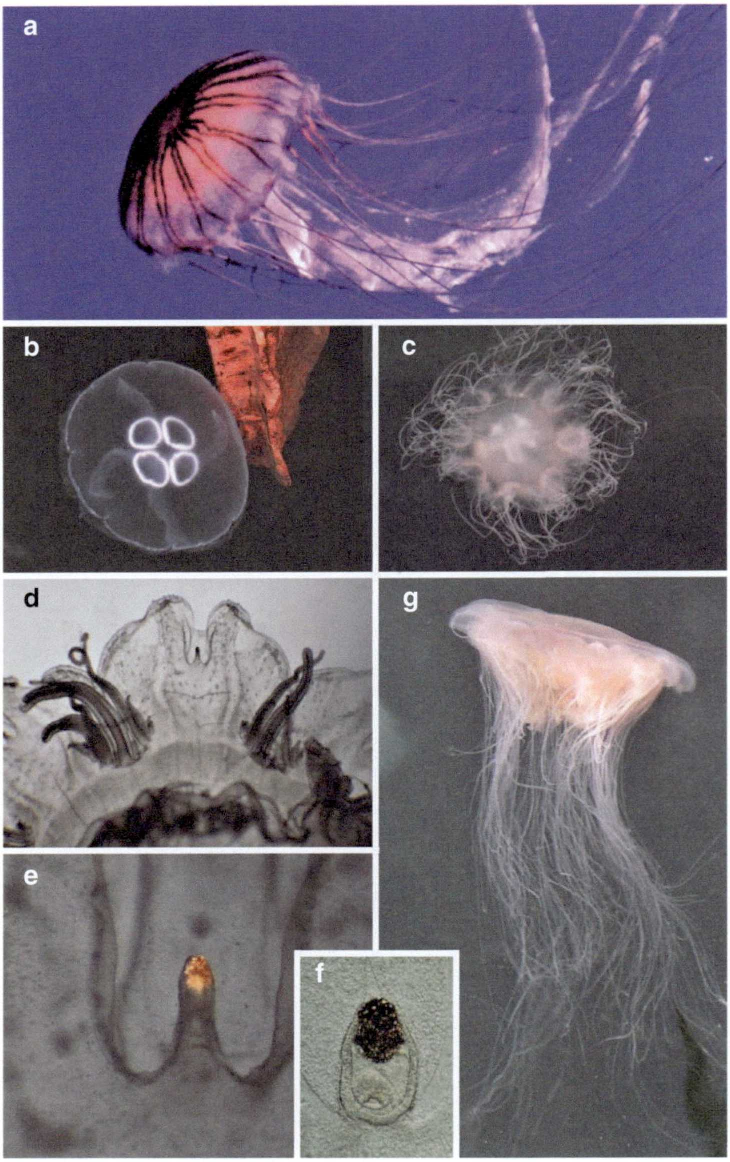

Plate 9 (**a**) *Chrysaora pacifica* (Cnidaria, Scyphozoa, Semaeostomae, Pelagiidae), diameter ca. 30 cm. Aquarium Valencia, Spain 2018. (**b**) *Aurelia aurita*, diameter ca 25 cm. Kiel, Germany 2018. (**c**) *Cyanea capillata* (Cnidaria, Scyphozoa, Semaeostomeae, Ulmaridae), diameter ca. 25 cm. Kiel, Germany 2018. (**d**) *Cyanea capillata* (Cnidaria, Scyphozoa, Semaeostomeae, Cyaneidae), diameter ca. 20 cm. Kiel, Germany 2018. (**e**) *Cyanea capillata* (Cnidaria, Scyphozoa, Semaeostomeae, Cyaneidae), diameter ca. 4 cm. Kiel, Germany 2018. (**f**) *Cyanea capillata* (Cnidaria, Scyphozoa, Semaeostomeae, Cyaneidae), rhopalium, length 3 mm. Helgoland, Germany 2017. (**g**) *Cyanea capillata* (Cnidaria, Scyphozoa, Semaeostomeae, Cyaneidae), rhopalium, length 3 mm. Helgoland, Germany 2017

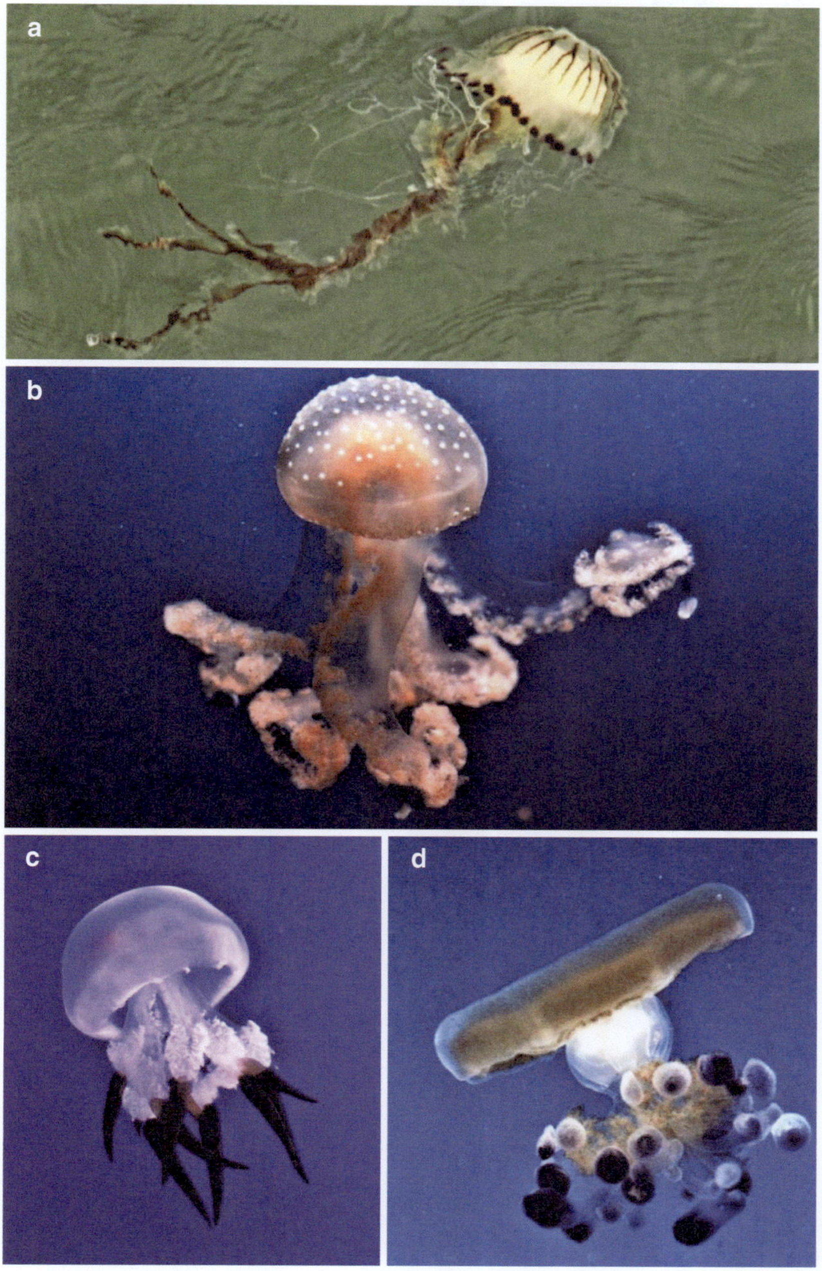

Plate 10 (**a**) *Chrysaora hyoscella* (Cnidaria, Scyphozoa, Semaeostomeae, Pelagiidae), diameter ca. 20 cm. Helgoland, Germany 2017. (**b**) *Phyllorhiza punctata* (Cnidaria, Scyphozoa, Semaeostomeae, Mastigiidae), diameter ca. 6 cm. Aquarium Valencia, Spain 2018. (**c**) *Rhopilema esculentum* (Cnidaria, Scyphozoa, Semaeostomeae, Rhizostomatidae), diameter ca. 6 cm. Aquarium Valencia, Spain 2018. (**d**) *Cassiopeia andromeda* (Cnidaria, Scyphozoa, Semaeostomeae, Cassiopeiidae), diameter ca. 4 cm. Aquarium Valencia, Spain 2018

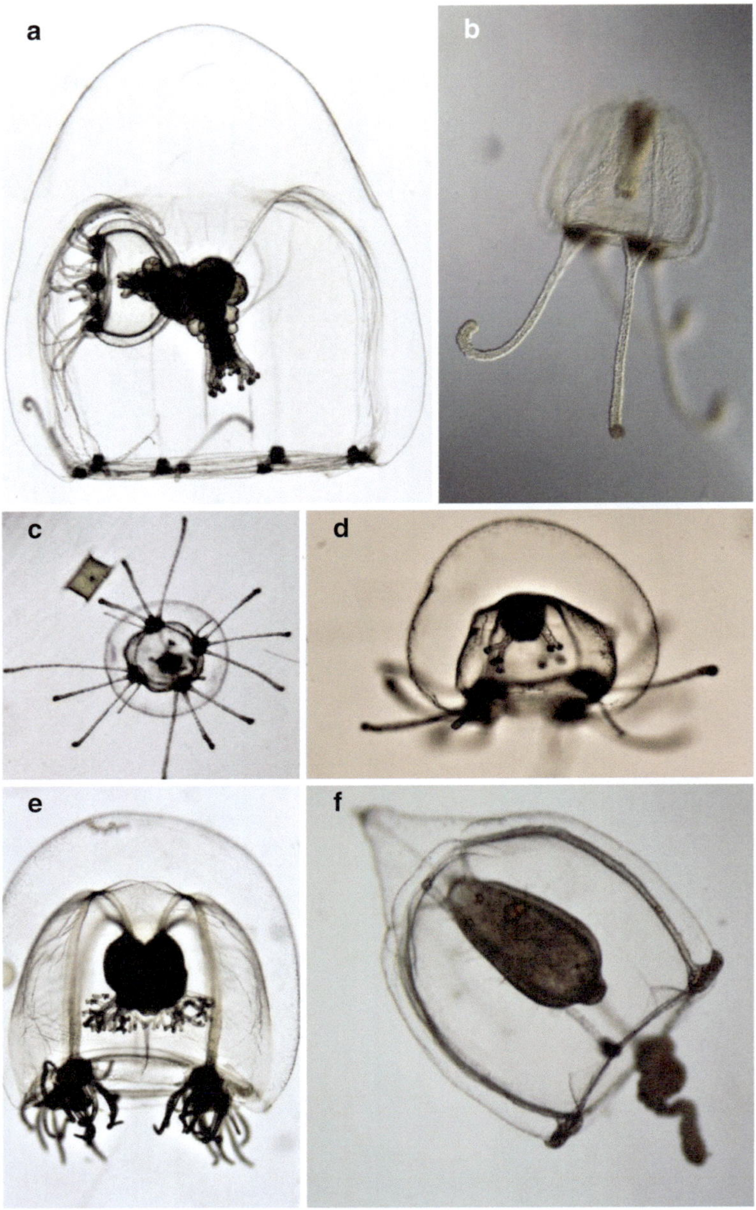

Plate 11 (**a**) *Rathkea octopunctata* (Cnidaria, Hydrozoa, Anthoathecata, Rathkeisae), three generations, diameter 3.2 mm. Sylt, Germany 2013. (**b**) *Clythia* sp. (Cnidaria, Hydrozoa, Leptothecata, Campanulariidae), diameter 420 µm. Marbella, Spain 2017. (**c**) *Bougainvillia britannica* (Cnidaria, Hydrozoa, Anthoathecata, Bougainvilliidae), oral view, diameter 920 µm. Helgoland, Germany 2013. (**d**) *Bougainvillia britannica* (Cnidaria, Hydrozoa, Anthoathecata, Bougainvilliidae), lateral view, diameter 920 µm. Helgoland, Germany 2013. (**e**) *Bougainvillia britannica* (Cnidaria, Hydrozoa, Anthoathecata, Bougainvilliidae), diameter 6 mm. Helgoland, Germany 2013. (**f**) *Corymorpha nutans* (Cnidaria, Hydrozoa, Anthoathecata, Corymorphidae), diameter 1.7 mm. Helgoland, Germany 2013

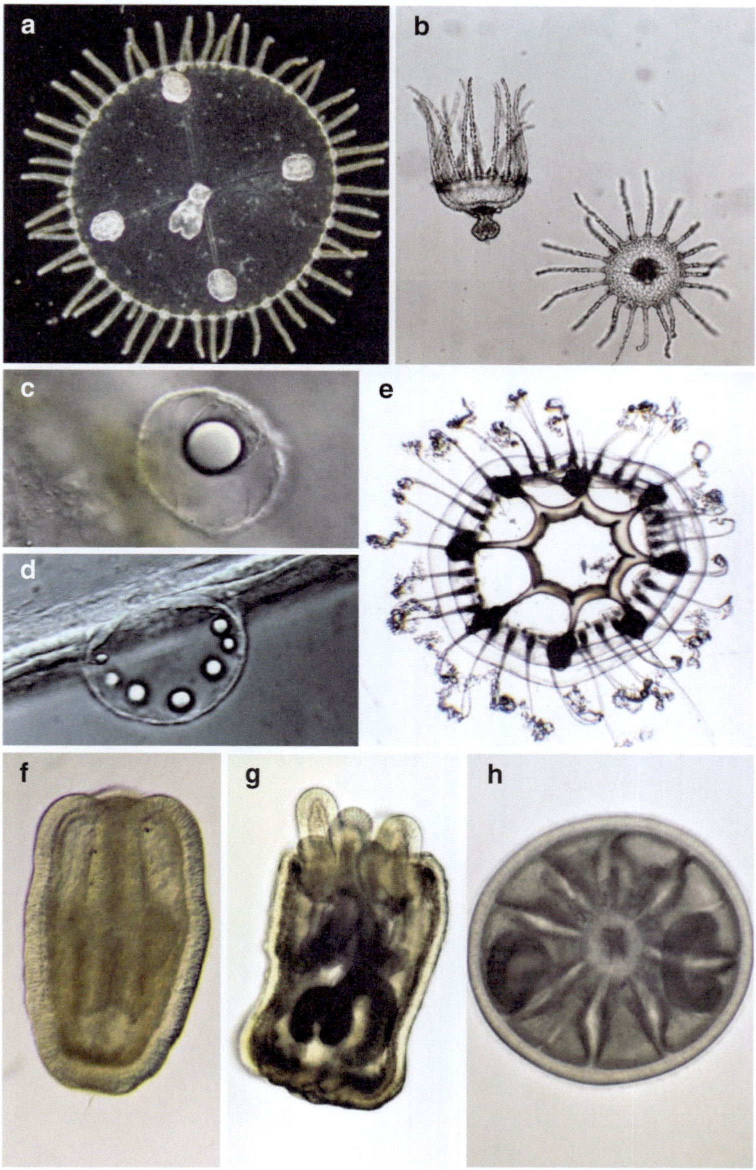

Plate 12 (**a**) *Obelia* sp. (Cnidaria, Hydrozoa, Leptothecata, Campanulariidae), diameter 1.8 mm. Helgoland, Germany 2017. (**b**) *Obelia* sp. (Cnidaria, Hydrozoa, Leptothecata, Campanulariidae), diameter 300 µm. Cape Town, South Africa 2011. (**c**) *Obelia* sp. (Cnidaria, Hydrozoa, Leptothecata, Campanulariidae), statocyst, diameter 28 µm. New Brunswick, Canada 2015. (**d**) Eirene sp. (Cnidaria, Hydrozoa, Leptothecata, Eirenidae), statocyst, diameter 95 µm. Helgoland, Germany 2013. (**e**) *Melicertum octocostatum* (Cnidaria, Hydrozoa, Leptothecata, Melicertidae), diameter 8 mm. Helgoland, Germany 2017. (**f**) Anthozoa, planula, length 550 µm. New Brunswick, Canada 2012. (**g**) Anthozoa, planula, metamorphosis, lateral view. Cape Town, South Africa 2011. (**h**) Anthozoa, planula, metamorphosis, oral view. Cape Town, South Africa 2011

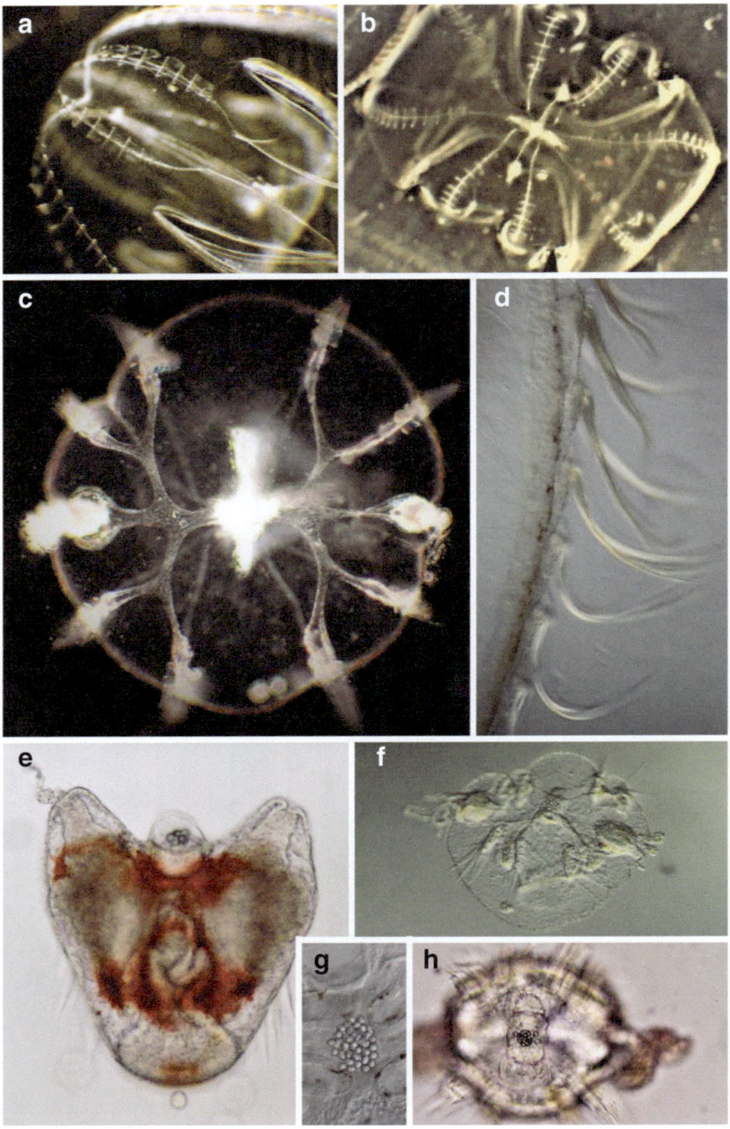

Plate 13 (**a**) *Mnemiopsis leidyi* (Ctenophorae, Tentaculata, Lobata, Bolinopsidae), lateral view, width 1.1 mm. Sylt, Germany 2016. (**b**) *Mnemiopsis leidyi* (Ctenophorae, Tentaculata, Lobata, Bolinopsidae), oral view, width 1.1 mm. Sylt, Germany 2016. (**c**) *Pleurobrachia pileus* (Ctenophorae, Tentaculata, Cydippida, Pleurobrachiidae), oral view, diameter 1.4 mm. Sylt, Germany 2016. (**d**) *Pleurobrachia pileus* (Ctenophorae, Tentaculata, Cydippida, Pleurobrachiidae), ctenes, length ca. 100 µm. Sylt, Germany 2015. (**e**) Ctenophora, indet. species, early stage, length 400 µm. Cape Town, South Africa 2011. (**f**) *Pleurobrachia pileus* (Ctenophorae, Tentaculata, Cydippida, Pleurobrachiidae), very early stage, diameter 450 µm. Sylt, Germany 2016. (**g**) *Beroe cucumis* (Ctenophorae, Nuda, Beroida, Beroidae), statocyst, diameter 60 µm. Sylt, Germany 2015. (**h**) *Pleurobrachia pileus* (Ctenophorae, Tentaculata, Cydippida, Pleurobrachiidae), very early stage, 300 µm. New Brunswick, Canada 2012

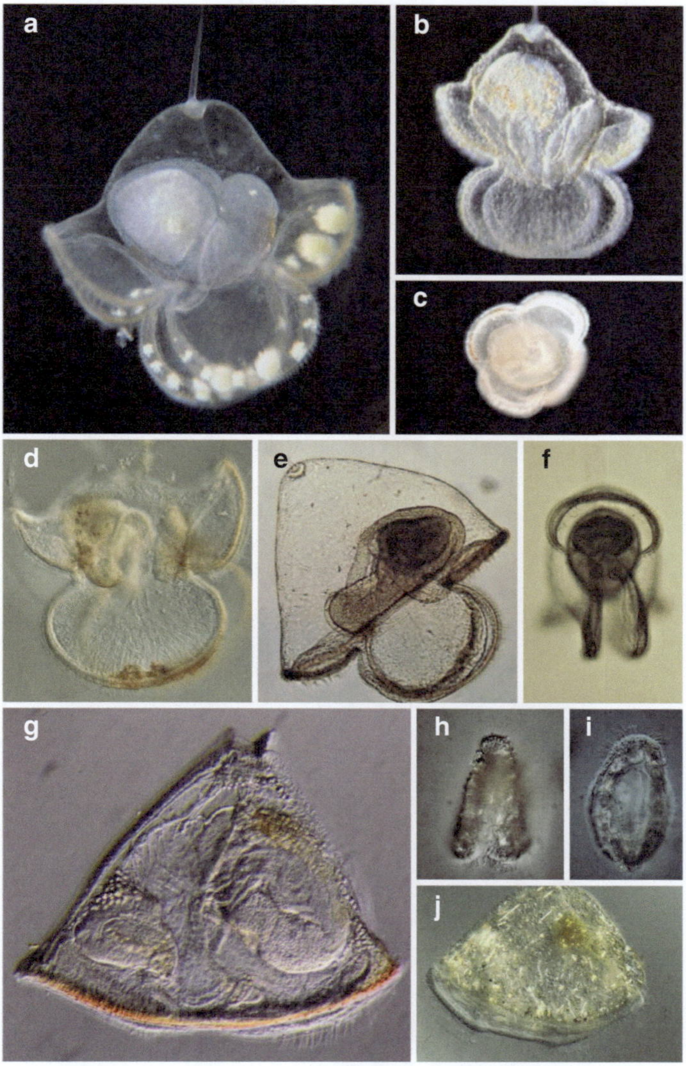

Plate 14 (**a**) *Lineus* sp. (Nemertea, Pilidiophora, Heteronemertea, Lineidae), pilidium larva, width 850 µm. Helgoland, Germany 2015. (**b**) C*erebratulus* sp. (Nemertea, Pilidiophora, Heteronemertea, Lineidae), pilidium larva, width 600 µm. New Brunswick, Canada 2012. (**c**) *Lineus* sp. (Nemertea, Pilidiophora, Heteronemertea, Lineidae), pilidium larva, width 800 µm, Helgoland, Germany 1017. (**d**) *Lineus* sp. (Nemertea, Pilidiophora, Heteronemertea, Lineidae), pilidium larva, width 450 µm. Helgoland, Germany 2021. (**e**) *Lineus* sp. (Nemertea, Pilidiophora, Heteronemertea, Lineidae), pilidium larva, width 800 µm. Helgoland, Germany 2017. (**f**) *Lineus* sp. (Nemertea, Pilidiophora, Heteronemertea, Lineidae), pilidium, frontal view, height 450 mm. Helgoland, Germany 2017. (**g**) *Membranipora* sp. (Bryozoa, Gymnolaemata, Cheilostomatida, Membraniopridae), cyphonautes larva, width 650 µm. New Brunswick, Canada 2012. (**h**) *Electra pilosa* (Bryozoa, Gymnolaemata, Cheilostomatida, Electridae), cyphonautes larva, frontal view, width 110 µm. Sylt, Germany 2015. (**i**) *Electra pilosa* (Bryozoa, Gymnolaemata, Cheilostomatida, Electridae), cyphonautes larva, oral view, height 110 µm. Sylt, Germany 2015. (**j**) *Electra pilosa* (Bryozoa, Gymnolaemata, Cheilostomatida, Electridae), cyphonautes larva, lateral view, height 110 µm. Sylt, Germany 2015

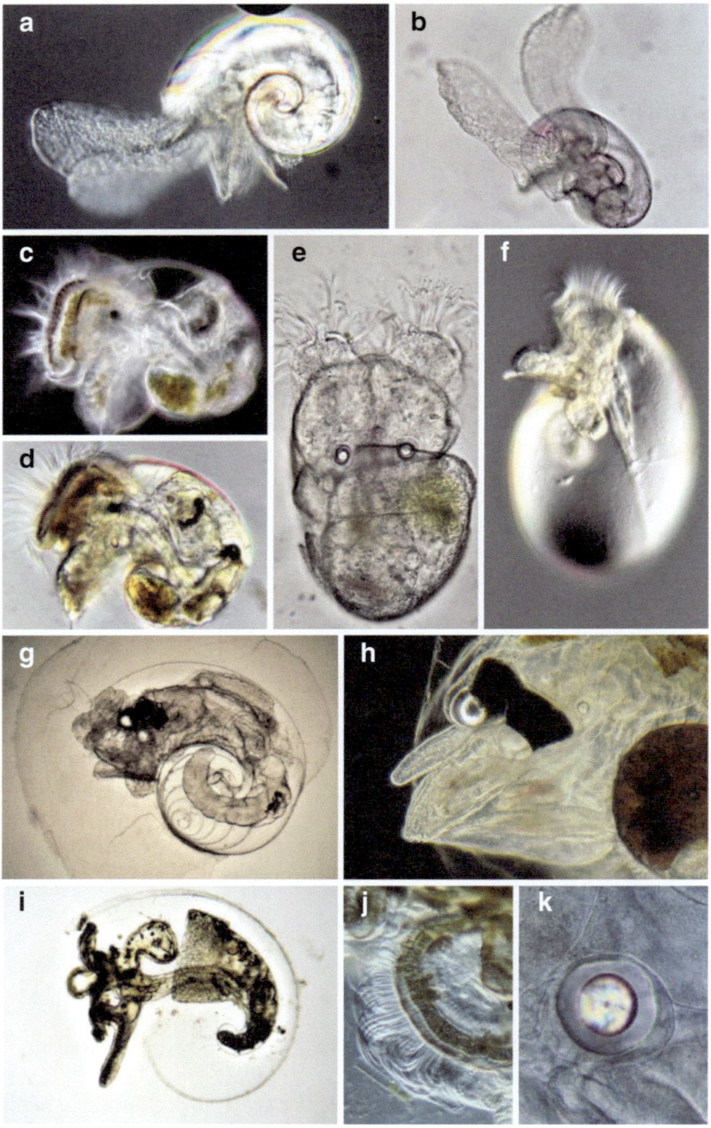

Plate 15 (**a**) *Limacina inflata* (Mollusca, Gastropoda, Pteropoda, Limacinidae), width 350 µm, Lanzarote, Spain 2019. (**b**) *Limacina inflata* (Mollusca, Gastropoda, Pteropoda, Limacinidae), width 350 µm, Lanzarote, Spain 2019. (**c**) *Littorina* sp. (Mollusca, Gastropoda, Littorinimorpha, Littorinidae), diameter 410 µm. New Brunswick, Canada 2012. (**d**) *Littorina* sp. (Mollusca, Gastropoda, Littorinimorpha, Littorinidae), diameter 410 µm. New Brunswick, Canada 2012. (**e**) Gastropoda, veliger, indet., height 210 µm. Helgoland, Germany 2010. (**f**) Gastropoda, Nudibranchia, undet. veliger larva, length 270 µm. New Brunswick, Canada 2012. (**g**) *Atlanta gibbosa* (Mollusca, Gastropoda, Littorinimorpha, Atlantidae), diameter 1.2 mm. Lanzarote, Spain 2022. (**h**) *Atlanta gibbosa* (Mollusca, Gastropoda, Littorinimorpha, Atlantidae), diameter 1.2 mm. Lanzarote, Spain 2022. (**i**) *Lamellaria* sp. (Mollusca, Gastropoda, Littorinimorpha, Velutinidae), shell diameter 850 µm. Helgoland, Germany 2013. (**j**) *Lamellaria* sp. (Mollusca, Gastropoda, Littorinimorpha, Velutinidae), cilia of velum. Helgoland, Germany 2013. (**k**) *Lamellaria* sp. (Mollusca, Gastropoda, Littorinimorpha, Velutinidae), statocyst. Helgoland, Germany 2013

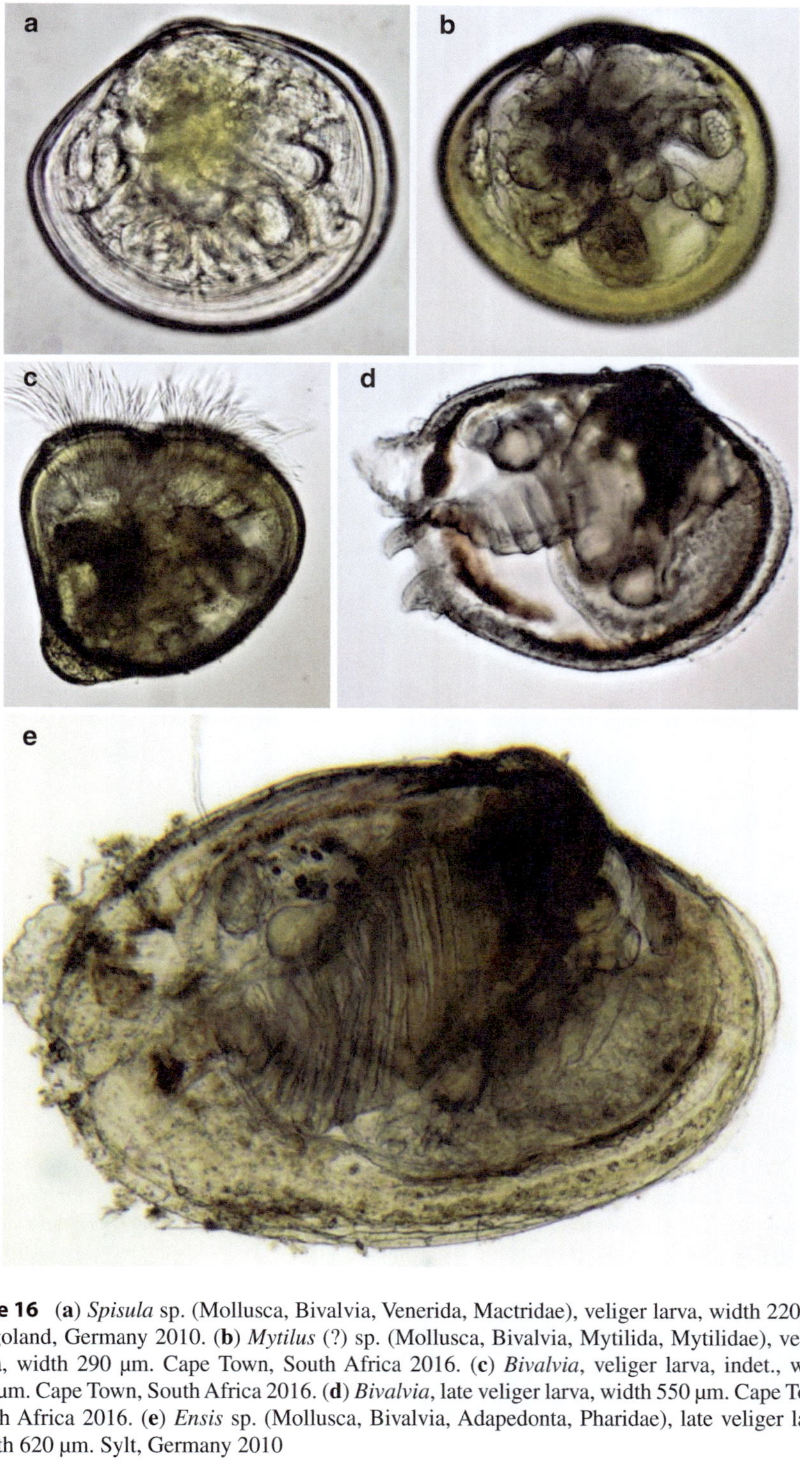

Plate 16 (**a**) *Spisula* sp. (Mollusca, Bivalvia, Venerida, Mactridae), veliger larva, width 220 µm. Helgoland, Germany 2010. (**b**) *Mytilus* (?) sp. (Mollusca, Bivalvia, Mytilida, Mytilidae), veliger larva, width 290 µm. Cape Town, South Africa 2016. (**c**) *Bivalvia*, veliger larva, indet., width 450 µm. Cape Town, South Africa 2016. (**d**) *Bivalvia*, late veliger larva, width 550 µm. Cape Town, South Africa 2016. (**e**) *Ensis* sp. (Mollusca, Bivalvia, Adapedonta, Pharidae), late veliger larva, length 620 µm. Sylt, Germany 2010

Plate 17 (**a**) *Tomopteris helgolandica* (Polychaeta, Phyllodocida, Tomopteridae), length 170 μm. Helgoland, Germany 2012. (**b**) *Tomopteris septentrionales* (Polychaeta, Phyllodocida, Tomopteridae), length 2 mm. Helgoland, Germany 2021. (**c**) *Tomopteris septentrionales* (Polychaeta, Phyllodocida, Tomopteridae), length 4.4 mm. Helgoland, Germany 2014. (**e**) *Myrianida prolifera* (Polychaeta, Phyllodocida), head of male, width of eyes 300 μm. Helgoland, Germany 2019. (**d**) *Myrianida prolifera* (Polychaeta, Phyllodocida), female with eggs, length 2.4 mm. Helgoland, Germany 2014. (**f**) *Syllidae* (Polychaeta, Phyllodocida), male, indet., length 950 μm. Palamos, Spain 2009. (**g**) Polynoidae (Polychaeta, Phyllodocida), larva, length 1.35 mm. Marbella, Spain 2017. (**h**) Nereidae (Polychaeta, Phyllodocida), nectochaeta, indet., length 450 μm. Helgoland, Germany 2006. (**i**) Nereidae (Polychaeta, Phyllodocida), nectochaeta, indet., length 700 μm. Sylt, Germany 2009. (**j**) Nereidae (Polychaeta, Phyllodocida), nectochaeta, length 850 μm. Marbella, Spain 2017. (**k**) *Eulalia* sp. (Polychaeta, Phyllodocida, Nereidae), trochophora larva, indet., diameter 160 μm. Marbella, Spain 2017. (**l**) Typhloscolecidae (Polychaeta, Phyllodocida), larva, length 1.05 mm. Lanzarote, 2022

Plate 18 (**a**) Sabellariidae (Polychaeta, Canalipalpata, Sabellariidae), trochophra larva, diameter 320 µm. Palamos, Spain 2009. (**b**) *Polygordius* sp. (Polychaeta, Polygordiida, Polygordiidae), metatrochophora, diameter 720 µm. Helgoland, Germany 2014. (**c**) *Chaetopterus variopedatus* (Polychaeta, Canalipalpata, Chaetopteridae), larva, length 800 µm. Helgoland, Germany 2012. (**d**) Spionidae (Polychaeta, Canalipalpata), larva, indet., length 800 µm. Palamos, Spain 2009. (**e**) *Megalone filiformis* (Polychaeta, Canalipalpata, Spionida, Spionidae), larva, length 1.1 mm. Helgoland, Germany 2014. (**f**) *Lanice conchilega* (Polychaeta, Canalipalpata, Terebellidae), aulophora larva I, length of tube 1.8 mm. Helgoland, Germany 2017. (**g**) *Pectinaria koreni* (Polychaeta, Canalipalpata, Pectinariidae), trochophora larva, diameter 190 µm. Sylt, Germany 2016. (**h**) *Pectinaria koreni* (Polychaeta, Canalipalpata, Pectinariidae), diameter 320 µm. Sylt, Germany 2014. (**i**) *Pectinaria koreni* (Polychaeta, Canalipalpata, Pectinariidae), tube, length 1.25 mm. Sylt, Germany 2014

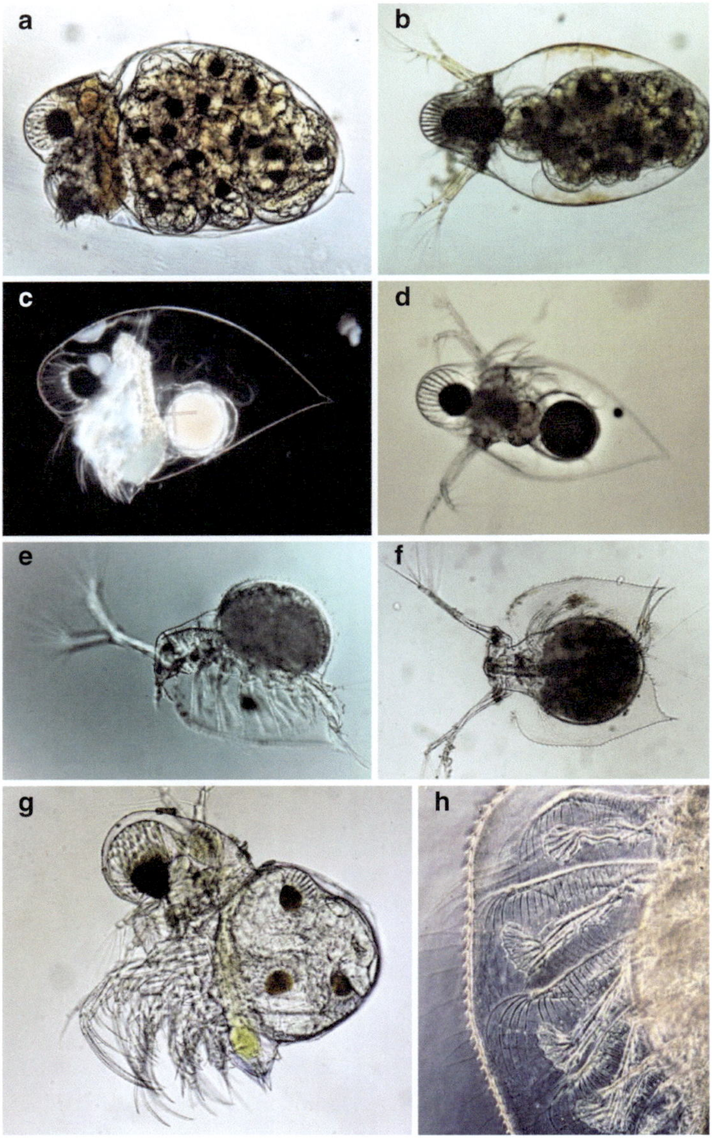

Plate 19 (**a**) *Evadne nordmanni* (Crustacea, Branchiopoda, Phyllopopda, Onychopoda, Podonidae), female with eggs, length 1 mm. Helgoland, Germany 2017. (**b**) *Evadne nordmanni* (Crustacea, Branchiopoda, Phyllopopda, Onychopoda, Podonidae), female with eggs, length 900 μm. Helgoland, Germany 2017. (**c**) *Evadne nordmanni* (Crustacea, Branchiopoda, Phyllopopda, Onychopoda, Podonidae), female with egg, length 700 μm. Helgoland, Germany 2017. (**d**) *Evadne nordmanni* (Crustacea, Branchiopoda, Phyllopopda, Onychopoda, Podonidae), female with eggs, length 700 μm. Helgoland, Germany 2017. (**e**) *Penilia avirostris* (Crustacea, Branchiopoda, Phyllopoda, Ctenopoda, Sididae), female with eggs, length 1.1 mm. Helgoland, Germany 2014. (**f**) *Penilia avirostris* (Crustacea, Branchiopoda, Phyllopoda, Ctenopoda, Sididae), female with eggs, length 1.1 mm. Helgoland, Germany 2017. (**g**) *Pleopis polyphemoides* (Crustacea, Branchiopoda, Phyllopoda, Onychopoda, Podonidae), female with eggs, length 600 μm. Cape Town, South Africa 2016. (**h**) *Penilia avirostris* (Crustacea, Branchiopoda, Phyllopoda, Ctenopoda, Sididae), legs for filter feeding, length of the photo-frame 900 μm. Lanzarote, Spain 2022

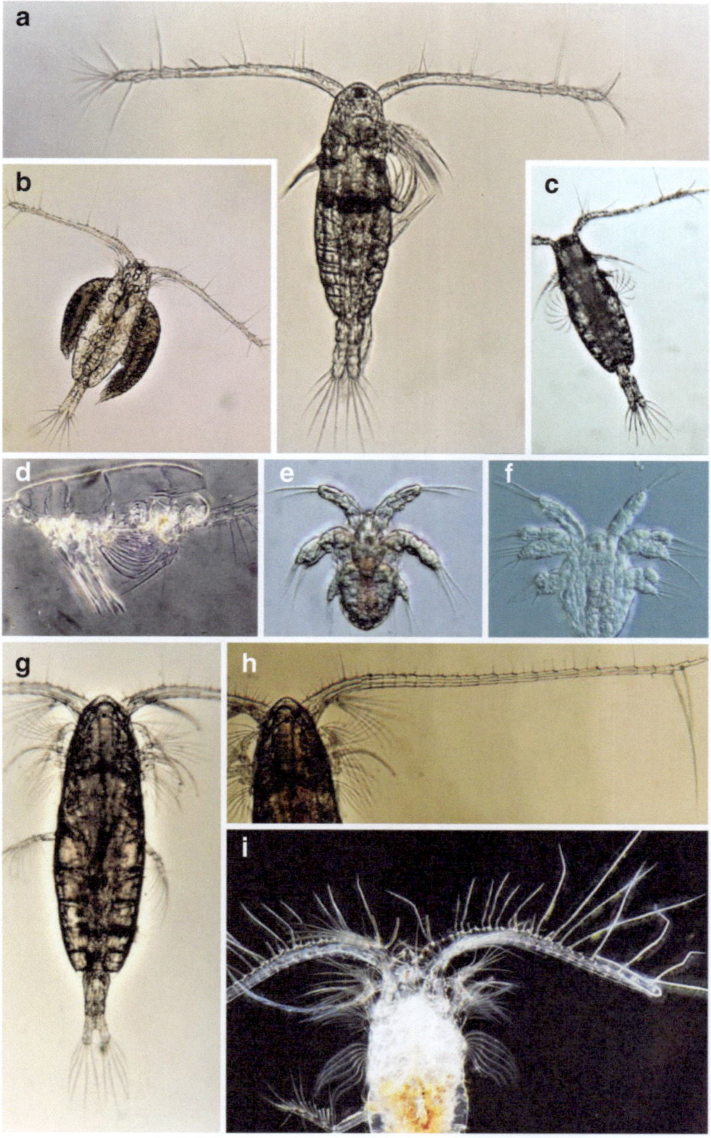

Plate 20 (**a**) *Acartia tonsa* (Crustacea, Copepoda, Calanoida, Acartiidae), female, length 1 mm. From Helgoland culture, 2021. (**b**) *Acartia clausi* (Crustacea, Copepoda, Calanoida, Acartiidae), with parasitic bopyrid (Crustacea, Malacostraca, Peracarida, Isopoda, Bopyridae), length 520 µm. Marbella, Spain 2017. (**c**) *Acartia clausi* (Crustacea, Copepoda, Calanoida, Acartiidae), with expanded filtering legs, length 1 mm. From Helgoland culture, 2021. (**d**) *Acartia clausi* (Crustacea, Copepoda, Calanoida, Acartiidae), exuvia, lateral view, length 900 µm. Lanzarote, Spain 1922. (**e**) *Acartia tonsa* (Crustacea, Copepoda, Calanoida, Acartiidae), nauplius, just hatched, length 53 µm. From Helgoland culture, 2021. (**f**) *Acartia tonsa* (Crustacea, Copepoda, Calanoida, Acartiidae), nauplius, just hatched, length 53 µm. From Helgoland culture, 2021. (**g**) *Calanus* sp. (Crustacea, Copepoda, Calanoida, Calanidae), length 1.2 mm. Helgoland, Germany 2021. (**h**) *Calanus* sp. (Crustacea, Copepoda, Calanoida, Calanidae), antenna, length 3 mm. Helgoland, Germany 2021. (**i**) *Calocalanus* sp. (Crustacea, Copepoda, Calanoida, Paracalanidae), length 1.3 mm. Lanzarote, Spain 2019

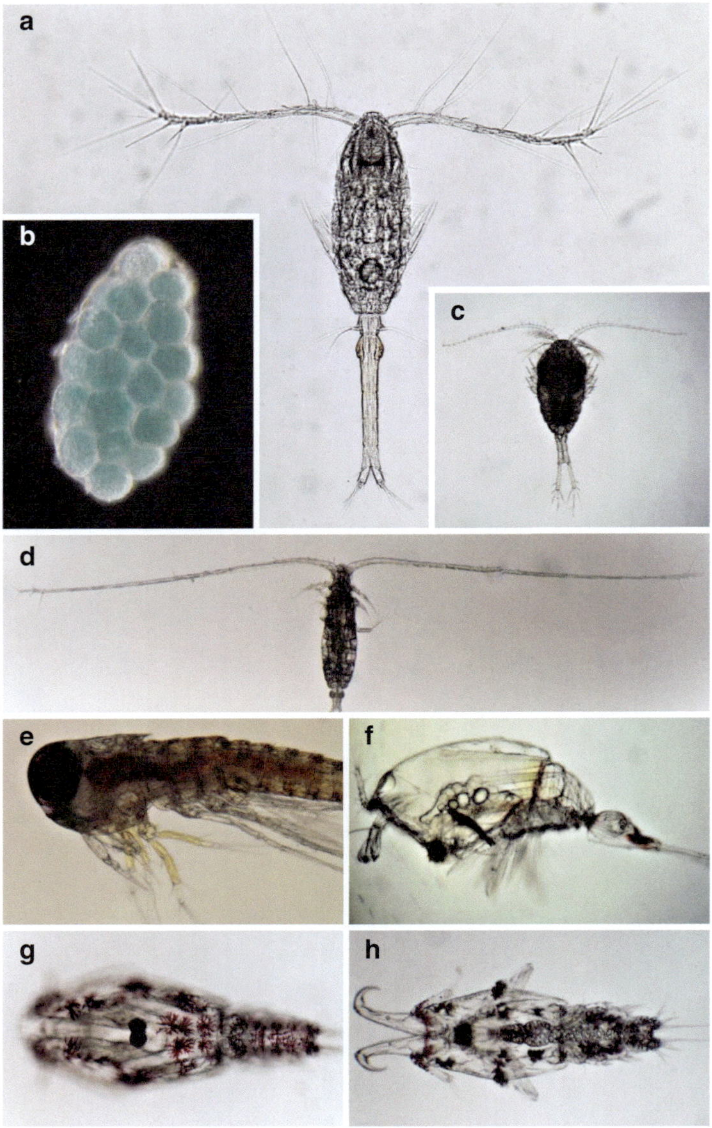

Plate 21 (**a**) *Oithona* sp. (Crustacea, Copepoda, Cyclopoida, Oithonidae), female, length 1.4 mm. Helgoland, Germany. (**b**) Egg sac of an indet. species (Crustacea, Copepoda), length 700 µm. Lanzarote, Spain 2019. (**c**) *Temora longicornis* (Crustacea, Copepoda, Calanoida, Temoridae), female, length of antenna 1.4 mm. Helgoland, Germany 2021. (**d**) *Mecynocera clausi* (Crustacea, Copepoda, Calanoida, Paracalaniae), body length 1.05 mm. Lanzarote, Spain 2019. (**e**) *Distioculus* sp. (Crustacea, Copepoda, Harpacticoida, Miraciidae), body length 750 µm. Lanzarote 2019. (**f**) *Ditrichocorycaeus* sp. (Crustacea, Copepoda, Cyclopoida, Corycaeidae), body length 1 mm, length of eyes 400 µm. Helgoland, Germany 2012. (**g**) Indet. ergasilid (Crustacea, Copepoda, Cyclopoida, Ergasilidae), body length 600 µm, Cape Town, South Africa 2011. (**h**) Indet. ergasilid (Crustacea, Copepoda, Cyclopoida, Ergasilidae), body length 600 µm, Cape Town, South Africa 2011

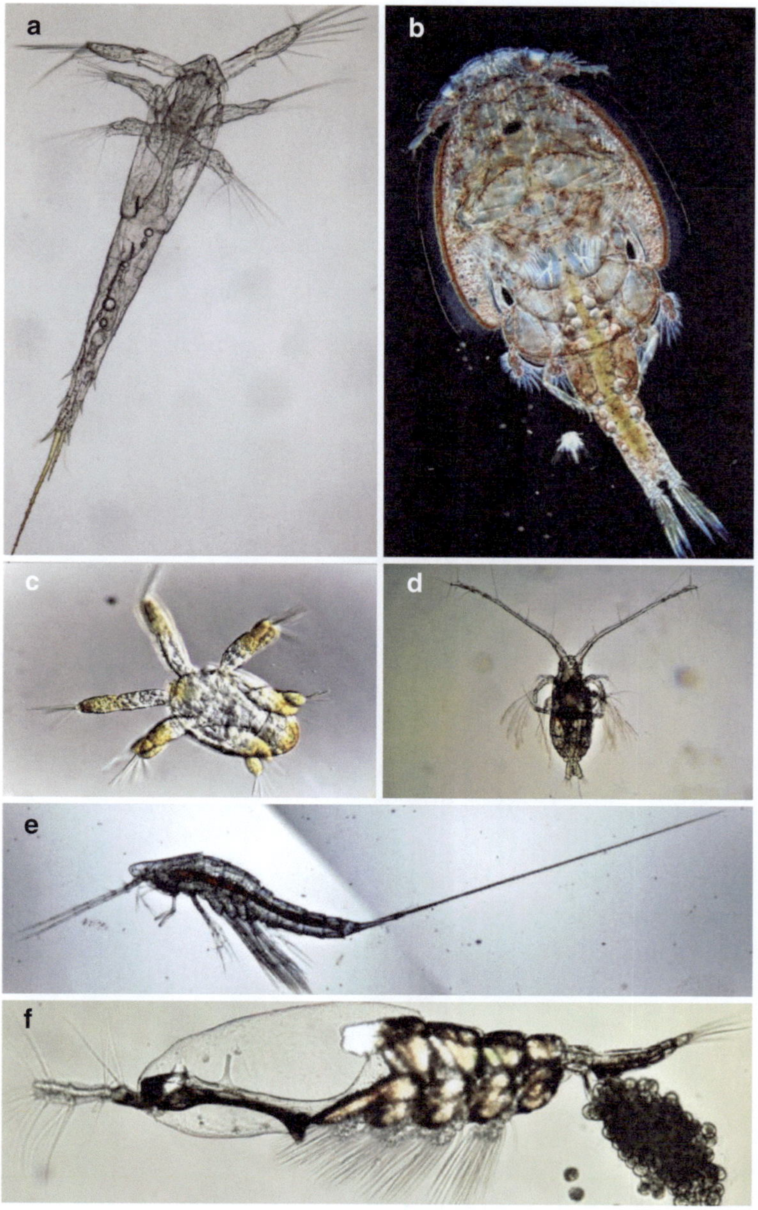

Plate 22 (**a**) Pontellid nauplius (Crustacea, Copepoda, Calanoida, Pontellidae), body length 1.9 mm. Lanzarote, Spain 2022. (**b**) *Caligus* sp. (Crustacea, Copepoda, Siphonostomatoida, Caligidae), body length 5 mm. Helgoland, Germany 2017. (**c**) Calanid nauplius (Crustacea, Copepoda, Caqlanoida, Calanidae), length 240 μm. Marbella, Spain 2017. (**d**) Naulius, indet. (Crustacea, Copepoda), body length 1.15 mm. Lanzarote, Spain 2019. (**e**) *Macrosetella gracilis* (Crustacea, Copepoda, Harpacticoida, Miraciidae), body length 1.4 mm. Lanzarote, Spain 2022. (**f**) *Monstrilla helgolandica* (Crustacea, Copepoda, Monstrilloida, Monstrillidae), female, length 1.8 mm. Helgoland, Germany 2011

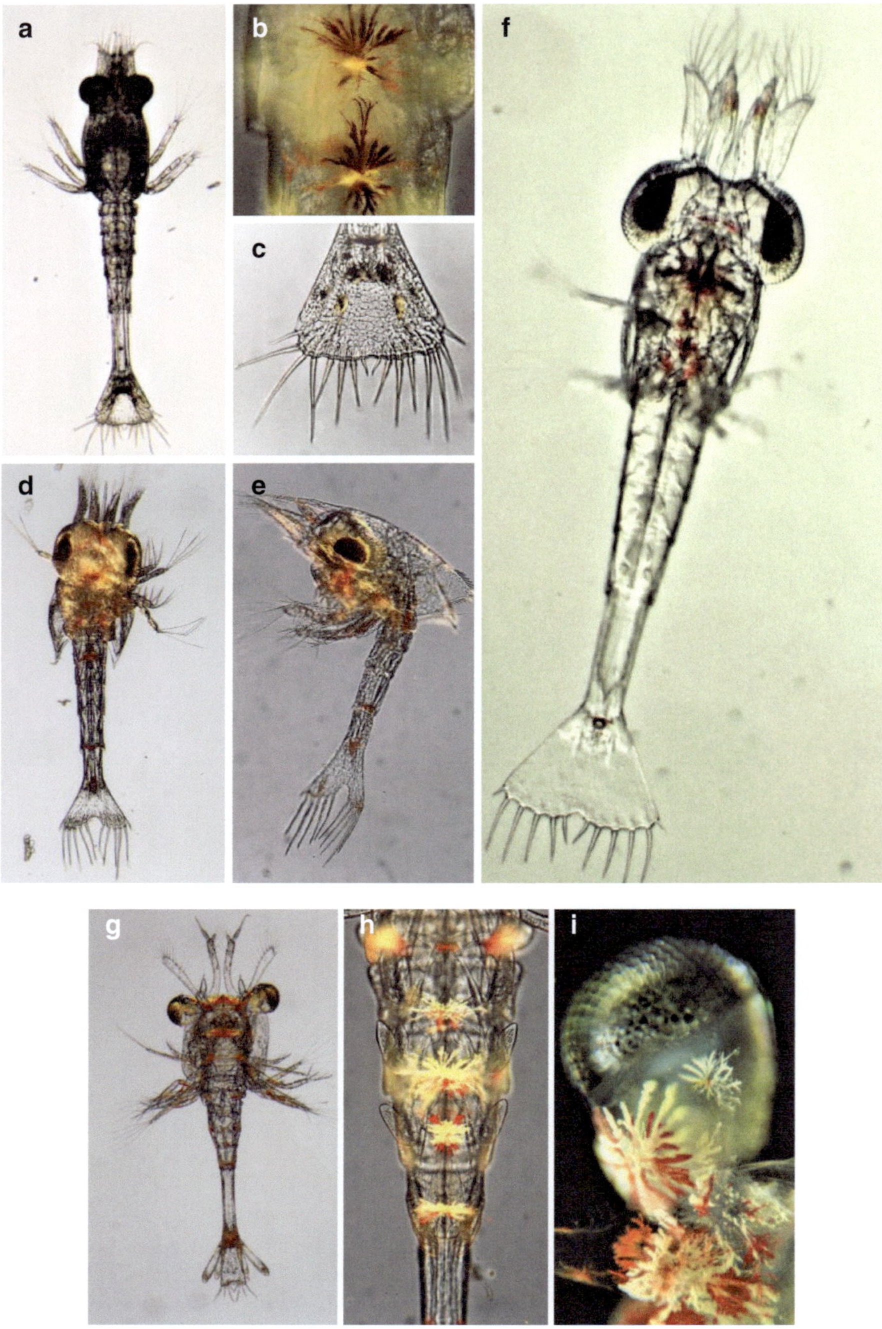
a
b
c
d
e
f
g
h
i

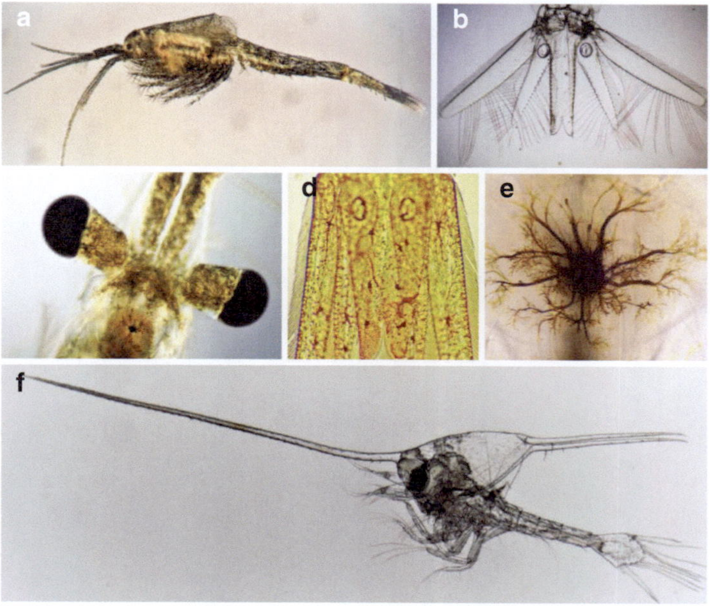

Plate 24 (**a**) *Mysis relicta* (Crustacea, Malacostraca, Mysida, Mysidae), lateral view, length of carapax 1 mm. Sylt, Germany 2009. (**b**) *Mysis* sp. (Crustacea, Malacostraca, Mysida, Mysidae), tail fan of an exuvia, width 1.3 cm. New Brunswick, Canada 2012. (**c**) *Mysis* sp. (Crustacea, Malacostraca, Mysida, Mysidae), head, width of eyes 3 mm, New Brunswick, Canada 2012. (**d**) *Mysis* sp. (Crustacea, Malacostraca, Mysida, Mysidae), tail fan, width of foto 1.5 mm. New Brunswick, Canada 2012. (**e**) *Mysis* sp. (Crustacea, Malacostraca, Mysida, Mysidae), abdominal chromatophore, diameter 900 µm. New Brunswick, Canada 2012. (**f**) *Porcellana* sp. (Crustacea, Malacostraca, Decapoda, Pleocyemata, Porcellanidae), zoea, length 5.3 mm. Helgoland, Germany 2012. (**g**) *Carcinus maenas* (Crustacea, Malacostraca, Decapoda, Pleocyemata, Brachyura, Carcinidae), zoea, body length 1.5 mm. Helgoland, Germany 1013. (**h**) *Liocarcinus holsatus* (Crustacea, Malacostraca, Decaopda, Pleocyemata, Brachyura, Polybiidae), megalopa larva, width of eyes 1.7 mm. Helgoland, Germany 2012. (**i**) *Corystes* sp. (Crustacea, Malacostraca, Decapoda, Pleocyemata, Brachyura, Corystidae), megalopa larva, body length 4.5 mm. Helgoland, Germany 2019

Plate 23 (**a**) *Crangon crangon* (Crustacea, Malacostraca, Decapoda, Crangonidae), zoea larva I, length 2.5 mm. Helgoland, Germany, 2012. (**b**) *Crangon crangon* (Crustacea, Malacostraca, Decapoda, Crangonidae), zoea öarva I, abdominal segments, width 120 µm. Helgoland, Germany 2012. (**c**) *Crangon crangon* (Crustacea, Malacostraca, Decapoda, Crangonidae), zoea larva I, width of tail segment 260 µm. Helgoland, Germany, 2012. (**d**) *Upogebia* sp. (Crustacea, Malacostraca, Decapoda, Upogebiidae), zoea, width of eyes 650 µm. Helgoland, Germany 2012. (**e**) *Galathea* sp. (Crustacea, Malacostraca, Decapoda, Galatheidae), zoea, length 1.2 mm. Helgoland, Germany, 2012. (**f**) *Galathea* sp. (Crustacea, Malacostraca, Decapoda, Galatheidae), zoea, length 1.2 mm. Helgoland, Germany, 2012. (**g**) *Processa* sp. (Crustacea, Malacostraca, Decapoda, Processidae), zoea, length 2.9 mm. Helgoland, Germany 2012. (**h**) *Processa* sp. (Crustacea, Malacostraca, Decapoda, Processidae), zoea, length 2.9 mm. Helgoland, Germany 2012. (**i**) *Processa* sp. (Crustacea, Malacostraca, Decapoda, Processidae), zoea, diameter of eye 450 µm. Helgoland, Germany 2007

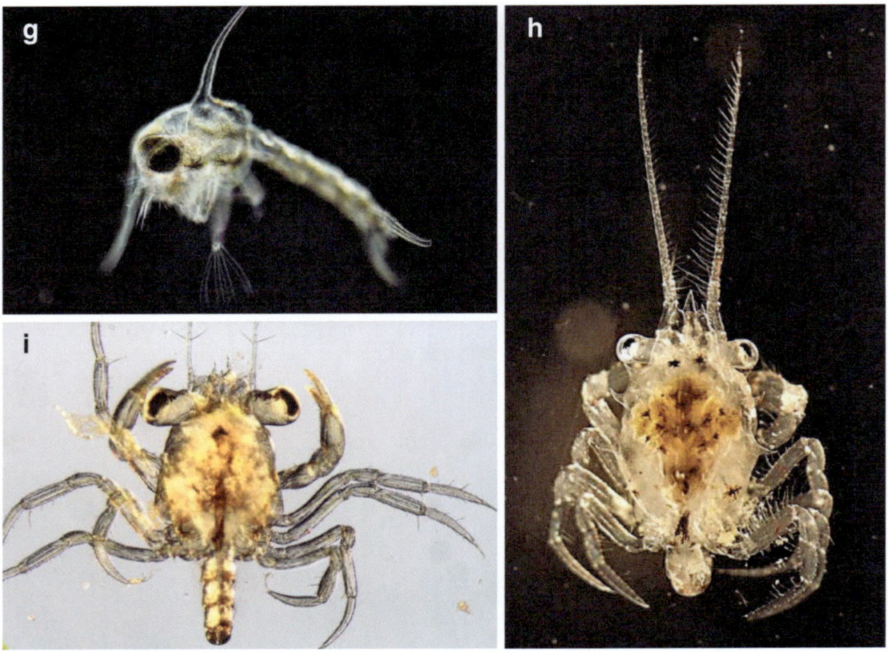

Plate 24 (continued)

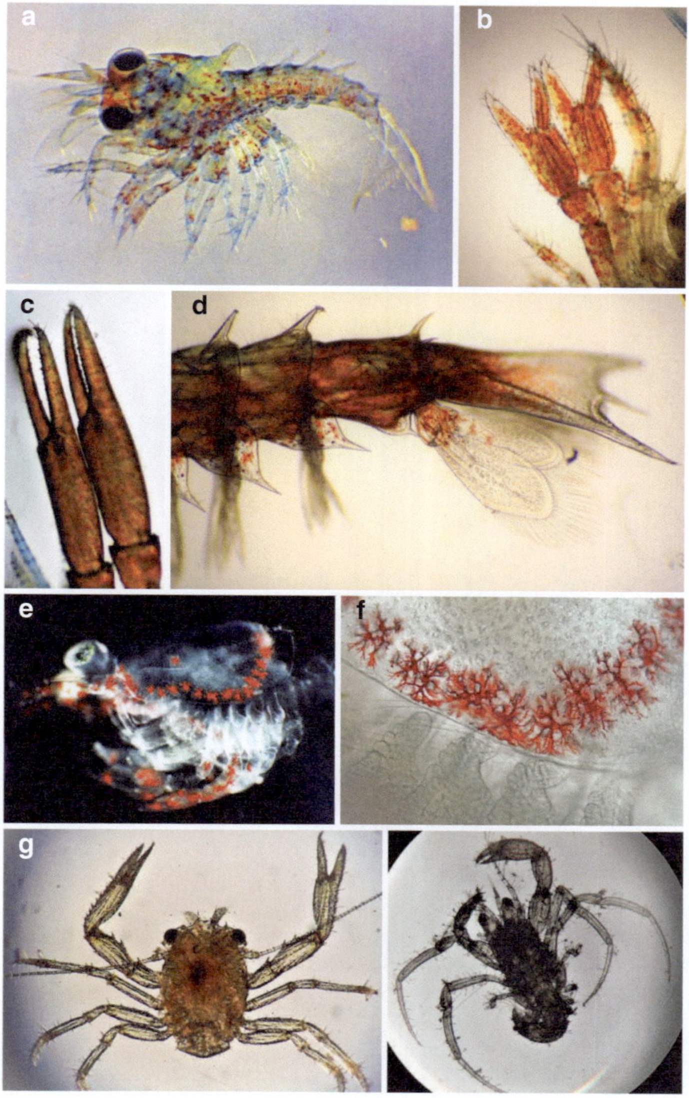

Plate 25 (**a**) *Homarus americanus* (Crustacea, Malacostraca, Decapoda, Pleocyemata, Astacidea, Nephropidae), first stage, body length 1.1 cm. New Brunswick, Canada 2012. (**b**) *Homarus americanus* (Crustacea, Malacostraca, Decapoda, Pleocyemata, Astacidea, Nephropidae), third stage, chelae, length 1.7 mm. New Brunswick, Canada 2012. (**c**) *Homarus americanus* (Crustacea, Malacostraca, Decapoda, Pleocyemata, Astacidea, Nephropidae), fourth stage, chelae, length 3.8 mm. New Brunswick, Canada 2012. (**d**) *Homarus americanus* (Crustacea, Malacostraca, Decapoda, Pleocyemata, Astacidea, Nephropidae), third stage, end of abdomen, length ca. 3.5 mm. New Brunswick, Canada 2012. (**e**) Caridea, indet. (Crustacea, Malacostraca, Decapoda, Pleocyemata), carapax with chromatophores, width of photo ca. 5 mm. Cape Town, South Africa 2012. (**f**) Caridea, indet. (Crustacea, Malacostraca, Decapoda, Pleocyemata), chromatophores and gills, width of photo ca. 2 cm. Cape Town, South Africa 2012. (**g**) *Pagurus bernhardus* (Crustacea, Malacostraca, Decaopda, Pleocyemata, Brachyura, Paguridae), megalopa, width of eyes 1.1 mm. Sylt, Germany, 2009. (**h**) *Pagurus bernhardus* (Crustacea, Malacostraca, Decaopda, Pleocyemata, Brachyura, Paguridae), late megalopa, body length 3.5 mm. Sylt, Germany 2009

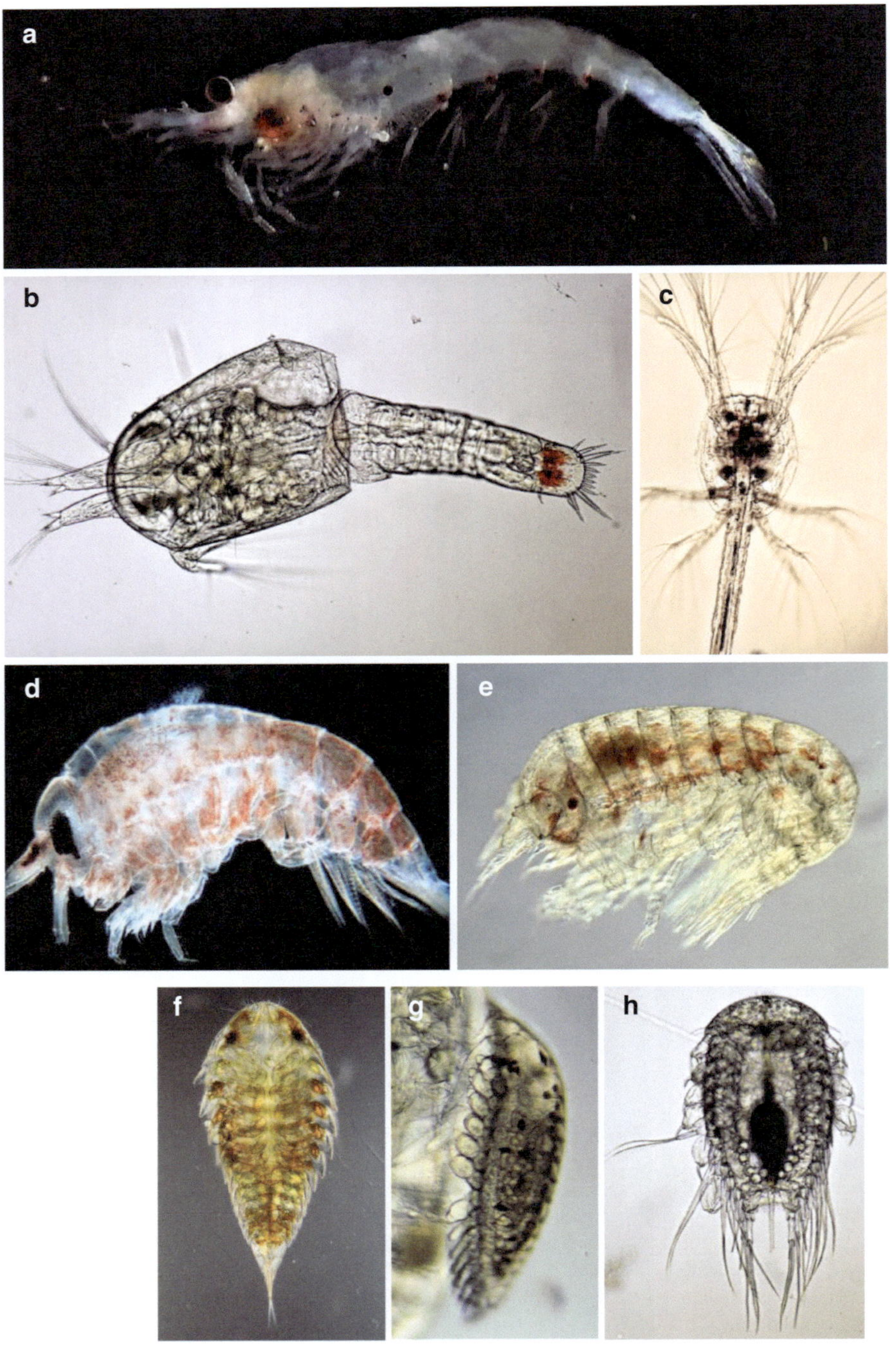
a
b
c
d
e
f
g
h

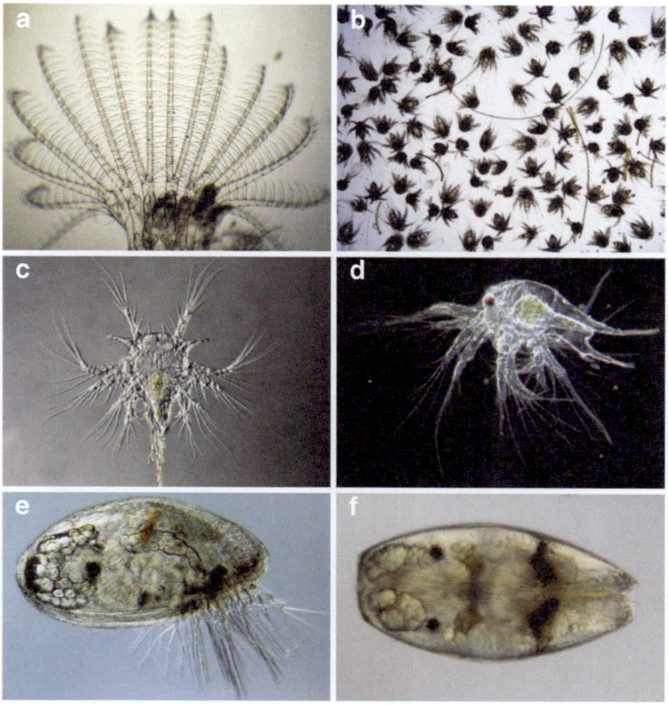

Plate 27 (**a**) Balanid exuvia, indet. (Crustacea, Thecostraca, Cirripedia, Balanomorpha, Balanidae), width 3.5 mm. Cape Town, South Africa 2016. (**b**) Balanid nauplii, indet. (Crustacea, Thecostraca, Cirripedia, Balanomorpha, Balanidae), just hatched nauplii, body length 300 µm. Sylt, Germany 2013. (**c**) Balanid nauplius, indet. (Crustacea, Thecostraca, Cirripedia, Balanomorpha, Balanidae), width 240 µm. Marbella, Spain 2017. (**d**) Balanid nauplius, indet. (Crustacea, Thecostraca, Cirripedia, Balanomorpha, Balanidae), lateral view, body length 500 µm. Sylt, Germany 2013. (**e**) Cypris larva, indet. (Crustacea, Thecostraca, Cirripedia, Balanomorpha, Balanidae), lateral view, length 720 µm. Sylt, Germany 2014. (**f**) Cypris larva, indet. (Crustacea, Thecostraca, Cirripedia, Balanomorpha, Balanidae), frontal view, body length 460 µm. Marbella, Spain 2017. (**g**) *Peltogaster* sp., nauplius. (Crustacea, Thecostraca, Cirripedia, Rhizocephala, Peltogastridae), length 360 µm. St. Cyr, France 2018. (**h**) *Lepas* sp., nauplius (Crustacea, Thecostraca, Cirripedia, Thoracica, Scalpellomorpha, Lepadidae), length 360 µm. Marbella, Spain 2017

Plate 26 (**a**) *Euphausia* sp. (Crustacea, Malacostraca, Euphausiacea, Euphausiidae), length ca. 5 mm, Lanzarote, Spain 2022. (**b**) Calotropis larva, indet. (Crustacea, Malacostraca, Euphausiacea, Euphausiidae). Cape Town, South Africa 2012. (**c**) Penaeid larva (Crustacea, Malacostraca, Decapoda, Penaeidae), body length ca. 1.4 mm. Villefranche, France 2000. (**d**) *Hyperia* sp. (Crustacea, Malacostraca, Amphipoda, Hyperiidae), length 2.5 mm. Helgoland, Germany 2019. (**e**) Amphipod, indet. (Crustacea, Malacostraca, Amphipoda), body length 900 µm. Lanzarote, Spain 2019. (**f**) Bopyrid, indet. (Crustacea, Malacostraca, Isopoda, Bopyridae), body length 720 µm. Helgoland, Germany 2012. (**g**) Bopyrid, indet. (Crustacea, Malacostraca, Isopoda, Bopyridae), body length 520 µm on calanid. Marbella, Spain 2017. (**h**) Isopod, parasitic, indet. (Crustacea, Malacostraca, Isopoda), body length 450 µm. Cape Town, South Africa 2016

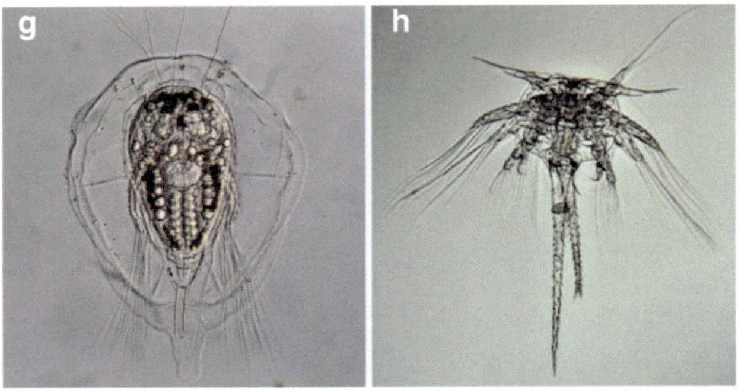

Plate 27 (continued)

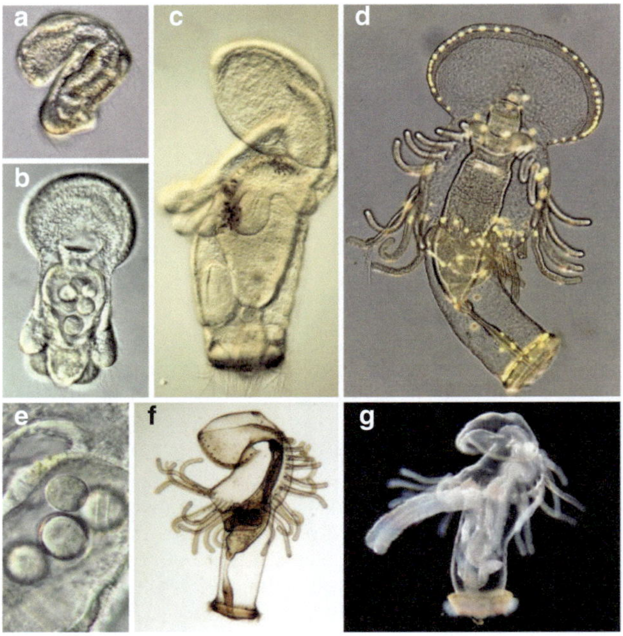

Plate 28 (**a**) *Phoronis* sp. (Phoronida, Actinotrochidae), early actinotrocha larva, length 120 µm. Marbella, Spain 2017. (**b**) *Phoronis* sp. (Phoronida, Actinotrochidae), early actinotrocha larva, length 220 µm. Marbella, Spain 2017. (**c**) *Phoronis* sp. (Phoronida, Actinotrochidae), actinotrocha larva, length 450 µm. Sylt, Germany, 2009. (**d**) *Phoronis* sp. (Phoronida, Actinotrochidae), late actinotrocha larva, width 750 µm. Sylt, Germany, 2009. (**e**) *Phoronis* sp. (Phoronida, Actinotrochidae), dinoflagellate in the gut of medium actinotrocha larva, diameter ca. 50 µm. Sylt, Germany, 2009. (**f**) *Phoronis* sp. (Phoronida, Actinotrochidae), actinotrocha larva, length 1 mm. Sylt, Germany, 2009. (**g**) *Phoronis* sp. (Phoronida, Actinotrochidae), actinotrocha larva, beginning metamorphosis, length 1.5 mm. Helgoland, Germany 2017. (**h**) *Sagitta* sp. (Chaetognatha, Sagittoidea, Aphragmophora, Sagittidae), "head", frontal view, length of chaetae 220 µm. Cape Town, South Africa 2016. (**i**) *Sagitta* sp. (Chaetognatha, Sagittoidea, Aphragmophora), diameter 600 µm. Helgoland, Germany 2012. (**j**) *Sagitta* sp. (Chaetognatha, Sagittoidea, Aphragmophora, Sagittidae), tail fin, width 1 mm. Helgoland, Germany 2021. (**k**) *Sagitta* sp. (Chaetognatha, Sagittoidea, Aphragmophora, Sagittidae), "head", lateral view, length of chaetae 220 µm. Cape Town, South Africa 2016

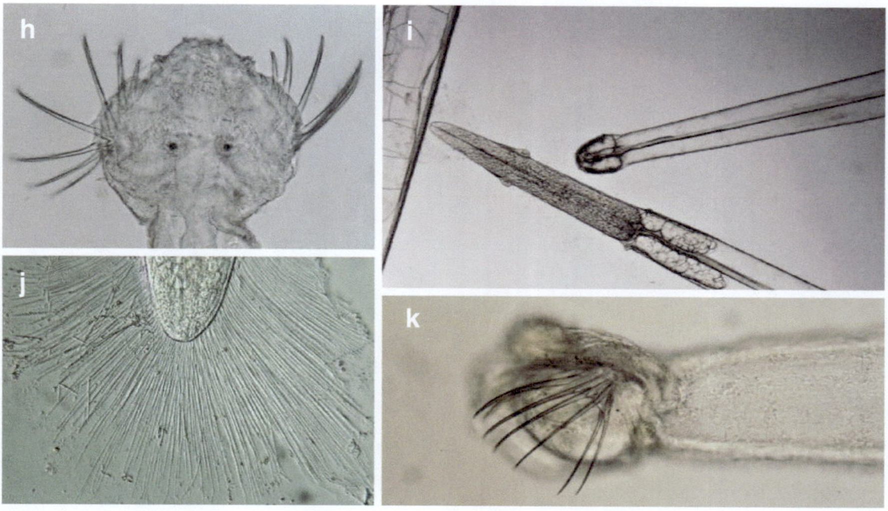

Plate 28 (continued)

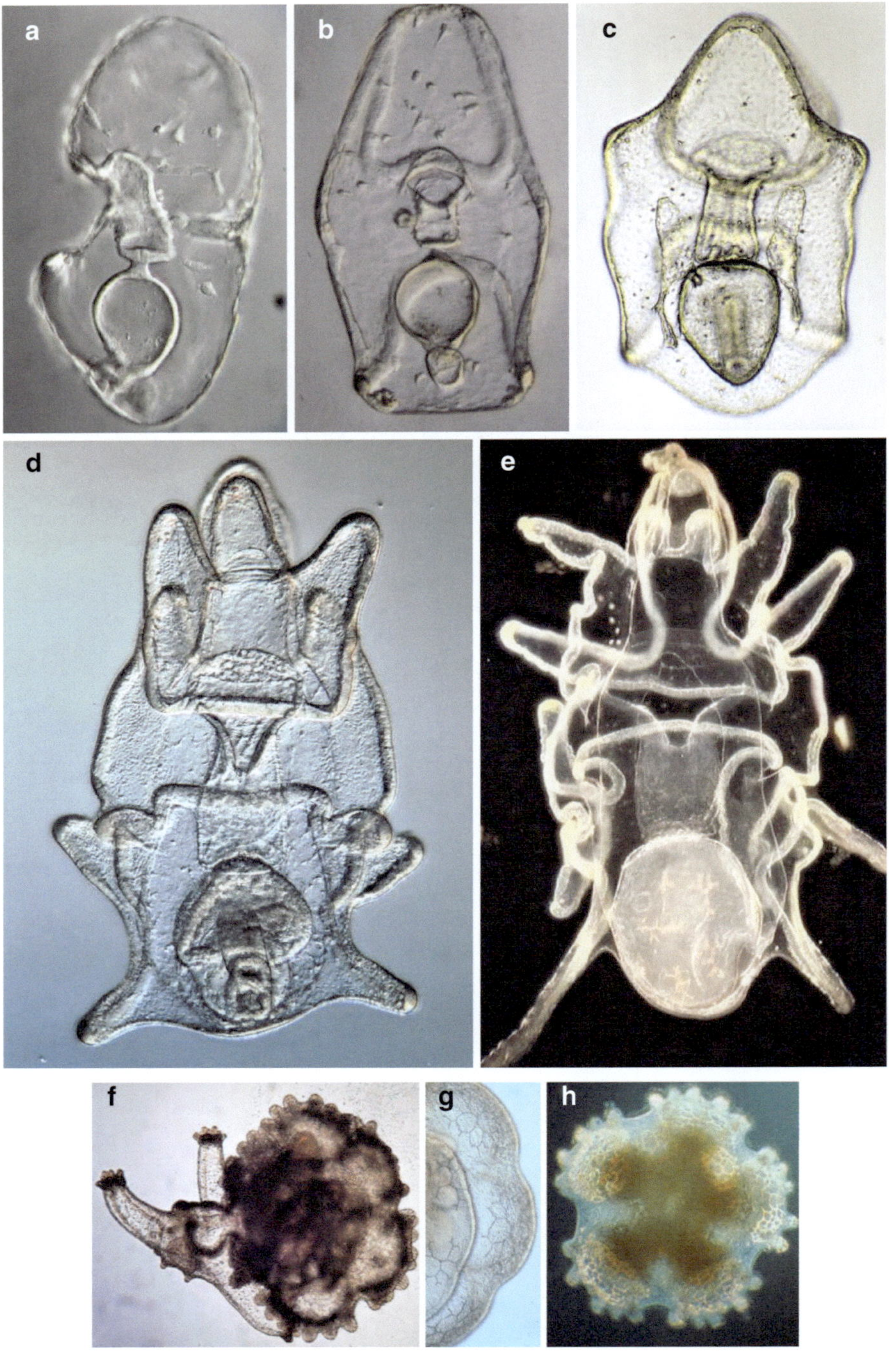
a
b
c
d
e
f
g
h

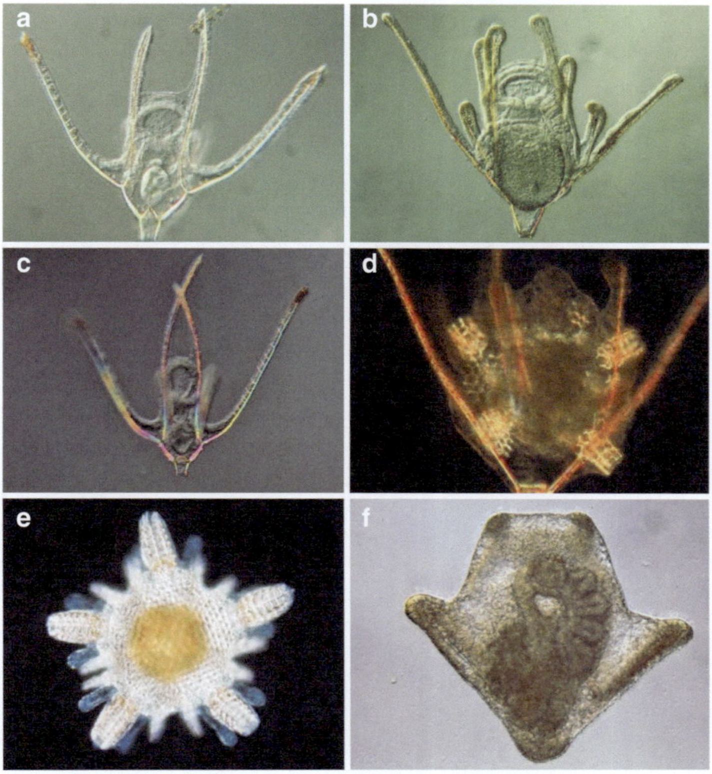

Plate 30 (**a**) *Ophiura albida* (Echinodermata, Ophiuroidea, Ophiurida, Ophiuridae), pluteus larva, width 600 µm. Helgoland, Germany 2021. (**b**) *Ophiura albida* (Echinodermata, Ophiuroidea, Ophiurida, Ophiuridae), pluteus larva, width 900 µm. Helgoland, Germany 2017. (**c**) *Ophiura albida* (Echinodermata, Ophiuroidea, Ophiurida, Ophiuridae), pluteus larva, width 600 µm. Helgoland, Germany 2021. (**d**) *Ophiura albida* (Echinodermata, Ophiuroidea, Ophiurida, Ophiuridae), pluteus larva, metamorphosis, width 700 µm. Helgoland, Germany 2017. (**e**) *Ophiura albida* (Echinodermata, Ophiuroidea, Ophiurida, Ophiuridae), early brittle star, diameter 800 µm.

Plate 29 (**a**) *Asterias* sp. (Echinodermata, Asteroidea, Forcipulatida, Asteriidae), early dipleurula larva, lateral view, length 400 µm. Marbella, Spain 2017. (**b**) *Asterias* sp. (Echinodermata, Asteroidea, Forcipulatida, Asteriidae), early dipleurula larva, frontal view, length 400 µm. Marbella, Spain 2017. (**c**) *Asterias* sp. (Echinodermata, Asteroidea, Forcipulatida, Asteriidae), early dipleurula larva, frontal view, length 630 µm. Cape Town, South Africa 2016. (**d**) *Asterias rubens* (Echinodermata, Asteroidea, Forcipulatida, Asteriidae), late dipleurula larva, length 1.05 mm. Helgoland, Germany 2015. (**e**) *Asterias rubens* (Echinodermata, Asteroidea, Forcipulatida, Asteriidae), brachiolaria larva, length 1.7 mm. Helgoland, Germany 2015. (**f**) *Asterias rubens* (Echinodermata, Asteroidea, Forcipulatida, Asteriidae), metamorphosis, diameter 1.6 mm. Helgoland, Germany 2018. (**g**) *Asterias rubens* (Echinodermata, Asteroidea, Forcipulatida, Asteriidae), first skeletal elements (ossicles), width of photo 2 mm. Helgoland, Germany 2015. (**h**) *Asterias rubens* (Echinodermata, Asteroidea, Forcipulatida, Asteriidae), end of metamorphosis diameter 1 mm. Helgoland, Germany 2019

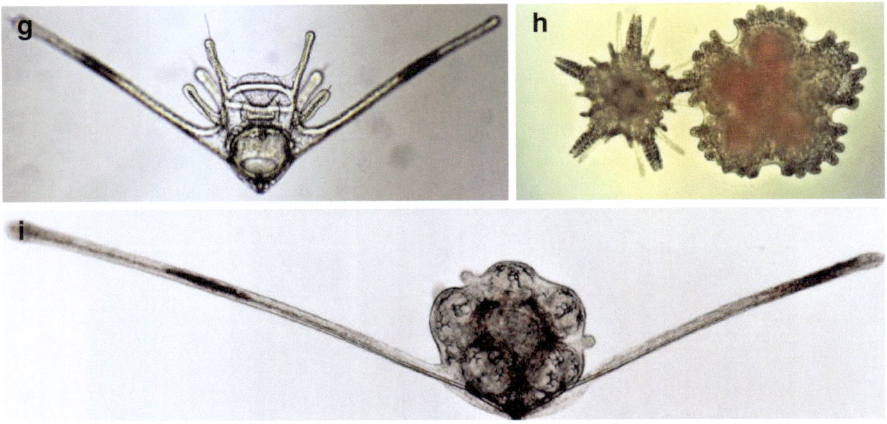

Plate 30 (continued) Helgoland, Germany 2017. (**f**) *Ophiurid pluteus*, indet. (Echinodermata, Ophiuroidea, Ophiurida), width 410 µm. Marbella, Spain 2017. (**g**) *Ophiurid pluteus*, indet. (Echinodermata, Ophiuroidea, Ophiurida), width 1.4 mm. Cape Town, South Africa 2016. (**h**) Brittle star (diameter 700 µm) and starfish (diameter 750) after metamorphosis. Helgoland, Germany 2018. (**i**) Ophiurid pluteus, indet. (Echinodermata, Ophiuroidea, Ophiurida), width 1.85 mm. Cape Town, South Africa 2016

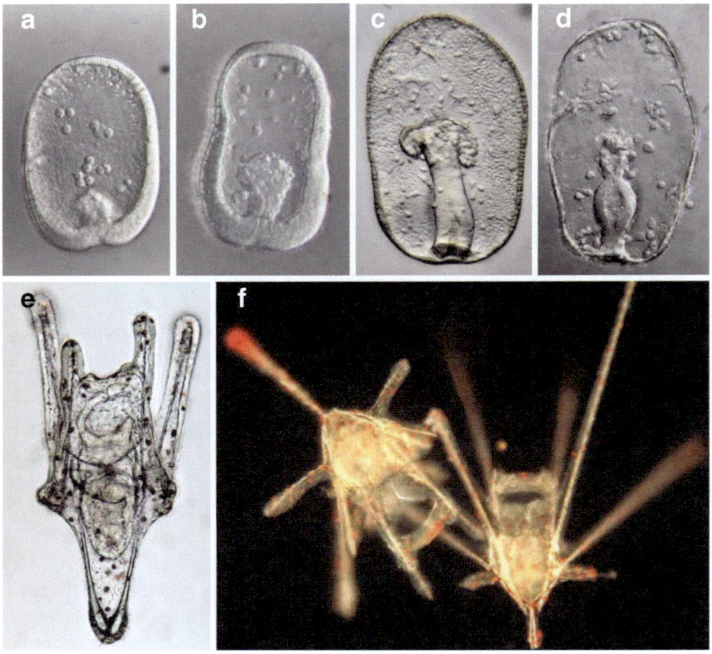

Plate 31 (**a**) Echinoid dipleurula larva (Echinodermata, Echinoidea, Echinidea), length 240 µm. Marbella, Spain 2017. (**b**) Echinoid dipleurula larva (Echinodermata, Echinoidea, Echinidea), length 210 µm. Marbella, Spain 2017. (**c**) Echinoid dipleurula larva (Echinodermata, Echinoidea, Echinidea), frontal, length 350 µm. Marbella, Spain 2017. (**d**) Echinoid dipleurula larva (Echinodermata, Echinoidea, Echinidea), frontal, length 340 µm. Marbella, Spain 2017. (**e**)

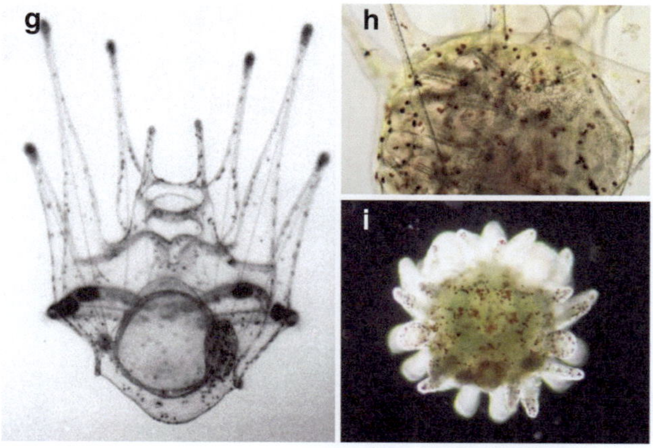

Plate 31 (continued) Echinoid, early pluteus (Echinodermata, Echinoidea, Echinidea), frontal, length 500 µm. Marbella, Spain 2017. (**f**) *Echinocardium cordatum*, plutei (Echinodermata, Echinoidea, Spatangoida, Loveniidae), longest terminal process 340 µm. Helgoland, Germany 2017. (**g**) *Psammechinus miliaris*, latr pluteus (Echinodermata, Echinoidea, Camarodonta, Parechinidae), width ca. 800 µm. Helgoland, Germany 1973. (**h**) *Echinocardium cordatum* (Echinodermata, Echinoidea, Spatangoida, Loveniidae), late metamorphosis, diameter 350 µm. Helgoland, Germany 2017. (**i**) Echinid, end of metamorphosis (Echinodermata, Echinoidea, Echinidea), diameter 630 µm. Marbella, Spain 2017

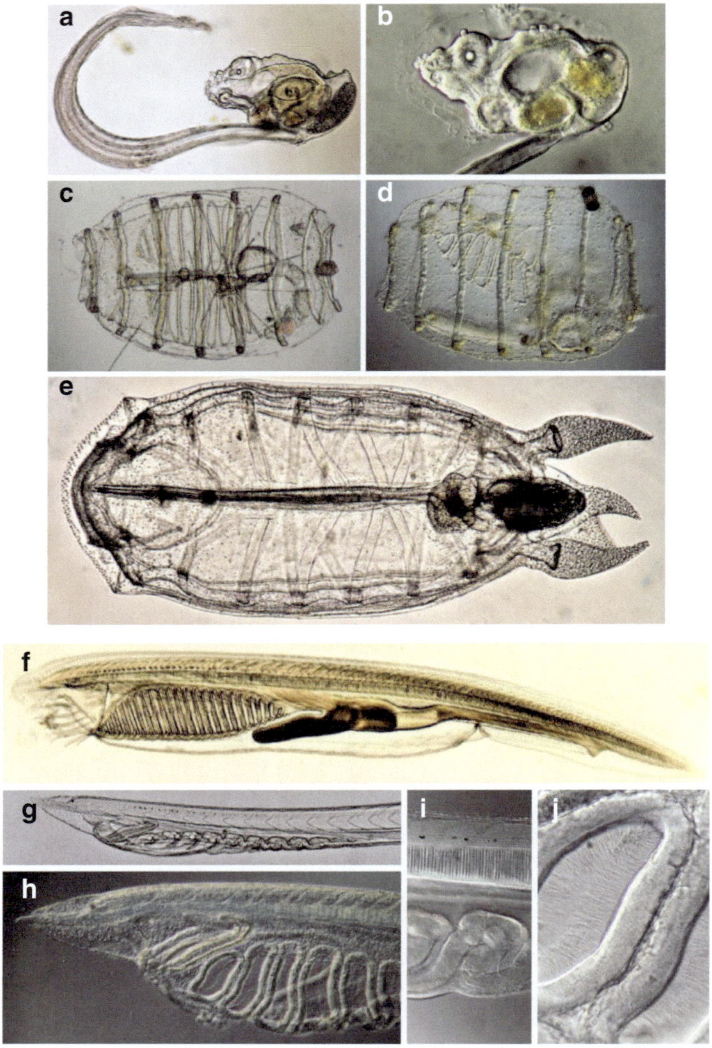

Plate 32 (**a**) *Oikopleura* sp. (Tunicata, Appendicularia, Oikopleuridae), length of "head" 370 µm. Cape Town, South Africa 2016. (**b**) *Oikopleura dioica* (Tunicata, Appendicularia, Oikopleuridae), length of "head" 190 µm. Sylt, Germany 2016. (**c**) *Doliolum* sp. (Tunicata, Thaliacea, Doliolida, Doliolidae), ventral view, length 1.4 mm. Sylt, Germany 2016. (**d**) *Doliolum* sp. (Tunicata, Thaliacea, Doliolida, Doliolidae), lateral view, length 920 µm. Marbella, Spain 2017. (**e**) *Salpa* sp. (Tunicata, Thaliacea, Salpida, Salpidae), frontal view, length 4 mm. St. Cyr, France 2018. (**f**) *Branchiostoma lanceolatum* (Chordata, Leptocardii, Branchiostomidae), young specimen, length ca. 10 mm. Sylt, Germany 2019. (**g**) *Branchiostoma lanceolatum* (Chordata, Leptocardii, Branchiostomidae), larva, length 4.2 mm. Helgoland, Germany 2012. (**h**) *Branchiostoma lanceolatum* (Chordata, Leptocardii, Branchiostomidae), front part of a larva, gills well developed, width of photo 1 mm. Helgoland, Germany 2012. (**i**) *Branchiostoma lanceolatum* (Chordata, Leptocardii, Branchiostomidae), part of body with gills, chorda and neural tube, height of photo 280 µm. Helgoland, Germany 2012. (**j**) *Branchiostoma lanceolatum* (Chordata, Leptocardii, Branchiostomidae), part of gills, width of photo 100 µm. Helgoland, Germany 2012

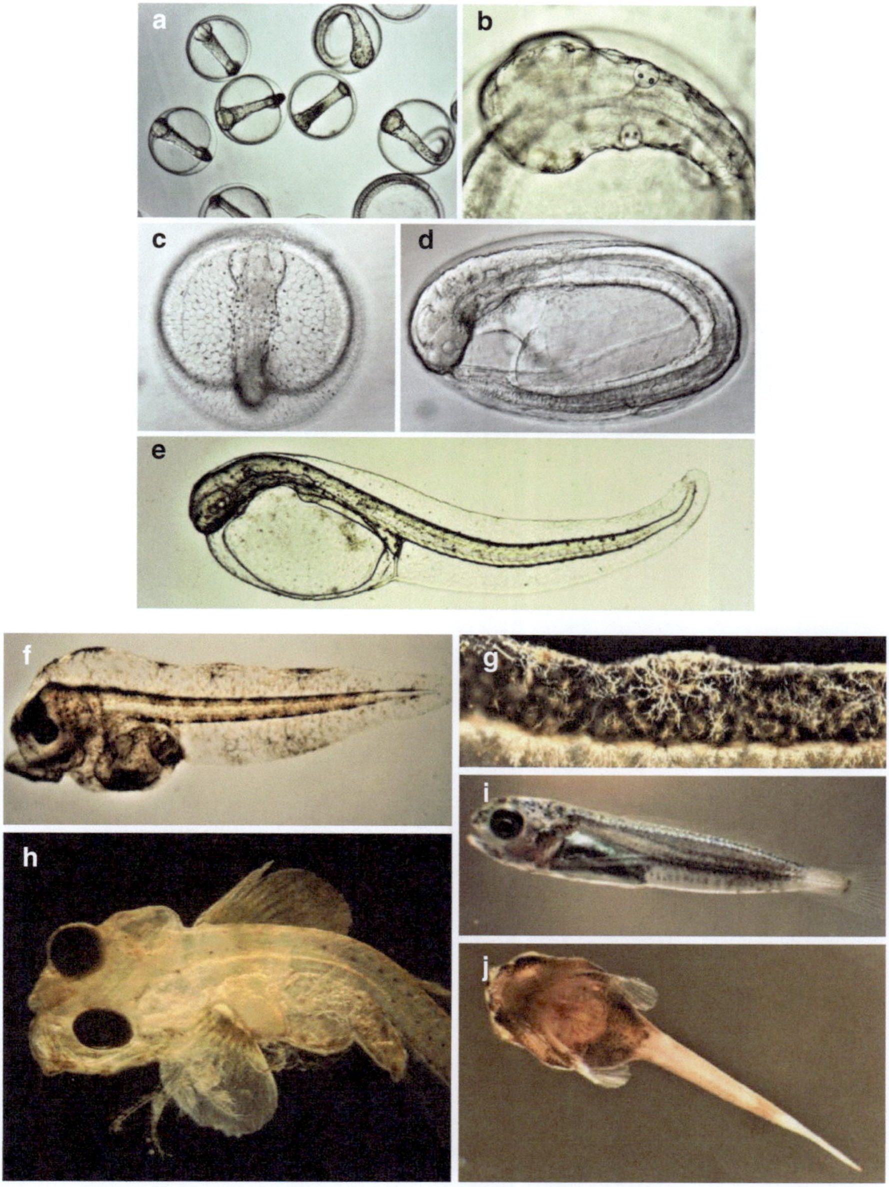

Plate 33 (**a**) Fish eggs in different stages and views (Chordata, Vertebrata, Osteichthyes, Teleostei), diameter 800 µm. Helgoland, Germany 2016. (**b**) Fish egg with embryo (Chordata, Vertebrata, Osteichthyes, Teleostei), diameter 800 µm. Helgoland, Germany 2016. (**c**) *Callionymus lyra*, egg (Chordata, Vertebrata, Osteichthyes, Teleostei, Callyonimidae), diameter 800 µm. Helgoland, Germany 2017. (**d**) Fish egg with embryo (Chordata, Vertebrata, Osteichthyes, Teleostei), length 1.4 mm. Helgoland, Germany 2017. (**e**) Fisch larva, just hatched (Chordata, Vertebrata, Osteichthyes, Teleostei), length 2.2 mm. Helgoland, Germany 2016.

Plate 33 (continued) (**f**) *Buglossidium luteum*, larva (Chordata, Vertebrata, Osteichthyes, Teleostei, Pleuronectiformes, Soleidae), length 5.5 mm. Helgoland, Germany 2016. (**g**) *Buglossidium luteum*, larva (Chordata, Vertebrata, Osteichthyes, Teleostei, Pleuronectiformes, Soleidae), dorsal part with chromatophores, width of photo 500 μm. Helgoland, Germany 2016. (**h**) *Agonus cataphractus* (Chordata, Vertebrata, Osteichthyes, Teleostei, Perciformes, Agonidae), length 6.2 mm. Helgoland, Germany 2016. (**i**) *Gadus morhua* (Chordata, Vertebrata, Osteichthyes, Teleostei, Gadiformes, Gadidae), length 1.3 cm. Helgoland, Germany 2016. (**j**) *Cyclopterus lumpus* (Chordata, Vertebrata, Osteichthyes, Teleostei, Scorpaeniformes, Cyclopteridae), length 5 mm. Helgoland, Germany 2017

MIX
Papier aus verantwortungsvollen Quellen
Paper from responsible sources
FSC® C105338

If you have any concerns about our products,
you can contact us on
ProductSafety@springernature.com

In case Publisher is established outside the EU,
the EU authorized representative is:
Springer Nature Customer Service Center GmbH
Europaplatz 3, 69115 Heidelberg, Germany

Printed by Libri Plureos GmbH
in Hamburg, Germany